Topical Outlines

Since most major topics appear more than once in the spiral, we show here at a glance the chapters or sections where several of the most important themes can be traced.

For more details, consult the index.

Second Edition

Musical Acoustics

Donald E. Hall
California State University, Sacramento

Brooks/Cole Publishing Company
Pacific Grove, California

Brooks/Cole Publishing Company
A Division of Wadsworth, Inc.

Printed in the United States of America
10 9 8 7 6

Library of Congress Cataloging-in-Publication Data
Hall, Donald E.
Musical acoustics/Donald E. Hall—2nd ed.
p. cm.
Includes bibliographical references and index.
ISBN 0-534-13248-0
1. Music—Acoustics and physics. I. Title.
ML3805.H153 1990
781.2—dc20

90-2374
CIP
MN

Sponsoring Editor: Harvey C. Pantzis
Editorial Assistant: Jennifer Kehr
Production Editor: Penelope Sky
Manuscript Editor: Steven M. Bailey
Permissions Editor: Carline Haga
Interior and Cover Design: Sharon L. Kinghan
Cover Photo: © Ohlmeyer & Coughlin/Fran Heyl Assocs.
Photo taken at St. Peter's Lutheran Church, NYC
Art Coordinator: Lisa Torri
Interior Illustration: Lotus Art
Photo Editor: Ruth Minerva
Typesetting: Polyglot Compositors
Cover Printing: Phoenix Color Corporation
Printing and Binding: Arcata Graphics/Martinsburg

For Karl and Kurt

Preface

In musical acoustics we have a unique opportunity to see science and art working together. Along the way they are sometimes friendly antagonists, but ultimately they are partners in teaching us what music is and how it works.

This book is for an introductory, nontechnical college course; it requires no background in either science or music. It is readable to the novice and at the same time interesting and significant to the experienced musician. I offer a balanced presentation of all aspects of musical acoustics. I connect traditional physical analyses (Chapters 8–16) to musical reality by explaining how our ears and brains interpret musical events (Chapters 17–19). I include a brief discussion of hi-fi components (Chapter 16), but the topic is not emphasized because it is peripheral to musical acoustics.

Because this study is a "science appreciation" experience for most readers, I have kept two purposes in mind: (1) to help you use simple physical concepts as tools for understanding how music works, and (2) to use your interest in music to motivate the study and appreciation of scientific methods. Any given chapter may challenge you with several points that are not obvious on first reading. I hope you will welcome these challenges as opportunities for real learning.

Features

Boxed material is as important as the main text (unless the title is preceded by an asterisk), and is generally not any more difficult; it is displayed separately so that the central discussion is not interrupted. *Starred* sections, including all of Chapter 16, are optional, and not prerequisite to later sections. Some present details that are not expected to interest the average reader; some require more background in science than is otherwise assumed.

Each chapter includes a number of study aids: a summary; references; a list of symbols, terms, and relations; exercises; and suggestions for extended projects. The exercises are extremely important, and I have taken care to make them realistic and interesting. Many of these problems are answered in the back of the book, but the ultimate solution is usually much less significant than a clear and careful presentation of the reasoning that leads to it. Starred exercises either depend on optional text material or are somewhat more difficult than the others.

Several features pertain to the book as a whole. In the glossary you can quickly check the definitions of many terms. The topical outlines inside the front cover show at a glance where each of several subjects is discussed. The illustration inside the back cover greatly simplifies the naming and identification of notes, intervals, harmonic series, and combination tones. Appendix A is a brief introduction to written music for those who do not read it already, and Appendix B contains necessary information about the metric system and scientific notation.

Instructors familiar with the first edition will notice that the major change in this revision is the reordering of Chapters 8–10. I also have taken this opportunity to incorporate recent research findings, to correct some errors and smooth out some rough writing, and to offer additional exercises. The *Instructor's Manual* contains numerous suggestions for films, lecture demonstrations, and laboratory activities, as well as complete solutions to the exercises.

Spiral Structure

The subject matter falls naturally into three major areas:

1. Production of sound by various sources, especially by musical instruments
2. Propagation of sound from source to listener, not only by direct transmission through the air but also through electronic reinforcement, or recording and reproduction
3. Perception and judgment of sound by the human brain, on the basis of nerve signals bringing information from the ears

Because a thorough study of any of these areas requires some knowledge of the other two, it is not practical simply to take each one in order. I have instead chosen a *spiral approach,* which is diagrammed.

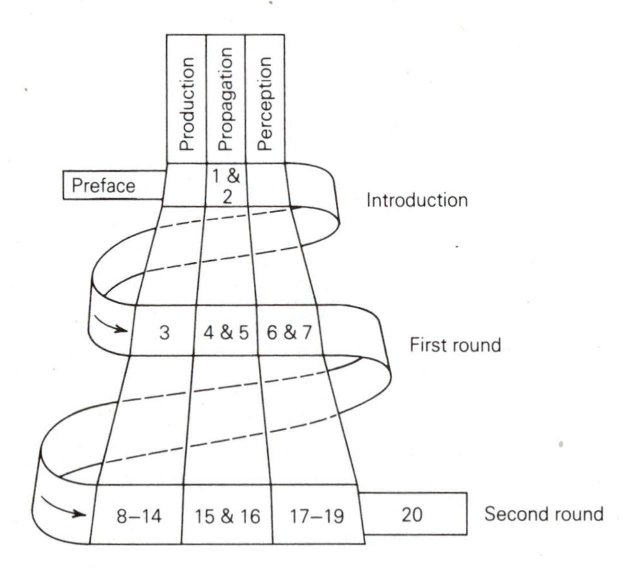

The first two chapters are a broad general introduction to acoustics and to such conceptual tools as analogies and graphs. With Chapters 3–7, we make the first full trip around the spiral, establishing a foundation in all three areas. The remaining chapters complete another, more leisurely, turn around the spiral, encompassing many more topics and concepts. This structure accommodates a variety of courses with differing emphases, and it is relatively easy for the instructor to designate some areas as optional. It is also possible for different students in the same course to concentrate a major part of their second-round efforts in different chapters, according to their interests and backgrounds. I recommend that the book be used this way, because it contains too much material to be covered completely in a one-semester course.

A further advantage of the spiral organization is that we can raise certain knotty questions and take time to mull them over before tackling the explanations. The famous inventor Charles F. Kettering persuaded a friend to hang an empty birdcage in his house, betting that this would force him to buy a bird. The friend soon grew so tired of having people ask, "When did your bird die?" that he did buy a bird, thereby losing the wager. Kettering believed that we deliberately hang puzzling questions in our minds, like empty birdcages. Their nagging presence will eventually drive us to find a solution to fill the void.

The study of musical acoustics is seriously limited when it does not balance psychological and artistic insights with the physics. I hope that as you proceed you will think often about the interplay among the three central points of view, and let each one contribute to your understanding of how music works.

Acknowledgments

The following people reviewed all or part of the manuscript for the first edition, and made many helpful suggestions: William T. Achor; R. Dean Ayers; John W. Coltman; Lowell Cross; Eric N. Koch; Edward L. Kottick; Mark Lindley; William R. Savage; John C. Schelleng; Gabriel Weinreich; Frederick L. Wrightman; and John S. Zetts. I thank consulting editor John Shonle for his invaluable help in clarifying and organizing my thoughts. More recently, William Hartmann, Anders Askenfelt, and William E. Baylis were especially kind in providing extensive critical comments that helped me improve the text in its current revision.

I am especially indebted to Ray Hefferlin (Southern College), Peter Sturrock (Stanford University), and Claude Barnett (Walla Walla College). Each accepted the crucial role of mentor during successive stages of my development as student, physicist, and teacher, and I cherish the innumerable lessons I learned from them.

It was a privilege to teach musical acoustics with Gaylen Hatton for several years. Department chairs Michael Shea and Edward Gibson deserve thanks for their support and encouragement for my teaching and writing. I also thank Mike Snell, Jenny Sill, Autumn Stanley, and Linda Hayes for helping me start the project.

I am grateful to my parents for encouraging my interest in learning and in music. And my wife, Carol Maxwell, has been very patient about all the time this book has taken.

Donald E. Hall

About the Author

After receiving his Ph.D. in physics at Stanford University in 1967, Don Hall taught at Walla Walla College, where he first offered a musical acoustics course in 1971. He earned an M.A. in music at the University of Iowa in 1973. After teaching at the University of Colorado in Denver, he came to California State University at Sacramento, where his course is sponsored by the music and physics departments.

Hall has been an active amateur and professional musician since childhood, playing many different instruments, but concentrating on the pipe organ. He is a member of the American Guild of Organists, has played for services at many churches, and has given several solo recitals. He has also built several harpsichords; the most recent two-manual model now dominates his living room.

Contents

Second Edition

Musical Acoustics

CHAPTER

1 The Nature of Sound

Perhaps you remember from childhood, as I do, the old riddle about whether a tree falling in some remote forest makes any sound. The answer depends upon how we define "sound." If no person is within 50 miles, a psychologist might say "no" because no human perception of sound has taken place. A biologist might ask whether a bird or a fox is close enough to hear. But a physicist would simply answer "yes," insisting that sound means a disturbance of air, which needs no living observer.

Your dictionary also will tell you that sound can mean either the sensation of hearing *or* the air disturbance that causes it. Clearly, then, to understand sound properly we must study both its physical and its psychological aspects and relate them to each other.

This chapter will lay a foundation for the others to follow by introducing some of the basic concepts of sound. After some sharpening of distinctions among the words we use in talking about sound and music, we will describe the division of our subject into three main areas: the creation of sound, its transmission through the air, and its detection by a listener. In Section 1.3 we begin to describe the basic properties of sound waves, taking advantage of helpful analogies with other kinds of waves. In the final parts of this chapter we will begin to study more closely such specifics as the speed of sound waves and their description in terms of sound pressure.

1.1 ACOUSTICS AND MUSIC

Acoustics is the science of sound, and traditionally the term has meant especially the study of the physical nature of sound. Acoustics is one of the main divisions of classical physics, along with motion (mechanics), heat (thermodynamics), light (optics), electricity, and magnetism. In this century, atomic and nuclear physics also have become major areas of investigation.

Although acoustics remains an important part of physics, its scope has increased. If you were to attend a meeting of the Acoustical Society of America today, you would find much of the time devoted to physiology and psychoacoustics, and even to reports of experiments in which the human ear can hear nothing. As technology has advanced, the meaning of acoustics has gradually broadened, as you can see in Box 1.1. It now merges into several other areas related either directly or indirectly to audible sound.

BOX 1.1 KINDS OF ACOUSTICS

Ultrasonics: vibrations too fast for the ear to detect, nevertheless useful for calling dogs, detecting burglars, and removing dirt from small machine parts. Solid-state physicists use ultrasonic waves to study details of crystal structure.
Infrasonics: vibrations too slow for the ear to detect, used by atmospheric physicists to study blast waves or weather systems.
Underwater sound: useful in detecting submerged objects, whether schools of fish or submarines, with sonar devices.
Structural vibration: building motion caused by wind, earthquake, or trampling of feet.
Physiological acoustics: mechanisms of ear and nerve operation and their pathology, often involving experiments on animals.
Psychological acoustics: human perception of sound; judgments, comparisons, and reactions to various sounds.
Speech and hearing: organization of sounds for human communication; strong emphasis on therapy for correcting such problems as stuttering, aphasia, and deafness.
Noise measurement and control: rapidly burgeoning activity in response to concerns about environmental noise, including aircraft, highway traffic, industrial machinery, and rock concerts.
Architectural acoustics: designs and materials for improving homes, offices, and concert halls.
Musical acoustics: mechanisms of sound production by musical instruments; effects of reproduction processes or room design on musical sounds; human perception of sound as music.

The first four of these areas lie outside our scope. But clearly the next five are relevant to our study, even though most of the experiments are not done for musical reasons. Relatively few scientists have studied musical acoustics for its own sake, but these few have given us many important insights.

Let us confine our attention now to audible sound and consider several contrasting types. *Music* includes those intentional combinations of sounds that we choose to hear for esthetic enjoyment and usually depends on an orderly pattern of sounds for its pleasing effect. *Speech* sounds have much in common with music but differ in their purpose; they communicate the entire range of human ideas through word symbols rather than by conveying emotions directly. *Noise* is a term sometimes used vaguely to encompass all other sounds, but it means especially those that are unorganized, unpleasant, or unwanted.

The boundary between music and noise, however, is not distinct. Each new generation of teenagers seems to like music that its elders hear as headache material. And many a piece now accepted as standard concert fare was considered outrageous when first performed. The riot over Stravinsky's *Rite of Spring* (Box 1.2) was an extreme case, but hostile receptions for new musical ideas are actually common. We will adopt the cautious attitude that almost any audible sound may reasonably appear in some composer's music.

***BOX 1.2** "...BUT IS IT MUSIC?"

On the evening of May 29, 1913, a new ballet titled *The Rite of Spring* was performed for the first time in Paris. The music was by Igor Stravinsky, whose earlier works (*The Firebird* and *Petrouchka*) already had made his reputation as a rebel and innovator. Partisan factions welcomed this new opportunity to demonstrate their passions, as later recounted by Carl Van Vechten:

> A certain part of the audience, thrilled by what it considered a blasphemous attempt to destroy music as an art, and swept away with wrath, began very soon after the rise of the curtain to offer audible suggestions as to how the performance should proceed. Others of us, who liked the music and felt that the principles of free speech were at stake, bellowed defiance. It was war over art for the rest of the evening and the orchestra played on unheard, except occasionally when a slight lull occurred. The figures on the stage danced in time to music they had to imagine they heard and beautifully out of rhythm with the uproar in the auditorium.

1.2 ORGANIZING OUR STUDY OF SOUND

Let us think about sound in three basic ways: (1) how it is created, (2) how it travels from one place to another, and (3) how it affects the senses and emotions of a listener. We can call these (1) *production*, (2) *propagation*, and (3) *perception*.

You may easily observe that sound production has something to do with vibration. Just rest your fingertips lightly against the lump in the front of your neck (the larynx, voice box, or "Adam's apple") and sing, hum, or talk. The vibration you feel is good evidence that sound originates when something moves back and forth very fast. The evidence is even more obvious when you pluck a guitar string: You can not only feel the vibration by touching the string but also see its blurred appearance. So we must study vibrating objects to understand sound production, and we will do this in Chapters 2–3 and 9–14.

FIGURE 1.1 Fine cork dust was originally spread evenly along the bottom of this tube. A very strong sound sent out continuously by a loudspeaker on one end, together with its echo reflected from the other end, has disturbed the air. Where the air motion was most vigorous, the dust was swept away. Little piles of dust have collected in other places where the air did not move very much. (Photo by Stephen Hamilton)

Skipping ahead for a moment to perception, it is also quite easy to deduce that your ears detect sound. But it is not at all obvious *how* they do it or how much your nerves and brain modify the sound information they receive. To your ears, two of the most prominent properties of a musical sound are its pitch and loudness. **Pitch*** is the sensation of how "high" or "low" a sound is, or how far toward the right or left end of the piano keyboard. The high-pitched sounds are often called *treble*, and the low ones *bass*. **Loudness** is the sensation of strength or weakness in a sound. We must take care not to assume that one sound wave carrying twice the physical energy of another would necessarily seem twice as loud to our ears; only actual measurements can tell. You will learn more about sound perception in Chapters 6–7 and 17–19.

As for propagation from source to detector, everyone knows that sound vibrations go through the air. Yet we cannot see or feel them moving. So we try to "see" sound with our imaginations, using an analogy. For example, we compare sound to water waves spreading over the surface of a pond. (This does not actually prove anything, but it does make it seem more plausible that a similar thing might happen in the air.) We also may bring technology to the aid of our senses—for instance, by exploring some space with a small microphone that detects a vibratory motion of the air at every point. Perhaps the most direct demonstration that air moves vigorously as sound passes is that shown in Figure 1.1.

It may surprise you to learn that propagation, seemingly the most abstract part of acoustics, is the simplest for physicists. Sound traveling through air obeys an especially simple type of equation that mathematicians call *linear*. The physical meaning of linearity is that many different sounds (say, from different instruments in an orchestra) can travel through the same space at the same time, and each will be unaffected by the others. Light waves also ordinarily have this property of traveling through the same space without

*Many important terms, including these in boldface type, are defined in the Glossary.

altering one another. Physicists find it much easier to calculate and explain linear phenomena than they do any others.

We will study sound propagation in Chapters 4–5 and 15–16. This leaves most of the book for explaining sound production and perception. As you will see, nonlinearities make these two areas a little harder to understand, so they require a larger share of our efforts.

In the rest of this chapter, and in Chapter 2, we will become acquainted with the basic nature of sound and with some general conceptual tools that make it possible to communicate ideas about acoustics more clearly and efficiently.

1.3 THE PHYSICAL NATURE OF SOUND

Sound entails a disturbance of the air through which it moves. But what specific kind of disturbance? Sound in air consists of longitudinal waves carrying energy outward from their source. We must now become acquainted with those waves.

Although we will abandon the distinction later, for now let us distinguish between vibrations and waves. We take *vibration* to mean a rapid back-and-forth movement of a single object or of a single small piece of some large object, such as the tip of one prong of a tuning fork. But a *wave* will mean a disturbance traveling outward in all directions from a vibrating source, like the expanding circular ripples around a point where stones are dropped one after another into a pool of water.

There is a close connection between the two concepts, because the passage of a wave through any region causes each little piece of material in that region, in turn, to vibrate. But that vibration does not carry any material very far, and after the wave has passed, each piece returns to its original position. Thus deep-ocean waves do not pile water higher and higher against the shore, and sound waves do *not* bring a steady stream of air from the speaker to your ear. It is not the air that travels across the room; it is the signal *in* the air.

Figure 1.2 suggests what a sound wave would look like if we could see the air. The areas of darkest shading are called *compressions;* here the density and pressure of the air are greater than they would be in the absence of the sound wave. The lightest areas are called *rarefactions,* where the density and pressure are reduced below their normal values. Each compression is created by temporarily moving air into that region from the adjoining rarefactions on both sides. A short time later this compressed air will reexpand, position A in Figure 1.2 will gradually be occupied by a rarefaction instead, and the compression will move on to B. Thus, as the pattern moves outward, the air at any given point repeatedly becomes more and then less dense.

That sound travels in the form of waves is an extremely important point, because all waves have certain properties in common. Sometimes we can

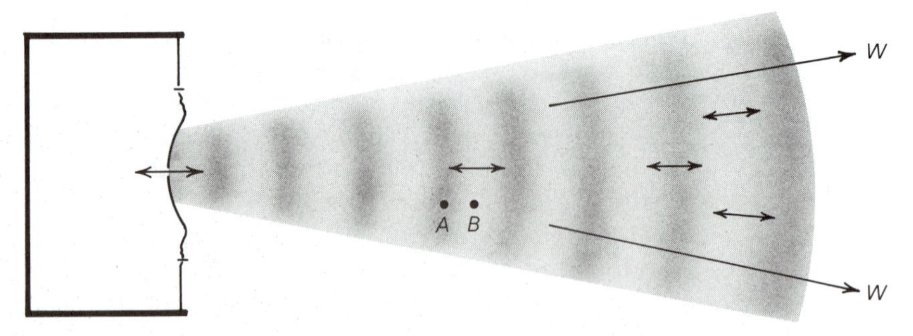

FIGURE 1.2 Sound, a longitudinal wave, travels outward from a vibrating loudspeaker cone. It is the wave pattern that moves continuously outward, as indicated by the arrows *W*. Individual parcels of air, as indicated by the short double-ended arrows, move back and forth parallel to the direction of wave travel but always remain in the same vicinity. The shading greatly exaggerates the changes in density, which are ordinarily only minute fractions (such as millionths) of the average density.

understand sound better by looking at analogous behavior in some other type of wave or vice versa (Box 1.3). Water surface waves are most familiar and easiest to picture; but there are also radio waves, light waves, and waves you can make by shaking one end of a taut rope or spring.

Some waves can be classified according to whether the local disturbance is transverse (crosswise) or longitudinal (lengthwise), with respect to their direction of travel. When you shake a horizontal stretched rope, you set up a transverse wave (Figure 1.3). Whether you shake side to side or up and

*BOX 1.3 THE VALUE OF WAVE ANALOGIES

The idea that different kinds of waves share many of their properties is one of the most powerful tools in physics. The scientist who has studied waves in general can always anticipate much of the behavior of any particular wave type, even one that is new to him.

There are many other types of waves besides those mentioned already. The stiffness of any solid material, such as a steel rod, enables it to transmit both longitudinal and transverse waves, and the two generally travel independently at different speeds. The term *sound* is usually associated only with the longitudinal waves; the others can be described as *flexural* (bending) or *torsional* (twisting) waves. The seismic waves caused by earthquakes include all these kinds.

Plasma waves occur in extremely hot gases and also come in both longitudinal and transverse types. They are important in such places as the earth's ionosphere and the solar wind that sweeps past the planets. Proper understanding and control of plasma waves is one of the main tasks in the effort to build controlled-fusion nuclear power plants, which may eventually replace our present fossil-fuel energy sources.

Wave analogies extend even to the *quantum waves* introduced by Louis de Broglie and Erwin Schrödinger in the 1920s. These describe the behavior of elementary particles at the submicroscopic level. The simple, naive picture of electrons or neutrons as particles (meaning miniature billiard balls) is incomplete, for experiments show that they also have wavelike properties. The concept of quantum waves is what makes possible most of our present understanding of atoms and their nuclei. We shall occasionally point out further aspects of wave analogies in later chapters.

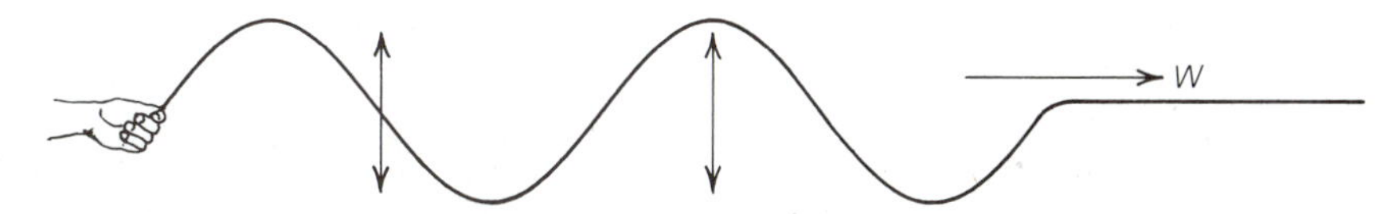

FIGURE 1.3 A transverse wave travels along a rope being shaken at one end. As the wave travels in direction *W*, the presently undisturbed portion of the rope on the right also will begin to move. It is only the wave pattern that moves to the right; each individual piece of the rope moves up and down as shown by the double-ended arrows. Because the wave motion and the particle motion are perpendicular to each other, this is called a *transverse wave*.

down, each piece of the rope vibrates similarly in a direction *perpendicular* to the length of the rope as the wave passes. Likewise, electromagnetic waves (which include radio, light, and X rays) have a disturbance at right angles to the direction in which the wave moves; all such waves are called *transverse*.

Sound in air, on the other hand, is a *longitudinal* wave. By this we mean that the vibration of each little parcel of air is aligned *parallel* to the direction of wave travel. ("Parcel" means we imagine the air divided into little cubes, each a millimeter across, for instance, and containing twenty million billion or so molecules.) Because these parcels do not move all in step with each other, they become crowded together in some places and more spread out in others, just like the spring coils in Figure 1.4. That is, the motion results in fluctuations in the density of the air. These, in turn, cause the air pressure to go above and below its normal value, and these pressure fluctuations are what most microphones (as well as your eardrums) actually detect.

Do not get confused here: Sound is rightly called a wave phenomenon in that it is a vibratory disturbance traveling away from its source, and it has

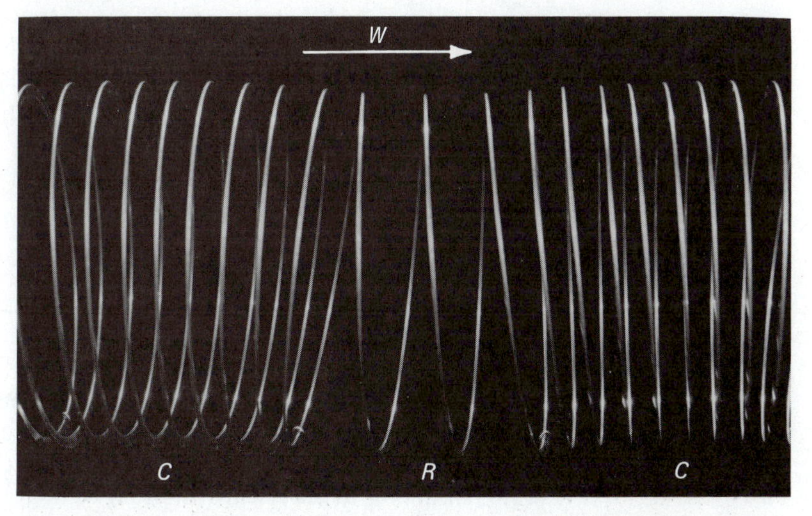

FIGURE 1.4 A longitudinal wave traveling along a "Slinky" spring. It has compressions *C* and rarefactions *R* where the individual coils are closer together or farther apart than average.

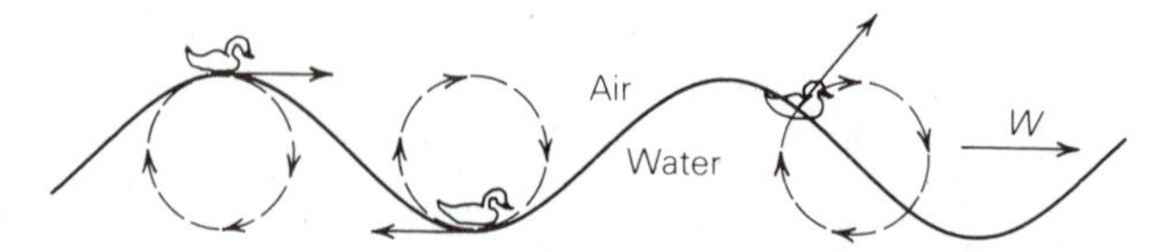

FIGURE 1.5 A wave travels across the surface of the ocean, disturbing a family of ducks. While the wave pattern moves always in the direction *W*, each individual parcel of water always remains in the same vicinity. Each duck helps us visualize what happens to the particular bit of water on which it floats. Water on a crest of the wave moves forward, but half a cycle later when that region becomes a trough, the same bit of water will be moving backward. Where a trough has just passed and a crest is coming, the water carries the duck upward. The complete motion of each bit of water takes it round and round a circular path. This wave cannot be called either longitudinal or transverse. When the wave gets into shallow water near shore, the picture is no longer quite this simple.

many properties in common with other waves. But it does *not* cause the kind of change of shape we intuitively call wavy-looking, as a transverse wave does.

Water surface waves do not fit either category; their motion involves both longitudinal and transverse components. In this respect they are more complicated than sound. They also are more complicated when they "break" as they approach a beach; that is analogous to a shock wave (sonic boom) rather than to ordinary sound. Try to think only of gentle swells or ripples resembling those in Figure 1.5.

We often will use a simplified, stylized representation of sound waves (Figure 1.6) as an alternative to the detailed picture of Figure 1.2. By analogy to water waves, we call the points of greatest compression *crests*. And by analogy to light rays, we draw a set of arrows to show the direction in which the waves are moving. These rays are always perpendicular to the wave crests. Either a series of crests or a set of rays can give a good idea what a sound wave is doing, and we may sometimes sketch one without the other.

The distance from one crest to the next along the direction of travel is customarily called the **wavelength** (although *wave spacing* would be more descriptive). Be sure to notice that wavelength is *not* measured along (parallel to) a crest. The Greek letter λ (lambda) is used to represent wavelength. Wavelengths for sound range from approximately $\lambda = 2$ cm for the highest audible pitches to $\lambda = 20$ m for the lowest.

1.4 THE SPEED OF SOUND

In ordinary conversation, sound seems to arrive at our ears the very instant each word is spoken. But at distances of 20 meters or so, you can begin to notice that it takes a little time for the sound to make the trip. Some people find it very distracting to sit toward the rear of a large concert hall, because

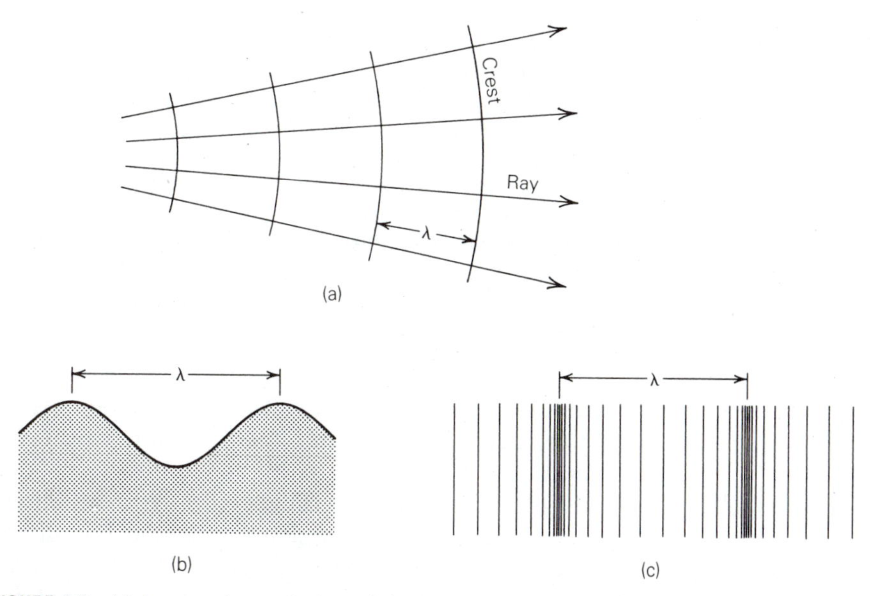

FIGURE 1.6 (a) A series of wave crests, and the corresponding rays showing their direction of travel. The distance from one crest to the next is the wavelength, λ. (b, c) More literal representations for water waves and sound waves, respectively.

the arriving sound of each chord travels much slower than does light, and so lags too much behind the visible motion of the conductor's arm. The lag is quite obvious if you watch a worker swinging a sledgehammer from a block away.

Still, sound travels fast enough that you must be a bit clever to measure just how fast. (Try the method described in Project 2 at the end of the chapter, or at least read about it.) The speed of sound in dry air at room temperature ($T = 20°C$) is $v_{20} = 344$ meters per second. This is often abbreviated as 344 m/s, and it is equivalent to 1130 feet per second, or 770 miles per hour. At this rate it takes about 3 seconds to travel 1 kilometer, so you can tell the distance of a thunderstorm in km by counting the number of seconds delay between lightning flash and thunder and dividing by three. This speed is very fast by everyday standards; few of us have yet ridden a supersonic airplane. But to physicists used to dealing with the speed of light (nearly a million times faster), sound seems quite slow.

It is fortunate for musical purposes that air is a *nondispersive medium*. This means that all sounds, whether high or low in pitch, travel through the air at the same speed. Think what it would be like to sit in the back of a concert hall if they did not: Treble and bass notes played at the same time by the musicians on stage would hit your ears at different times, so any chordal music would be hopelessly scrambled!

*This constant speed for all sounds is a property not necessarily shared by other types of waves. All colors of light travel at the same speed through a vacuum and very nearly so through air. But they travel at quite different speeds through glass; this is why they separate when they go through a prism. Water waves do not all travel at the same speed either; the water surface is a *dispersive medium* on which long waves travel faster and short waves slower. You can see this easily if you look down from a bridge at the wake of a boat passing underneath. If you live near the ocean, you may be familiar with another example. A storm far out at sea creates both long and short waves. The long waves arrive first, and it may take another day or two before the short ones make it to shore.

If the air temperature changes, so does the speed of sound. This is reasonable, because temperature is a measure of the violence of molecular agitation. If the temperature is higher, the random molecular motions are faster, neighboring molecules collide more often, and they can pass the sound disturbance faster from one region to another. For all musical purposes, we can say that the sound speed increases about 0.6 m/s for each degree of temperature rise on the Celsius scale. (In old units, this is 1.1 ft/s for each degree Fahrenheit.) For instance, in a warm room at 30°C (86°F), the speed is 350 m/s, nearly 2 percent faster than at 20°C. This is more than enough to wreak havoc with the sound from a pipe organ that was tuned at 20°: The organ's reed pipes would stay at about the same pitch, while the other pipes would go sharp by about a third of a semitone (as we shall understand better after Chapters 12 and 13). This is also why wind instrument players sometimes blow gently and soundlessly through their instruments to warm them up (that is, fill them with air at operating temperature) before making an important entrance during performance.

*The speed of sound also depends on what medium it passes through. At room temperature, sound travels 1000 m/s through helium but only 270 m/s in pure carbon dioxide. Sound generally travels faster in liquid or solid substances than in gases. Its approximate speed is 1500 m/s through water, for example, and 6000 m/s in steel.

1.5 PRESSURE AND SOUND AMPLITUDE

We need to have precise measures not only for the wavelength of sound and its speed of travel but also for its strength. One possible measure would be the distance each bit of air moves to either side of its normal position during

*As explained in the Preface, this smaller type is used throughout the book to denote optional material.

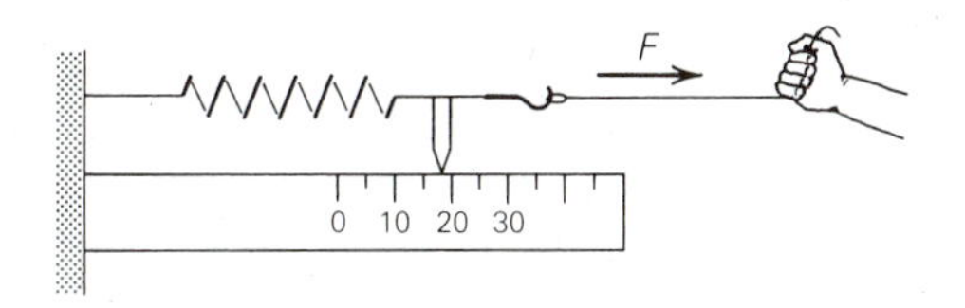

FIGURE 1.7 One way to measure a force.

its vibration. This is called the *displacement amplitude*. For ordinary sound waves it is very small—on the order of millionths of a meter or less. Because it is extremely difficult to measure such motions directly, displacement amplitude is rarely used.

Instead, it is far more common to specify *pressure amplitude*. This is defined as the maximum increase of air pressure (above normal atmospheric pressure) in a sound wave compression. Sound pressure amplitudes also are typically very small (for example, millionths of atmospheric pressure), but they are easily measured because microphone diaphragms respond directly to the pressure fluctuations.

We have relied thus far on your intuitive notion of air pressure. You know, at least in the case of a balloon or a bicycle tire, that air can push outward against its surroundings. But we should take care now to explain precisely what words such as *force* and *pressure* mean in physics.

Whenever you pull or push upon any object, you are exerting a **force** on it. We cannot be content just to speak of large and small forces; we must have some way of measuring the exact strength of a force. This is most easily done by applying the force in question to a spring gauge. The stronger the force, the more the spring is stretched, and opposite a pointer (such as that in Figure 1.7) we can read a number that tells how strong. (We will not discuss here the question of how the spring gauge was initially calibrated.)

The metric unit of force is the *newton* (N); approximately $4\frac{1}{2}$ N equal 1 pound of force. That is about how much force you exert to pick up a flute. It takes a force of several thousand newtons to lift a piano.

No force is completely described unless we give its direction as well as its magnitude. (Physical quantities having both direction and magnitude often are called *vectors*.) A force of 20 N to the right is not the same as a force of 20 N to the left; it will cause a different motion. When two or more forces act at the same time, their directions must be considered in calculating the total force. If two people each push with a force of 300 N in the same direction on a stalled car on level ground (Figure 1.8a), the total force of 600 N will (if we neglect friction) speed up the car twice as fast as either one would alone. But if one person is at the front and the other at the back, pushing in opposite directions (Figure 1.8b), the total force on the car is zero and it goes nowhere.

There is an exact mathematical procedure for adding vectors when they point in any direction. It will suffice here to use your intuitive understanding that if you push north on a piano while I push east, each with a force of 200 N (Figure 1.8c), the net result will be the same as if one of us alone had pushed

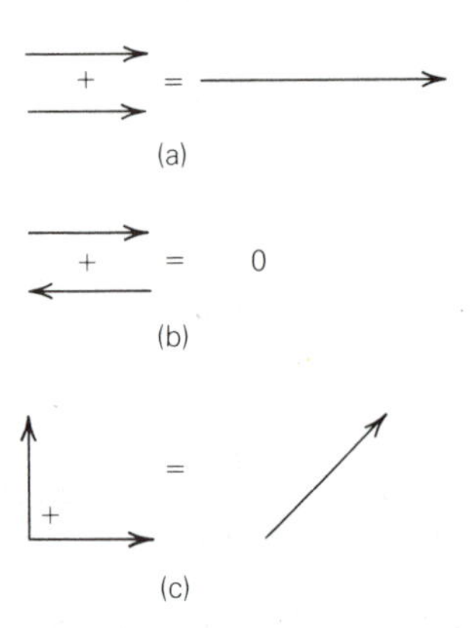

FIGURE 1.8 Examples of the combined effect of two forces.

northeast with a force considerably less than 400 N (the exact figure happens to be 283 N).

So far we have treated the concept of force as if an entire force were applied at a single point. For some purposes such an approximation is entirely adequate; for instance, we usually care only about the total force (called *tension*) applied to a guitar string by the tuning peg and not about the way that force is distributed over the small cross section of the string.

But in real life many forces are applied over an entire surface, not just at one point, and it can make a great difference how large the area is. Pressure is a measure of how much force is concentrated on each part of a surface. Here is the formal, precise definition: **Pressure** means force per unit area, $p = F/S$. (We will use the symbol S for surface area to avoid confusion later with A for amplitude.) Appropriate units for pressure are newtons per square meter, N/m^2, as illustrated in Figure 1.9.

Consider, for example, the weight of a woman supported on a thin high heel—say, 500 N distributed over an area of $.0002\ m^2$ (that is, 2 square centimeters). This can dent a floor more effectively than the step of an elephant (say, 10,000 N distributed over $0.1\ m^2$). In the second case, the pressure is $(10^4\ N)/(10^{-1}\ m^2) = 10^5\ N/m^2$ (see Appendix B for an explanation

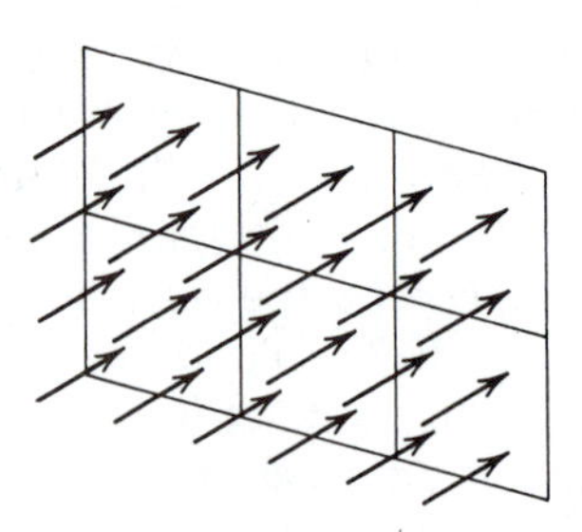

FIGURE 1.9 A total force of 24 N exerted over an area of $6\ m^2$, for a pressure $p = F/S = 4\ N/m^2$. This diagram is only schematic, because each individual newton, in turn, actually is spread over every part of its quarter of a square meter.

of this scientific notation). For the woman, the corresponding figure is $(5 \times 10^2 \text{ N})/(2 \times 10^{-4} \text{ m}^2) = 2.5 \times 10^6 \text{ N/m}^2$. Although only one small bit of floor is underneath the heel, that bit has just as much load as if a total of 2.5 million newtons were distributed over every square meter. That is 25 times as much pressure as for the elephant.

We are particularly interested in the pressure exerted by the air upon every surface that it contacts. Under normal conditions, this is approximately 10^5 N/m^2 (or 14.7 lb/in^2, or 1000 millibars on a weather map). This amount of pressure is called *one atmosphere* and often is used as an alternative unit to measure pressure; for instance, $5 \times 10^6 \text{ N/m}^2 = 50$ atm.

Because the atmosphere exerts 10^5 N on each square meter of any surface exposed to it, a storefront plate glass window with area $S = 10 \text{ m}^2$ is subjected to a total force $F = pS$ of approximately 10^6 N from the air outside pushing inward. Why does the glass not break under such a huge force? It is because the pressure of the air on the inside pushes outward with an equal and opposite force; the net force on the glass is zero. What if the pressure inside stayed the same while the pressure outside dropped 2 percent? (That can happen in a tornado if the store is shut up tightly so that air cannot escape.) There would be a net outward force of $1.00 \times 10^6 - 0.98 \times 10^6 = 0.02 \times 10^6 = 2 \times 10^4$ N, and the window would immediately explode outward.

Air is springy; if you try to squeeze it into a smaller volume it will push back with increased pressure. (Think of squeezing an inflated rubber balloon or pushing the handle of a bicycle tire pump.) By squeezing the chest cavity with certain muscles, we can make the air pressure inside the lungs greater than outside. Then the air in a trumpet mouthpiece held to the lips is acted on by an outward force from the pressure of the adjoining air in the mouth that is greater than the inward force from the adjoining air on the outside. The net outward force causes the steady outward streaming motion of the air in the mouthpiece.

The springiness of air is essential to sound wave motion—the increased pressure in each compression pushes outward in all directions on the surrounding air parcels where the pressure is lower, moving them aside so that the compressed air can reexpand.

The pressure amplitudes of sound waves at comfortable listening levels range from about 0.01 to 1 N/m^2, or 10^{-7} to 10^{-5} atm. A pressure amplitude of 10^{-5} atm means the pressure has a maximum value of 1.00001 atm in the compressions and a minimum of 0.99999 atm in the rarefactions.

SUMMARY

Psychologists are interested in the human perception of sounds. Physicists concentrate on understanding the physical mechanisms by which sound is produced and transmitted through the air. They regard sound as one of many kinds of wave phenomena and realize that soundlike vibrations also exist beyond the reach of our hearing. Musicians choose and arrange certain sounds for emotional effect, and they may be able both to compose and to perform more effectively if they understand what psychologists and physicists tell them about these sounds and their production.

Sound waves are longitudinal and travel through air with a speed of 344 m/s at room temperature. They are described in terms of their wavelength and amplitude. The wavelengths (crest-to-crest distances) of audible sound lie between a few centimeters and several meters. The pressure amplitudes (greatest increases above normal atmospheric pressure in the compressions) are extremely small fractions of 1 atm of pressure.

REFERENCES

Occasionally it is useful for the student to compare how someone else expresses the ideas in this book with my own presentation. For that purpose I recommend two of the many texts written at this level, those by Backus and by Rossing.* Two other excellent books for supplementary reference, which cover somewhat less ground but in greater depth, are those by Benade and by Roederer.* Benade discusses in detail the physical mechanisms of sound production by musical instruments, and still with little use of mathematics. Roederer delves more deeply into some of the psychophysical phenomena. We shall refer directly to particular sections of both books in later chapters.

Journals in which you will find current research include the *Journal of the Acoustical Society of America (JASA), Acustica,* the *Journal of the Audio Engineering Society,* and *Music Perception.* To delve beyond the references mentioned in this book, you may find it helpful to consult the two Resource Letters by Rossing, published in the *American Journal of Physics, 43,* 944 (1975) and *55,* 589 (1987). Each provides literally hundreds of leads in all areas of musical acoustics.

With regard to the critical reception of new music, you may find it amusing to look at the *Lexicon of Musical Invective* compiled by Nicholas Slonimsky (New York: Coleman-Ross, 1953). (I am conservative enough to feel that a significant fraction of the invective still is justified.)

SYMBOLS, TERMS, AND RELATIONS

v speed
T temperature
λ wavelength
F force
S area
p pressure
$v_{20} = 344$ m/s (20°C)
$v_T = 344 + 0.6\,(T - 20°)$ m/s

$p = F/S$
$v = d/t$
sound
vibration
wave
longitudinal
transverse

amplitude
compression
rarefaction
pitch
loudness
treble
bass

*See the list inside the front cover for complete bibliographic information.

EXERCISES

1. Are the simple categories of music, speech, and noise adequate to classify a cat's meow? A mockingbird's song? Wind whistling in the trees?

2. How far does sound travel during 0.002 second if the temperature is 20°C?

3. How long does it take for sound to arrive at the back of an auditorium if the temperature is 30°C and the stage is 35 m away?

4. It is possible for changes in sound speed as little as $\frac{1}{2}$% (1.7 m/s) to be important in certain demanding musical situations. How many degrees, then, would be an allowable range of temperature variation?

5. Suppose a xylophone mallet with a hard-rubber head remains in contact with the wooden bar for only a few milliseconds, but during that short time it exerts a force of 500 N. If that force is concentrated in a contact area of only 5 square millimeters (that is, 5×10^{-6} m^2), how much pressure is being exerted on that part of the bar? Express the answer both in N/m^2 and in atm. Does this suggest that the wood is deformed during impact? If you have any lingering doubts, take a close look at the surface of a xylophone bar some time!

6. Imagine that you are lying on your back on the floor. Estimate the surface area of your chest in m^2 and use this to calculate the total force with which the atmosphere pushes down upon you. How many people, weighing 500 N each, must stand on you to exert as much force? Why doesn't this outside air pressure immediately crush your chest?

7. Suppose a sound wave has pressure amplitude 0.3 N/m^2. What is this amplitude expressed as a number of atmospheres? What are the maximum and minimum pressures (in atm) in the presence of this wave?

*8. A common gauge pressure (excess of inside over outside pressure) for a bicycle tire is 60 lb/in^2. About how many atm is this? How many N/m^2? Estimate (either in in^2 or in cm^2) how much tire area makes contact with the road to support a person weighing 700 N (157 lb) on a 100-N (22-lb) bicycle.

9. Imagine for a moment that treble sounds could travel at 400 m/s, but bass sounds at only 300 m/s. If sound actually behaved that way, and you were listening to a brass band from a distance of 120 m, how much delay would you hear between trumpet and tuba notes when they begin a chord together?

10. You direct the marching band at the University of Southern North Dakota. During the last game of the season, the temperature in the stadium is −10°C (14°F). What is the speed of sound? How long does it take the music to travel from the band to a spectator sitting 65 m away?

11. Suppose a particular sound wave momentarily creates an extra pressure of 10^{-4} atm upon a microphone diaphragm that has an area of 1 cm^2. What total force in newtons does this make on the diaphragm? (Hint: To deal with the mixed units in this problem, you may want to study Appendix B.)

12. If the rarefactions of a particular sound wave reduce the pressure to 0.997 atm, what is the pressure in the compressions? What is the amplitude of this wave, expressed first in atm and then in N/m^2? Is this a soft or a loud sound?

PROJECTS

1. Experiment with waves using a Slinky or stretched rope. Observe and measure where possible, even if only approximately, such things as longitudinal and transverse waves, the speed with which they travel, the way they reflect when they reach the end, and so forth. What do you have to do to get short wavelengths? To get long wavelengths? Do all wavelengths seem to travel at the same speed?

2. Find a large building from which you can stand 30 or 40 m away and hear a good echo when you clap your hands. Find a clapping rate for which each echo splits the time between two consecutive claps in half. Use your wristwatch to find how long it takes for 10 or 20 such claps. (Explain why this is more accurate than just measuring the time from one clap to the next.) Pace off the distance from your clapping position to the wall. Remember the definition:

$$\text{Speed (of anything)} = \frac{\text{distance traveled}}{\text{time elapsed}}.$$

Use it to compute the speed of sound from your distance and time data, and see whether the answer is as close to 344 m/s as you could reasonably expect for this technique. If your answer seems more like a half or a fourth of this, think carefully about how the sound makes a round trip and how this takes only half the time between successive claps.

CHAPTER

2 Waves and Vibrations

Sound waves are dynamic; they evolve in time. We must add some precise description of these time processes to the "snapshots" of spatial variation on which we concentrated in Chapter 1. We will describe the precise counting of vibrations in Section 2.1 and the determination of the exact nature of each individual vibration in Section 2.2.

These pressure changes are too rapid to comprehend without artificial aid, so we must ask what helpful tools and concepts physics can provide. We shall see that modern electronic devices make it easy to dissect sound down to thousandths of a second. There turns out to be a great variety of different sound waveforms. To understand this wealth of information clearly, we must learn how to use its graphical representation, which is presented in Section 2.3.

The question of which waveform is most elementary leads us to consider the simplest type of vibrating source for these waves in Section 2.4. These "simple harmonic oscillators" later will prove useful in understanding complex vibrations as well. The study of simple vibrations presents the opportunity to state clearly such important physical concepts as inertia, restoring force, natural frequency, and (in Section 2.5) work and energy.

2.1 THE TIME ELEMENT IN SOUND

Unlike a painting, a piece of music does not sit still for our contemplation; change in time is part of music's essential nature. We may study a musical score at leisure, but study can never fully replace the actual hearing of the music. If we try to stretch out a musical experience to discern more detail, we change its quality. If we rush through, we again destroy one of the most important properties of the performance. This is true not only of the succession of musical notes in a composition but also of the dynamic nature of sound itself within every note.

There are several characteristic time scales for musical events. The longest is the time for a piece to begin, exhibit its overall structure, and end—anywhere from a minute to an hour or more. In the case of a symphony, the overall organization may be quite complex. After dividing it into movements, sections, and phrases, we come down to individual chords and notes; this is like dividing a book into chapters, paragraphs, sentences, words, and letters.

The individual notes typically last around a second, although some may last several seconds and others only a small fraction of a second; let us call this range of durations the medium time scale. We are directly conscious of both medium-and long-scale events, and they represent a predominantly psychological aspect of acoustics. We shall say more about them in Chapters 7 and 19.

Each musical note contains many individual sound vibrations, and their rate is too rapid for us to recognize them separately. This short time scale involves hundredths, thousandths, or even smaller fractions of a second. Eventually, we would like to know the exact nature of each vibration, but first we should have some way of telling how long a time it takes.

Before the age of electronics it was difficult to measure such details accurately. Nevertheless, clever experimenters did gain some information. In the early seventeenth century, Galileo Galilei made sound by rubbing a card along the serrated edge of a coin. You can do the same thing, perhaps more easily, with the teeth of a comb. If you count the number of teeth and estimate the length of time used in the movement, you will have an idea how often the vibrations repeat. With attentive listening you may be able to identify the pitch of this sound by comparing it to various notes on a piano. The faster you move the card along the comb, the higher the pitch. (On one trial, I estimated that I heard a pitch somewhere around middle C while snapping 40 teeth in very roughly one-fifth of a second. This would indicate vibration at the rate of 200 per second, about 20 percent off the correct answer—not surprising in view of the difficulty of estimating such a short time interval.)

In the nineteenth century, several people made accurate determinations of the vibration rate for notes of various pitch, both with toothed wheels that could be rotated at a known rate and with siren discs (Figure 2.1). The results generally are described in terms of the **frequency** of the vibrations, that is, their rate of repetition. We shall use the symbol f for frequency, which is measured in vibrations per second (also sometimes called cycles per second or hertz, abbreviated Hz). For example, the pitch to which most orchestras tune, by international agreement, has $f = 440$ Hz and is called A440. It causes your eardrums to vibrate 440 times each second.

We sometimes also speak of the **period** P of a vibration, or the length of time taken for a single complete cycle of motion. Because this is the number of seconds per cycle, and frequency is the number of cycles per second, we must have the simple relation $P = 1/f$. For example, this says that each vibration from an A440 tuning fork last $\frac{1}{440}$ of a second, or 0.00227 s. This also could be written as $P = 2.27$ ms, where ms stands for *millisecond*, or one-thousandth of a second. (See Appendix B for a summary of information about metric units and prefixes.)

If we are interested only in finding the frequency of vibration, we can do so with a stroboscope. The strobe illuminates the vibrating object with a series of brief flashes of light, leaving it in darkness most of the time. The strobe principle has long been known, but xenon flashlamps make it easier to use.

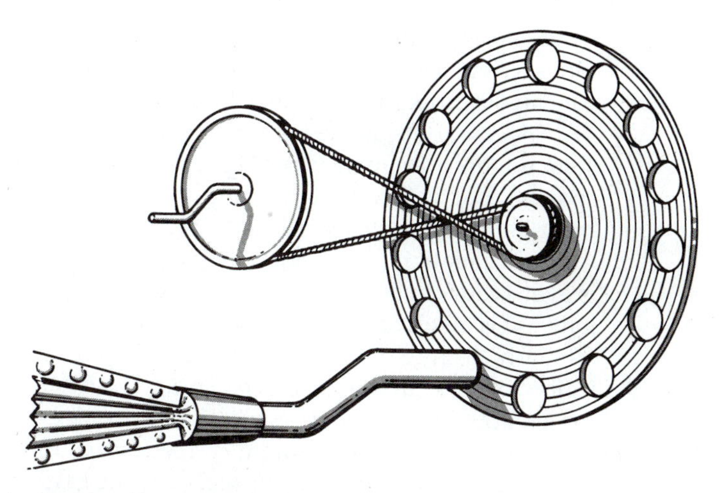

FIGURE 2.1 As a siren disc turns, each hole allows a little puff of air from the bellows to pass through. Counting the number of holes around the circle and the number of times the crank is turned each second determines the frequency of sound vibration generated.

On a modern strobe lamp, we can turn a dial to adjust the flash frequency. When this matches the frequency of the vibrating object, each flash catches it at the same point in its motion, and thus the object appears as if it were stationary. If the frequencies are slightly different, the object appears to move very slowly. Electronic frequency counters are now readily available, too; upon receiving input from a microphone, they automatically count and digitally display either the period or the frequency of the signal.

The range of audible frequencies covers about 10 octaves, and is conveniently remembered as extending from approximately 20 Hz up to 20,000 Hz (or 20 KHz, where KHz stands for kilohertz and means 1000 cycles per second). We perceive low and high frequencies as having low and high pitches, respectively.

The frequencies of radio waves used in commercial broadcasting are much higher. AM radio uses a band around 1 MHz (megahertz, a million cycles per second) and FM at around 100 MHz; television extends up to several hundred MHz.

We now have three fundamental physical quantities that describe any repetitive wave: the speed v, the wavelength λ, and the frequency f. There is an extremely important relation among them, expressed by the equation $v = f\lambda$. In any situation in which we know two of the three quantities, this equation always makes it possible to find the third. Let us see why the equation is true.

Suppose an A440 tuning fork is radiating sound waves outdoors (where we can avoid confusing reflected waves) with the temperature at 20°C. During 1 second, 440 compressions move outward, one after another. By the time the 441st compression is created, the first one has moved 344 meters away. If the sound were visible, a snapshot at that moment would show 440 complete waves spread evenly over 344 meters, so that the distance from each crest to the next must be 344 m divided by 440. This is 0.782 m, or 78.2 cm, and is what we defined in Chapter 1 to be the wavelength λ.

The same argument can be applied in every case. During 1 second a source whose frequency is f Hz will send out f complete vibrations, and at speed v they will spread over a distance of v meters. Thus, the crest spacing must be $\lambda = v/f$.

This equation is true not only for sound but also for every kind of wave. It is even true for complex waveforms (such as those in Figures 2.4 and 2.5), as long as f is taken to mean the rate of repetition of the entire complex vibration. Similarly, λ must mean not just the distance from one crest to another subsidiary one but to the next *corresponding* crest where the entire pattern begins to repeat.

Sometimes it will be more convenient to use the equation in the forms $f = v/\lambda$ or $v = f\lambda$. It is easy to see with the last form that the units of these physical quantities reinforce the argument:

$$v\left(\frac{\text{meters}}{\text{second}}\right) = f\left(\frac{\cancel{\text{cycles}}}{\text{second}}\right) \times \lambda\left(\frac{\text{meters}}{\cancel{\text{cycle}}}\right).$$

(It helps here to think of λ as "meters per cycle.") If you ever forget the equation, you can always reconstruct it correctly just by insisting that the units come out right.

2.2 WAVEFORMS

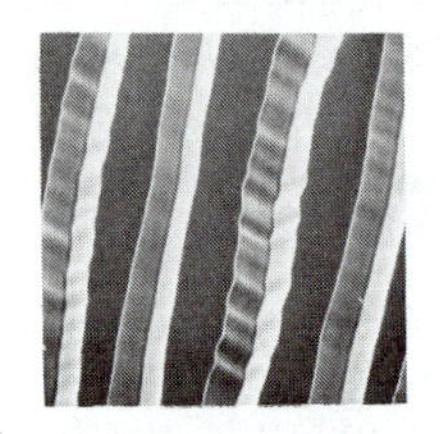

FIGURE 2.2
Magnified view of a phonograph record groove. (Courtesy of Csaba Hunyar.)

Scientists have devised many different ways to visualize the details of sound vibrations. At first, their efforts took a great deal of ingenuity. After phonograph records were invented, an ordinary microscope could be used to inspect a groove (Figure 2.2). Here, preserved in wax, was a clear record of how the original sound drove the cutting stylus back and forth.

Modern technology now makes it easy to get such information. A key role is played by the oscilloscope (Box 2.1), a device for displaying how an electrical signal changes with time. We need only take a sensor (such as a microphone) that converts acoustic to electrical signals and attach it to an oscilloscope to immediately see every little detail of a complex vibration (Figure 2.4).

There is an endless variety of audible waveforms (Figure 2.5). The wave shape is closely related to our perception of the quality or "color" of a tone. But before we can describe this relationship clearly, we need to develop several other concepts. So the full story of tone color will emerge gradually later on.

Even a single musical instrument can generate many different wave shapes when it is played in different ways. In later chapters we shall study the microstructure of sounds from various sources in greater detail. This will include percussive sounds that quickly die away (Chapters 9, 10); artificially

BOX 2.1 THE UBIQUITOUS OSCILLOSCOPE

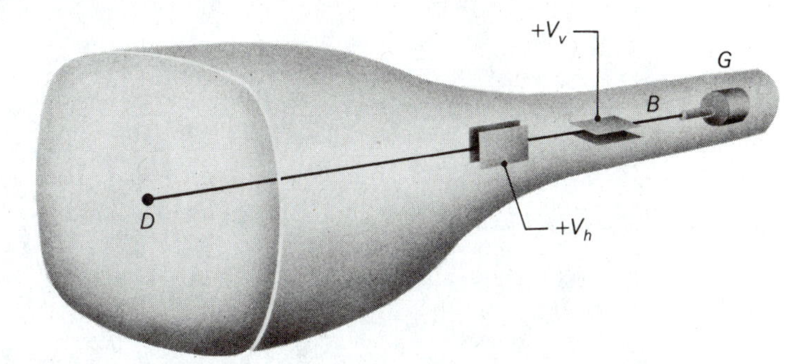

FIGURE 2.3 Idealized sketch of a cathode-ray tube. The electron gun G sends out a beam B of fast-moving electrons, making a bright dot D when they hit the phosphorescent screen. A positive voltage V_v applied to the top plate of the vertical-deflection pair will attract the (negatively charged) electrons toward it as they pass, moving the dot upward on the screen. The repulsion of a negative V_v will move the dot down. Similarly, positive and negative V_h on the right plate of the horizontal-deflection pair will move the dot right and left on the screen. The second plate in each pair is electrically grounded. The deflecting voltages are provided by amplifying circuits controlled by knob settings and by external input signals (for example, the antenna pickup for a TV set or a microphone input to an oscilloscope).

The oscilloscope is extremely versatile, and is an indispensable tool in many fields of science. Oscilloscopes are common not only in physics laboratories and TV repair shops but also in hospital intensive-care units and sound recording studios and even in musical instrument repair work. Every serious student of acoustics should have some experience using an oscilloscope.

How does an oscilloscope work? It is a close cousin to a television set: Each has a picture tube and each has electronic circuitry to control what will be shown on the screen. The picture tube (sometimes called a cathode-ray tube, or CRT) has a phosphorescent coating that glows whenever it is struck by a stream of electrons that originates in the neck of the tube.

The electron beam can be steered right or left and up or down by deflectors. The TV picture tube has magnetic deflectors; two pairs of current-carrying coils act as electromagnets. The same purpose is accomplished in most oscilloscopes with electrostatic deflectors (Figure 2.3). The electron beam passes between two pairs of flat plates; electric charges placed on these plates can attract or repel the electrons in the beam and steer them in any direction.

In the usual mode of operation, the horizontal deflection plates are controlled by circuits that cause the electron beam to move across the screen from left to right at a uniform rate; after each trip the beam jumps back to the left edge and begins again. By turning a knob you can control this rate (the "sweep speed"). When it is slow (for example, 1 second for each sweep), you can easily follow the motion of the bright dot with your eye. When it is fast (for example, 1 millisecond for each sweep), your persistence of vision (aided by some afterglow of the phosphor coating) makes it seem as if there is a permanent bright streak.

Meanwhile, some external signal (such as the voltage from a microphone) is being amplified and used to control the vertical deflection of the beam. An increase in air pressure upon the microphone, for instance, can cause a positive voltage that moves the bright dot upward when applied to the vertical deflection plates. And an air pressure below normal can cause a negative voltage and move the dot downward. Then the repeated fluctuations of a sound wave cause the dot to move up and down repeatedly. Because the dot also is moving across at the same time, the total effect is to make it follow a wavelike path across the screen.

It can be very helpful to see this visual representation of a sound wave at the same time you hear it. There are many sounds for which the eye can detect changes in the waveform that escape the ear's attention.

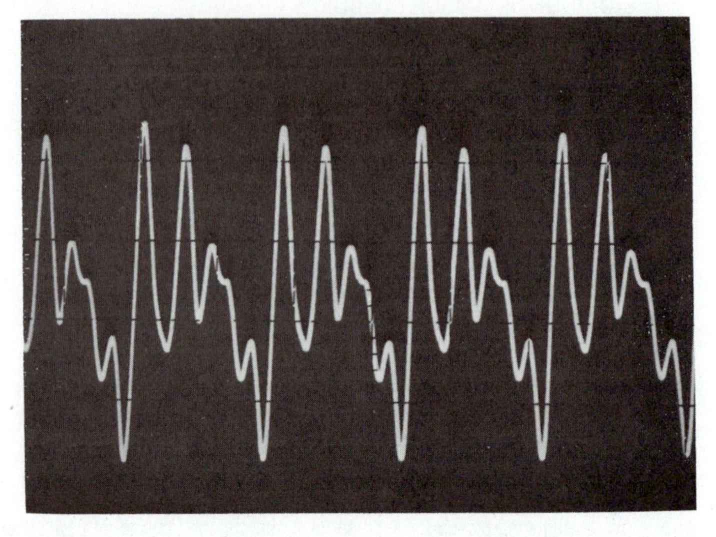

FIGURE 2.4 Oscilloscope trace generated by the author's voice, singing the vowel $\bar{o}$ at a pitch near A_2 ($f \simeq 110$ Hz). As will be explained in Section 2.3, this tells how the air pressure upon the microphone varies (vertically) as time passes (horizontally). The segment shown here lasted only 45 ms.

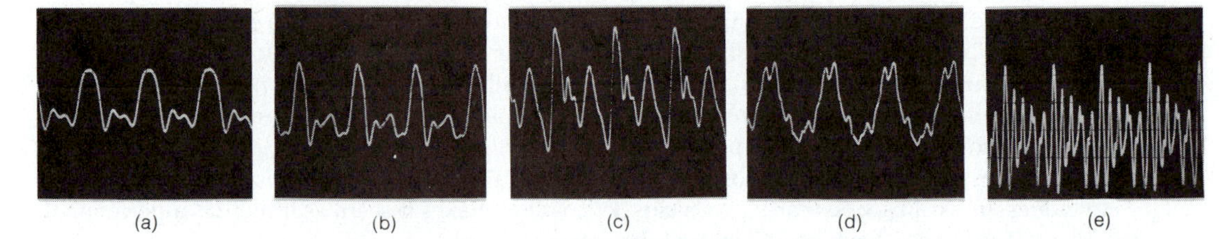

FIGURE 2.5 Sample waveforms produced by different sound sources. (a) A flute, (b) a trumpet, (c) a soprano saxophone, (d) a violin, all playing A_4 ($f = 440$ Hz); these $3\frac{1}{2}$ cycles last approximately 8 ms. (e) A bassoon playing A_2 ($f = 110$ Hz), for which these $4\frac{1}{2}$ cycles take approximately 40 ms. Do not attach too much absolute significance to these waveforms, because some details depend on where the microphone was located.

produced, perfectly steady sounds that keep repeating the same vibrational pattern indefinitely (Chapter 8); and tones from string and wind instruments that are approximately steady but also change in crucial ways as they start and stop (Chapters 11–14).

There is one important point we can make here about waveforms. It takes only a little experimentation with a microphone and oscilloscope to see that some sound vibrations are inherently more complicated than others. Certainly those waves that have several subsidiary peaks within each cycle (such as in Figure 2.4) deserve to be called complex.

This still leaves many possible forms for relatively simple waves, such as those in Figure 2.6. But can we single out any one of these as the simplest of all? We shall answer that question in Section 2.4. First we will pause to make sure we understand clearly what all these waveform pictures mean.

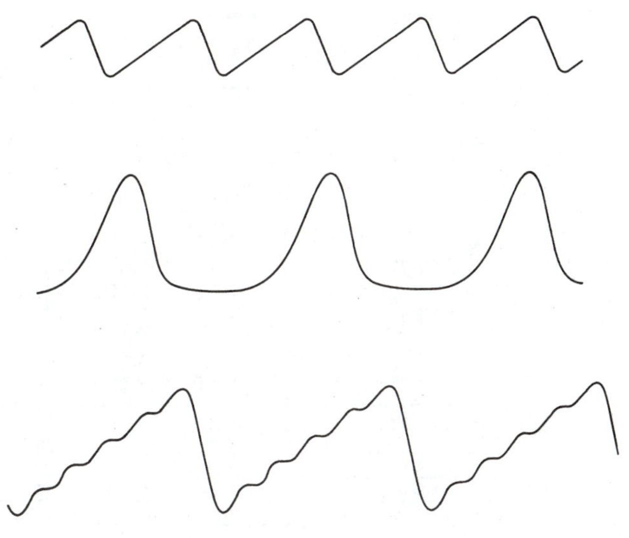

FIGURE 2.6 Several possible forms for relatively simple waves, with only a single maximum and a single minimum within each cycle.

2.3 FUNCTIONAL RELATIONS

Figures 2.4–2.6 show what we may actually see in certain experimental observations. They are pretty pictures and are quite suggestive of wavelike behavior. But what is their precise meaning?

These are examples of a general concept called *functional relationship* by mathematicians. These relationships involve *variables* or quantities that may have several different values under different circumstances. For instance, the amount of income tax that I pay is a variable; it might be many different amounts, depending on my income, on my dependents, and on my itemized deductions. My income also is a variable, differing from one year to another according to whether I have changed jobs, gotten a raise, or made an astute investment.

A *functional relation* is merely an expression of how one variable is related to another. Mathematicians may study functional relations as abstractions for their own sake, but scientists are always concerned with underlying physical relationships, especially of cause and effect. The speed of a rocket ship is related to the length of time its engine has operated, because the force generated during that operation is the very thing that has brought that speed into being. The strength of a sensation of loudness in your head is related to the intensity of the sound waves reaching your eardrums, because the sensation is the result (through several intermediate steps) of the action of those waves.

Functional relationships pervade our everyday lives; they are often easy to recognize and to describe qualitatively. We may say, for example, "Tax is an

AUGUST 1987 WEATHER CALENDAR

DAYS	SUN Rise (PDT) AM	SUN Set (PDT) PM	LENGTH OF DAY HRS.	LENGTH OF DAY MIN.	MOON (PDT) Rise	MOON (PDT) Set	TEMPERATURE DATA Record MAX.	Record Year	Record MIN.	Record Year	Normal MAX.	Normal MIN.	Normal AVER.
1	6:07	8:17	14	10	12:45p	11:35p	106	1980	50	1887	93	60	77
2	6:08	8:16	14	08	1:53p	—	107	1946	50	1887	93	60	77
3	6:09	8:15	14	06	3:04p	12:05a	107	1969	51	1919	93	60	77
4	6:20	8:14	14	04	4:18p	12:41a	106	1966	50	1897	93	60	76
5	6:11	8:12	14	01	5:29p	1:28a	107	1978	50	1950	93	60	76
6	6:12	8:11	13	59	6:33p	2:27a	108	1978	50	1891	93	60	76
7	6:13	8:10	13	57	7:27p	3:38a	108	1913	50	1931	92	60	76
8	6:13	8:09	13	56	8:11p	4:58a	108	1984	50	1919	92	60	76
9	6:14	8:08	13	54	8:47p	6:19a	108	1984	50	1931	92	60	76
10	6:15	8:07	13	52	9:16p	7:39a	108	1971	50	1919	92	60	76
11	6:16	8:05	13	49	9:42p	8:59a	110	1898	49	1910	92	60	76
12	6:17	8:04	13	47	10:07p	10:06a	106	1898	50	1910	92	60	76
13	6:18	8:03	13	45	10:32p	11:16a	111	1933	48	1921	92	60	76
14	6:19	8:02	13	43	10:59p	12:24p	107	1920	49	1887	92	60	76
15	6:20	8:00	13	40	11:29p	1:31p	108	1920	51	1955	92	60	76
16	6:21	7:59	13	38	—	2:37p	105	1920	50	1955	91	60	76
17	6:22	7:58	13	36	12:04a	3:40p	106	1967	51	1917	91	60	75
18	6:22	7:57	13	35	12:46a	4:38p	107	1950	52	1894	91	60	75
19	6:23	7:55	13	32	1:34a	5:27p	108	1950	51	1890	91	60	75
20	6:24	7:54	13	30	2:28a	6:08p	106	1950	48	1914	91	59	75
21	6:25	7:52	13	27	3:27a	6:50p	102	1982	49	1910	91	59	75
22	6:26	7:51	13	25	4:28a	7:21p	106	1891	50	1901	91	59	75
23	6:27	7:50	13	23	5:30a	7:48p	109	1913	50	1908	91	59	75
24	6:28	7:48	13	20	6:31a	8:11p	108	1931	50	1887	91	59	75
25	6:29	7:47	13	18	7:32a	8:33p	104	1931	52	1887	91	59	75
26	6:30	7:45	13	15	8:33a	8:15p	106	1894	50	1929	90	59	75
27	6:30	7:44	13	14	9:35a	9:15p	108	1894	51	1952	90	59	75
28	6:31	7:42	13	11	10:38a	9:39p	105	1915	50	1910	90	59	75
29	6:32	7:41	13	09	11:44a	10:06p	104	1976	49	1880	90	59	75
30	6:33	7:39	13	06	12:53p	10:39p	106	1976	48	1887	90	59	75
31	6:34	7:38	13	04	2:04p	11:21p	108	1976	51	1914	90	59	75

FIGURE 2.7 A table of information about daylight hours and weather in Sacramento. If you believe sunrise times do not jump about randomly, then you can spot a typographical error in that column.

increasing function of income," meaning that if the income increases then so will the tax. But words are an inefficient way for scientists to express these relations with precision, so they resort to several other means of communication.

One way to express these relations is by listing various pairs of corresponding values of the variables in a table with two (or more) columns, as in Figure 2.7. (This example appears in my local newspaper each month.) In similar fashion, we shall use such tables several times below. Table 7.1 (page 115), for example, indicates a relation between one variable measured in numbers and another variable with qualitative values identified by code words or symbols. The scientist more commonly deals with cases in which both variables are quantitative; we will see an example of this in Table 5.1 (page 76). The important point about all such tables is that whenever we find one particular value of either variable to be of special interest, we can consult the table to find the corresponding value of the other variable. For instance, as I write this paragraph on August 6, I can see that the sun rose this morning at 6:12 and will set tonight at 8:11. By examining many such individual cases, we may develop some understanding of the underlying relationship.

A second way to represent the same information is by encoding it in the shorthand of an algebraic formula. For example, we could immediately convey all the information in Table 5.1 to a physicist merely by stating "$LD = 10 \log IR$." Another example from everyday life is $t = d/v$, where t is the amount of time it takes to drive your car a distance d at average speed v. It is a concise way of saying, "If you divide the number of kilometers to your destination by the speed you expect to maintain in km per hour, you will get the number of hours the trip will take." The formula that produces the day lengths shown in Figure 2.7 is

$$L = 12\,\text{hr}[1 + (2/\pi)\arcsin(\tan N \tan D)] + C,$$

which is far more complicated than anything you will be asked to do in this book. But it illustrates the general point that the sun's position D on any day of the year can be used to calculate the day length L.

The great power of any symbolic formula is that it encompasses every specific case. You should *not* assume that all such formulas will mystify you. Many of them are well within the grasp of every reader of this book, and the main barrier to understanding them is neglecting to think carefully about the physical meaning behind each symbol. Even so, we will use such mathematical shorthand only sparingly.

But there is a third way to represent functional relations, one that we will use repeatedly. This is *graphical representation*, and it bears out the truth of the saying, "A picture is worth a thousand words." Consider the curve labeled "Sacramento" in Figure 2.8 as a representation of the relation between day length and date. This shows the same information that we would find in a complete series of tables like Figure 2.7, because that is precisely the information that is used to construct the graph. For example, taking the data for August 6, you can locate that date on the horizontal axis and then move upward until nearly even with 14 hours on the vertical axis and make a dot at point A. After doing this with many pairs, you finally would be able to sketch a smooth curve passing through all those points.

Once the graph is plotted, anyone may use it to find pairs of corresponding values without having the original table or formula. For example, you can answer the question "How long will the sun be up in Sacramento on February 1?" by following the arrows through point B. The other arrows through C will answer "On what date is the sun up for 13 hours?"

Figure 2.9 gives you further opportunities to see the variety of information that can be conveyed quickly by a graph. After thinking carefully about the information shown in each graph here, you should be prepared to use all the other graphs that lie ahead. In particular, we now can say precisely what is represented in the earlier graphs in this chapter. The oscilloscope traces are

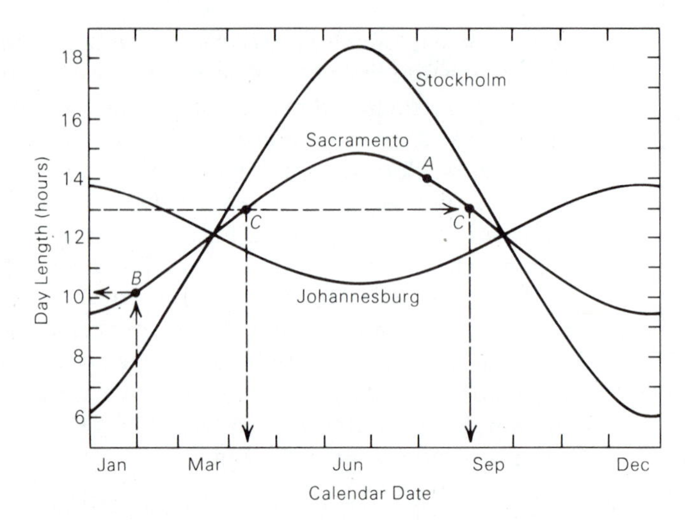

FIGURE 2.8 A graph representing information such as that found in Figure 2.7 for three cities at different latitudes. The answers to the questions posed in the text are (*B*) approximately 10 hr 10 min and (*C*) either April 11 or September 2.

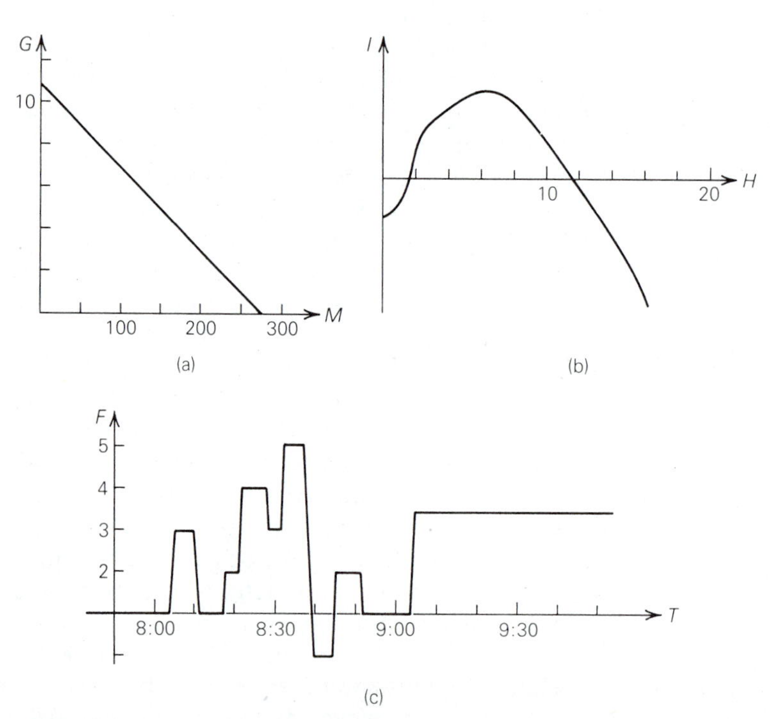

FIGURE 2.9 Examples of graphs. (a) Number of gallons of gasoline *G* remaining in a car's fuel tank as a function of distance traveled in miles *M*. (b) Improvement *I* in a musical performance as a function of the number of hours per day *H* spent practicing in preparation. (c) Height above ground in floors *F* of an elevator as a function of time *T*. For further interpretation, see Exercise 11.

graphs of voltage from a microphone versus time; because the voltage is directly proportional to the sound pressure caused by the voice or instrument, we also may regard the graphs as showing how this sound pressure varied over time.

One additional point is illustrated by considering how long the sun is up for other cities at various distances north or south of the equator. We must admit that now there are more than two interdependent variables involved; in the formula above it means we are concerned with different possible values of N. We could handle that by making a whole set of graphs (or tables), each representing day lengths for a different city. But as long as we take only a few examples, we can imagine them drawn on transparent plastic and superimposed as in Figure 2.8. Similarly in Chapter 6, we will want to show the relationship between a sound's physical strength and the corresponding perceived loudness. Again, there is actually a third variable, for different results are obtained when the experiment is carried out with tones of high, medium, or low pitch. We will deal with this in the same way we did here by showing a whole family of curves (Figure 6.10, page 102).

2.4 SIMPLE HARMONIC OSCILLATION

Now let us return to the question of what type of wave motion should rank as the simplest of all. The simplest sound waveforms come from those sound sources that vibrate in the simplest way. So what are the essential requirements for any object to vibrate?

First, there must be some shape or position of **equilibrium** in which the object could remain at rest. For the examples of Figure 2.10, these are denoted by E. Second, any displacement of the object from equilibrium must bring

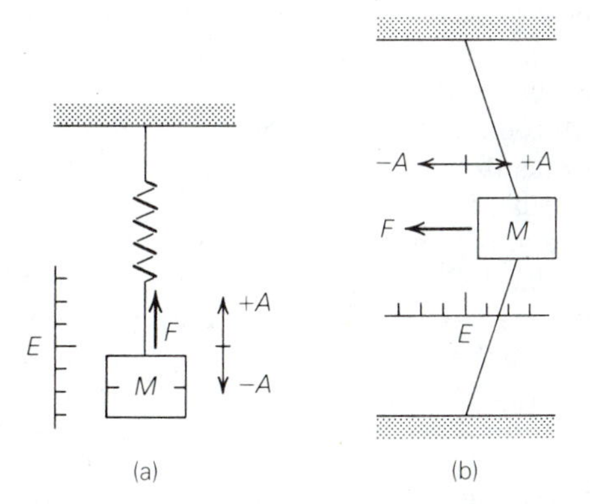

FIGURE 2.10 Two examples of simple vibrating systems. (a) Stretched spring provides restoring force F for mass M. (b) Tension in both support cables provides restoring force. E represents equilibrium position; A represents amplitude of motion.

into play a **restoring force** that tends to pull or push it back toward the equilibrium position. This restoring force generally can be thought of as "stiffness" or "springiness." Examples include both the return pull of a stretched spring and the extra outward push from a lump of air squeezed by the motion of the adjoining lumps.

Third, to keep the object moving *through* the equilibrium point, there must be **inertia**. This is a name for the tendency of any body to continue whatever motion it already has; only the action of forces can change that motion. Thus, a pendulum does not just stop when it reaches the bottom of its swing; its inertia causes it to overshoot, and some force must act to slow it down and bring it back again from the other side. Inertia is a manifestation of the object's mass (Box 2.2).

The vibration process then works like this: Suppose the object is displaced

BOX 2.2 MASS AND WEIGHT

Forces can be exerted in many different ways—by ropes, sticks, fingers, wind, magnets, and so on. One especially familiar force is that of gravity: The Earth pulls downward on any and every object in its vicinity (whether touching it or not) with a force called the weight of that object. Note carefully: *Weight is not the same thing as mass!*

Mass tells how much matter an object is made of and is measured in kilograms; it is an *intrinsic* property of the object. Specifically, "how much matter" means how many of the elementary particles called protons and neutrons. This is essentially without regard to how tightly they may be packed together in different kinds of atoms or molecules. A kilogram of feathers may occupy a much larger volume of space than a kilogram of lead, but within a margin of less than 1 percent both contain the same total number of protons and neutrons.

Mass is responsible for the property of inertia; it is a measure of how hard it is to change the motion of an object. This is expressed by Isaac Newton's famous law of motion, $F = Ma$. Its meaning is clearer in the form $a = F/M$, which says that when identical forces are exerted on two different objects, the more massive one will show less acceleration (change in velocity).

Weight is the *force* with which a mass is attracted to the Earth (or whatever other planet it may be near), and is measured in newtons; it is an *extrinsic* property. The relation between weight and mass is given by weight $= Mg$; g (the acceleration of gravity) tells how rapidly an object will accelerate in free fall, and is determined by the mass of the Earth and by how close we are to it. The value of g on the Earth's surface is approximately 9.8 m/s^2; that is, for every second it remains in free fall (neglecting air resistance), a dropped object will increase its downward speed another 9.8 m/s over what it was before.

A typical male student in your class might have a mass of 70 kg and a weight of $70 \times 9.8 =$ 686 N. (His weight would be called 154 pounds in the old English system.) If the same student lived on the moon, his mass would still be the same. But because g on the moon is only about 1.6 m/s^2, his weight would be only a sixth as much, around 114 N (or 26 lb). If he were on a space trip, far from any planet or moon, his mass would still be the same, but his weight would be practically zero. It is most unfortunate that food-processing companies are careless about this distinction. You can pick up a can of peas in the grocery store that says, "Net weight 454 grams," yet there is no such thing. It is the net *mass* of 0.454 kg that you pay for; its weight is about $0.454 \times 9.8 = 4.45$ N here on Earth, but it would be different if you took the can to another planet.

a distance A toward the right as in Figure 2.10b, perhaps because someone has pulled it aside and then let go. The restoring force pulls toward the left and produces motion. As long as M is to the right of E, F continues to make it go faster and faster toward the left. When M reaches E, its inertia keeps it going. Once past E, the force F pulls toward the right, gradually reducing the speed until M comes to rest a distance A to the left. F continues to act toward the right, and M begins to move toward E again. Once past E, the force F reverses direction and slows down M until it comes to rest at its original starting point. Then the whole sequence begins over again. The maximum displacement (A) to either side of equilibrium is the **amplitude** of the vibration.

The special type of vibration called **simple harmonic motion (SHM)** occurs whenever the restoring force is of the uniquely simple kind called *linear*. This means the restoring force is directly proportional to the displacement; that is, moving the object twice as far from its equilibrium configuration will result in precisely twice as much restoring force (Figure 2.11). For physicists, SHM is especially important because sufficiently small vibrations of natural systems usually are of this kind. It is convenient for us that the restoring forces in violin strings and air-filled pipes are very nearly linear.

The frequency of a simple oscillation always is determined by the strength of its restoring force and by its inertia. For example, take an ideal spring of stiffness K, meaning that to stretch the spring out a small distance y requires a force $F = Ky$ pulling on it, as in Figure 2.11a. Then a mass M hanging on the end of the spring will bounce up and down in SHM with frequency $f = (1/2\,\pi)\sqrt{K/M}$, where $\pi \cong 3.1416$. Notice that this formula says (in an exact way) that the vibration will be more rapid on a stiffer spring (one with larger K) but slower if the mass is increased. One of the most remarkable things about the formula is that A does not appear—it says the frequency of SHM does *not* depend on whether the amplitude is large or small. This is a property we would very much like musical oscillators to have so that the pitch of each note will remain steady whether it is played loudly or softly.

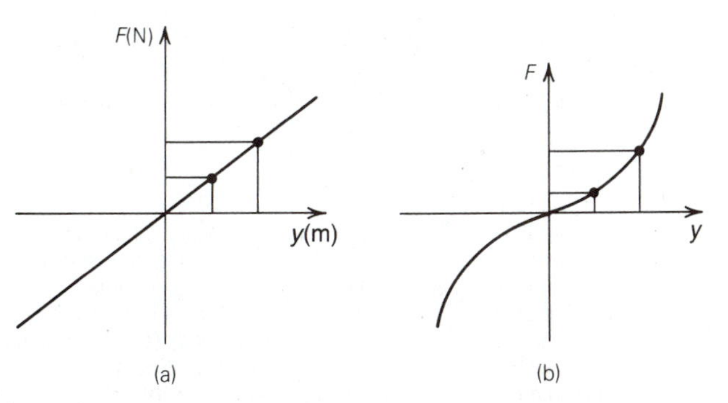

FIGURE 2.11 Dependence of restoring force F upon displacement y for (a) linear and (b) nonlinear restoring forces. The distinctive property of the linear case is that a doubling of displacement always causes a precise doubling of restoring force.

The importance of the formula extends beyond this particular example, for in *every* case of SHM we always can identify something playing the role of stiffness or elasticity and something playing the role of inertia or mass. The frequency of vibration will always depend on these quantities in just the same way it does on K and M for the spring. Several musical examples of similar square-root formulas (for drumheads, guitar strings, and so on) occur in Chapters 9 and 10.

An equivalent term for SHM is *sinusoidal motion*. Figure 2.12a shows a graph of this motion. An object vibrating with SHM creates a sound wave in

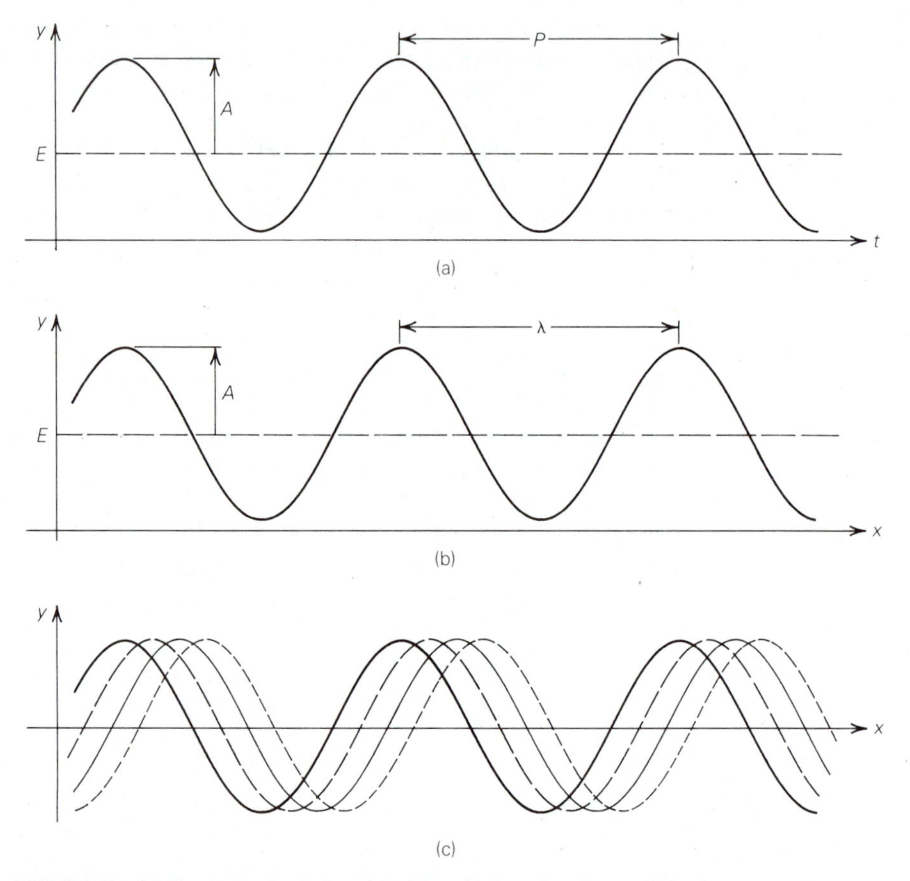

FIGURE 2.12 (a) Graph of a simple harmonic (sinusoidal) motion. The vertical axis y represents displacement (such as height of a mass hanging on a spring or horizontal position of a pendulum bob), while the horizontal axis t indicates the passage of time. Note the amplitude (maximum displacement) A and the period P in time. (b) A sinusoidal wave disturbance as a function of position x in space. The y axis represents some aspect of the disturbance (such as height of water surface, electric field strength of a radio wave, or air pressure in a sound wave). Note the amplitude A and wavelength λ. This graph is like a snapshot of the wave, and we can imagine a series of such snapshots (c) forming a moving picture in which we see the waveform travel across the page. The time history for the SHM of each single piece of the wave-carrying medium is a graph like (a), although for different pieces the peaks occur at different times.

which each bit of air also undergoes SHM. This is called a simple harmonic wave, or sinusoidal wave, or just sine wave for short (Figure 2.12b,c). You can make an approximately sinusoidal sound wave by whistling.

The full importance of SHM in acoustics will become apparent only when we see (in Chapter 8 and onward) how other, complex motions can be understood as combinations of several simple harmonic motions.

2.5 WORK, ENERGY, AND RESONANCE

For a deeper understanding of vibrating systems, we should go beyond a mere description of the possibility of SHM. We should look further into the relation between an oscillator and its surroundings, and especially into what must be received from the surroundings to get the vibrations started.

The key concepts here are work and energy. These two words are used in a much more precise and limited way in physics than in most people's everyday use. **Work** is done whenever one object exerts some force upon another *while it moves* in the direction of that force. If there is no motion, there is no work, however great the force may be. I may push all day against a brick wall and go home tired, but as long as the wall does not yield I have done no work at all as physics understands it. It is also important that only force components in the same direction as the motion count for doing work. An object that moves only horizontally, for instance, has no work done on it by the force of gravity, which is vertical; only if it moves up or down can the Earth's attraction do any work upon it.

The measure of the amount of work is the force multiplied by the distance through which it acts: $W = F \times D$. To lift a 10 kg mass, I must pull upward with a force equal to its weight, which is 98 N (about 22 lb). If I lift it through a vertical distance of 2 meters, the amount of work I have done is $98\ \text{N} \times 2\ \text{m} = 196\ \text{N} \cdot \text{m}$; this also is called 196 joules (J). (A 100-watt lightbulb uses this much energy in about 2 seconds.) For another example, take a violinist using a force $F = 0.6$ N to pull the bow across the strings. The work done in pulling the bow a distance $D = 0.3$ m is $W = 0.6\ \text{N} \times 0.3\ \text{m} = 0.18$ J. Some small portion of this work eventually goes to move the air around the violin.

Energy is an intangible property gained by anything upon which we do work. It can be defined as a measure of the net amount of work done upon some object in the past or, equivalently, the amount of work it could do in the future upon other objects. Energy, then, is a quantity that is transferred from one body to another by the process of doing work. It may be stored, like money in the bank, and withdrawn later. Energy (the balance in the account) is measured in the same units, joules, as work (the deposit or withdrawal).

A moving billiard ball has energy (called kinetic energy), because if it is stopped in a collision with another ball it can do work upon that ball; that is, the first ball can set the second in motion, thereby transferring energy to it. Water stored behind a dam also has energy (called potential energy) by virtue of its altitude, because if it falls to a lower level through a turbine it does work. This work may turn a generator to supply electrical energy to nearby homes. A stretched spring also stores potential energy, because it can do work by pulling a massive object along as it returns to its relaxed position.

Energy does not magically appear out of nowhere. According to a fundamental law of nature, we must always have *conservation of energy*. Energy may be transferred from one body to another, or it may change from one form to another, but the grand total *never* changes. When you give kinetic energy to a ball by throwing it, that is at the expense of the chemical energy you stored in your body when you last ate.

Sound waves carry energy. The vibrating source does work upon the adjacent air molecules when it sets them in motion. The energy is passed on from neighbor to neighbor, and finally those molecules next to your eardrum push it aside; thus they do work upon it and deliver the energy to its destination in the form of vibrations of the internal structure of the ear.

The continual competition between restoring force and inertia in any harmonic oscillator can also be described as an interplay of potential and kinetic energy. As the mass moves toward the equilibrium position (as at times t_1 or t_3 in Figure 2.13), the restoring force does work on it, thus increasing its kinetic energy. This energy is withdrawn from the reservoir of potential energy stored in the spring. As the mass moves beyond the equilib-

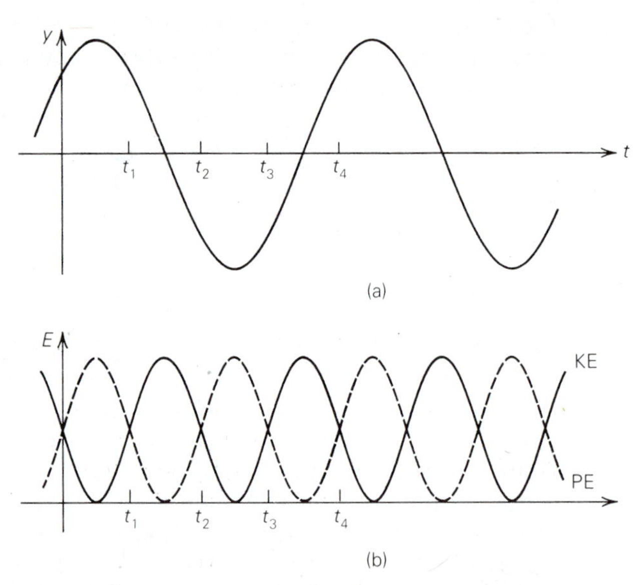

FIGURE 2.13 (a) Displacement of an oscillator as a function of time. Positive and negative y may represent points above and below equilibrium for the mass in Figure 2.10a or to the right and left in Figure 2.10b. (b) Kinetic energy KE (solid line) and potential energy PE (dashed line) for the same oscillator. At times t_1 and t_3, PE is being converted to KE, and vice versa at t_2 and t_4. Notice that for this idealized frictionless oscillator, the total energy in PE and KE combined remains constant throughout the motion.

rium position and slows down, the restoring force is in the direction opposite to its motion. This means the restoring force does negative work on the mass, reducing its kinetic energy (times t_2 and t_4 in Figure 2.13). Another way to say this is that now it is the mass that is doing work on the spring and delivering energy to it. Thus, during vibration, energy shuttles back and forth between kinetic and potential forms, but the total energy remains always the same. In real life, of course, this total oscillation energy may gradually decrease as friction or radiation drains it off as heat or sound energy.

Where does the oscillation energy originally come from? A violin string does not spontaneously begin to vibrate—it must be plucked, struck, or bowed. We have here two general categories of excitation—impulsive and continuous. Plucking and striking both require some external agent (such as a finger) to push the string aside. That is, the finger does work on the string, delivering in a single impulse the supply of energy for the subsequent free vibration of the string.

In continuous excitation, an external agent (such as the violinist's arm pulling the bow) does more and more work on the string while it vibrates. (Exactly how this is accomplished is somewhat complicated and will be explained in Chapter 11.) This continuing energy input makes up for the losses to friction and radiation and thus maintains the vibration energy at a constant level.

An especially efficient way to deliver energy continuously to an oscillator is through **resonance**, which occurs when the driving force cooperates with the oscillator by alternating at about the same frequency that the oscillator would naturally prefer. This enables the driving force to push to the right when the mass moves to the right and to the left when the motion is toward the left, thus always doing positive work and delivering more energy to the oscillator. To understand what that means, think of pushing a child in a swing. Only when you time your pushes to occur at the same frequency as the swing's own natural motion can you cause a large-amplitude oscillation by repeated gentle pushes.

The famous tides in the Bay of Fundy provide another example of resonance. They are much higher (10 or 15 meters) than tides in other bays, because the size of the bay and its narrow outlet give it a natural period of water inflow and outflow of approximately 12 hours. As the Earth rotates, the tidal forces from the moon pull first one way and then the other that often and thus excite a resonant response.

For the time being we will use the term *resonance* rather loosely to indicate strong oscillation occurring at a naturally preferred frequency. For example, you may hear sound from resonant oscillation of the air in an empty gallon jug or soft-drink bottle when you blow across its mouth. (These and other acoustic oscillators with large air reservoirs and narrow necks sometimes are called *Helmholtz resonators*.) We will develop a more formal definition of resonance and explain more about how it can be achieved in Chapters 11–13.

SUMMARY

Change and movement are indispensable elements of music on all time scales. Modern devices, especially the oscilloscope, enable us to study many details of the individual vibrations in a sound, which occur at rates of hundreds or thousands per second. The fullest understanding of the significance of what we see on an oscilloscope screen is gained only when we think about the general meaning of functional relationships and about how they are represented by tables and graphs.

The frequency of a wave tells how often its main crests pass a fixed point in space, whereas the speed tells how fast it moves from one place to another. The wavelength tells the distance in space from one crest to the next similar crest on a snapshot that freezes the motion everywhere at the same instant in time. These three quantities are related through the important equation $v = f\lambda$. Out of many possible waveforms, it is the sinusoidal or simple harmonic wave that is the key to understanding many features of sound production and perception.

Sinusoidal waves come from sources vibrating in simple harmonic motion. Such oscillations result from the action of linear restoring forces, which vary in direct proportion to the displacement from equilibrium. Every simple harmonic oscillator has its own natural frequency, determined jointly by its stiffness and its inertia. Oscillatory motion involves both potential and kinetic energy. Resonance provides an especially efficient mechanism to supply an oscillator with large amounts of energy.

REFERENCES

Anyone interested in the beginnings of musical acoustics as a modern science should become acquainted with Hermann von Helmholtz's *On the Sensations of Tone* (translation by A. Ellis from the German edition of 1877, reprinted by Dover, 1954). Before his time there were a few scattered observations, but Helmholtz carried out an extensive program of research over many years. His investigations were aided by his expertise in physiology as well as in physics; he was the first to measure the velocity of nerve impulses, in 1850. His appreciation of the uniquely musical elements in musical acoustics still provides a high standard not always matched by modern writers.

SYMBOLS, TERMS, AND RELATIONS

f frequency
P period
λ wavelength

v speed of sound
K stiffness
M mass

Mg weight
$\pi \cong 3.1416$
$g = 9.8\ \mathrm{m/s^2}$

A amplitude
a acceleration
SHM simple harmonic motion
F force
D distance
W work

$P = 1/f$
$v = f\lambda$
$f = (1/2\pi)\sqrt{K/M}$
$W = FD$
hertz
equilibrium

inertia
restoring force
sine wave
functional relation
kinetic energy
potential energy

EXERCISES

1. If the frequency of a middle-C sound wave is 262 Hz, what is the period of each vibration?

2. If the frequency of a certain FM radio wave is 100 MHz, what is the period of each individual vibration?

3. If $P = 1/f$, then also $f = 1/P$. What is the frequency of a sound wave whose vibrations occur 5 ms apart?

4. The marks on the dial of my wristwatch indicate that the hand jumps once every 0.2 s. What is the frequency with which ticks are emitted? (This is not to be confused with the much higher frequency of the vibrations that make up the sound of each tick.)

5. Suppose you measured both frequency and wavelength of a sound wave, and found them to be 175 Hz and 2 m, respectively. What speed of sound does that imply? Try to think of two different reasons why the answer may not be 344 m/s.

6. If some ocean swells are spaced 25 m apart, and as they sweep past your small boat it bobs up and down 15 times per minute, at what speed are these waves traveling? What distance would they cover in 24 hours?

7. If a tuning fork vibrates at 1000 Hz, what is the wavelength of the sound waves produced? (Assume $T = 20°C$.)

8. If a violin string vibrates at 1720 Hz, what is the wavelength of the sound produced in the surrounding air? (Assume $T = 20°C$.)

9. If an organ pipe is built to produce sound of wavelength 4 m, what frequency will this sound have? (Assume $T = 20°C$.)

10. If a trumpet produces waves of wavelength 0.5 m in air with $T = 30°C$, what is their frequency?

11. For each of the three graphs of Figure 2.9, describe as completely as possible the information conveyed by the graph. Include, for example, a speculation upon what type of car may be involved in part (a) on the basis of the number of miles per gallon it is getting and possible reasons why the elevator in part (c) did not go anywhere after 9:05 A.M. Discuss also the lack of any quantitative units for the I axis in part (b).

12. Identify another situation from everyday life that illustrates a functional relationship and sketch a graph that represents it.

13. You are watching water waves pass a tide gauge fastened to the end of a pier. At their highest the crests reach a mark labeled 3.8 m, and at the low point of the troughs you can just see the 3.2 m mark. What is the amplitude of these waves?

14. The mass of 1 cubic meter of air at sea level is about 1.3 kg. How much does it weigh in newtons? How much would a stack of 8000 such cubes weigh? If this total weight were all applied to an area of 1 m^2, how much pressure would it produce on that surface? The answer should look familiar. (Although the atmosphere actually gets less dense with increasing altitude, the total amount of air is the same as if these sea-level cubes were stacked some 8 km high.)

15. If a student weighs 490 N, what is her mass in kg?

*16. If a steady force of 2 N applied to a certain spring stretches it 0.2 m, what is its stiffness, K? If a mass of 0.1 kg is hung on this spring, what will be its natural frequency of oscillation?

*17. You blow over the top of a jug and make note of the pitch you hear. Now you partially fill the jug with water, which reduces the volume available to the air, and blow again. Is the pitch, and thus the natural frequency of oscillation, higher or lower than before? Do you attribute this mainly to a change in stiffness or in inertia? To decide, think about the air inside the body of the jug versus the small plug of air in the neck. Which moves most during the vibration and therefore is mainly responsible for the vibrational inertia? Was this changed by adding water? Which is squeezed and thus is mainly responsible for the vibrational stiffness? Was it changed by adding water?

18. If you push against a car with a force of 400 N but the brakes are set and the car goes nowhere, how much work are you doing? If the brakes are released and you push this hard while the car moves a distance of 30 m, how much work have you done? Ideally, when you push the car you would like all the work to appear as an increase of kinetic energy (a faster motion) of the car. In real life much of this energy is wasted in what other form?

19. Suppose a hydroelectric generating station takes 100 cubic meters of water per second from its storage reservoir and lets it drop a vertical distance of 30 m to get up the speed to run a turbine. Given that each cubic meter has a mass of 10^3 kg, what is the total mass of 100 m^3? Using Mg, calculate the weight of those 100 m^3. (Keep the arithmetic simple by using the approximation $g \simeq 10\ m/s^2$.) How much work is done upon 100 m^3 of water by this force as it falls 30 meters? The answer tells you how many joules of energy are made available every second to run the generator. If the generator has 85% efficiency (that is, it wastes 15% of that energy by converting it into heat, noise, and so on), how many joules per second are being made available in electrical form? Because 1 J/s is the same as 1 watt, how many 100-watt lightbulbs could be kept burning by this generator?

20. The speed of radio waves is 3×10^8 m/s, the same as the speed of light. If a spacecraft transmits a radio signal with frequency 3×10^9 Hz, what is the wavelength of those waves?

21. Typical advertising by makers of hi-fi turntables includes claims such as "Our tone arm will track accurately with a tracking force of less than one gram." That statement is technically incorrect; explain why and what the makers really mean.

22. If the microphone diaphragm of Exercise 11 in Chapter 1 moved 10^{-5} m under the influence of that 10^{-3} N force, how much work did the force do?

23. Examine Figure 2.5. Which of these five instruments would you say operates most nearly like an ideal simple harmonic oscillator? Which least?

24. If you fasten lumps of clay on the prongs of a tuning fork, does this change the amount of moving mass? Does it change the stiffness of the prongs? Explain then whether the fork's natural frequency of oscillation is increased or decreased by adding the clay.

PROJECTS

1. Take a comb, preferably a long one, and choose some way of rapidly snapping its teeth to hear a tone whose pitch can be identified by comparison with a piano or other instrument. Estimate the frequency of the vibration produced (as described in Section 2.1), and discuss whether this is in reasonable agreement with the frequency of the comparison note as found on the inside of the back cover. Do not be surprised if you are off at first by an octave (a factor of 2); this is rather difficult for the ear to judge.

2. Any television set is a crude stroboscope with a fixed frequency of 30 Hz (that is how often the electron beam paints a picture on the entire screen). Explain in these terms how the spokes in a moving wagon wheel sometimes appear on TV to be stationary or even move backward. Wave a pencil in front of a TV screen in a darkened room; count the multiple images and time the period of your waving and see if you can verify that the strobe frequency is 30 Hz. Can you find some object around the house that will vibrate near enough to 30 Hz (or 60 or 90 or . . .) to have its motion frozen by the strobe effect when illuminated by the TV screen? A spatula or a hacksaw blade or a plastic foot ruler are all good candidates when held firmly against something solid at one end and plucked at the other. A spinning toy gyroscope with a stripe painted across its disc also would be good.

3. If you have access to a microphone and oscilloscope, learn how to connect them properly and then observe the signals displayed for different sound sources. Make as wide a variety of sounds as you can. Describe the differences in terms of amplitudes, frequencies, and complexity of waveforms.

CHAPTER

3 Sources of Sound

A cricket chirps in the night. A distant train whistles. The heroine's contented humming turns to a shriek of terror as she hears the intruder ascending the creaky stairs. . . .

Sounds stimulate a host of familiar images in our minds, and we could easily list hundreds of things or processes that produce sound. But a list alone provides little insight into how those sounds are created. For a scientific study of sound, we should try to recognize those key elements in our examples that make possible a unified explanation of the origin of all sounds.

We already have noted that vibrating objects produce sounds as they disturb the surrounding air. So we now should study (1) how objects can be set in vibration, (2) how vibrations can be sustained or made to continue for a long time, and (3) how well the energy of vibrations can be transferred into the air. It is a combination of these factors that determines the strength of the resulting sound waves.

In this chapter we will describe several ways of generating sound, especially the basic processes that are used in most musical instruments. This will be only an introduction to the processes of sound generation, and many more details will follow in Chapters 8–14. We will find that we need not have those details, however, to explain certain general features, such as the tendency for larger instruments to produce lower notes.

3.1 CLASSIFYING SOUND SOURCES

Before discussing the details of any particular type of sound production, let us think about several ways of dividing sound sources into different categories.

First, we might consider *natural* versus *artificial* sounds. Nature provides many sounds in our environment, some of which are musically useful. But most music consists of sound deliberately produced through consciously controlled processes, and we want to identify and understand these processes.

Second, we are not limited to *original* sounds, but now have the opportunity to hear *reproduced* sounds as well. We are interested in the faithfulness with which the replica mimics the original and also in the physical processes of recording, storing, and reproducing the sound. We will wait until Chap-

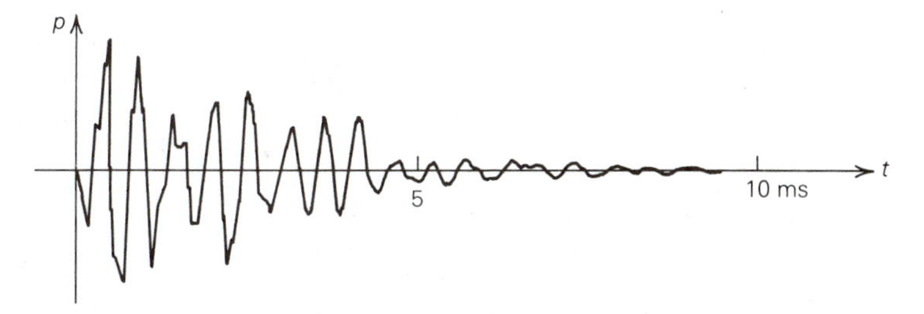

FIGURE 3.1 Oscilloscope trace of the transient "clack" sound from a short wooden stick struck by a metal rod. Compare this with the steady sounds of Figures 2.4 and 2.5 (page 22), whose waveforms repeat over and over again with the same amplitude.

ter 16 to look at those processes in detail. You may think also of the electronic modification of sounds, on which Section 3.6 offers comments.

Third, there are fundamental differences between *transient* and *steady* sounds. Impulsive or transient sounds are temporary and quickly die away (Figure 3.1). They occur when sources are set in vibration at one moment (such as by plucking a guitar string) but left alone thereafter. Steady sounds continue at the same level as long as we choose to sustain them (for example, as long as you are willing or able to continue blowing at the same rate into a trumpet). In such cases we need to understand the mechanisms by which a steady input of energy (from the air pressure in your lungs, for instance) somehow maintains a continuing vibration of the sound source.

Finally, we may sometimes classify sounds according to the means of production; that is, according to *families* of commonly used devices. For instance, we refer to wind instruments or string instruments. But both of these families must be subdivided further; the flute and the oboe are both wind instruments, yet there are large differences in the way they work.

As with any classification, reality does not necessarily adapt itself to our pigeonholes. The piano is a string instrument, yet in some respects it belongs to the percussion family because its strings are struck by hammers. These classifications also may cut across one another. For example, a violin may produce either transient or steady sounds, depending on whether it is plucked or bowed.

3.2 PERCUSSION INSTRUMENTS

You can make a sound by striking any hard object against another. What happens when you do this? While they are in contact at the point of impact, each object is exerting a strong force upon the other and causing some distortion—such as a dent. If the object is highly elastic (which means highly insistent on returning to its original shape when released—like steel, but not putty), the dent is not permanent; the distorted part springs back out when the force is removed.

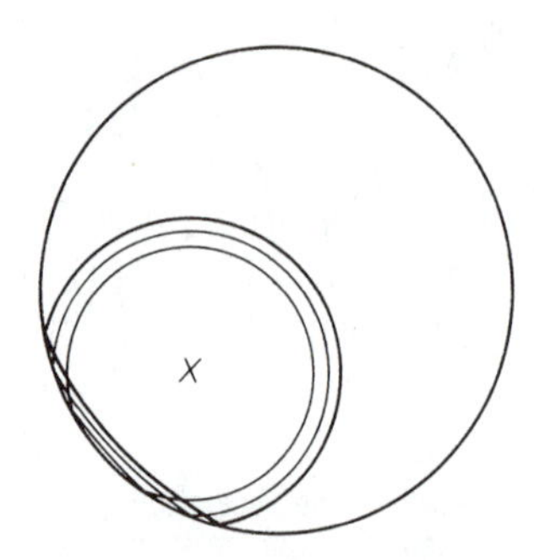

FIGURE 3.2 Wave disturbance traveling on a drumhead shortly after it was struck at point *X*. You may watch similar waves by releasing drops from an eyedropper into a full teacup.

But meanwhile, to relieve some of the stress, part of the deformation has been shared with some adjoining material that was not actually touched by the striking object. And that, in turn, is passed on, in much the same way as sound waves travel through air, and parts of the object farther and farther away from the striking point also are deformed. This is probably easiest to visualize in the case of a drumhead (Figure 3.2), but practically the same phenomenon happens to a gong, a cowbell, or a wooden bar on a xylophone.

In other words, the solid material is capable of carrying a wave disturbance back and forth within itself; this includes the possibility of both longitudinal and transverse waves. As this disturbance makes the surface of the object vibrate, the adjoining air will be disturbed, too, creating a sound wave that can travel to your ear. What are the essential properties of this air vibration near a struck object?

1. The sound may be loud or soft, and this depends on the combination of two factors: the amplitude of the vibrations of the solid object and the surface area that is vibrating. A harder blow creates larger-amplitude vibrations and correspondingly louder sound; as the vibration amplitude dies away, the sound also fades. And if a tuning fork and a bass drumhead both vibrate with the same amplitude (say, 3 mm), the drumhead's large surface will move a much greater amount of air and create a much louder sound.

2. The sound is transient; it soon dies away. This may happen in a fraction of a second (for example, a xylophone bar or snare drum), a second or two (a church bell or a large bass drum), or many seconds (a large gong or a tuning fork). What has happened to the energy that was originally present in the vibration? One possibility is that it has been turned into heat by the flexing of an object that is not highly elastic. This certainly accounts for much of the energy imparted to the xylophone bar when it was struck and perhaps also for an appreciable share of energy in the case of a drumhead.

But as the surface of the vibrating object pushes on the surrounding air, it delivers energy to the air; so the sound also carries away some of the original energy. This can be the major loss mechanism in some cases. A tuning fork, for instance, will continue to vibrate for a long time when held by itself,

FIGURE 3.3 Symphonic percussion instruments. Two sets of tympani are in the background. Notice several triangles just below the glockenspiel in the center. To the right of center is a vibraharp with metal bars and a set of orchestral chimes. To the left of center are a marimba and (next to the bass drum) a xylophone, both with wooden bars. The long pipes toward the left end of the marimba and vibraharp are purely for visual symmetry; they are plugged near the top, and the resonating portion gets uniformly shorter toward the treble end just as for the xylophone. (Courtesy of Ludwig Drum Co.)

because it is creating only a weak sound wave. But if you hold the handle of the tuning fork against a table top, the table vibrates, too. Its larger surface moves more air, creating a much louder sound wave. This carries away energy at a greater rate, and the vibration of the tuning fork dies away much faster as its energy supply is depleted.

3. In most cases, the sound does not produce a clear sensation of pitch; that is, you cannot easily match it to one particular note on the piano. Only for such special cases as drumheads placed over specially shaped "kettles" to make tympani, metal bells cast in a certain shape, or wooden bars undercut in just the right way can we play definite notes in a musical scale (see Figure 3.3). In Chapter 9 we will begin to understand what is so special about those shapes. We also will be able to see then why the "color" or quality of the tone depends strongly on which part of the object is struck.

4. Whether or not the impression of pitch is very definite, the sound at least gives some feeling of being high or low, and this must mean that both the object and the surrounding air are vibrating at a correspondingly high or low frequency. This frequency is determined by the size of the object and by the speed at which vibrations move through it; that speed, in turn, is determined by the stiffness or hardness of the material and by its mass or inertia.

Thus, if you strike two bars of different length, both made of the same material, it takes the disturbance a longer time to make a round trip between the ends of the longer bar; the longer bar passes on to the air a vibration of longer period and lower frequency. And if you strike two bars of identical size and shape, one of steel and one of cast iron, the steel bar produces sound of higher frequency (shorter period) because it is stiffer and so carries disturbances from one end to the other and back faster than the cast iron. We will show in Chapter 8 formulas that tell precisely how these frequencies are related to the material and dimensions of a bar.

A great variety of standard percussion instruments are commonly used in modern orchestras and bands. We can divide them into classes that depend on vibration of a flexible membrane mounted on some kind of rigid frame or vibration of metallic or of wooden objects. If you like fancy names, you can call these *membranophones, metallophones,* and *xylophones* (the prefix *xylo-* simply means "wooden").

Drums most commonly have a uniform membrane fastened on a circular hoop; this is pulled down over another smaller hoop to put the membrane under the desired amount of tension. In the case of modern tympani, there usually is a pedal mechanism with which the player can easily and quickly change the tension for retuning to another pitch. Historically, the membranes were of animal skin, but plastic is now far more common in the United States; it is cheaper, stronger, of uniform thickness, and much less subject to shrinking and swelling with changes in humidity. Bass and snare drums have two heads on a cylindrical frame, with the air between transmitting the vibration from one to the other; their sound has no definite pitch.

Metallophones may also have either definite or indefinite pitch. The former would include vibraphones, bells, and chimes, and the latter triangles, cymbals, and gongs. The high elasticity of hard metals generally is associated with low internal damping and thus with relatively little loss of energy in heating. That is, the springiness of the metal stands in contrast to the "mushy" or "creaky" nature of lossy materials with greater internal damping such as skin or wood. Thus, vibrations in metallophones generally die away slowly unless they are deliberately damped out by contact with some soft object (such as a hand).

The xylophone class includes, of course, the instrument commonly called by that name and its close cousin the marimba; both have several dozen bars mounted in a pattern like that of the piano keyboard, with a resonator pipe underneath each bar to reinforce its sound. There also are hollow wooden instruments, such as temple blocks.

3.3 STRING INSTRUMENTS

A stretched string hit with a hammer could be called just another metallophone. A long, thin string, however, has very special properties that give its sound a much more definite pitch than most percussion instruments. The

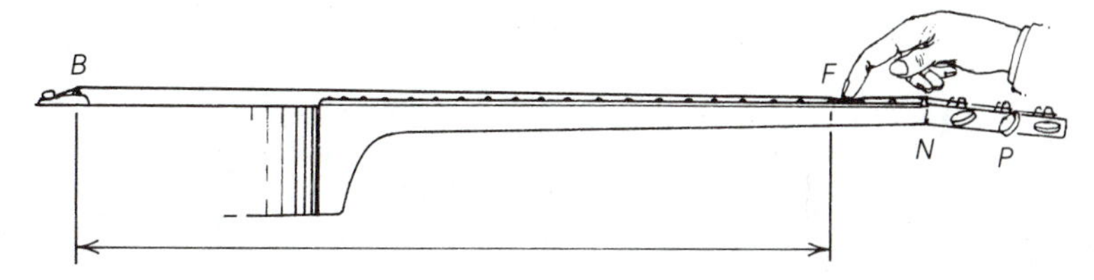

FIGURE 3.4 A guitar string is held above the neck by a bridge *B* on the body and another bridge *N* (customarily called the nut) on the neck near the tuning peg *P*. The arrows indicate the length of string allowed to vibrate when the string is held against one of the frets *F*.

piano, which is a collection of about 200 such strings hit by felt-covered hammers, is used musically for melody and harmony, not primarily for rhythm; it usually is classified with the strings rather than with the percussion instruments.

We can get a somewhat different quality of sound from the same strings by plucking instead of striking them. This is done directly with the fingers on a harp and via a key mechanism on the harpsichord, which has a keyboard like the piano's. All of these instruments require one or more separate strings for each different note that is to be played.

An entirely different approach is to have only a few strings, but to obtain several notes from each by using different portions of its length. We can do this by simply pressing the string at one point against the neck of the instrument so that only part is left free to vibrate (Figure 3.4). There may be thin wooden, metal, or plastic strips called *frets* mounted across the fingerboard, serving as guides and well-defined points against which to press the strings. The shorter the active string length, the higher pitch the note will have. As with the percussion instruments, this is simply because a transverse wave requires less time to go from one end to the other and back on a shorter string.

Such instruments are popular in many cultures, and in ours the most common is the guitar. All the same acoustical considerations apply to lutes, mandolins, balalaikas, vihuelas, and the like. A 12-string guitar produces all the same notes as a six-string guitar but with a richer sound, because each note is being produced by a pair of strings tuned the same (or sometimes an octave apart) and sounding together.

Once we set the strings vibrating, their energy must be passed on to the air. You may wonder if the moving string does this directly by pushing on the adjacent air, but that process is extremely inefficient. This is because the string is so thin: only a small bit of air need move slightly aside to let the string pass (Figure 3.5). You may verify how little the air is disturbed by the string alone by playing an electric guitar with the amplifier turned off. You can easily understand this problem with an analogy: "You can't fan a fire with a knitting needle."

To create a loud sound we must move larger amounts of air, and this is best done by pushing against it with larger surfaces. The electric guitar accomplishes this by taking a very weak signal directly from the string,

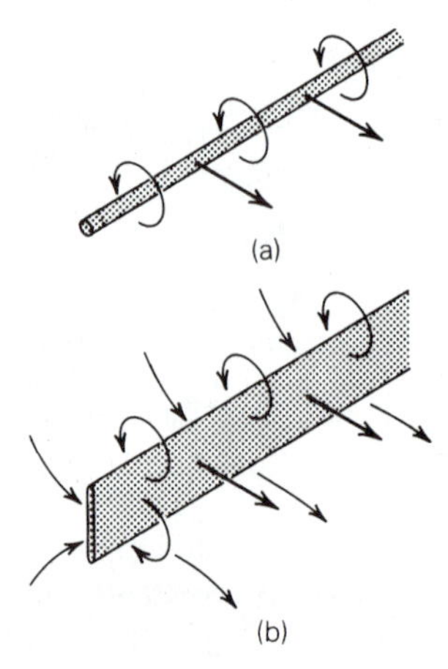

FIGURE 3.5 (a) As a thin string moves through the air, only a small amount of air needs to move out of the way and then fill in again when the string has passed. (b) A wide ribbon of material moving in the same way causes a much larger disturbance of the surrounding air.

amplifying it electronically, and driving a large speaker cone with the amplified signal. The acoustic guitar (and the harp, the piano, and instruments of the violin family) use a mechanical resonator; the motion of the strings is transmitted to a box or soundboard on which they are mounted, and the vibration of this surface accounts for nearly all the sound you hear. The intermediate step does not give you something for nothing; a large-amplitude motion of the string causes only a small-amplitude motion of the (much heavier) soundboard, which nevertheless causes louder sound because of its greater surface area. As with the tuning fork and table top (in Section 3.2), the enhancement of radiation by the soundboard simply drains away the energy supplied by the original pluck more quickly; compare the duration of sound from a plucked violin string with that from an electric guitar.

When we consider the viol and violin families, we face the question of how to sustain the string oscillation indefinitely instead of letting it die away. To say merely that a bundle of horsehair sticky with rosin is pulled across the string does not provide much insight. We must understand *how* the bow hair pulls on the string to understand the particular kind of vibration that results and understand, in turn, the particular tone quality of the violin. This we will do in Chapter 11.

Violin family instruments do not have frets on their fingerboards, so the player may produce a note of any pitch and is not limited to notes in some predetermined scale. This greater range of possibilities is accompanied by greater difficulty in playing, for finger positioning must be much more accurate than on a viol or a guitar.

3.4 WIND INSTRUMENTS

A continuous airstream is another means of supplying energy to sustain a steady vibration. Again we must eventually consider the details to under-

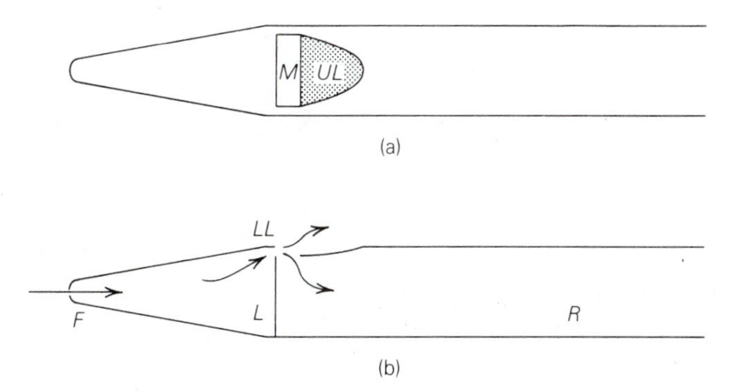

FIGURE 3.6 (a) Front view of an organ flue pipe. (b) Cutaway side view of the same pipe. Air from the wind chest enters the foot of the pipe *F*, which acts as a secondary reservoir from which a narrow stream escapes through the slit between the languid *L* and the lower lip *LL*. The airstream crosses the mouth *M* and is split as it hits the upper lip *UL*, where an edgetone is created. The remainder of the pipe *R* is simply a resonator whose length is adjusted to modify the edgetone and reinforce the desired frequencies.

stand what kind of sound is produced; for now we will introduce only the two main possibilities to see how they influence the classification of wind instruments.

A narrow airstream directed against a sharp-edged rigid obstacle at the right speed and angle will not just flow smoothly past the edge, it will flow rhythmically, first one way and then the other; this disturbance creates sound waves that propagate outward. The resulting **edgetone** has a definite pitch, corresponding to the frequency of airstream oscillation, but the pitch may be obscured by the accompanying hissing sound. (You can make an edgetone by blowing with pursed lips to direct a narrow jet of air against the edge of a playing card or even a sheet of paper.) If there is an adjacent, nearly enclosed air reservoir of proper size and shape, it may respond strongly to the vibrations at one particular frequency. This resonance modifies and reinforces the edgetone vibrations, giving a much louder sound and one with more easily identifiable pitch. A crude example is provided by blowing over the mouth of a jug or pop bottle.

Most organ pipes (Figure 3.6) are simply tubular resonators with a stream of air directed against a sharp edge at one end; each pipe produces only a single note. A recorder (Figure 3.7) is essentially the same, but fingerholes along the side of the tube allow the player to change its effective length so that it will resonate at different frequencies to play different notes. The modern transverse flute also operates on the same principle, but the airstream comes directly from the player's mouth instead of through a mouthpiece windway. This gives the flute player much more flexibility in the strength and quality of sounds produced but at the price of making the flute more difficult to play than the recorder.

The other wind instruments also involve resonating tubes; it does not really matter whether the tube is straight as in a clarinet or coiled up as in a French horn. But while the flute family uses an edgetone, the reed and brass families make use of airflow through a narrow opening of variable width. As the airstream leaves the mouth, variations in air pressure can encourage the

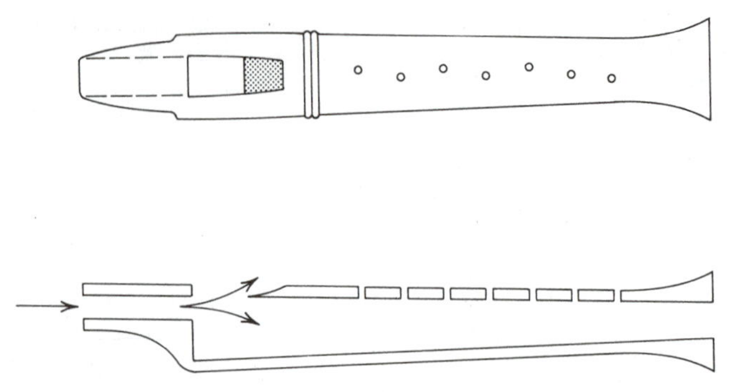

FIGURE 3.7 Top and side cutaway views of a recorder. Note the similarities to Figure 3.6. The amount of taper is variable from one instrument to another. The flared bell is mainly for visual effect; it is only the unflared inner bore that matters acoustically.

flexible edges of the opening to vibrate and alternately reduce or increase airflow as the opening is narrowed or widened. (We will show how this happens in Chapter 13.)

The flexible boundary may be a thin piece of cane reed, in which case we have the *reed woodwind* instruments. (The term woodwind usually includes the flute as well, but as far as acoustics is concerned it would be better unused, because quite different physical mechanisms are involved.) The clarinet and saxophone both come in several sizes and have a single reed fastened against the open side of a hollow mouthpiece (Figure 3.8a). The difference between clarinet and saxophone is in the main body of the instrument; the clarinet has a cylindrical barrel, while the saxophone is conical. The oboe, English horn, and bassoon form another family, using conical bores and double reeds (Figure 3.8b).

Although all the reeds mentioned above are placed in the mouth, it also is possible to have the reed enclosed in a small chamber (Figure 3.8c), with the player blowing (or, in the case of bagpipes, squeezing) only to keep the pressure above atmospheric in that chamber. The krummhorn is a capped-reed instrument that was popular in the Renaissance and is sometimes heard now in concerts of early music. As with the contrast between flute and recorder, the modern oboe with its reed in the player's mouth can be better controlled to provide a wider range of sounds than can a capped reed.

In all these instruments the reed itself can be made to sound at many different pitches; this is easily demonstrated by blowing through the reed (or, for a clarinet, the reed and mouthpiece together) detached from the rest of the instrument. It is the attached resonating pipe that determines the pitch at which the entire instrument will sound, just as it does for the flute family instruments.

It also is possible to let a vibrating reed be the principal determiner of pitch. This is what happens in a harmonica and in the reed pipes of a pipe organ. Because these reeds need to have a very strong preference for one frequency over all others, they are made of metal rather than cane.

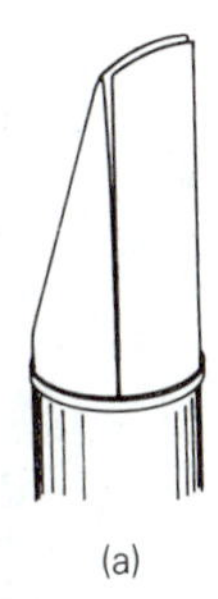

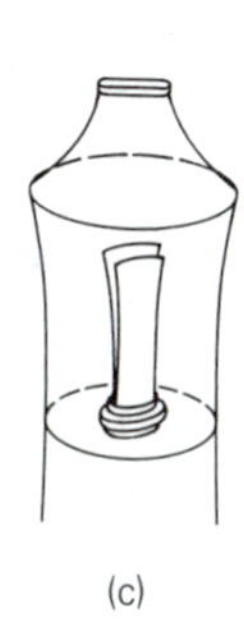

FIGURE 3.8 (a) Single reed and mouthpiece for clarinet or saxophone. (b) Double reed, fastened together by winding with wire and attached directly to a bassoon or oboe. (c) Capped double reed enclosed inside a mouthpiece, as used on the krummhorn.

The *brass* instruments also have a flexible opening that controls their air input, but in this case it is formed by the player's lips. Like a cane reed, the lips alone (or with a detached trumpet mouthpiece) can vibrate at any of a wide range of frequencies, and their buzzing sound is not very musical. When the mouthpiece is attached to the instrument, sound waves travel down the tube and back. Upon returning to the mouthpiece, each wave helps control the lip motion that launches the next wave. It is this resonant action of the long tube back upon the lips that forces them to vibrate in a much more regular (and therefore more musical) way, as well as making it practically impossible to buzz except at a few special "privileged" frequencies. We will show why this is so in Chapter 13.

The human voice operates somewhat like a brass instrument. The vocal cords buzz just like a trumpet player's lips. But they are not attached to a long, thin, highly resonant tube, so they can buzz at any of a wide range of frequencies, like lips in a mouthpiece alone. The mouth cavity alters the sound somewhat as it passes through, but without imposing any limitations on the frequency of vocal cord vibration; this is like attaching a trumpet mouthpiece to a shoebox instead of a long, thin tube. We will present more detail on the voice in Chapter 14.

3.5 SOURCE SIZE

From your own experience you probably find it easy to agree that larger or smaller instruments tend to produce sounds of lower or higher pitch, respectively. But we should state this more carefully and recognize certain limitations.

First, we need to introduce a word that may be unfamiliar to some readers. An **octave** is a natural unit of measurement for the separation of two notes in pitch; for this particular separation they sound so much alike that both are called by the same name even though one is obviously higher than the other.

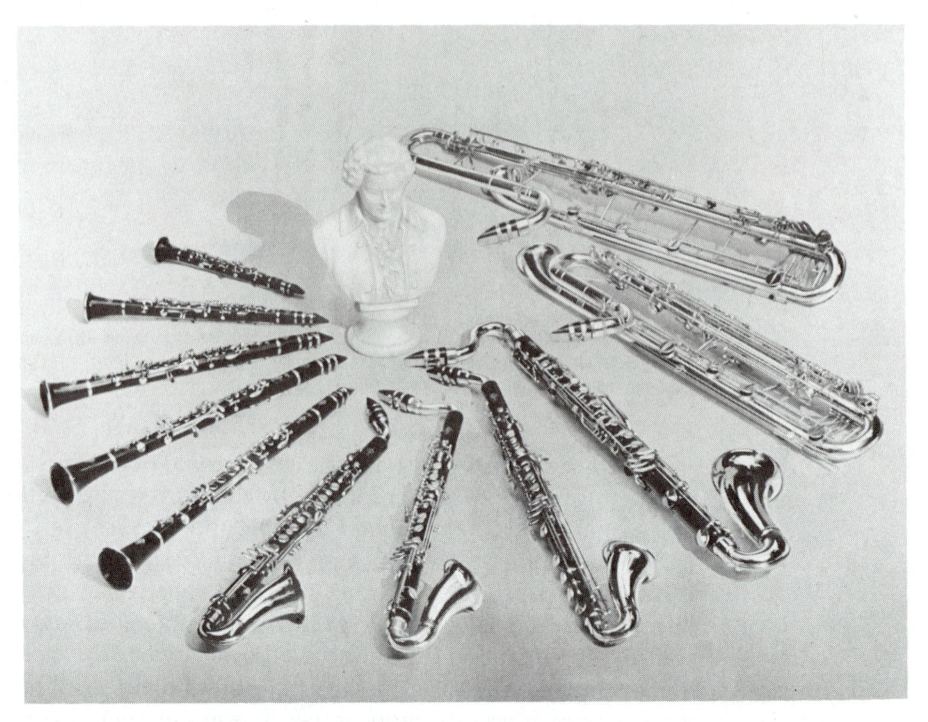

FIGURE 3.9 A family gathering of clarinets, from A♭ sopranino through B♭ contrabass. (Courtesy of G. Leblanc Corporation.)

A demonstration is in order here, as the special nature of the octave is much more obvious from simply hearing it than from any description. (See Exercise 6 for directions on how to do this for yourself with a piano.)

If we take a single family of instruments (as in Figure 3.9), all using the same basic sound-producing mechanism, we find in some cases a definite relation between size and pitch. The trombone, for instance, plays an octave below the trumpet and does so by having twice as long a tube. There is a similar relation between the flute and piccolo and between tenor and soprano saxophones. (Box 3.1 offers comments on the way music is written for members of some instrument families.)

This suggests one of the most fundamental connections between numbers and music: If we double the length of a tube, sound waves will take twice as long to go down and back, and whatever regular oscillations may occur will have twice as long a period as before, or half the frequency. If we observe that this corresponds to an octave difference in pitch of the sound produced, we can speculate that the reason for the important special musical significance of the octave is that one of the tones has precisely twice the frequency of the other. It still remains (for Chapter 18) to explain why a doubling of frequency is so special in terms of how it affects the inner ear and the brain.

If we compare instruments from different families, the relationships are more complicated. A flute and a clarinet have similar length, yet the clarinet

*BOX 3.1 TRANSPOSING INSTRUMENTS

Many wind instruments often are referred to by musicians as *transposing instruments,* but the term is somewhat misleading. It does not refer to any special capability of the instruments, only to an agreement about how the music will be *written* for them. (The note names used here can be related to musical staff notation with the help of Figure E, found on the back endpaper.)

To see why this is done, consider a tenor saxophone player, who has practiced a certain sequence of fingering patterns that gives an F-major scale. Suppose he wishes occasionally to play an alto saxophone; now the same sequence of fingerings produces a B♭ major scale.

He sees now a written note that ordinarily leads to playing a C_4 and presses the keys that would produce this on the tenor sax. But on the alto, F_4 comes out instead. One way to avoid this mistake is to give the brain completely separate training for each instrument. Another is to learn the technical skill of transposition or rapid mental translation of each note upward or downward on the staff to a different note, which then is allowed to determine the fingering pattern.

But a third way, which requires no special preparation or training for the performer, is for the transposition to be included before the music is written down. Thus, all alto sax music is written so that whenever C_4 appears on the staff it means "press the same combination of keys you would if you saw C_4 while playing music written for tenor sax." Because the written music already has allowed for the transposition, the player need not even think about which note actually is sounding.

In the case of saxophones, the entire family is transposing; the players are trained so that a written C_4 will result in a sounded $E_2^\flat$, $B_2^\flat$, $E_3^\flat$, $B_3^\flat$ on the baritone, tenor, alto, and soprano sax, respectively; it is because of this that they are called E♭-alto, and so on. Put the other way around, if you want a tenor and alto player both to produce the same sound, say, C_4 on the piano, you must show the tenor player a written D_5 but the alto player a written A_4.

The common trumpet today is in B♭, meaning that when C_5 appears on the staff the player will leave all the valves up and produce the sound of $B_4^\flat$ on the piano. Other size trumpets (for example, in C, D, or E♭) were common in the eighteenth century and still are used occasionally in some orchestral music. The common clarinet is the B♭-soprano, and other members of the family (Figure 3.9) in B♭ and E♭ are used in bands. Orchestral music often calls for clarinet in A; then the player can just attach the mouthpiece to a slightly longer barrel and let the fingering patterns still correspond to the same written notes.

Although this notational trick makes it easy for wind players to switch around among several different instruments, it also has two serious drawbacks. The first is the enormous confusion it creates for a conductor. If the score shows each part as it is being seen by the player of that part, then the conductor must somehow learn to tolerate the appearance that various parts of the orchestra are playing in several different keys at the same time. On the other hand, if the score is all in concert pitch (so that it could be played on the piano as written), then every time the conductor communicates with a player about a mistake, he must mentally transpose what he sees (for example, a G_4) to call it by the name the player is expecting (for example, A_4 for the trumpet).

The other drawback is that some people with "absolute pitch" ability find it intolerable to play transposing instruments. As a child, I took one lesson on the trumpet, found that I could not bear always hearing a note coming out a step lower than the one I knew I saw on the page, and promptly gave it up. Ever since, although I have played many instruments, my experience has been strictly confined to the nontransposing ones.

plays notes nearly an octave lower; we shall see (in Chapter 13) how this is caused by the clarinet reed closing off the air column at one end, whereas the flute is open at both ends. Again, the oboe and clarinet are both reed instruments of similar length, yet the clarinet can play over half an octave lower; in this case our explanation will invoke the difference between a conical and a cylindrical bore.

Comparison between a violin and a clarinet is an entirely different matter. The size of the clarinet cannot be changed without changing its musical range, because the speed of sound through the air inside the instrument remains always the same. But the size of a violin or cello is determined partly by what is convenient for the player to hold on the arm or between the knees. This is because the original vibration is not in an air column but in a string, and we can make the speed of waves traveling on the string be as fast or as slow as we like by choosing thicker or thinner strings or putting them under more or less tension.

This point is further illustrated by regarding the four strings on a violin as four separate sound-producing devices that just happen to be mounted on the same box. Because all four are the same length, we get sounds of lower (or higher) frequency and pitch by using heavier (or lighter) strings that will carry vibrations back and forth at slower (or faster) speed. Increased tension when the strings are tightened with the tuning pegs also increases the speed. The four strings together can produce a much wider range of notes than would be practical with a single string.

Again, even though a cello plays in a range beginning an octave below a viola, it is generally somewhat less than twice as big. And, unlike wind instruments, string instruments can be made in half and three-quarter sizes for easier handling by children. But the smaller versions have two drawbacks: To produce the same pitches, they must use heavier strings or less tension, either of which leads to a duller tone; and the smaller body is less helpful in radiating the sound into the air.

There is another complication with percussion instruments. Although membranes are like strings in that doubling the size of a tympani head (still made of the same material and under the same tension) would lower its pitch one octave, that is not true of metallophones or xylophones. If you cut a long uniform bar in half, you will find that the pitch you get by striking the shorter bar is not one but two octaves higher than for the original bar. This is because transverse waves of different wavelength do not all travel at the same speed in a solid bar: halve the wavelength and you also double the speed, so that complete cycles of vibration in the half-length bar take only one-fourth the time they did in the original bar.

3.6 SOUND FROM THE NATURAL ENVIRONMENT

Composers have long wanted to include naturally occurring sounds in their music. But in past centuries they were limited to imitating these sounds with musical instruments; for example, by writing a bird call into a flute part or using tympani for rolling thunder or muted violins for wind sighing in the trees. After all, it is rather difficult to make a real bird or breeze perform on command.

In the 1920s, Respighi directed that a phonograph recording of a nightingale's song be played as part of his symphonic poem, *The Pines of Rome*. Many contemporary composers take advantage of tape recorders to collect snippets of natural sound that can be played back on cue at any time. The imagination of the collector is the only limitation, for by striking or rubbing whatever objects are available, you can produce a great many different sounds. By playing these back and rerecording at different speeds, through filters, or backward, you can achieve an even greater variety.

If a musical composition consists primarily of natural sounds and their derivatives, it is sometimes called by the French term *musique concrète*, meaning "the music of real objects" (as opposed to abstract music). Currently, the composer who has the inclination and the recording equipment to undertake *musique concrète* probably also has electronic oscillators and can produce artificial music without using any natural source or traditional musical instrument. You are more likely to hear *concrète* effects as partial ingredients in a piece of electronic music. For an early example, listen to Edgar Varèse's *Poeme Electronique* (1958).

These distinctions between natural and artificial are not especially important to our study of acoustics, because the same concepts we use in understanding the production of sound by traditional instruments are the ones we would use to analyze sound from any other source. There is also no definite dividing line between *musical instrument* and *musique concrète* in the progression from (1) playing a piano normally through (2) reaching inside and plucking the strings, (3) beating on the strings with a drumstick, and (4) hammering on all parts of the piano to (5) hammering on a nicely resonant bridge railing. If you find this expansion of musical resources appealing, you should delve into the works of Henry Cowell and John Cage.

SUMMARY

Plucking or striking any object produces a transient sound. The percussion instruments that are most musically useful depend on one or more of these effects: (1) particular shapes that produce sound with a definite pitch, especially the long thin string as found in the piano or hammered dulcimer; (2) springy, highly elastic material (metal) for longer-lasting sounds or lossy materials (wood, plastic, leather) for relatively short, dull thumps; (3) smaller, lighter, or stiffer objects or objects under greater tension to produce higher-pitched sounds; (4) a larger surface area or an associated resonator to produce louder sounds.

Steady sounds require a continuing supply of energy, and there are two common ways of doing this in musical instruments: (1) with the work expended in pulling a bow across a string and (2) with the work done in blowing a steady stream of air through a narrow opening, either against a sharp edge or past a flexible reed or lip. If we include electronic oscillators, then we can say that a third way to produce continuous sound is to supply the energy electrically. In later chapters we will explain just how the oscillations are sustained and how their exact form is determined.

Wind instrument sizes are related directly to the frequencies of sound they produce; within a given family, halving the length doubles the frequency and thus raises the pitch one octave. String instrument size has only a qualitatively similar relation to pitch, because the frequency also is influenced by string diameter and tension.

Tape recording makes it practical for composers to use the unlimited variety of sounds made by sources other than traditional musical instruments. Full exploitation of the possibilities includes electronic modification of the sounds.

REFERENCES

The fundamental characteristics of transient and steady sounds are discussed at much greater length in Benade, especially Chapters 2 and 4. Additional references on the acoustics of specific instruments will be suggested in later chapters.

For further accounts of the instruments of the orchestra and their musical uses, you may consult typical "Introduction to Music" texts as well as various musical encyclopedias. (See the references at the end of Chapter 7.) For more detail, check your library for separate books about different instruments or families.

SYMBOLS, TERMS, AND RELATIONS

percussion
winds
strings
transient
damping
frets
musique concrète
edgetone
reed
octave

EXERCISES

1. How would you classify the instrument and the sound when a violin is played *col legno?* (That means the bow is flipped over and its wooden shaft is used to strike the strings.) What if the score calls for the player to rap his knuckles against the body of the violin? Or to shatter it over the head of the neighboring player?
2. Explain why placing your finger exactly in the middle of a guitar or violin string and pressing it against the neck should give a tone about an octave higher than the open (full-length) string. Explain (in terms of the tension in the string) why you may in reality get a difference slightly greater than an octave, especially

if the bridges are high and hold the string very far above the fingerboard.

3. If a soprano saxophone is 69 cm long and the tenor sax plays an octave lower, how long should the tenor sax be?

4. Take a flexible tape measure and find the total length of tubing (not counting the extra tubing associated with the valves) of a trumpet, a trombone, and a tuba. Do your results verify the claim that doubling the length lowers the pitch an octave? Measure a French horn, too; in this case, you may find the results puzzling until you study Chapter 13.

5. If you have never played a brass instrument, ask a friend to let you blow on one. Blow and buzz your lips with the mouthpiece alone, and verify that you can produce an entire range of pitches by varying the air pressure and the lip tension. Then attach the mouthpiece to the instrument, and experience for yourself the feeling that the instrument will let you produce a few special pitches but make it much harder to produce any others in between.

6. If you are not familiar with the concept and the sound of octaves, experiment with playing pairs of notes at random on the piano. Discover for yourself that whenever one note in the pair is the twelfth one above or below the other (counting both white and black keys), there is a very special relation between their sounds. Describe in your own words how an octave relation sounds different to your ears than any other pair.

7. What frequency will sound to your ears one octave below 400 Hz? What frequency will be an octave above 400 Hz? Two octaves above 400 Hz?

8. In a certain set of organ pipes, the speaking length for the note C_5 ($f = 524$ Hz) is 30 cm. How long a pipe do you expect for C_4 ($f = 262$ Hz)? What frequency and what pipe length for C_3, one more octave lower?

9. When an organist practices in an unheated church in January the temperature is at 8°C, but when the organ was tuned the temperature was 20°C. Explain whether the pipes will speak with higher or lower than normal frequency. What percentage of change has there been in the speed of sound? What percentage difference does this cause in the speaking frequency of most of the pipes?

10. When the metal bars on a vibraphone are struck by themselves, the sound is rather weak. But when a tubular resonator of the proper length is immediately under the bar, the sound is much louder. Rossing reports vibration durations of approximately 30 to 40 seconds in the former case, but less then 10 seconds in the latter. Explain the connection.

PROJECTS

1. Hit, tap, rub, scratch, and blow on many different objects. Compare the resulting sounds as to duration, loudness, musical quality, and high or low pitch. Try to explain these comparisons in terms of the size and material of the objects, and the methods used to excite them.

2. Find a wide variety of sounds, as in Project 1, and record them ($\frac{1}{4}$-inch, reel-to-reel tape). Use a second recorder to copy at different speeds. Chop, splice, turn backward, and so on, and reassemble to make a *musique concrète* composition.

CHAPTER

4 Sound Propagation

We tend to take sound propagation for granted, but it has several significant features that deserve our closer attention.

Sound may change its direction of travel for several reasons. One is reflection from hard surfaces, which is familiar as the cause of echoes. Less well known are refraction and diffraction, which describe bending and spreading effects. Diffraction is extremely important in musical acoustics. After describing each of these effects individually, we will comment in Section 4.3 on how they influence the quality of music played outdoors and how acoustical problems in amphitheaters can be handled.

After an optional section on frequency shifts for moving sound sources, we will consider interference effects in Section 4.5. These result from the way one sound wave may sometimes cooperate with another, while at other times they may oppose or even cancel each other. All the effects introduced qualitatively here can eventually be described quantitatively with the measurement criteria that follow in Chapter 5.

4.1 REFLECTION AND REFRACTION

In real life, sound waves can never travel forever in the same direction. When waves encounter an obstacle we expect them to be reflected; we have all heard echoes near a cliff or a large building. Echoes are noticed most easily at distances of 50 or 100 meters from a large, flat, hard surface. Such echoes seem as if they come from a point behind the reflecting surface, just as light waves reflected from a mirror seem to come from an image behind the mirror.

Sound will reflect from surfaces of any shape or size. If the surface is smooth, the reflection is regular and orderly (Figure 4.1a). But a rough surface causes irregular or *diffuse* reflection (Figure 4.1b). The criterion that distinguishes rough from smooth is whether the bumps on the surface are as large as the wavelength of the waves. A good optical mirror should not have bumps larger than a thousandth of a millimeter, because the light waves are even smaller than that. But sound waves are unaffected by such small detail, and a textured wall that appears rough to your eyes may reflect sound like a

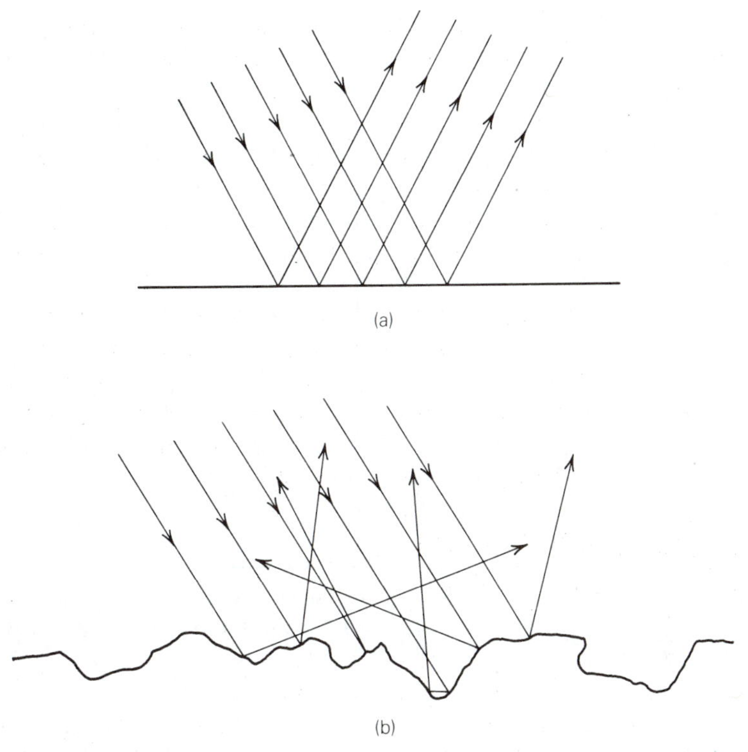

FIGURE 4.1 (a) A group of waves all traveling in the same direction (like a light beam) undergoes regular or *specular* reflection at a sufficiently smooth surface. (b) Diffuse reflection occurs when waves encounter a rough surface. The wave energy is scattered in many different directions by different parts of the surface. A reflecting wall will not seem this rough to an A440 sound wave unless it has bumps a meter or more across.

smooth mirror. It takes irregularities several centimeters in size to diffuse high treble notes, and any bumps less than 1 or 2 meters in size will seem quite smooth to the lowest bass notes.

The reflected wave is always weaker than the original, because part of the sound energy is *absorbed* at the reflecting surface. The fraction absorbed is much higher for a soft surface such as heavy drapery than for a hard bare plaster wall. Reflection and absorption are both important in determining whether a room is a pleasant place to listen to music, and we shall consider them at length in Chapter 15. Repeated reflection, in particular, accounts for the lingering reverberation of sound in an enclosed room.

Multiple reflection also makes it possible for sound to travel from one room to another (Figure 4.2). This is why you can often hear people talking in another room even though you cannot see them.

A situation in which the speed of sound changes with position may cause sound energy to follow curved paths. This is called **refraction**. Any type of waves can be refracted. Desert mirages, for instance, result from light rays curving as they pass through regions of differing air density.

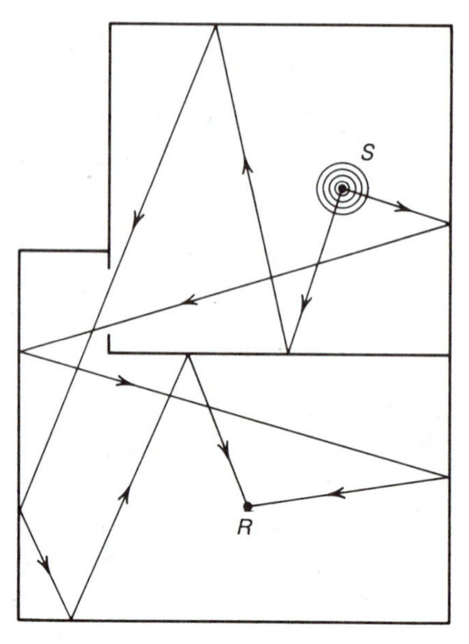

FIGURE 4.2 Multiple reflections allow sound waves from the source *S* to be received at *R* in a different room. These two paths are only representative of the many that are possible. There is one, not shown, that involves only a single reflection.

You may encounter some interesting effects of sound refraction; we will describe these only briefly, as they are seldom of any musical importance. First, on an extremely still day (or, more likely, night) you may hear sounds (say, the ringing of a church bell) from a mile or more away that surprise you by being much louder than usual. The weather condition called inversion favors this (along with smog), for it means that cold air is close to the ground and warmer air above it. Because the speed of sound is less in cold air, the wave fronts behave as in Figure 4.3, and more sound energy is steered back toward the ground. In more ordinary circumstances (especially on a clear, sunny day), the ground and adjacent air are warmest, and the air temperature decreases with increasing height. Then the sound closest to the ground travels fastest (Figure 4.4), the wave crests are distorted upward, and you receive practically no sound from sources that are not nearby.

Wind can produce similar effects. Specifically, if the air near the ground is nearly still and the wind speed increases with height, the wind carries along higher parts of the wave fronts. When you are downwind from the source, the wind brings the sound higher up toward you more quickly; the wave fronts bend over somewhat as in Figure 4.3, and you hear louder sound. When you are upwind from the source, the higher parts of the wave fronts are held back, they bend upward as in Figure 4.4, and you hear very little from a distant source.

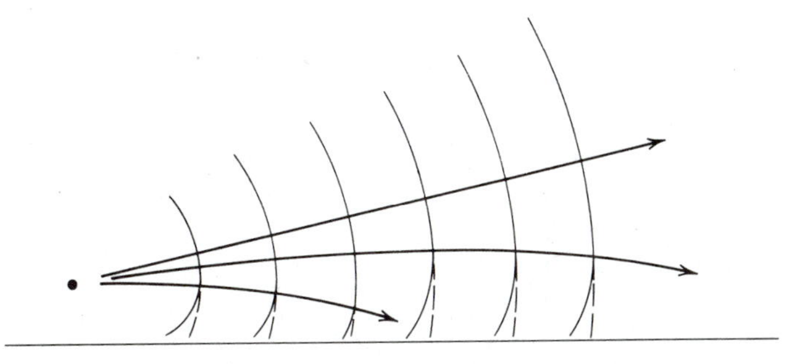

FIGURE 4.3 Downward bending (refraction) of sound waves in any situation where the sound speed is reduced toward the bottom of the picture. Dashed lines show where wave crests would be if sound speed were the same everywhere. This picture applies if air is cooler near the ground or if a wind is blowing from left to right.

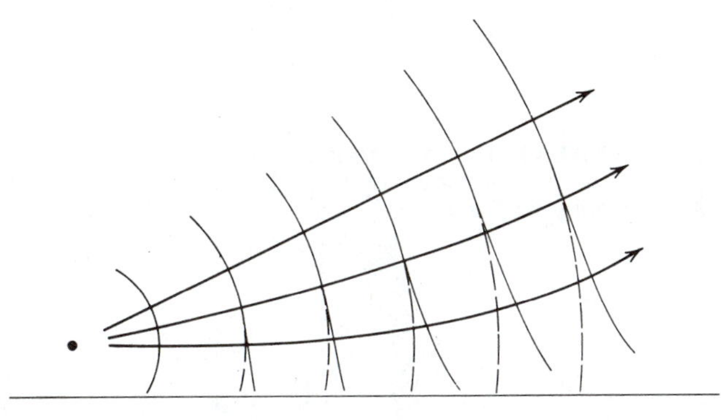

FIGURE 4.4 Sound refraction in any situation where the speed of sound increases toward the ground, such as for hotter air below or wind blowing from right to left. As wave fronts are steered away from the ground, a dead region is left at the lower right where practically nothing from this source will be heard.

4.2 DIFFRACTION

Sound can travel around corners. When indoors, we usually explain this away as merely a result of multiple reflections. But outdoors, even with no other buildings nearby to provide reflections, you can still hear a siren getting louder before you can see the ambulance approaching on the cross street as in Figure 4.5.

As an analogy to help understand this, think of long ocean swells passing through a narrow opening in a breakwater into a harbor. Although they may remain strongest in their original direction, they also spread out and reach into every part of the harbor (Figure 4.6). The same diagram could represent sound waves coming through a doorway and spreading throughout a room. (They would have to be bass notes to spread that well.)

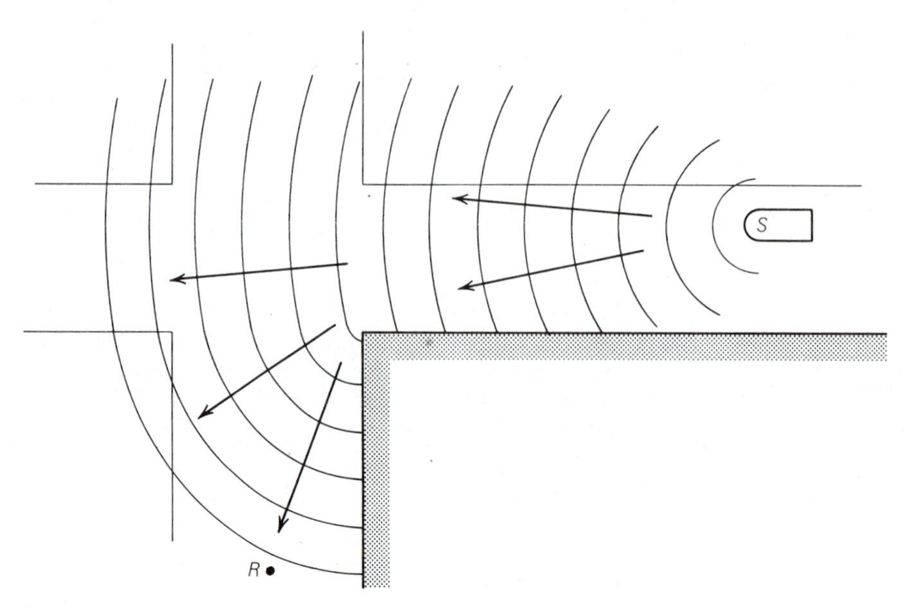

FIGURE 4.5 Even in the absence of reflections, the source of sound S, such as an ambulance siren, can be heard around the corner of a building by the receiver R. The waves spread around the corner by the process of diffraction.

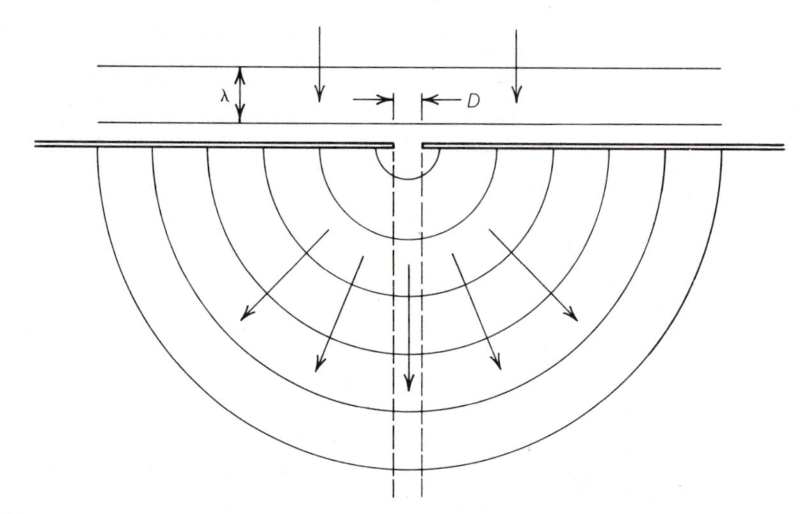

FIGURE 4.6 Waves coming through a narrow opening spread out almost equally in all directions. If waves acted like bullets, we would expect them to reach only the narrow region between the two dashed lines, leaving a shadow zone on either side. But these waves do *not* leave such shadows. It is equally true that waves passing a narrow obstacle will very effectively fill in behind it, as long water waves do when encountering a piling.

*BOX 4.1 THE WAVE NATURE OF MATTER

Certain types of behavior are exhibited by all waves. We also can turn this around and say that any phenomenon exhibiting such wavelike behavior deserves to be considered a wave phenomenon.

Thus we claim that the fundamental properties of matter itself indicate a wavelike nature. If we knew only that electrons sometimes are reflected, we might still think of them as bulletlike particles. After all, a stream of tennis balls can be reflected from a wall. But electrons tend to spread like waves throughout a region instead of being localized at any one point. We also observe refraction, diffraction, and interference with electron beams. (Because electron wavelengths are typically 10^{-9} m or less, diffraction and interference are manifest only in experiments where electrons pass through incredibly small openings, such as those between neighboring atoms in a crystal.)

Although this evidence requires that electrons be considered as waves, they show definite particlelike properties in still other experiments. Apparently it is incorrect to try forcing a choice that electrons must be classified *either* as particles *or* as waves. Instead, it is the fundamental nature of matter to have a duality, an ability to be both wavelike and particlelike.

This spreading is called **diffraction** and is a general property of all waves (Box 4.1). How much waves are diffracted depends on the relation between the wavelength λ and the other distances involved. In general, if λ is about the same size as the width D of an opening, or if it is larger, the waves spread with comparable strength in all directions. But if λ is much smaller than D, as in Figure 4.7, there is only a little spreading, and that part of the wave going into the "shadow" region is extremely weak.

You may object that you have never noticed light waves behaving this way. The headlight beams of the ambulance in Figure 4.5 can point down the cross street, and they do not come around the corner of the building to light

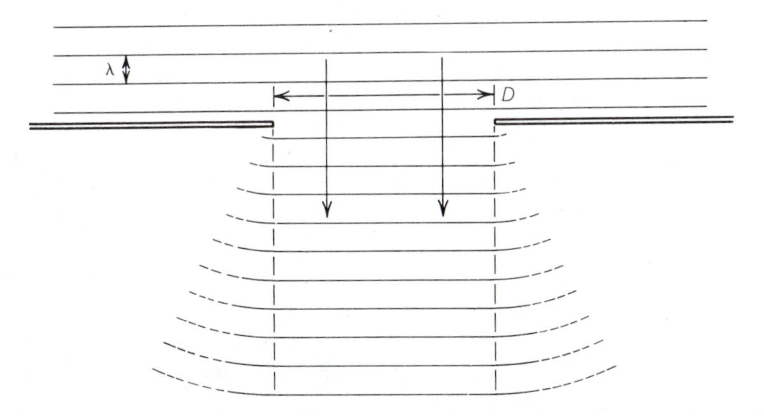

FIGURE 4.7 Waves coming through an opening that is large compared to their wavelength λ remain largely confined to a beam continuing in the original direction. The penetration into the shadow zones outside the dashed lines is very weak. It is equally true that waves passing a large obstacle will leave a shadow region behind it largely undisturbed.

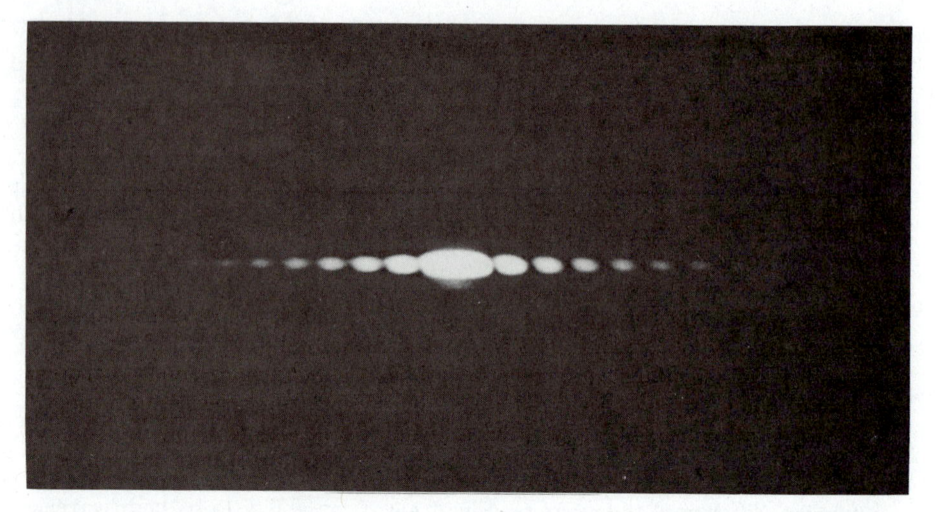

FIGURE 4.8 A diffraction pattern made with visible light. Imagine the photographic film being placed along the bottom of Figure 4.7. If it were not for diffraction, there would be a sharp boundary between light and shadow. (Photo by Stephen Hamilton.)

the street where you stand. The truth is that light *does* spread, but not enough that we notice it under everyday conditions. Because the wavelengths of visible light are less than a thousandth of a millimeter, they are very small compared to most everyday objects and openings, and the diffraction of light is so weak you do not ordinarily notice it. But send light in the right way through a hairline opening into a darkened room and you can see a beautiful diffraction pattern (Figure 4.8).

Sound wavelengths are much longer, ranging from approximately 10–15 m for the lowest audible notes down to 2 cm for the highest. Thus, we can expect strong diffraction of sound around objects of ordinary size. And we can understand why bass notes travel around corners or obstacles more effectively than do treble notes.

Using the idea of diffraction, we can explain several additional interesting effects:

1. You can talk to another person and be understood, even though not directly facing her. Sound emerging from your mouth spreads out to both sides, because it comes through an opening only a few centimeters wide. This is small compared to the wavelength of most speech sounds. Similar comments apply to sound coming out the end of a trumpet.
2. Bass sounds from a loudspeaker spread out evenly, while treble sounds from the same speaker are more nearly confined to a narrow cone in the forward direction. The diffraction effect here is the same as if the sounds were coming through a speaker-size opening in the wall from another room. One of the reasons that treble sounds usually are sent through a smaller speaker is so that diffraction will spread them more evenly in all directions.
3. Your left ear can hear sounds that come from a source on your right.

Bass notes have wavelengths much larger than your head, and they diffract so well that their strength is nearly the same at both ears. But high treble notes form more of a "shadow" and may be much weaker at one ear than the other. This is one of several clues you use when judging where a sound comes from.

4.3 OUTDOOR MUSIC

Although music is more often performed indoors, it is easier to analyze the spreading of sound energy in outdoor situations. We will wait until Chapter 15 to discuss sound in rooms, where repeated reflections from the walls are important in determining the sound levels.

The first and most obvious generalization is that any given source sounds weaker outside; whatever sound misses your ears as it first moves outward is lost forever, whereas an enclosed room will trap it and send it by repeatedly. The practical consequence of this is that outdoor musical performances require sources that can put out a lot of energy—large groups rather than soloists, brass bands rather than string quartets, or aid from electronic amplifiers. Second, and perhaps less obvious, is the "deadness" of outdoor sound. It has the same cause: The first sound you receive directly from the source is not followed by a series of overlapping reflected sounds, and thus it lacks the reverberation and lasting warmth of indoor sound. Third, again for the same reason, it is hard to distribute sound evenly over a large audience outdoors. The reflections inside a room come from many different directions, and will average out in a well-designed hall so that all listeners receive nearly the same sound levels. But outdoors the single outgoing wave inevitably weakens rapidly as you go farther from the source. Perhaps for all these reasons, the composer Hector Berlioz went so far as to say, "There is no such thing as music in the open air."

Several strategems can reduce these problems and steer the sound where it is most needed. Foremost is to have some reflecting structure or partial enclosure near the performers redirect some of the sound that is otherwise wasted going in directions where there are no listeners. This takes its simplest form where the audience is concentrated on one side so that a shell can be put on the opposite side (Figure 4.9); a famous example is the Hollywood Bowl. A well-designed shell can deliver 10 or more times as much sound energy to the more distant parts of the audience as they would otherwise receive, and greatly reduce the difference between front and back rows. The location of the performer is important; if too far out front, or if the shell is too strongly curved, the result can be to focus the sound in a way that makes the nonuniform distribution worse instead of better. A good rule of thumb is that the performer should be a little less than halfway out from the shell to its center of curvature (Figure 4.10). Shells are also sometimes useful indoors; for example, behind a small group of musicians using only the front part of a large stage.

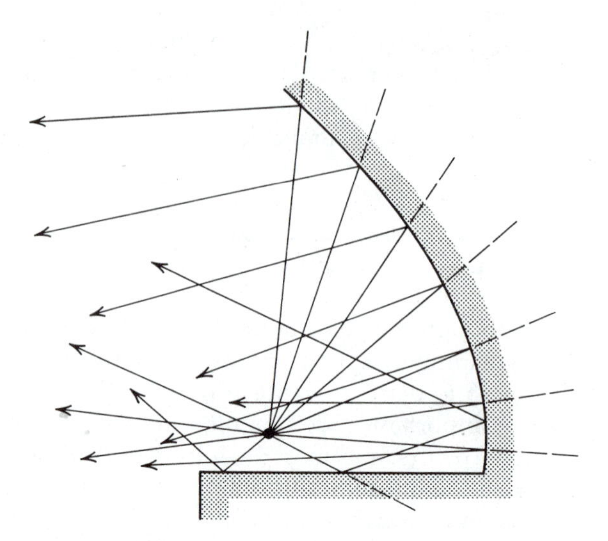

FIGURE 4.9 Section through a stage (side view) backed by an acoustical shell; audience is on the left.

If the audience is to be on all sides, it is still possible to redirect some of the sound that would otherwise escape upward. Think of the bandstands that are found in many city parks, such as those diagrammed in Figure 4.11 (where these ray paths are valid only for the mid- to high-frequency range).

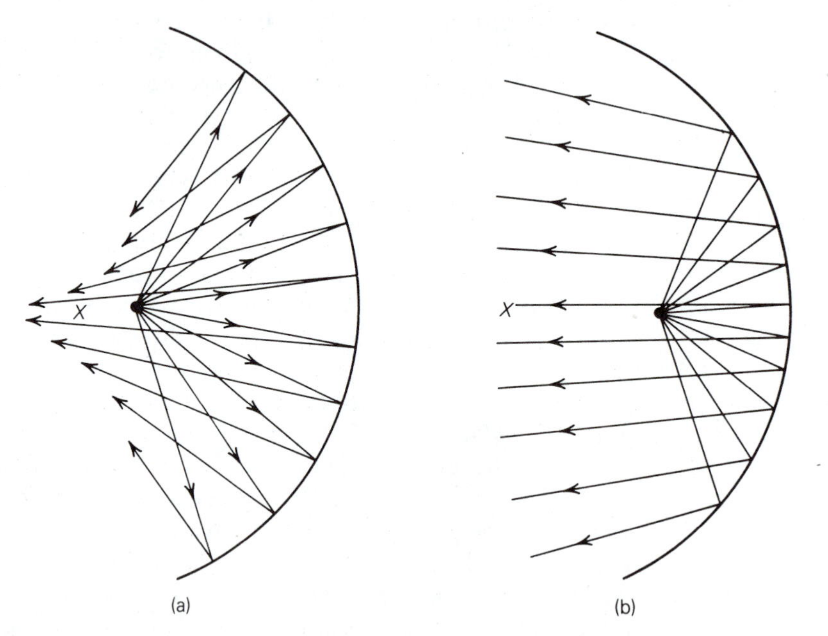

FIGURE 4.10 Reflection of sound by a concave wall (top view), with the center of curvature located at *X*. (a) With the performer too far out from the wall, sound is focused largely on one small part of the audience. (b) Proper placement is a little closer than half the radius of curvature and spreads sound nearly uniformly.

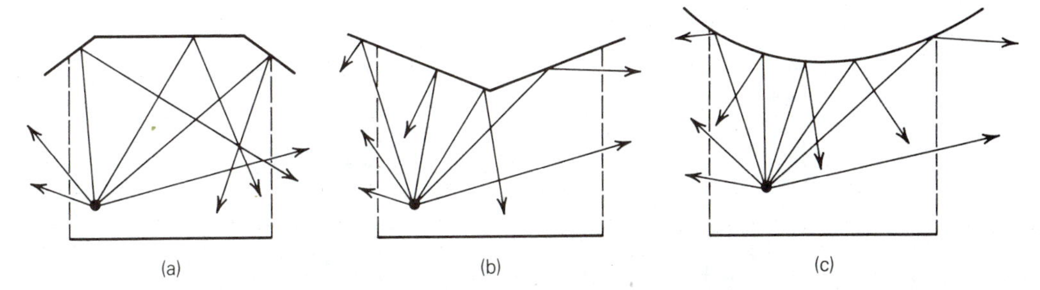

FIGURE 4.11 Cross section (side view) of some possible roof shapes for outdoor pavilions and a few typical paths followed by reflected sound. Shape (a) traps a lot of sound so that performers can hear each other better (maybe even uncomfortably loud), but it is not very helpful to the audience outside. Shape (b) helps the audience, but half the band may hardly hear the other half. Shape (c) is a compromise that serves both purposes.

Notice that a good design not only helps the audience but also makes it easier for the musicians to hear one another. Tiered seating for the audience helps, too, both visually and acoustically, and has been exploited ever since the Greek amphitheaters. The raised seating area intercepts more of the total sound output than if the listeners were on flat ground.

All of these aids, together with electronic amplification, can make for satisfactory open-air music performances. But these must be judged as such and should never be expected to sound even nearly the same as they would inside a good enclosed auditorium.

*4.4 THE DOPPLER EFFECT

Think about listening to a jet plane flying low overhead or a speeding truck passing as you stand beside a highway. You will recall that a relatively high-pitched sound from the approaching source changed to a lower pitch as the source receded. This change of pitch because of motion is called the Doppler effect. It has almost no musical importance, but it is another interesting illustration of the behavior of waves in general and is very easily understood.

In Figure 4.12, the source S is moving toward the right at half the speed of sound. As each circular wavefront moves outward, it remains centered on the point where it was first emitted, even though the source is no longer there. This snapshot of how the waves are distributed in space at a particular time makes it clear that the waves are crowded closer together on the right than they would be if the source were stationary, and spread farther apart on the left. The shorter wavelength on the right suggests that observer A receives wave crests at higher frequency, and the longer wavelength on the left suggests that observer B receives a lower frequency. You can further convince yourself of this by saying that all the wave crests travel through the air with the same speed, but the more recently emitted ones did not have as far to go

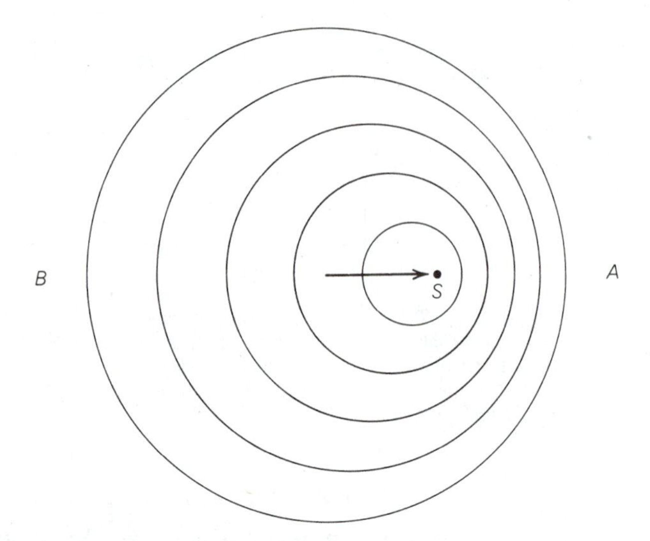

FIGURE 4.12 Wave crests emitted by a moving source *S*. The outermost wave was emitted when the source was at the left end of the arrow; a sixth crest is just now being created. Observer *A* receives crests crowded closer together than if the source were stationary; for observer *B* they are spread farther apart.

to reach *A*, so the rate of arrival of wave crests there is greater than the rate at which *S* emits them. Similarly, the later waves have farther to go to get to *B*, so arrive less often.

There also is Doppler effect if the source is stationary and the observer is moving. This, too, is easily understood; in the case where the observer is receding from the source, the later crests must travel a greater distance to catch up with him, and the frequency received is less than the frequency emitted.

If V, the relative speed of the source and observer, is much less than v, the speed of the signal, the amount of frequency shift is given by the simple formula $(f_1 - f_0)/f_0 \cong V/v$. Here f_0 is the frequency emitted and f_1 the frequency received, and V is considered positive for approach and negative for separation. This same formula is used by astronomers to deduce the speed of recession of distant galaxies when they observe Doppler shift of the light waves received from those galaxies. In that case, of course, v must be the speed of light rather than sound. The shift in frequency is manifest as a change in color of the light, which is analogous to a change in pitch of sound.

4.5 INTERFERENCE AND BEATS

What is the combined effect when similar sound waves arrive from several directions at once? Surprisingly, this sometimes results in a sound weaker than any one of the signals would be by itself. The way they may either cooperate or cancel out is called **interference**.

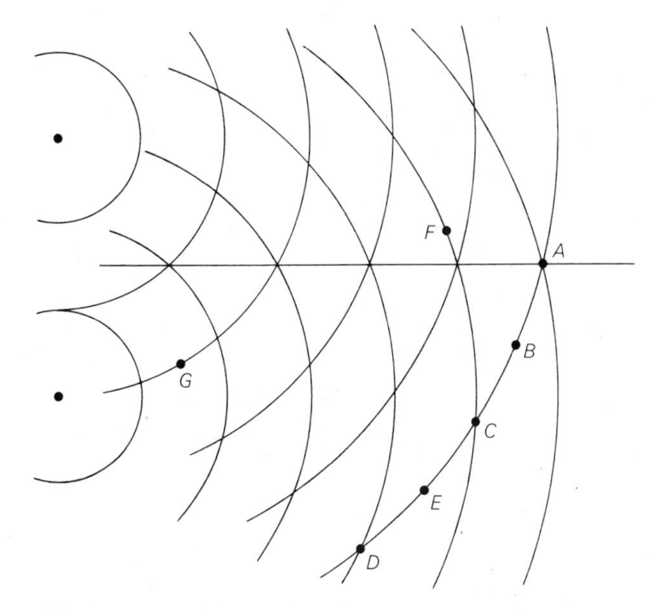

FIGURE 4.13 Wave crests of the same frequency being sent out from two sources. You should think of this as only a snapshot of a dynamic process in which these crests are continually moving outward. Constructive interference occurs at *A* (as well as all points along the straight line drawn through *A*) and also at many other points such as *C* and *D*. At many points, such as *B* and *E*, the crest of one wave always coincides with a trough of the other so that there is no net disturbance (destructive interference). See Figure 4.14 for another view of the waves arriving at *A* and *B*.

Let us study the case of two sources putting out identical sounds; for example, a pair of speakers both driven by the same amplifier with a monaural (not stereo) signal. Consider point *A* in Figure 4.13, which is equidistant from two sources of identical sine waves of wavelength λ. Every time a crest from one speaker reaches *A*, so does a crest from the other speaker, and the two signals are said to be *in phase* at *A* (Figure 4.14a). The total disturbance at *A* then has twice the amplitude of either wave alone (Figure 4.15a).

The situation is quite different at other points such as *B*, which is half a wavelength farther from one speaker than from the other. The travel times differ by half a vibration period, so by the time a given crest arrives from one speaker, the corresponding crest from the other speaker already has passed and the following trough is at *B* instead (Figure 4.14b). One wave's compression cancels the other's rarefaction (they are *out of phase*), and the net result is no disturbance at all (Figure 4.15b). This suggests that one would avoid ever driving two speakers in the same room with the same signal; in most rooms, however, waves reflected from the walls tend to even out the distribution of sound compared to these idealized pictures.

The enhanced intensity at *A* is called *constructive interference*; it also occurs at points such as *C* or *D* where there are one or more full wavelengths difference in distance. In general, when L_1 and L_2 are the distances from the two sources to the observation point, what matters is how many wavelengths

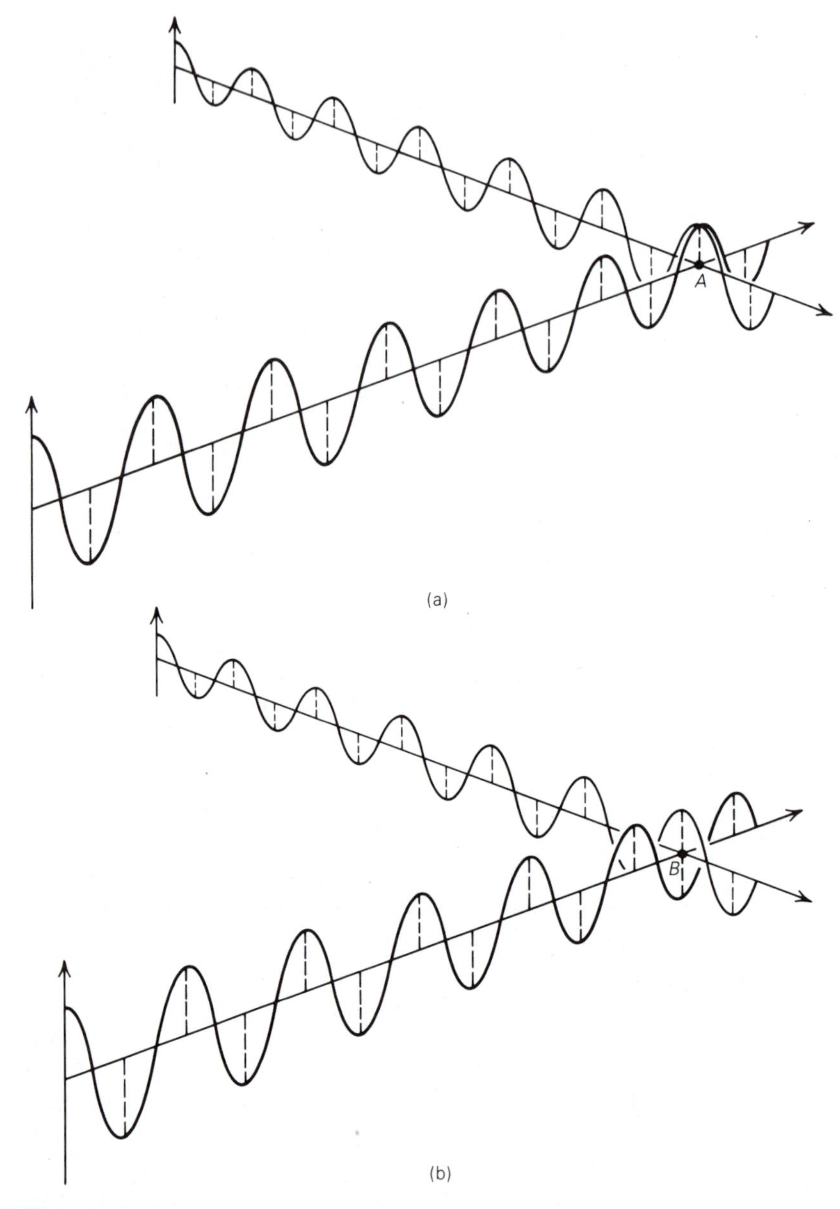

FIGURE 4.14 Graphs of the waves along some of their lines of travel in Figure 4.13. Waves arrive at point *A* in phase and cooperate to produce a large disturbance. But because of the extra half-wavelength distance, the wave from the upper speaker opposes the other one at *B*. See also Figure 4.15.

occupy the extra distance, so we must always examine the relation $n\lambda = L_1 - L_2$. Whenever n is any whole number (positive, negative, or zero), the condition for constructive interference is satisfied. But whenever n is an integer plus an extra half (for example, 1.5 at point E), the waves are out of phase and we have *destructive interference.*

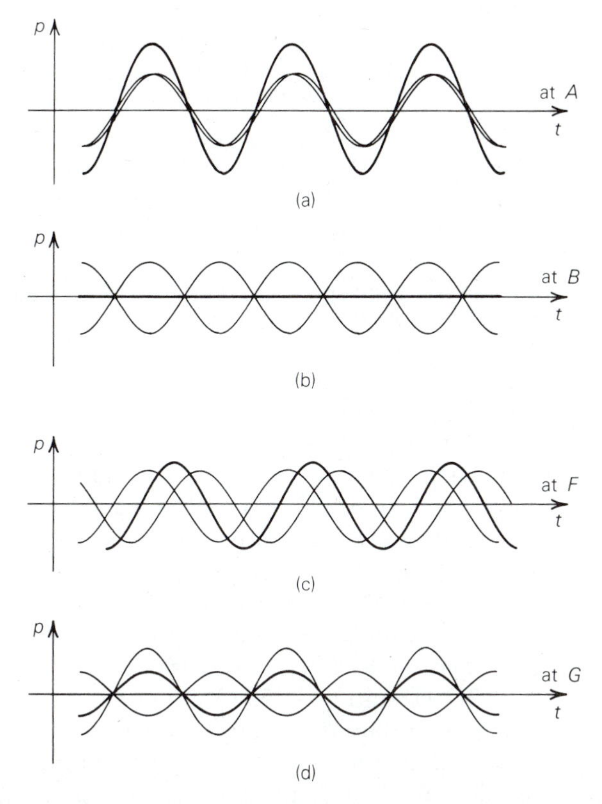

FIGURE 4.15 Time history graphs of waves passing particular points in Figure 4.13. (a) At point *A* both waves always do the same thing at the same time (they are drawn slightly displaced to remind you that there are two), and the total disturbance (darker line) has just twice the amplitude of either wave alone. (b) At point *B* contributing waves are out of phase, and total disturbance is zero. (c) Point *F* has one wave about a quarter-cycle behind the other, and the total disturbance is only a little larger than either alone. (d) At point *G* the two waves are out of phase but have considerably different amplitudes, so the destructive interference is not complete.

For equal-strength waves, a continuous variation in phase results in the net amplitude taking on all intermediate values between the extremes of zero and twice that of one wave alone (Figure 4.15c). But there also can be interference between two waves whose individual strengths are not equal (as would occur for point *G* in Figure 4.13 because it is much closer to one speaker than to the other), and in such cases there cannot be complete cancellation (Figure 4.15d). For example, let one wave alone have five times the amplitude of the other. Then the amplitude of the combined wave must be between four times (out of phase) and six times (in phase) that of the smaller wave alone.

What happens if waves from two sources have almost the same frequency but not exactly? Now, regardless of which point in space we consider, we will have the phenomenon of **beats** shown in Figure 4.16. For a while, the two waves stay nearly in phase, and there will be constructive interference. But one wave gradually gets further and further out of step with the other, and after a while the interference becomes destructive. The listener will not

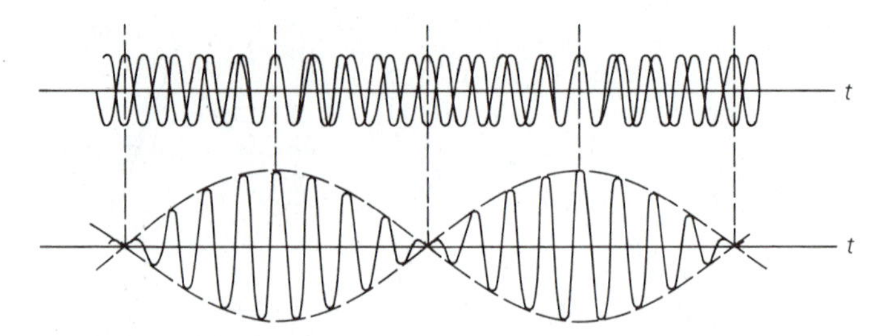

FIGURE 4.16 Time history graphs showing (top) two sine waves of slightly different frequency and (bottom) the total disturbance when both are present at the same time. Notice how maximum and minimum amplitudes for the combined beating wave correspond to in-phase and out-of-phase combinations.

perceive two separate sounds but one whose intensity seems to slowly and repeatedly rise and fall. This pulsation of the sound is described by telling how many times per second the maxima or beats occur.

To see how the beat frequency is related to the two original frequencies, consider a 1-second interval that begins at a moment when the two waves are exactly in phase. During this second, there will be f_1 cycles of one wave and f_2 cycles of the other. Supposing f_1 to be the larger of these, the first wave has then made $f_1 - f_2$ extra cycles compared to the other. For example, if $f_1 = 261$ Hz and $f_2 = 256$ Hz, the first wave makes five more complete cycles per second than the other. When 1 second is past, they are just coming into phase again for the fifth time, so there must be five beats per second. In general, the beat frequency is always given by $f_b = f_1 - f_2$.

Beats often are used by musicians in tuning their instruments. The more rapid the beats you hear when your instrument and another play together (or from a piano string and a tuning fork), the greater the difference in their frequencies. If after adjustments you hear no beats at all, the two sources are in tune with each other. In practice, it is difficult to hear beats at a rate much slower than one in 2 or 3 seconds, so matching frequencies to within about 0.5 Hz will pass as good tuning. At the other extreme, beats can no longer be heard individually at rates faster than 30 or 40 per second; the combined sound has instead a rough quality for frequency differences up to 100 or 200 Hz. We will relate this to the properties of the inner ear in Chapter 17.

SUMMARY

Sound waves, like all other kinds, are subject to reflection, diffraction, interference, refraction, and Doppler frequency shifts. Of these effects, the first three are very important in musical acoustics.

Multiple reflections are a major factor in sound leakage from one room to another and in room reverberation. Waves become weaker at each reflection, because part of their energy is lost to absorption.

Diffraction is the spreading that occurs whenever waves go around corners, or especially through narrow openings or around small obstacles. Here *narrow* and *small* mean "in comparison to the sound wavelength."

Interference means that similar waves coming from two or more spatially separated sources may either aid or oppose each other. The interference is constructive or destructive according to whether the waves arrive in or out of phase. Two waves of slightly different frequency will go repeatedly in and out of phase with each other to produce beats. You may like to think of beats as an interference pattern whose alternating maxima and minima occur in time rather than in space.

Sounds produced outdoors generally impress the listener as weaker, deader, and more strongly dependent on location than indoor sounds. These problems can be overcome partially with tiered seating and reflective shells behind the performers.

REFERENCES

An article on the acoustics of Greek amphitheaters by R. S. Shankland in *Physics Today*, 26, 30 (October 1973), is informative, although the emphasis is more on speech clarity than on music.

SYMBOLS, TERMS, AND RELATIONS

λ wavelength
f frequency
n number of cycles
$D \lesssim \lambda$ for strong diffraction
$f_b = f_1 - f_2$
$L_1 - L_2 = n\lambda$
reflection
absorption
refraction
diffraction
Doppler shift
interference
beats

EXERCISES

1. Sketch a picture of what happens when ocean waves with (a) λ = 10 m and (b) λ = 200 m encounter a small island of width 100 m. Discuss the amount of shelter you will find in each case by taking your small boat to the lee side of the island.

2. Discuss how much the sound is blocked off if you sit behind a large pillar in a concert hall. Is the answer different for treble and for bass? Is the problem alleviated by reflections from the walls?

3. Explain in terms of diffraction how a megaphone enables you to project your voice mainly in one direction.

4. A partially open window presents an

opening 20 cm wide. What kind of sounds from outside will go across the room in a well-defined beam? What kind will spread evenly throughout the room? Roughly what frequency marks the boundary between the two cases?

*5. Suppose an observer were standing at some distance off to the side in Figure 4.12; that is, at the bottom of the page, so that the source is moving neither toward nor away from him but across his line of sight. How much Doppler shift would there be for him? Explain.

*6. A musician standing close beside a railroad track hears the whistle blowing while a train passes her. She reports that the pitch dropped by the musical interval called a major third. As we shall learn later, this means the received frequency must have been about 12% higher as the train approached (and 12% lower as it receded) than the f_0 that would have been heard with the train at rest. How fast was the train going?

7. Sound of frequency $f = 688$ Hz is sent through two speakers as in Figure 4.13. What are several values for path-length difference $L_1 - L_2$ that will lead to constructive interference? What are several that will give destructive interference?

8. Suppose you listen to sound from two loudspeakers, as in Figure 4.13, at a distance of 6 m from one and 4.8 m from the other. What are several wavelengths and the corresponding frequencies for which you will experience constructive interference? What are several wavelengths (and frequencies) for which the interference is destructive?

9. The speaker S in Figure 4.17 can send sound to the listener L not only directly but also by reflection from the hard wall W. (Suppose the walls not shown are

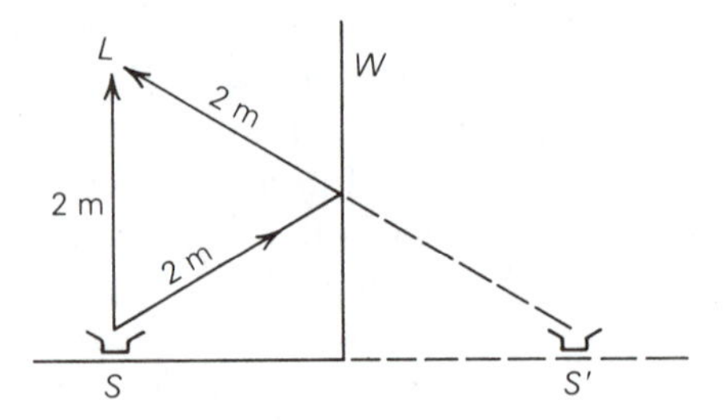

FIGURE 4.17

covered with heavy draperies, the ceiling with acoustic plaster, and the floor with thick carpet, so that the problem is not complicated by any other reflections.) The combined sound is just the same as if there were a mirror image speaker S', with both sending out identical signals. For frequency $f_1 = 172$ Hz, is there constructive or destructive interference at L? What kind of interference for $f_2 = 86$ Hz? Would those conditions change if the listener moved to other points in the room? Explain (in terms of how the two-signal path-length difference changes) why this situation would be improved by moving the speaker into the corner.

10. Two organ pipes speak at frequencies 523.0 and 520.6 Hz. What is the frequency of beats heard when both sound together?

11. A certain piano string and an A440 tuning fork heard together give three beats per second. What are the two possible values for the frequency of the string vibration?

12. If three instruments play together, with frequencies 440, 438, and 443 Hz, what beat frequencies will result?

13. The diameter of a trumpet bell is roughly 10 cm. When the trumpet plays middle C ($f = 262$ Hz), does the sound come out in a narrow beam or spread in all direc-

tions? Give clear physical justification for your answer.

14. The radio waves of Exercise 20 in Chapter 2 have wavelength 0.1 m. If we want to send this signal in a well-defined beam from the spacecraft toward Earth, not wasting energy by allowing it to spread in all directions, roughly how large a transmitting antenna must the spacecraft carry? Explain your reasoning.

15. Airplanes can be located by radar, which detects radio waves reflected back toward a transmitter. In analogous fashion, bats can locate insects by making high-pitched squeaks and listening for echoes. Using concepts from this chapter, can you explain why radar works much better with microwaves (typical wavelength 10 cm) than it would in the AM radio band (wavelengths more than 100 m)? Can you explain similarly why a bat can succeed using sound at ultrahigh frequencies (80 KHz, for example), but would fail miserably with 800 Hz signals?

PROJECTS

1. Find out as much as you can about waves from experiments with water. Drop pebbles into a pond; watch the wake from a boat or make your own by dragging a stick through the water; make waves in a shallow layer of water covering the flat bottom of a large glass kitchen dish. Try to verify as many as you can of the statements about wave behavior in this chapter.

2. Search your memory and explain whether your own experiences agree with the claim that bass notes go around corners more effectively than do treble. Better yet, design and carry out an experiment (for example, with a tuba and a piccolo) to test this. Better yet, do not rely on your ears to judge, but measure both sources with a sound level meter (which is easy to use here, even though not explained until Chapter 5). You can define a degree of attenuation for each sound as the number of decibels difference between one reading directly along a line of sight and another at the same distance from the source but around the corner.

CHAPTER

5 Sound Intensity and Its Measurement

Some sounds are barely audible, and others are so loud as to be painful. What is the physical property of sound that causes these sensations, and how can we measure it exactly? How wide a range of sound levels is useful in music, and how do they compare with other familiar sounds? Can we explain the variation in sound strength when we move around, toward, or away from its sources? These are some of the questions we will consider in this chapter.

You must keep clearly in mind that we are talking now about the physical measurement of sound strength. The human perception of sound strength is *not* the same thing; we reserve the term *loudness* for this psychological sensation, which will be studied in Chapter 6.

As we will explain in Section 5.1, even the physical aspect of sound strength may be described by two different measures; namely, amplitude and intensity. It is intensity that will be used most in subsequent chapters. But we also will learn in Section 5.2 how the decibel scale is used to specify these intensities in an alternative form called *sound level*. In the remainder of the chapter, we study typical sound levels found around us and try to understand some of the reasons for their variation. We will show in particular how to compute the sound levels expected when two or more sounds are combined.

5.1 AMPLITUDE, ENERGY, AND INTENSITY

Let us lay aside questions of frequency, waveform, or duration for a while and focus our attention on sound strength alone. What is the appropriate physical measure of sound strength or weakness?

One possibility is the amplitude of the sound wave. More specifically, we could think about the displacement amplitude (the greatest distance that each little packet of air moves to either side of its normal position) or the velocity amplitude (the greatest speed that packet achieves during its oscillation). But both of these are quite difficult to measure directly; as we mentioned earlier, it is much easier to measure pressure amplitude (the greatest variation of pressure above and below atmospheric). Because all three represent essentially the same information, let us use the term *amplitude* from here on to mean pressure amplitude whenever referring to sound.

Another possibility is to work directly and primarily with the *energy* carried by the sound wave. This energy is related to the amplitude, but they are not the same thing. To understand how they are related, think about a mass attached to a spring and sliding on a frictionless horizontal surface. If you pull the mass aside 3 cm from its equilibrium position and release it, it oscillates with an amplitude of 3 cm. The work you originally did by pulling the mass aside is manifest in the energy of the vibration.

Now suppose you start over again, pulling the mass aside 6 cm and releasing it to get a vibration of 6 cm amplitude. Not only must you pull the mass twice as far, but also you must pull twice as hard. Because the amount of work done depends on the force multiplied by the distance, you must do four times as much work as in the first case, and so you provide four times as much energy for the resulting vibration. If you pulled three times as far (and three times as hard), there would be nine times as much energy; 10 times as far (and 10 times as hard) would make 100 times as much energy. We can encompass all cases with the general statement that *the energy of oscillation is proportional to the square of the amplitude.* This is true not only for the mass on the spring but also for *any* simple harmonic oscillation, including light and sound waves.

Any statement about the strength of a sound wave is imprecise unless we specify whether we mean amplitude or energy. Consider three different waves called X, Y, and Z; let Y have twice the amplitude of X, and Z twice the energy of X. Then Y and Z are *not* of equal strength; Y carries $(2)^2 = 4$ times as much energy as X, which is twice the energy of Z. Energy is usually of more direct interest than amplitude, so let us concentrate on how to describe and measure the energy in a sound wave.

This energy is neither all in one place nor sitting still. Because the energy is traveling and is spread throughout the region where the waves exist, we do not attempt to measure the total amount. It is more appropriate to measure the *energy flow* at each location. Consider the total energy E received by a sound detector over a length of time t. For a steady sound, the continuous flow of energy means that any increase in t makes a proportionate increase in E; the longer we wait, the greater the accumulation of received energy. We are interested only in the *rate* of energy transfer; that is, the energy received *per unit time.* This quantity $P = E/t$ is called **power**, and is measured in watts (1 W = 1 J/s). A 100 W lightbulb, for instance, uses 100 joules of energy for each second it stays on.

But even the power received by our detector still depends on the size of the sensitive area S with which it intercepts the sound waves. We do not want our measure of sound strength to depend on detector size, so we ask for the *power per unit area:* $I = P/S = E/St$. I is called the **intensity** and is measured in watts per square meter. A total power $P = 10$ W spread evenly over a surface of area $S = 5\ \mathrm{m}^2$, for example, means an intensity $I = 2\ \mathrm{W/m}^2$ falling on every part of that surface.

Because the intensity is the rate of energy flow, it also is proportional to the square of the wave amplitude. This can be written mathematically as

$I \propto A^2$ or, equivalently, $I = bA^2$, where b is some constant. We can nearly always phrase our problems in terms of comparing one sound with another, so that we will have no need to know b. We can instead use ratios: $(I_1/I_2) = (A_1/A_2)^2$.

*For sine waves, $I = A^2/2\rho c$, where A is the pressure amplitude, ρ is the mass density, and c the sound speed. Thus, the constant b for ordinary atmospheric conditions is .0012 $(\text{W/m}^2)/(\text{N/m}^2)^2$.

5.2 SOUND LEVEL AND THE DECIBEL SCALE

The loudest sounds we encounter rarely exceed 1 W/m^2 in intensity. Yet they may involve a trillion times as much energy as the softest sounds we can hear. This makes it somewhat inconvenient to use directly the intensity in W/m^2; if we did, we would find ourselves nearly always referring to tiny fractions of a W/m^2. It has become customary instead to use the **sound level** scale, which is labeled in **decibels** (abbreviated dB). Practical sound measurements are routinely made with sound level meters (see Box 5.1) that give readouts in decibels.

You can think of the *sound intensity level* (*SIL*) as a code; there is always a correspondence between this code and the intensity, but they are not the same thing. It is easiest to decipher the code if we realize that a decibel simply means one-tenth of a bel (named after Alexander Graham Bell). The bel, in turn, is defined to represent a *ratio* of 10 to 1 between two intensities. Let us repeat that for emphasis: A bel is *not* an *amount* of sound; it is a *relation* between two sounds.

Thus, if sound Y carries 10 times as much energy as sound X, we say its level is 1 bel, or 10 dB, higher. That is, $I_Y/I_X = 10$ means $SIL_Y - SIL_X = 10$ dB. If a third sound Z has in turn 10 times the intensity of Y, its level is another 10 dB higher. Notice very carefully how this means that sound Z carries 100 times as much energy as X (10 times as much as Y, which itself had 10 times as much as X), but its sound level is merely 20 dB higher (10 dB plus another 10 dB). That is, $I_Z/I_X = 100$ corresponds to $SIL_Z - SIL_X = 20$ dB.

In general, suppose we take a case where it requires a 1 followed by n zeros to express how many times as much energy one sound carries compared to another; that is, $I_1/I_2 = 10^n$. By considering a series of intermediate sounds each 10 times as intense as the one before, we can see that the first sound level must be n bels, or $10n$ dB, higher than the other; that is, $SIL_1 - SIL_2 = 10n$ dB. Thus, if one sound has 10 million ($10{,}000{,}000 = 10^7$; see Appendix B) times the intensity of another, we must say that the first sound level is 7 bels, or 70 dB, higher than the second.

BOX 5.1 THE SOUND LEVEL METER

Although our ears can make rough qualitative comparisons of sound strength, they are not adequate for precise measurement of intensity. There are two reasons for this. First, it is too awkward to make allowances for the variation in judgment from one person to another or even from one time to another for the same person. Second, even if by repeated trials and use of statistics we were to obtain very accurate average judgments of loudness, these estimates still would not directly measure the physical intensity.

The *sound level meter* is an electronic instrument designed to measure this intensity. The meters range in price from under $100 for one with limited capabilities to several thousand dollars for highly sensitive and versatile models (Figure 5.1). They are usually powered by batteries so as to be completely portable. They have three main parts: a built-in microphone, amplifying circuits, and a readout display. The microphone should have a uniform response to all audible frequencies. Newer models may have the sound level displayed in digital form, but many still have the traditional needle pointer that moves along a numbered scale calibrated in decibels. Some meters can preferentially measure different parts of the sound, as explained in Section 5.4.

FIGURE 5.1 A precision sound level meter. (Courtesy of GenRad, Inc.)

If the ratio is not a simple power of 10, we may consult Table 5.1, which will always enable us to approximately encode or decode this information. Some of the numbers in the left-hand column are not exact; they have been rounded off to the nearest tenth. The intensity ratio for 1 dB, for instance, is actually close to 1.26 rather than to 1.30. To see how these numbers arise, do Exercise 4.

Because a table including every possible case would take too much space, we must learn to make combinations of these entries. For example, if we encounter a level difference of 36 decibels, we consider first the 30 and then the 6. We know that 30 dB corresponds to three powers of 10, or 1000, and we see in the table that another 6 dB corresponds to a further ratio of about 4. Thus, altogether the ratio of intensities corresponding to 36 dB must be about $1000 \times 4 = 4000$. Similarly, if we encounter an intensity ratio of 300, we write

TABLE 5.1 The ratio of sound intensities corresponding to various sound-level differences in decibels. Fractions of dB may be estimated in between these entries, although for musical purposes it is generally adequate just to take the nearest dB. In making combinations, the intensity ratios must always be multiplied when the level differences are added.

Intensity Ratio	Level Difference
$I_1/I_2 = 1.0$	0 dB $= SIL_1 - SIL_2$
1.3	1 dB
1.6	2 dB
2.0	3 dB
2.5	4 dB
3.2	5 dB
4.0	6 dB
5.0	7 dB
6.3	8 dB
7.9	9 dB
10.0	10 dB
100	20 dB
1000	30 dB
...	...
10^n	$10n$ dB

300 as 3×100 and see that the corresponding level difference is just a little less than 5 plus 20, or 25 dB.

Here is the cardinal rule you must always follow in this coding process: *When intensity ratios are multiplied, level differences in dB are added.* If (as in

TABLE 5.2 Approximate sound levels that might be encountered in various environments (left) and in musical performance (right). These are only typical figures; individual examples might well give readings 10 dB higher or lower.

Sound Source	Sound Level (dB)	I(W/m^2)	Reaction
Jet engine at 10 m	150	10^3	Unbearable
	140		
	130		
SST takeoff at 500 m	120	1	Painful
Amplified rock music	110		
Machine shop	100		
Subway train	90	10^{-3}	
Factory	80		
City traffic	70		Musically useful
Quiet conversation	60	10^{-6}	
Quiet auto interior	50		
Library	40		
Empty auditorium	30	10^{-9}	
Whisper at 1 m	20		
Falling pin	10		
	0	10^{-12}	Inaudible

BOX 5.2 SOUND LEVELS IN MUSIC

Sound levels below 50 dB are seldom useful in music, because they require that background noise (from adjoining rooms, audience movement, or ventilating systems) be kept even lower. Levels above 100 dB are not only unpleasantly loud, but also damaging to the ear, which progressively worsens with prolonged exposure. One must look to nonmusical reasons to explain why so many people voluntarily expose themselves to sound levels as high as 115 dB at popular music concerts.

It is tempting to try to associate definite sound levels with each dynamic marking in classical music. (If you are not familiar with the following terminology, look at Appendix A.) Some books suggest 70 dB as typical for medium loudness (*mf* or *mp*); 60, 50, and 40 dB for the soft markings (*p*, *pp*, and *ppp*); and 80, 90, and 100 dB for the loud markings (*f*, *ff*, and *fff*). But those round numbers exaggerate the level differences found in actual performance.

I measured levels during a Sacramento Symphony concert from one of the best seats in the auditorium, and found that they were seldom outside the range of 60 to 85 dB. When 90 dB was attained, it was unquestionably *fff*. I believe the orchestra is capable of producing 95 or 100 dB at my seat (although they never did in that particular program), but that happens rarely and would be considered *ffff*. Similarly, only a soft passage by a solo instrument was as low as 50 dB, and that was definitely *ppp*; 40 dB would be so nearly inaudible as to be practically useless.

There is another flaw in trying to associate sound levels uniquely with dynamic markings. Single instruments are in many cases simply incapable of even a 40-dB dynamic range. Patterson (see the chapter references) suggests that woodwind players especially may exhibit as little as 10 dB difference between their loudest and softest playing. Does the ear then interpret this to mean that such an instrument only plays *p* or *f*, never *pp* or *ff*? Probably not, because the listener will make some allowance for the instrument's capabilities; whatever greatest extremes are reached probably will be considered at least *pp* and *ff*, even if only 10 or 15 dB apart.

The player also can convey dynamic impressions in other ways, such as phrasing and articulation (see Chapter 7) or visual cues. If the player is visibly straining, your impression of loudness may be enhanced; if he appears relaxed, it may seem softer. So you could be led to think that one passage was *forte* and another *piano* when the sound level in decibels was the same in both cases.

This is a good example of one of the pitfalls awaiting every would-be scientist. It is the scientist's style to rely on objective measurements with a sound level meter, trying to connect these in some absolute, invariant way with musical dynamic markings. But music in reality involves the human element, which makes some judgments on a flexible rather than an absolute scale. A scientist must take care to identify all the variables in the problem.

Exercise 11) you have an intensity ratio of 130, you must not try to analyze it as $100 + 30$ (because that is not multiplication); you must call it 1.3×100.

When you encounter a statement such as, "The sound from that passing truck reached a level of 75 dB," it does not seem at first as if any ratio of two sound intensities is involved. But there actually is a ratio, for by convention we compare all sounds (such as those in Table 5.2) to a certain standard called I_0. This standard is such a soft sound that only very good human ears can detect it under ideal conditions. It is an intensity $I_0 = 0.000000000001 = 10^{-12}$ W/m^2. So, finally, consider a reading on our sound level meter of 90 dB—a level sometimes attained in musical performance (see Box 5.2).

This means a sound whose intensity is 10^9 times greater than the standard I_0. Thus, the intensity of the 90-dB sound is 10^{-3} W/m^2. For comparison, the intensity of direct sunlight on a clear day is roughly 10^{+3} W/m^2, or a million times as much. So a 90-dB sound carries only a minute amount of energy, even though we would call it a loud sound; our ears are extremely sensitive detectors!

---

*Those familiar with logarithms will recognize that all this is summarized in the formula $SIL = 10 \log (I/I_0)$. Technically, there is a flaw in what we have said above, because the microphone is directly sensitive to pressure rather than to intensity. What the sound level meter really measures is not the sound intensity level *SIL*, but the *sound pressure level*, $SPL = 20 \log (p/p_0)$, where the reference pressure is taken as $p_0 = .00002$ N/m^2 (rms). If we measured only single trains of waves traveling in one direction, the *SIL* and *SPL* would correspond directly to within approximately 1 dB. For sounds reverberating in closed rooms, however, the *SIL* and *SPL* may differ several dB from each other. Fortunately, the distinction is not crucial in most musical situations, and we will not concern ourselves with it in this book. But it would be important if we were making precise measurements in a research project.

5.3 THE INVERSE-SQUARE LAW

Before making any measurements with our sound level meter, we can anticipate one observable result. As we move farther away from a steady source of sound, we expect the sound level reading to diminish. But in exactly what way will it decrease? If we are inside a room, the answer may be complicated; indeed, in a highly reverberant room so many echoes may contribute that the level hardly changes at all from one end to the other. But if we listen to sound created in the middle of a large, flat, grassy field, we have a situation in which the sound moves out uniformly in all directions and never comes back. In that case, we can appeal to the law of conservation of energy. Imagine two large hemispherical surfaces centered at the source (Figure 5.2). Whatever energy passes across the inner surface, we must have exactly the same amount of energy crossing the outer surface a little later. The reason the intensity is less on the outer surface is merely that the same total energy has been spread over a larger area.

If the outer sphere has twice the radius of the inner, the energy that passes through 1 m^2 of the inner surface must be spread out over 4 m^2 by the time it gets to the outer surface. The intensity must be only one-fourth as large, and the sound level has decreased 6 dB. Every doubling of the distance lowers the sound level another 6 dB. If you measured 84 dB when 10 m from the source, you could anticipate that this would drop to 78 dB at 20 m, 72 dB at 40 m, 66 dB at 80 m, and so on.

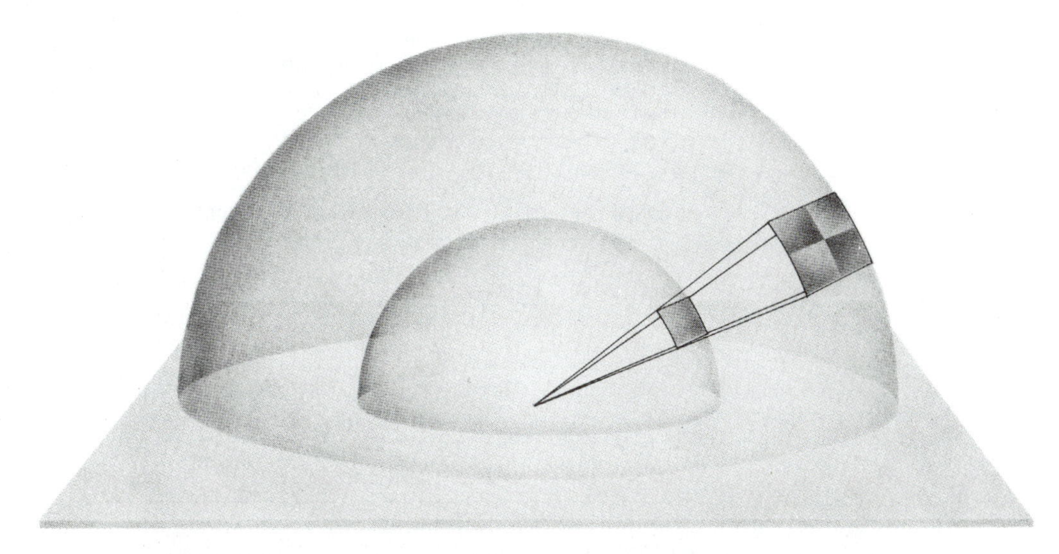

FIGURE 5.2 Sound spreading outward from a small source into unobstructed space. The energy crossing the patch on the inner sphere must be spread over the large patch on the outer sphere. Because the outer surface has twice the radius of the inner, the large patch equals four of the smaller ones, illustrating the geometrical basis for the inverse-square law.

A more general expression of this result is in the formula $I_2/I_1 = (r_1/r_2)^2$, or $SIL_2 - SIL_1 = 20 \log (r_1/r_2)$, known as the *inverse-square law.* For the same purely geometrical reason (that something spreads out and the surface area of a sphere increases as the square of its radius) this law also applies in optics, as well as in electromagnetism and gravitation. It is generally used much more in optics than in acoustics, because it is much easier there to achieve the necessary conditions of a small source radiating outward in all directions with no other complications. Even on a level, unobstructed athletic field, the sound level will tend to drop off faster than 6 dB per doubling of distance, because absorption by the grass steals a little of the energy.

*5.4 ENVIRONMENTAL NOISE

There has been much interest in recent years in the measurement and control of unwanted noise. Many communities have enacted ordinances governing permissible noise levels from air and highway traffic and from neighboring buildings. These ordinances often restrict the sound levels of outdoor rock concerts, too. The federal Occupational Safety and Health Administration (OSHA) also has set many standards for noise in factories, offices, and other work environments (Table 5.3).

These standards and measurements practically never tell the total amount of sound power put out by a source; instead they tell how strong the intensity is at some specific distance away from the source. If the distance were

TABLE 5.3 U.S. limits on permissible daily occupational noise exposure and more conservative limits as suggested by OSHA for avoidable exposure. Some other countries allow only a 3-dB level increase for each halving of exposure time.

	Maximum 24-Hour Exposure	
Sound Level (dBA)	**Occupational**	**Nonoccupational**
80		4 hr
85		2 hr
90	8 hr	1 hr
95	4 hr	30 min
100	2 hr	15 min
105	1 hr	8 min
110	30 min	4 min
115	15 min	2 min
120	0 min	0 min

different, the reading would be different. It is implicit in Table 5.2 that we all agree upon a "reasonable" distance in each case, such as half a kilometer from the end of the runway for an airplane takeoff, or 1 to 2 meters from the speaker for conversation.

When we use our sound level meter to measure various sounds in everyday life, we encounter complications. One of these is that most sound sources are not steady, so we cannot completely describe how loud they are with a single number. Unless we want to present detailed statistics on how often each different sound level occurred, we must sacrifice some of the information and hope that what remains is useful. One approach is to quote the highest reading observed and ignore all the rest. When community noise standards are written in terms of the maximum level, they can be extremely hard to achieve. If one large truck with a faulty muffler backfires in front of a school, such a standard may be violated regardless of whether that street is perfectly quiet the rest of the day.

Another approach is to use an average sound level; practical measurement of an average (usually over a 24-hour period) requires additional equipment to process the readings from the sound level meter. There are many different ways to take an average, so it is necessary to specify which will be used. One of the most popular is the equivalent level L_{eq}, which we obtain by imagining that the same total amount of received energy has been spread uniformly over the 24 hours. Another popular average is the day–night level L_{dn}, which is computed like L_{eq} except that all sounds between 10 P.M. and 7 A.M. are penalized by pretending that they were 10 dB higher than their actual levels. L_{dn} correlates well with long-term annoyance (Figure 5.3). These averages may lead to quite different conclusions from those based simply on the maximum level, because no matter how loud some sound is, if it lasts a short enough time it will not raise the average much. For example, an

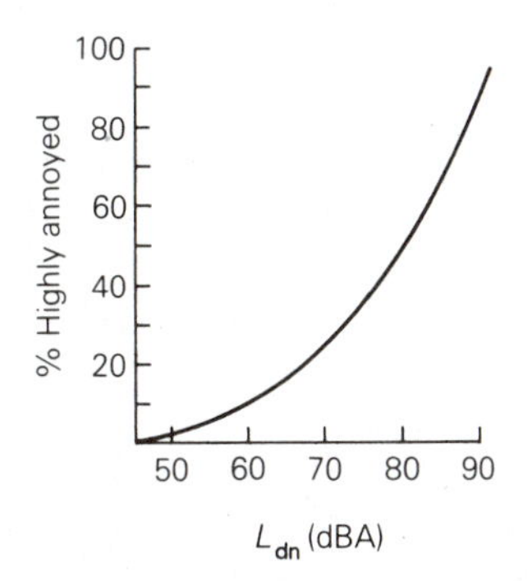

FIGURE 5.3 Relation between average environmental noise level and the resulting degree of annoyance. This graph is an average of the results of numerous surveys as reported by T. J. Schultz, *JASA*, *64*, 377 (1978).

explosion causing 110 dB for only 1 second and silence the rest of the time could give a 24-hour average of $L_{eq} = 60$ dB; yet that 1 second may be something we are unwilling to tolerate.

Still another approach is to find that level which is exceeded only 10 percent of the time (this is known as L_{10}), and require that it not be above some legal limit. For example, federal regulations require that protective measures such as walls or buffer zones be used wherever the L_{10} of freeway noise would be above 70 dB on adjoining residential property. Where highways pass schools, hospitals, or libraries, even stricter standards must be met.

---

*It can be rather difficult to achieve noise reductions of 10 dB or more in problem situations. Suppose, for example, you have a large compressor as part of the air-conditioning system for a commercial building. I have encountered such a machine, old and in poor repair, that was generating 75 dB at the boundary of an adjoining residential lot. The standard set by the county was 60 dB. It may seem as if a 15-dB reduction would not be difficult, until we calculate that this means *30 times* lower intensity. A modest wall in back of the compressor would not do, for it does not take much diffraction to have more than 3% of the former sound still sneaking over the wall. In this particular case, the problem solved itself when the compressor broke down and had to be replaced before the recommendations for its enclosure were completed. Much more attention is being paid these days to choosing quieter equipment for original installation.

Most sound level meters give us a choice of several methods of measurement. The discussion in Section 5.1 was presented as if we would measure all sound strictly on the basis of its energy, regardless of whether it was at high, medium, or low frequency. This would certainly be the physicist's first inclination, and it corresponds to the "linear" (or "flat" or "unweighted") scale on the meter, usually chosen by pushing a button or turning a knob.

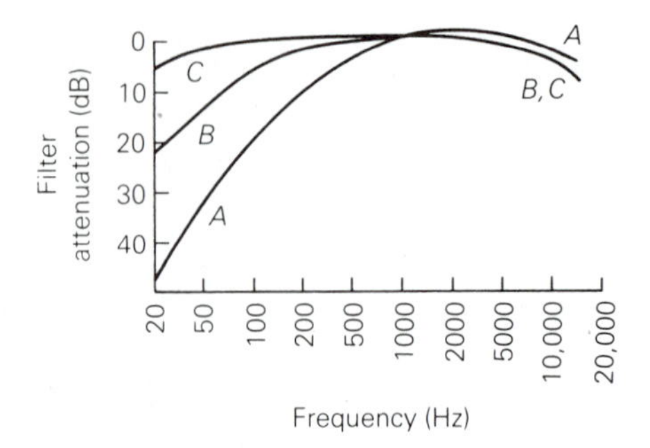

FIGURE 5.4 Relative sensitivity of a sound level meter at various frequencies for three standard weighting filters. The A, B, and C scales were designed to approximate human response at 40-, 70-, and 100-dB levels, respectively.

But what if the frequency is so high (or so low) as to be inaudible? Unless we go beyond the limits of its microphone, the sound level meter continues to give a reading, even though we hear nothing! Physicists may be interested in using the linear scale, but psychologists and noise-enforcement officers certainly are not. So our meter allows us to choose instead the A scale, where the microphone signal is sent through a filter circuit before being translated into a meter reading. This circuit passes on for measurement the full strength of signals with the medium-high frequencies to which the ear is most sensitive. But it cuts down somewhat the strength of the very high frequencies and greatly reduces the strength of the low frequencies to which the ear is less sensitive, and it cuts out altogether those that the ear cannot hear. Thus three different sources of sound at three widely differing frequencies, each giving equal intensity and equal readings on the linear scale, can give very different readings on the A scale. Sources of very high or especially of very low frequency will give low readings on the A scale, corresponding at least qualitatively to the way our ears would judge those sounds.

The A is always added as a suffix to remind us when any measurement was done with the A scale; for example, "My meter registered a maximum level of 93 dBA when that airplane flew over." Expensive meters also may give you a choice of B, C, and D scales; they differ in details, but all share with A the same general idea of approximating the human response (Figure 5.4). Even though B, C, and D are preferable sometimes in a technical sense, noise standards nearly always are specified in terms of dBA. This is partly because the A scale correlates well with risk of hearing loss, but also simply because the A scale is so widely available, even on old or cheap sound level meters.

Sound level measurements in dBA are a confusing mixture of physical and psychological elements. They are not the straightforward measurement of all energy on an equal basis; yet they are not direct human judgments either. They are physical measurements (made by a machine, not a person), but with a physical apparatus (the filter circuit) that makes allowances so that the results will be *similar in some ways* to human judgments. We will struggle with this distinction further in Chapter 6.

5.5 COMBINED SOUND LEVELS AND INTERFERENCE

What happens when two sources are both creating sound at the same time? *If* these are two independent sources putting out different sounds (specifically, different frequencies), the answer is simple. Each source continues to supply energy in the same way, just as if the other were not there. You can simply add the intensities that each source would make alone and get the intensity of the combined sound.

For example, consider a flute playing the note G_5, and a clarinet on B_4. Suppose the flute alone causes your meter to read $SIL_f = 60$ dB, and the clarinet alone gives $SIL_c = 63$ dB. What is the sound level when they play together? First of all, resist any temptation to add these two numbers together; the level of the combined sound is nowhere near 123 dB! Now $SIL_f = 60$ dB means the intensity of the flute sound is $I_f = 10^6 I_0 = 10^{-6}\,\mathrm{W/m^2}$; similarly for the clarinet, $I_c = 2 \times 10^{-6}\,\mathrm{W/m^2}$. The combined intensity is $3 \times 10^{-6}\,\mathrm{W/m^2}$, and so the combined sound level is about 65 dB. But we need not actually use all those numbers. The important thing is that $I_c = 2\,I_f$ (from Table 5.1, 3 dB higher means twice the intensity). So the combined intensity must be three times that of the flute alone, and the combined sound level is thus 5 dB higher than for the flute alone or 65 dB. (For another example, see Box 5.3.)

Matters become more complicated when we consider two sources whose frequencies are the same or nearly so. This is because the waves now maintain the same phase relation over a long period of time, and the total disturbance is entirely different depending on whether the interference is constructive or destructive.

Take, for example, two waves of the same frequency, with one having three times the amplitude of the other; that is, $A_1 = 3\,A_2$. If each were heard individually, the intensities would be related by $I_1 = (3)^2 I_2 = 9\,I_2$, and so $SIL_1 \cong SIL_2 + 9.5$ dB. Suppose you pick a point in the interference pattern where these waves are in phase with each other. There the combined amplitude is $A_c = A_1 + A_2 = 3\,A_2 + A_2 = 4\,A_2$. So the combined intensity and sound level are $I_c = (4)^2 I_2 = 16\,I_2$ and $SIL_c \cong SIL_2 + 12$ dB.

Now suppose you move to another location where these two waves arrive out of phase. There you will have destructive interference: $A_d = A_1 - A_2 = 3\,A_2 - A_2 = 2\,A_2$, $I_d = (2)^2\,I_2 = 4\,I_2$, and $SIL_d \cong SIL_2 + 6$ dB. If the separate sound levels were $SIL_2 = 77$ dB and $SIL_1 = 86.5$ dB, for instance, the combined level would always be somewhere between a maximum of $SIL_c = 89$ dB and a minimum of $SIL_d = 83$ dB. (Another example is worked out in Box 5.3.)

What happens if you have 10 violins carefully tuned, and instruct all 10 players to produce the same note? It is tempting, but incorrect, to think that we should add amplitudes and conclude that the resultant sound would have 10 times the amplitude, and thus 100 times the intensity, of one violin playing alone. Perhaps your next thought is that we must know the exact positions of all 10, as well as that of the observer (and any reflecting walls, too), to

*BOX 5.3 COMBINING AMPLITUDES, INTENSITIES, AND SOUND LEVELS

On first reading this chapter, you may feel unsure about how and when to use amplitudes, intensities, or sound intensity levels (*SIL*s). Sound strengths usually are expressed as *SIL*s (in dB), yet these always must be converted to another form before combining. Let us summarize several rules and techniques for finding how to describe the strength of a combination of two or more sounds.

1. You *never* just add *SIL*s.
2. You carefully refrain from writing such things as $10^{-7}\text{W/m}^2 = 50$ dB, which is a misuse of the equal sign. If it were truly an equality, we could do any legal arithmetic operation to both sides, for example, multiplying by 2; yet in this case that would give utter nonsense. *Corresponds to* is very different from *equals*!
3. If you combine sounds of quite different frequency (or if you care only about a long-term average, not beats), do not worry about amplitudes. Just add intensities, and if necessary convert the intensity of the combination to an *SIL* by using Table 5.1.
4. There are two ways to add intensities. The straightforward way is to express each one explicitly as a number of watts per square meter. Example: $SIL_1 = 69$ dB and $SIL_2 = 73$ dB (at different frequencies) mean $I_1 = 8 \times 10^6\ I_0 = 8 \times 10^{-6}\,\text{W/m}^2 = 0.000008\,\text{W/m}^2$ and $I_2 = 2 \times 10^7\ I_0 = 2 \times 10^{-5}\,\text{W/m}^2 = 0.000020\,\text{W/m}^2$, so $I_{comb} = 0.000028\,\text{W/m}^2 = 2.8 \times 10^7\ I_0$, which means $SIL_{comb} = 74.5$ dB. The clever way, which avoids the bother of converting to W/m^2 and back, is to keep everything in terms of ratios and comparisons. Same example: SIL_2 is 4 dB above SIL_1, so $I_2 = 2.5\,I_1$, so $I_{comb} = 3.5\,I_1$, so SIL_{comb} is 5.5 dB above SIL_1, which is 74.5 dB. You may choose whichever you find easier; perhaps sometimes you will want to do it both ways as a double-check.
5. If you combine sounds of the same frequency (or nearly enough the same to be concerned with slow beats), you must first convert information about their individual *SIL*s or intensities into information about their amplitudes. You then may add or subtract those amplitudes, according to whether the waves are in or out of phase, to find the amplitude of the combination. Finally, you may use the combined amplitude to find the intensity or the *SIL* of the combination.
6. The clever method of doing it all in ratios is especially advantageous when you have to use amplitudes; it avoids bothering to find each individual amplitude in N/m^2. Example: $SIL_1 = 69$ dB and $SIL_2 = 73$ dB as above (but now at nearly the same frequency). The long way finds the individual intensities $I_1 = 0.000008\,\text{W/m}^2$ and $I_2 = 0.00002\,\text{W/m}^2$, converts them to amplitudes $A_1 = 0.082\,\text{N/m}^2$ and $A_2 = 0.129\,\text{N/m}^2$, and combines these to $0.21\,\text{N/m}^2$ in phase or $0.047\,\text{N/m}^2$ out of phase, which corresponds to combined intensity $0.000054\,\text{W/m}^2$ or $0.0000026\,\text{W/m}^2$ and so to combined *SIL*s of 77 dB or 64 dB. The clever way merely says $I_2/I_1 = 2.5$, thus $A_2/A_1 = \sqrt{2.5} = 1.58$, thus $A_{comb}/A_1 = 2.58$ or 0.58, $I_{comb}/I_1 = (2.58)^2 = 6.7$ or $(0.58)^2 = 0.34$, and $SIL_{comb} = 8.2$ above or 4.7 below 69 dB; that is, 77 dB maximum and 64 dB minimum.

All such conversions are applications of this rule:

$$\left(\frac{A_x}{A_y}\right)^2 = \frac{I_x}{I_y} \xleftrightarrow{\text{Table 5.1}} SIL_x - SIL_y$$

compute phase relationships and account for interference that is partially constructive and partially destructive.

But in real life, the 10 violinists cannot all produce precisely the same frequency, no matter how hard they may try. Each pair of slightly different frequencies produces a slow beat; the interference is sometimes constructive and sometimes destructive. Ten violins together produce many different beat rates simultaneously, some slower and some faster and all gradually changing, because it is impossible even on one violin to keep the frequency exactly constant.

The continual ebb and flow of these multiple beats when three or more instruments play in unison is called *chorus effect* and adds a sort of warmth to the sound. It is important for composers to realize that having several instruments play together (or several voices sing together) does not merely produce louder sound; it also changes the quality of that sound. Although the intensity of the sound is not steady when chorus effect is present, the long-term average intensity is just the sum of the separate intensities, the same as if the frequencies had been quite different. In our example, the average intensity produced by the 10 violins is just 10 (not 100) times that for one violin alone.

SUMMARY

In describing the strength of a sound wave, it is important to distinguish among amplitude, intensity, and loudness. The first two are physical quantities, the last psychological. The intensity means the rate of energy flow (energy crossing unit area per unit time, W/m^2) and is proportional to the square of the amplitude.

For every intensity ratio there is a corresponding sound intensity level difference in decibels. Every *additional* decibel represents a *multiplication* of intensity by about 1.3 (more precisely, 1.26). A sound level $SIL = 0$ dB corresponds to the intensity $I_0 = 10^{-12}\ W/m^2$.

Under rare ideal conditions, sound levels drop 6 dB for every doubling of distance from a source. If human response to the sound is of prime importance, it is likely to be measured in dBA, which gives greatest weight to those sounds that our ears perceive best. Community noise ordinances generally are written in terms of maximum allowable sound levels or maximum allowable average levels, measured in dBA.

When two sounds of different frequency are combined, their intensities should be added to determine the resulting sound level. But for two sounds at the same frequency (or near enough the same to produce slow audible beats), it is the amplitudes that must be added or subtracted, according to whether they are in or out of phase, to determine maximum and minimum combined sound levels.

Multiple beat patterns from several instruments playing in unison produce the chorus effect, so that their sound is not the same as that of a single instrument amplified. The long-term average sound level in the presence of beats or chorus effect corresponds to simple addition of intensities, with no concern for phase relations.

REFERENCES

Blake Patterson wrote a very interesting article on dynamic levels of solo musicians in *Scientific American* (November 1974), p 78. Patterson draws some of his data from the article by M. Clark and D. Luce in *Journal of the Audio Engineering Society, 13*, 151 (1965).

The latter part of Chapter 3 in Backus treats interference, beats, and intensity, and includes an interesting table of measured greatest power outputs in watts, ranging from 0.05 W for a clarinet up to 67 W for an entire orchestra.

SYMBOLS, TERMS, AND RELATIONS

A pressure amplitude
E energy
P power
S area
$P = E/t$
$I = P/S$
$I_1/I_2 = (A_1/A_2)^2$
I intensity
SIL sound level
$I_0 = 10^{-12}$ W/m^2
I ratio ↔ *SIL* difference: Table 5.1
decibel
inverse-square law
chorus effect

EXERCISES

1. If a total energy $E = 200$ J is received uniformly during a time $t = 4$ s and spread over an area $S = 5$ m^2, what is the intensity?

2. If sound of intensity $I = 0.01$ W/m^2 falls on a window of area $S = 3$ m^2, what is the total power received? If this continues for an hour, what total accumulated energy arrives? Show that this amount of energy would be enough to lift a 1 kg mass upward about 11 m.

3. Suppose a sound of intensity $I = 10^{-6}$ W/m^2 falls on a detector of area $S = 7 \times 10^{-5}$ m^2. (That is about the size of an eardrum, two-thirds of a square centimeter.) What total power P in watts is being received by this detector? If the sound continues for 10 s, what total amount of energy E in joules is received?

4. Use a pocket calculator to multiply 1 by 1.26; multiply the answer by 1.26; multiply that answer by 1.26 again, and so on. Compare the answers with the numbers in Table 5.1. By the time you have multiplied by 1.26 ten times, does the answer come out to precisely 10.00? If not, find by trial and error a number slightly different from 1.26 whose tenth power will

be 10; that is the intensity ratio that corresponds to 1 decibel.

5. To get a sound intensity of 10^3 W/m^2 (that is, one carrying as much energy as bright sunlight), what sound level *SIL* (in dB) must you have?

6. What is the intensity *I* if the sound level is $SIL = 40$ dB? For 85 dB?

7. What is the sound level in dB if the intensity *I* is 10^{-10} W/m^2? If $I = 4 \times 10^{-7}$ W/m^2?

8. If one violin produces a reading of 75 dB on your sound level meter, show that you should get 78 dB from two violins playing together under the same conditions. What reading do you expect from 3 violins together? From 4, 5, and 10? How many violins would it take to produce a reading of 95 dB? How many for 105 dB?

9. Consider a fixed sound of intensity level $SIL_1 = 70$ dB and another (of different frequency) whose intensity level takes on the series of values $SIL_2 = 50$, 60, 70, 80, and 90 dB. To the nearest dB, what is the level of the combined sound in each case? Make a general statement about the combined level for any two sounds when one is much stronger than the other.

10. If two sources produce sound levels of 53 and 66 dB, what is the intensity ratio I_2/I_1?

11. If the maximum possible power output from a trombone is approximately 130 times that from a clarinet, and the clarinet produces a reading of 76 dB on your meter, what reading will the trombone produce at the same distance?

12. A singer is outdoors, on level ground, with no reflecting structures nearby. About how many dB difference in sound level would you expect between front and back rows if the audience is located between 5 m and 40 m from the source? How much difference do you suppose might be acceptable? Can a shell help solve this problem? How much will the shell help the deadness?

13. Two electronic oscillators are making sine waves, at frequencies $f_1 = 882$ Hz and $f_2 = 880$ Hz. At a certain point, the sound level from one alone would be $SIL_1 = 76$ dB, and from the other $SIL_2 = 82$ dB. How often do beats occur? What is the amplitude ratio A_2/A_1? How many times A_1 is the combined amplitude when the waves are out of phase and when they are in phase? What are the minimum and maximum levels (in dB) of the combined sound? What would be the combined level if the frequencies were widely different?

14. You might encounter sound levels of 75 dB in city traffic and 115 dB at a rock concert. What is the intensity ratio for these two sounds? What is their amplitude ratio?

15. Meyer and Angster report (in *Acustica*, *49*, 192, 1981) that violinists playing scales or arpeggios produced average sound levels that increased approximately 4 or 5 dB for each dynamic step when instructed to played at *pp*, *p*, *mf*, *f*, and *ff* levels. Describe what change occurred in the acoustic power output at each step. About how many times as much power was generated at *ff* as at *pp*?

16. If you attend a rock concert wearing earplugs that provide a reduction of 13 dB, what percentage of the sound energy do they block out?

17. Suppose two instruments, both at the same distance from you, produce readings of 82 and 65 dB on your sound level

meter when played separately. How many times as much power were you receiving from one as from the other? Without doing any calculations, can you make a close estimate of what the meter will read with both instruments playing together?

18. An outdoor air-raid siren produces a sound level of 115 dB at a distance of 10 m. How far away must you go to find a level of 85 dB? (Assume no reflecting surfaces or other complications to undermine the inverse-square law.)

PROJECTS

1. Ask a friend (or several) to play musical passages on some instrument while you measure with a sound level meter. What are typical levels for *pp*, *p*, *f*, *ff*, and so on? If at first you ask only casually for different dynamic levels, and then later you urge the player to use the greatest possible extremes, can you get him or her to increase the dynamic range? Compare your findings with Patterson's (see the chapter references).

2. Check out a sound level meter from your instructor, carry it around the campus or around town, and measure many different sounds, both musical and nonmusical. Try to include very loud and very soft sounds. Hold the meter away from you when measuring to minimize distortion of the readings by reflections from your body. When you report your results, be careful to specify for every measurement how far you were from the source.

CHAPTER

6 The Human Ear and Its Response

Loud, soft . . . high, low . . . smooth, harsh . . . hollow, full . . . In how many ways can you describe the sounds you hear? Is there no end to the nuances, or is there some definite number of qualities that will suffice to describe every sound? Is it fair to try measuring these qualities with numbers, or does that cause us to lose their essential meaning?

These are important questions in musical acoustics, for they are all ways of asking a deeper question: Just how far can science usefully go in analyzing and describing musical sounds? For that matter, how far can science go in connecting the physical and psychological aspects of *any* of our sensory perceptions? There always have been people who feel that physics must be confined to the objective study of matter and energy in nonhuman phenomena, while human perception is essentially subjective, and "never the twain shall meet."

But others feel that a science of human behavior is possible, and they have developed the discipline of psychology. Specialists in *psychophysics* set out to make an objective and quantitative study of how physical stimuli are related to human sensory perceptions. Their attitude is that these perceptions are not entirely vague or subjective after all. Although we study only auditory psychophysics here, human response to light, smell, taste, heat, and touch also can be studied with similar methods.

After a brief introduction to the anatomy and physiology of the human ear, we will consider in Section 6.2 the limits of its ability to detect and judge sounds. Section 6.3 spells out what is required for a complete description of both the physical and the psychological attributes of a sound and how they are connected. Simplest on the list of perceived sound properties are loudness and pitch, which we consider in detail in Sections 6.4 and 6.5. In the spirit of scientific analysis, these sections attempt to treat each property in isolation from all others. But reality is not so simple, and we must consider in Section 6.6 how loudness, pitch, intensity, and frequency are all tied together. The chapter closes with preliminary remarks on tone color and instrument recognition.

6.1 THE MECHANISM OF THE HUMAN EAR

Our ears are surprisingly sophisticated devices, and all the working parts are quite small. They are hidden from view and protected from damage by their location in recesses in the temporal bones (Figure 6.1). It is the task of the ear to convert incoming air-pressure fluctuations into electrical nerve impulses for processing by the brain. Let us learn about the ear by tracing this conversion step by step. As we proceed, refer both to the realistic picture of the ear and to the simplified schematic diagram (Figure 6.2) that clarifies what happens inside.

The *outer ear* consists of the pinna and auditory canal and ends at the eardrum. The *pinna* (or *auricle*) is the visible part standing out from your head. It is the inner portion of the pinna (about 3 cm across) that helps somewhat in funneling short-wavelength sound in toward the eardrum. The *auditory canal* (or *meatus*) is nearly 1 cm across and 2.5 to 3 cm long and allows the delicate eardrum to be set in where it is protected from dirt and sharp objects. The *eardrum* (or *tympanic membrane*) is a thin disc of fibrous tissue that bulges in and out in response to alternations in air pressure.

The *middle ear* is the enclosed chamber immediately behind the eardrum. It is connected to the throat by the *Eustachian tube*, through which air must move to equalize the pressure on both sides of the eardrum, such as happens when you climb a mountain. Unequal pressures distend the eardrum so that it cannot vibrate as well, reducing your hearing acuity, especially for low frequencies. Communication between the middle and inner ears is limited to two small membrane-covered openings called the *oval* and *round windows*, which confine the liquid called *perilymph* in the inner ear.

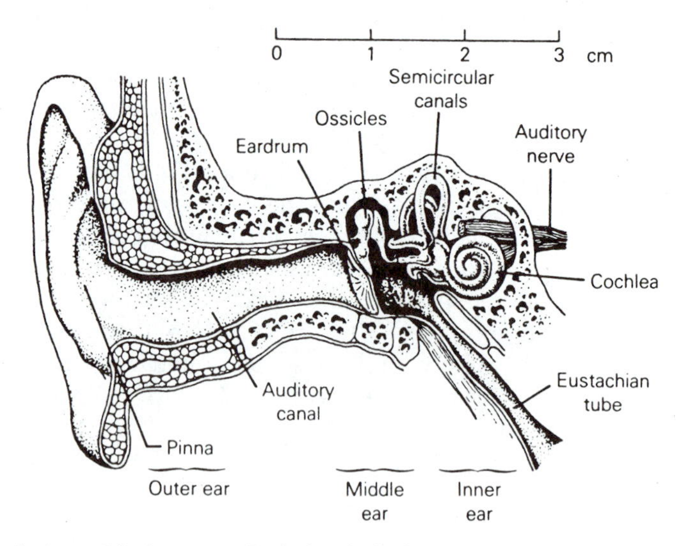

FIGURE 6.1 Anatomy of the human ear. For further detail of the middle and inner ear, see Figures 6.2 and 6.4.

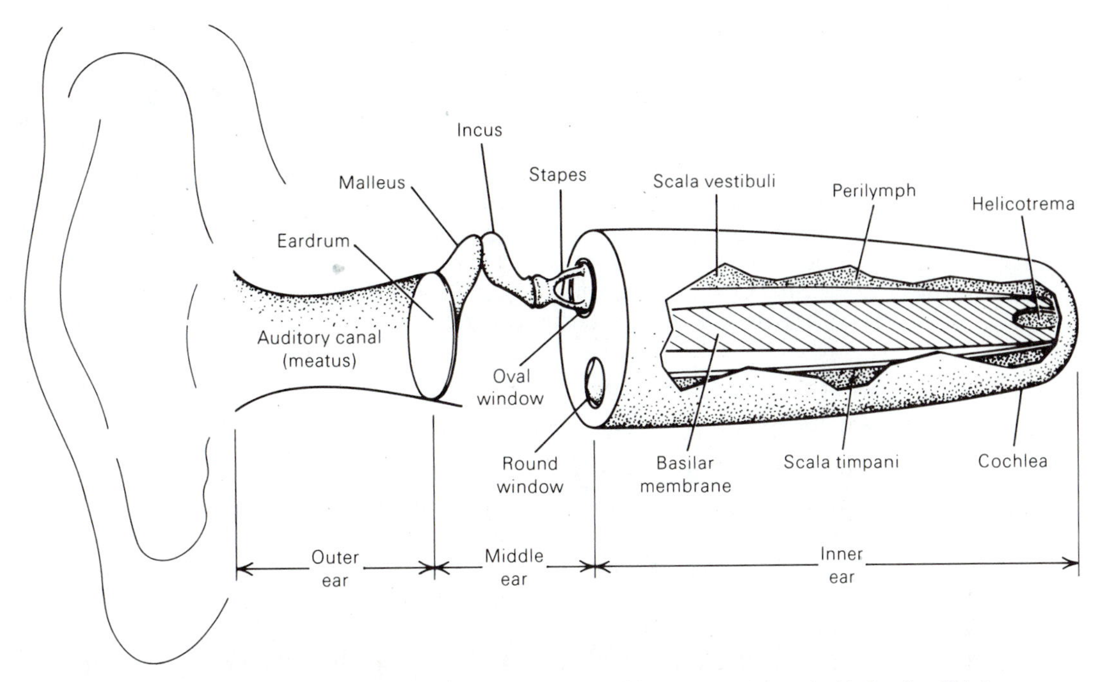

FIGURE 6.2 Schematic version of the ear, emphasizing parts most important in hearing. This is drawn as if the cochlea were uncoiled.

The middle ear contains three bones called the *malleus* (hammer), *incus* (anvil), and *stapes* (stirrup) or, collectively, the *ossicles*. They act as a lever system to transfer mechanical vibrations from the eardrum to the inner ear. The handle of the malleus is mounted on the eardrum, and the foot of the stapes on the oval window. For moderate-amplitude sounds, the malleus moves approximately 1.3 times as far as the stapes. As with any lever, a smaller force must move through a larger distance on one end while a larger force moves through a smaller distance on the other, so that both account for the same amount of work done. Thus, the stapes exerts approximately 1.3 times as much force on the oval window as the eardrum does on the malleus (Figure 6.3).

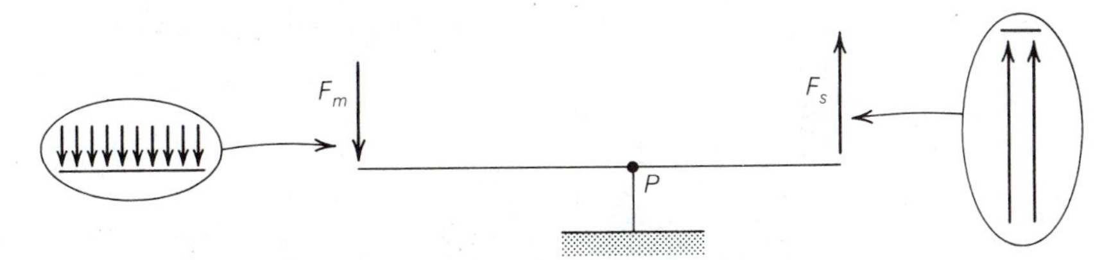

FIGURE 6.3 This simple lever, pivoted at P, is roughly equivalent to the middle ear. The total force F_m on the malleus is the result of a small acoustic pressure acting over the entire area of the eardrum. The somewhat larger total force F_s of the stapes upon the oval window produces a much larger pressure on that smaller area.

---

*Because the total force F_s from the stapes is applied to the very small area S_{ow} of the oval window, it produces a large pressure p_p in the perilymph: $p_p = F_s/S_{ow} = 1.3\ F_m/S_{ow} = 1.3\ (p_e S_e)/S_{ow}$. (Subscripts m and e refer to malleus and eardrum.) Because the area of the eardrum is approximately 20–25 times larger than that of the oval window (which is only about 3 mm^2), the pressure variations produced in the perilymph will be some 30 times greater than those on the eardrum. Intensity depends on the square of amplitude, so this suggests that we can hear sounds approximately 1000 times less intense (30 dB lower in sound level) than if the oval window received its vibrations directly from the air. This calculation is for an ideal lever system and does not account for frictional losses in the ligaments that hold the ossicles in place. According to von Békésy, the actual pressure ratio is 10 to 20 rather than 30, and thus the gain in sensitivity is only 20–25 dB.

The middle ear also contains two small muscles. The *tensor tympani* is attached to the malleus and increases the tension in the eardrum. The *stapedius* pulls the stapes sideways and reduces the mobility of the ossicle chain. Both actions reduce sound transmission through the middle ear, protecting the inner ear from damage by extremely loud sounds. The stapedius is activated by the *acoustic reflex* within about 10–20 ms for sound levels above 90–100 dB. The tensor tympani contracts as part of a more general reflex reaction taking about 10 times as long. Together, they provide up to 20 dB or more of attenuation for frequencies below 1 KHz.

The *inner ear* occupies a labyrinth of passages in the temporal bone; these include the *semicircular canals*, whose role is to sense changes in orientation of the head. The *cochlea* is a tapering tube coiled around roughly three times like a snail shell; if unrolled it would be about 3.5 cm long.

Inside the cochlea is an astonishingly complex structure called the cochlear partition (Figure 6.4). It separates two regions called the *scala vestibuli* and *scala tympani* along the entire length of the cochlea except for a small opening called the *helicotrema* at the end farthest from the middle ear; these regions are filled with perilymph. The partition has a hollow center called the *cochlear duct* or *scala media* filled with a more viscous fluid called *endolymph*. Because the *vestibular membrane* (or *Reissner's membrane*) is so thin and light, it just rides along with motions of the adjoining fluid and is acoustically unimportant. It is the **basilar membrane** whose vibrations determine what we hear. The *Organ of Corti* rides loosely on the inner part of this membrane and contains more than 20,000 *hair cells*. Different hair cells are disturbed when different parts of the basilar membrane move, and they initiate signals on the individual fibers of the *auditory nerve*.

The motion of the stapes must be passed on to the fluid in the scala vestibuli. If the oval window were the only outlet from the cochlea, the incompressibility of the perilymph would mean the stapes could hardly move at all. But the round window allows considerable motion by bulging out whenever the oval window moves in. For low frequencies there is enough time during each vibration for the fluid to move all the way up the scala

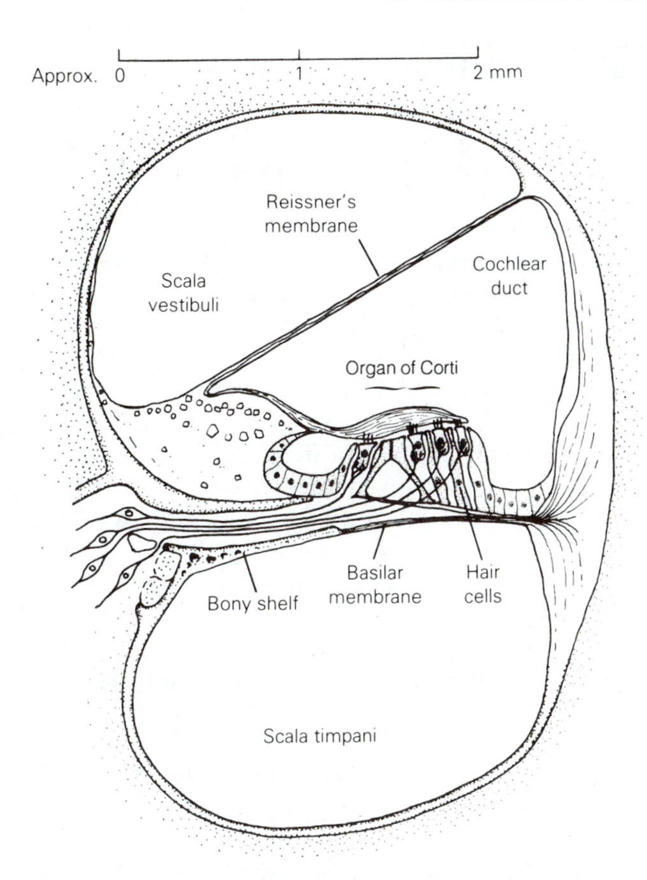

FIGURE 6.4 A cross section through the uncoiled cochlea. The center of the coil would be at the left. The width of the basilar membrane varies approximately from 0.08 mm near the oval window to 0.5 mm near the helicotrema.

vestibuli, through the helicotrema, and back down the scala tympani. But for high frequencies the fluid near the oval window must push down on the basilar membrane to communicate its motion quickly enough to the round window. The tendency for higher-frequency vibrations to disturb the basilar membrane closer to the windows is further encouraged by the fact that this end is relatively narrow, stiff, and light, while the end near the helicotrema is wider, more lax, and massive. The membrane width increases from less than 0.1 mm at the windows to around 0.5 mm at the other end.

None of the auditory nerve fibers from the inner ear goes directly "all the way to the top." Signals are mixed and partially processed at several way stations before finally being presented to the *auditory cortex* in the higher parts of the brain for conscious interpretation. Our sound perceptions depend just as much on the action of the brain as they do on the physiology of the inner ear. Unfortunately, our understanding of how the brain transforms and interprets information about sound is still rather rudimentary.

6.2 LIMITS OF AUDIBILITY AND DISCRIMINATION

Human ears do not respond to sounds that are too weak or too high or low in pitch. We are most sensitive to sounds with frequencies of approximately 2 to 5 KHz. A sound level of 0 dB (intensity $I_0 = 10^{-12}\ \mathrm{W/m^2}$) often is used as an easily remembered approximation to the faintest audible sound. But most of us require levels of 10 or 20 dB, and even more for lower and higher frequencies. We will introduce a diagram in Section 6.6 that easily summarizes this information.

We suggested in Chapter 1 that the audible range of frequencies extends from 20 Hz to 20 KHz, partly because those round numbers are easy to remember. There are, in fact, no sharp boundaries. It is a fairly unusual person who can hear frequencies as high as 20 KHz; a more common limit for a young person in good health is 17 or 18 KHz. Once into adulthood, the upper limit gradually decreases, and by retirement age it is likely to be below 12 KHz (for a woman) or 5 KHz (for a man). Fortunately, this leaves your capacity for musical enjoyment largely intact; it just removes a little of the brightness from the tone quality. (For further comments on hearing loss, see Box 6.1.)

At the lower end, sounds below 30 Hz are rather difficult to hear. But under proper conditions (high intensity and isolation from other distracting sounds), you may well be able to detect sine waves down to 20 or even 15 Hz. There have been reports of tracing this threshold all the way down to 2 Hz, but that is not really of musical interest. For one thing, below 20 Hz the quality of sensation changes from one of "hearing" to one of "feeling." For another, the thresholds are well above 100 dB. Finally, real music never uses pure sine waves at low frequencies anyhow. Although a large pipe organ may have notes as low as C_0 (16 Hz) and an exceptionally good bass speaker may be able to reproduce them cleanly, the musical usefulness of such notes depends almost entirely on the overtones that are present (for example, 32, 48, 64 Hz, and so on). We will study them in Chapters 8–14 and 17.

* Incidentally, you should *not* casually experiment with infrasound. High intensities at frequencies around 7–10 Hz can cause nausea (by disturbing the semicircular canals) and internal bleeding from bodily organs rubbing against one another.

Consider now this question: Among those sounds we *can* hear, how much must one sound differ from another so that you can detect the difference? This question arises for other perceptions as well: What are the minimum discriminable differences in color or brightness of lights, in sweetness of sugar solutions, in heaviness of objects lifted by hand? All these questions are answered with a classic procedure in psychophysics, the determination of **just noticeable differences** (*JNDs*, sometimes also called *difference limens*).

*BOX 6.1 HEARING LOSS

There are two basic kinds of deafness. *Conduction deafness* occurs when sound is not properly conducted from the eardrum to the inner ear. It most commonly results from repeated infections of the middle ear, during which the growth of extra fibrous tissue reduces the mobility of the ossicles. *Nerve deafness* means that the nerves fail to transmit signals to the brain even though there are normal vibrations in the inner ear; it may result from deterioration either of the hair cells or of the nerves themselves.

Older people normally experience a mild type of nerve deafness called *presbycusis,* in which they progressively lose their acuity for high frequencies while still hearing lower frequencies quite well (Figure 6.5). Prolonged exposure to high levels of environmental noise hastens this natural aging process.

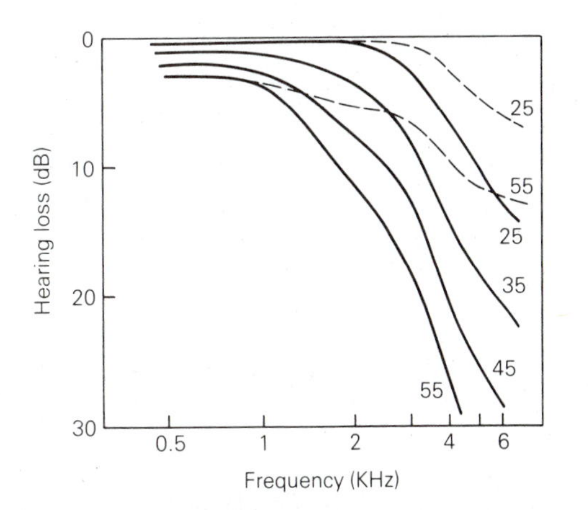

FIGURE 6.5 Average threshold shifts for 100 people with no history of industrial noise exposure (adapted from Berger, Royster, and Thomas, *JASA, 64,* 192, 1978). Men at various ages in years (solid curves) show much greater loss of high-frequency acuity than do women (dashed curves).

The two types of deafness can be distinguished by holding a vibrating object (such as a tuning fork handle) against the head. A victim of nerve deafness will hear no more than before. But a person with conduction deafness now can hear better because the middle ear is being bypassed by *bone conduction,* with the perilymph set in motion not by the oval window but by vibration of the entire bony structure surrounding the cochlea. Bone conduction can transmit vibrations directly from the mouth and nose to the inner ear, and that is why your own voice sounds different to you than to everyone else.

In acoustics we want to determine minimum detectable changes in both sound level and frequency for sine waves. Let us describe first how to determine *JND*s in sound level. This requires two oscillators to provide signals *X* and *Y*, both sine waves of the same frequency but at slightly different levels. The experimenter sets the output of one oscillator (*X*) at a fixed level (for example, 40 dB) at the subject's ears but changes the other oscillator output (*Y*) from time to time. The subject hears *X* and *Y* alternately, and is required to tell after each pair which one seemed louder. Clearly, if one sound level is 40 dB and the other 60, it will be easy to make a correct judgment 100% of the time. If both levels are 40 dB, the subject may want to say they are the same, but that is not allowed. In this "two-alternative forced-choice" scheme (commonly called a 2AFC experiment in psychologists' jargon), either *X* or *Y* must be picked as louder. So when they actually are equal, the choice will be random, 50% for each.

But what happens if the level of *Y* is 45 dB, or 41, or 40.1? Somewhere between 60 and 40 there must be a transition from total assurance to complete

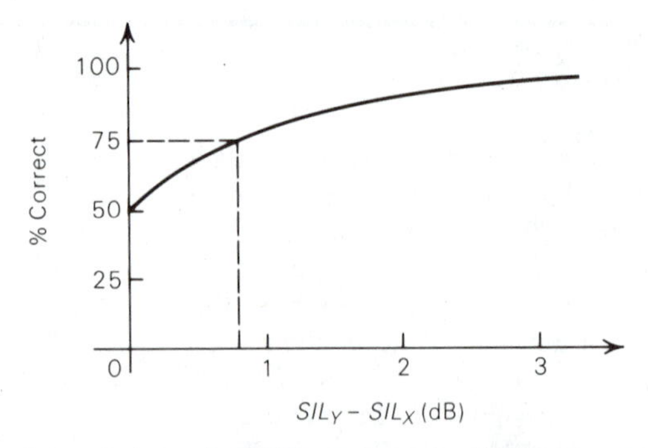

FIGURE 6.6 Proportion of correct choices in a *JND* experiment for 1000-Hz sine waves at 40-dB sound level. As the difference between sound levels *X* and *Y* becomes very small, correct determinations of which is louder approach the random 50% level. As the difference becomes large, performance approaches 100% correct. Dashed lines represent *JND* as defined by a simple 75%-correct criterion.

uncertainty, and that transition is not sudden (Figure 6.6). The experimenter explores both small and large differences, and determines that borderline case where the reliability of judgments is changing most rapidly. For $f = 1000$ Hz, a typical result would be $SIL_Y = 40.8$ dB for 75% correct choices, indicating that half the choices were random but the other half were based on actually perceiving a difference. We then would say that the *JND* in sound level is 0.8 dB for 1000-Hz sine waves at 40 dB.

This procedure can be repeated with the fixed oscillator reset to 50 dB, then 60, 70, and so on, to establish a *JND* for each level. When these data are graphed, we get the curve for 1000 Hz shown in Figure 6.7. Then the entire

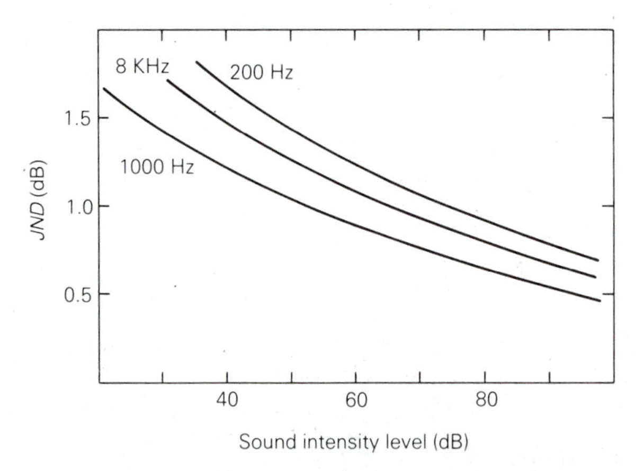

FIGURE 6.7 Just noticeable difference (*JND*) in sound level, as a function of initial sound level, for sine waves of several frequencies (after Jesteadt, Wier, and Green, *JASA*, *61*, 169, 1977). The 1000-Hz curve provides a fair approximation for all frequencies between about 500 and 4000 Hz. Results vary somewhat from one individual to another and also are sensitive to details of experimental procedure; these should be regarded as typical of what can be achieved with care by a good ear.

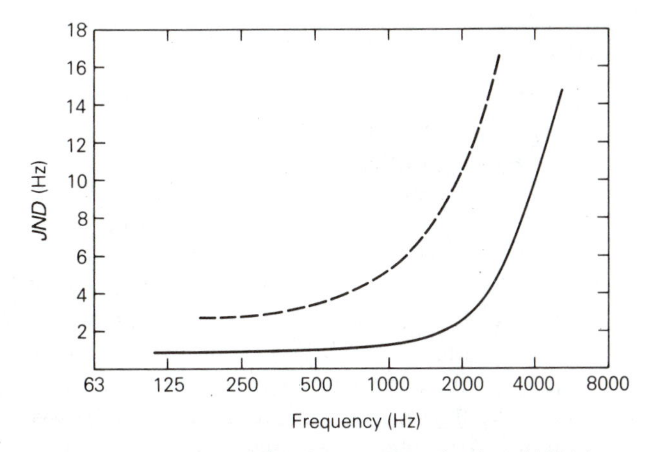

FIGURE 6.8 Just noticeable differences (*JND*) in frequency, as a function of initial frequency, for sine waves (after Wier, Jesteadt, and Green, *JASA*, *61*, 178, 1977). The dashed curve represents barely audible levels; the solid curve represents musically useful levels, roughly 60–90 dB. See J. D. Harris (*JASA*, *24*, 750, 1952) for discussion of why much larger values (for example, 3 Hz *JND*s at low and medium frequency) often quoted in the literature result from experimental designs that do not fully test pitch discrimination ability.

experiment can be repeated using other frequencies, giving a family of curves that covers all possible cases.

Although the *JND*s tend to be a little larger for lower frequencies and lower intensities, it is a fair approximation for most sounds of musical interest to say that the *JND* for sound level is between half a decibel and 1 decibel. That is, whatever intensity you already have, it must be changed by approximately 15–30% before you will notice the difference.

An analogous procedure will determine *JND*s in frequency: The subject listens to alternating tones of equal loudness with one frequency kept constant. The other frequency is varied to determine what minimum frequency difference is needed to judge reliably which of the two tones has higher pitch. This is repeated for other values of the fixed frequency to obtain Figure 6.8. The results may be summarized by saying that the *JND* in frequency for sine waves is around 1 Hz for all frequencies below 1 KHz, and begins to increase above that. Beyond 5 KHz, the *JND* rises rapidly as our pitch judgment becomes quite poor; above 10 KHz, pitch discrimination ability practically vanishes. It is no accident that the piano keyboard stops near 4 KHz. For complex waveforms, *JNDs* may be as little as 0.1 Hz at low frequencies; we will find later that this may be nicely explained by the presence of overtones.

As interesting as *JND* experiments may be, they are limited to deciding whether two sounds are the same or different; they do not determine *how much* different. Thus, in the next section we must prepare for more sophisticated approaches by carefully defining a framework for our judgments of sound.

6.3 CHARACTERISTICS OF STEADY SINGLE TONES

Although we are capable of perceiving several sounds simultaneously, any scientific study must begin with one tone at a time. Furthermore, we keep things as simple as possible by considering at first only tones that continue the same for a long time; that is, sounds whose waveform as shown by an oscilloscope repeats exactly the same shape over and over. Duration is obviously an extremely important quality of a musical note, but for sufficiently long notes it merely provides external boundaries, whereas we wish to dwell on those *intrinsic* properties of the tone that are independent of its beginning or end. We must return to the subject of duration in Chapters 7 and 17, for it does have important effects on our perception when it is brief.

What would constitute a complete *physical* description of a continuous tone? We already have discussed three things: amplitude (or, alternatively, intensity), vibration frequency, and wave shape. I claim that these completely determine the nature of the sound wave; if I have told you these three things, then there is nothing more to be said. You should think carefully about whether you believe that claim; try to imagine whether there is anything else to be said physically about the wave that does not just repeat some of this information.

What would constitute a complete description of our psychological *perception* of that wave? It certainly must include loudness and pitch, but how much more? If we think of smooth/rough, hollow/full, trumpetlike/violinlike, and so on, we cannot see any definite end to the list. But we can at least create a category to include everything other than loudness and pitch; it is called **timbre**, or tone color or tone quality. Then we can say that a sound is described by precisely three characteristics: loudness, pitch, and timbre. The price we pay for that neat package is that we have only postponed our need for a real understanding of what constitutes timbre.

Our definition of timbre suggests that it is quite different from the other two properties: Loudness and pitch each constitute a simple one-dimensional continuum (there is only one direction in which either can change), but timbre is complex and multidimensional (it can change in several different and independent ways).

Now how are the psychological perceptions related to the physical stimuli? A simple picture suggests itself: Perhaps intensity determines loudness, frequency determines pitch, and waveform determines timbre, each independently of the others (Figure 6.9). This correspondence is at least appropriate in that intensity and frequency are each one-dimensional (specified by a single number), while waveform is complex (we have not yet shown how to efficiently describe it). A little experimentation with an electronic oscillator and oscilloscope will quickly show at least some truth in this model. Turn up the amplitude and only the loudness changes while the pitch stays the same; turn the frequency knob and the most striking perceptual change is in pitch; switch waveforms and the timbre changes but the pitch does not.

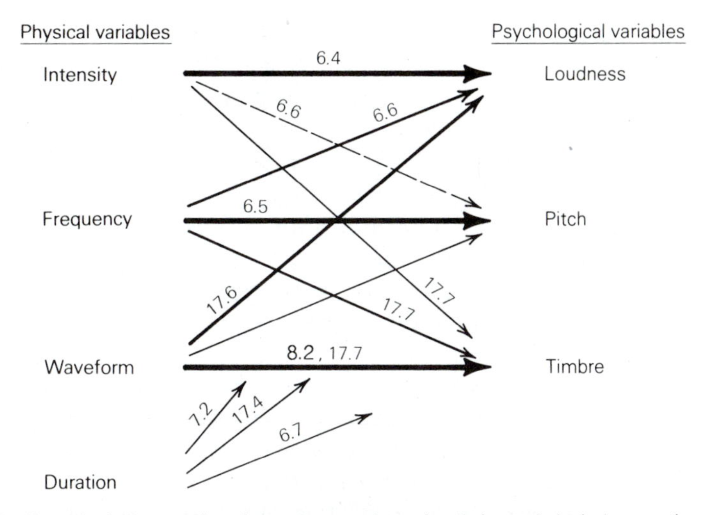

FIGURE 6.9 How physical properties of sound waves determine their psychological perception. Numbers on the arrows tell the chapters and sections in which each of these cause and effect relations is discussed.

If you pay careful attention in such an experiment, however, you also will find definite signs that this simple picture is not the whole truth. Large changes in frequency do produce changes in loudness even when the amplitude stays constant; sufficiently different waveforms of the same amplitude and frequency may have quite different loudness. There are also some cases where waveforms that look quite different on an oscilloscope nevertheless have timbres indistinguishable to your ears.

Our simple model, then, is only a first approximation, but it does suggest where to begin. It corresponds to the heavy lines in Figure 6.9. The lighter lines show additional causal relations that we must consider before we are through. Let us now begin at the top with the relation between intensity and loudness.

6.4 LOUDNESS AND INTENSITY

It is reasonable that loudness should depend primarily on intensity. The mechanism is clear: Larger air-pressure variations cause larger amplitude motions of the eardrum, middle-ear bones, oval window, perilymph, and basilar membrane. More violent motion of the basilar membrane means more stimulation of the hair cells and thus more nerve impulses sent to the brain.

It is interesting to know some typical numbers. For a quiet conversational level of 60 dB, the eardrum moves approximately 10^{-8} m, only some 100 times the diameter of a single atom. For an extremely loud sound of 120 dB, this motion is still no more than 10^{-5} m (a hundredth of a mm), and for an extremely faint sound of 0 dB, it is 10^{-11} m (a fraction of an atomic diameter). In each case the motion of the basilar membrane is perhaps 10 times smaller.

Nerve cells do *not* carry weak or strong voltages depending on the strength of their stimulation. They are on/off devices; they either send a pulse or not, and all pulses are the same size. Stronger stimulation results in *more* pulses, both from single nerves firing more often (up to 1000 times per second) and from more nerve endings reacting. The sensation of loudness is an interpretation by the brain of the average rate of arrival of pulses on the auditory nerve.

Can we establish a quantitative measure of loudness? Let us consider for now only the loudness of sine waves. That may seem like a stringent limitation, but we will find later that we can use this information on sine waves to explain most of our reactions to other waveforms. And let us begin with an experiment in which we use sine waves of fixed frequency (say 1000 Hz), allowing only their amplitude to change.

Now we may face the temptation to set up a measure of the loudness of a sound by just counting how many *JND*s it is above the threshold of bare audibility. We would find that it takes roughly 40 *JND*s to reach 60 dB at 1 KHz and another 40 *JND*s to get up to 85 or 90 dB, which suggests that the loudness at 85–90 dB is twice that at 60 dB. Unfortunately, *this does not work*; if you ask a large number of people, their overwhelming opinion will be that 85 dB is much more than twice as loud as 60 dB. So, whatever interest *JND*s have for their own sake, they do not serve to define a quantitative scale of loudness (Box 6.2).

Our object must be to listen to many different amplitudes, and judge *directly* how much loudness corresponds to each. That is, we want to determine by observation the functional relationship between loudness and intensity for 1000-Hz sine waves (and also eventually for other frequencies).

There are several kinds of experiments in which subjects directly determine a number to describe loudness. *Ratio estimation* requires listening to one sound, then another, then filling in the blank in the statement "Sound *Y* was_____times as loud as sound *X*." *Ratio production* requires the subject to listen to *XYXY*... while he adjusts the intensity of one of them until satisfied that it meets a stated requirement such as "Make sound *Y* twice (or five times, or half, etc.) as loud as *X*." In *magnitude estimation* the subject listens and assigns numbers to represent the relative loudness of a whole series of different sounds, rather than just one pair at a time.

Many such experiments have been done, and analyzed extensively by S. S. Stevens and his collaborators. Considering the human element involved, it can be claimed that they agree tolerably with one another when done with sufficient care. Thus, there is a useful scale of loudness on which many people can agree fairly well; it is not just an individual matter.

The unit of measurement for loudness is the **sone**. By convention, a loudness of 1 sone is defined to be that of a 1000-Hz sine wave at 40-dB sound

BOX 6.2 THE PSYCHOPHYSICAL LAW

It would be most satisfying to discover that what we learn about one human perception would apply just as well to all others. That is, if we establish a quantitative relation between intensity and loudness of sound, might not that same relation exist between other stimulus-judgment pairs such as intensity and brightness of light or weight and heaviness of an object held in the hand? After all, might not different parts of our sensory nervous system all work in much the same way? To some extent this is true, but there is often confusion as to the actual form of the functional relation.

The desire to establish this relation has sometimes led to oversimplification and to some people "seeing what they wanted to see" in the data. There is a piece of folklore dating back to the late nineteenth century, called the "psychophysical law," which still survives in many books. It often is associated with the names of Weber and Fechner, who were founding fathers of the science of psychophysics. Its esthetic appeal is so great that many people believe it even to this day; but, in fact, experiments show that it is not even approximately true.

The mathematical statement of this "law" is that the perceived psychological magnitude is proportional to the logarithm of the physical stimulus; that is, similar multiplications of the stimulus always produce similar additions to the response. In the case of sound, this simply means that our judgments of loudness would closely match the decibel scale; 80 dB would sound twice as loud as 40 dB, every 10-dB increase would cause an equal *addition* to the loudness, and so forth. *If* *JND*s were always the same number of dB and *if* numbers of *JND*s corresponded to perceived loudness, that would be true; but, in fact, neither of those conditions is met.

According to S. S. Stevens, most sensory perceptions correspond more nearly to power laws than to logarithms. That means equal stimulus ratios produce equal judgment ratios; that is, similar multiplications of the stimulus correspond to *multiplications* of the response rather than additions to it. Loudness of sound on the sone scale (at 1000 Hz and above 40 dB) is approximately proportional to the $\frac{1}{3}$ power (the cube root) of the intensity; this corresponds roughly to our statement that a 10-dB increase in sound level leads to a doubling of loudness. At frequencies below 100 Hz, the relation is more nearly $\frac{1}{2}$ power (square root), which means an approximate tripling of loudness for every 10-dB increase in intensity.

level, and all others are compared to that. Figure 6.10 shows how many sones of perceived loudness correspond to each different intensity level. Because the ear is not equally sensitive to all frequencies, any frequency other than 1000 Hz has its own loudness-intensity relationship; some of these are shown also in Figure 6.10. The graph says, for instance, that a 50-dB, 1000-Hz sine wave (point *A*) has a loudness of about 2.3 sones; that is, 2.3 times as much as the 40-dB standard. For a 50-dB, 100-Hz signal, however, the loudness is approximately only 0.7 sones (point *B*).

For the frequencies and intensities of musical interest (above 50 dB), it is a fair approximation to say that every 10-dB increase in sound level (that is, every multiplication of intensity by 10) will double the loudness in sones. This suggests that a group of 10 people all singing the same note will sound about twice as loud as a soloist, and a chorus of 100 twice again as loud; that is, four times as loud as the soloist.

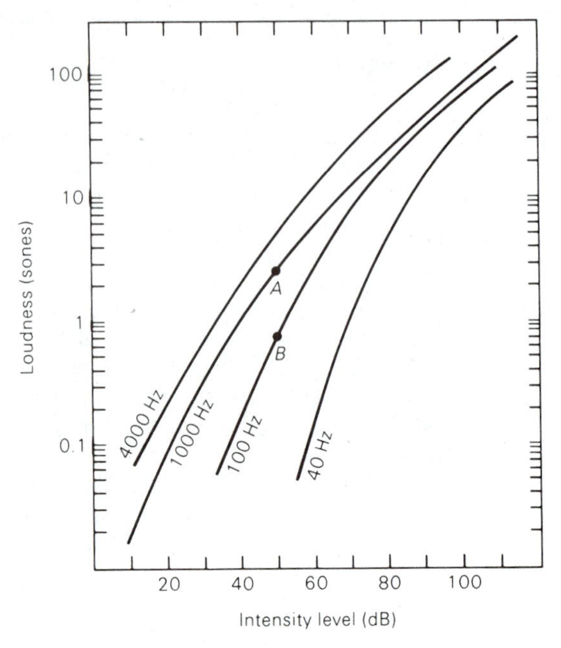

FIGURE 6.10 Perceived loudness as a function of intensity level, for sine waves of several frequencies. Points *A* and *B* represent examples discussed in the text.

---

*The preceding conclusions represent the generally accepted wisdom of the early 1970s. But it has been challenged in the work of Warren, who claims that, after certain subtle biases are removed from the experimental procedures, the data show that loudness doubles every 6 dB rather than 10 dB. This suggests the interesting possibility that our minds subconsciously try to equate halved loudness with doubled distance. Only time, and further refined experiments, can tell whether other scientists will become convinced that Warren is right instead of Stevens.

6.5 PITCH AND FREQUENCY

Some sounds have a clear and definite pitch; others are hard to judge. When a strong sensation of pitch exists, it is determined primarily by frequency. Let us again concentrate on the pitch of sine waves and postpone complications about pitch perception of complex waves until Chapter 17. It will be enough for now to say that (aside from odd exceptional cases) our ears assign to any *periodic* waveform (one that repeats without change) very nearly the same pitch as a sine wave with the same repetition frequency.

We suggested in Section 6.1 that different frequencies cause different parts of the basilar membrane to vibrate so that the brain receives pulses from different nerve endings, and these are interpreted as different pitches. This picture cannot explain all aspects of pitch perception, but it does have a large element of truth and will serve for the time being.

There is a fundamental way in which pitch/frequency differs from anything we have said about loudness/intensity. Loudness (like brightness, heaviness, and so on) is a matter of large or small, more or less, including the possibility of zero. But pitch is not large or small; it is here or there, high or low, left or right, and the idea of zero pitch is meaningless. (In technical terms, this distinction is indicated by saying that loudness is measured on a *prothetic* scale, while pitch is *metathetic*.) So we will make no attempt to say how big a pitch is, only how much different it is from other pitches.

What unit shall we use to measure pitch differences precisely? You will find in some books a discussion of the mel scale, in which a 1000-Hz tone is defined to be 1000 mels and all other tones are assigned a number of mels telling how they sound when compared to 1000. Notice the implication that other pitches are to be regarded as larger or smaller compared to 1000. This attempted analogy to sones for loudness is possibly of some interest to experimental psychologists, but it is irrelevant to musical acoustics. This is because (1) it is based on the misconception that musical pitch can be measured on a prothetic scale, and (2) a true mel scale can be obtained only by judging pitch *outside* any musical framework, preferably with subjects who have no musical experience whatever.

For our purposes, it is the judgments of well-trained musicians with keen ears that are important. The natural unit for these judgments is unquestionably the octave. Every octave interval is considered to represent an equal change in musical pitch throughout the range of pitches that can be judged reasonably well and thus be musically important. And within the octave other musical intervals (to be introduced in Chapter 7) will serve as invariant standards of pitch separation.

Recall how we originally introduced the octave with rather casual remarks that this peculiar sensation corresponds to a doubling of frequency. We now are saying more specifically that the octave has the remarkable distinction of being an appropriate unit for *both* a physical quantity (frequency ratio) and a psychological quantity (pitch separation). (Intensity and loudness had no such close connection.) We would carefully avoid giving the same word two distinct meanings except for the *experimental* observation that a psychological octave (a special pitch sensation) always corresponds to a physical octave (a frequency doubling) when tested under appropriate conditions.

Notice the similarity to the decibel scale: Every additional bel represents a multiplication of intensity by 10; every additional octave represents a multiplication of frequency by 2. This corresponds to the mathematical statement that pitch varies in proportion to the logarithm of frequency.

6.6 PITCH AND LOUDNESS TOGETHER

Now it is time to admit that pitch is not entirely determined by frequency and loudness is not determined by intensity alone. Let us examine here the

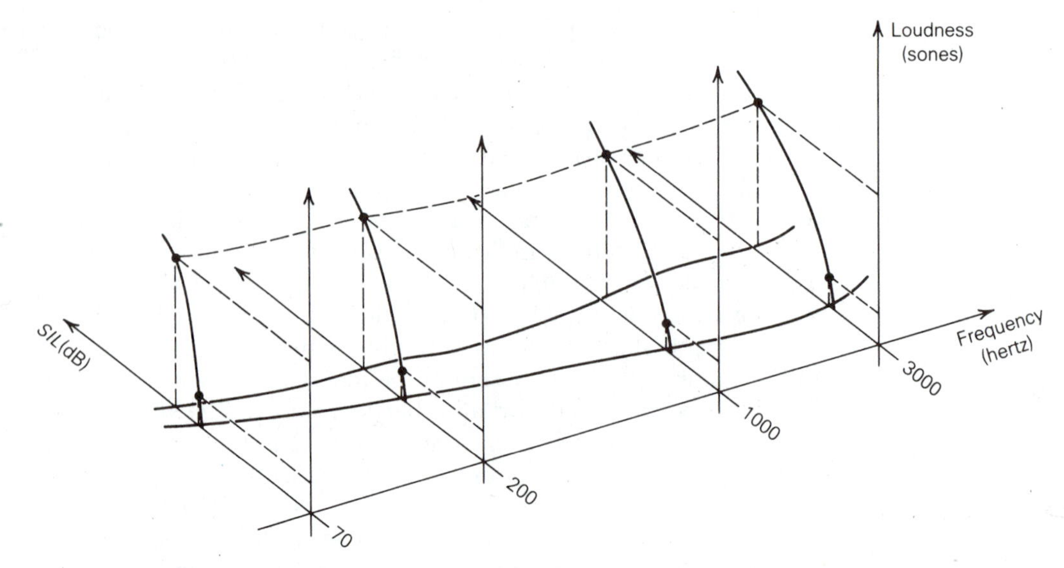

FIGURE 6.11 Dependence of loudness upon intensity level and frequency, shown as a collection of graphs in three dimensions. An observer stationed off the page to the upper right and looking back toward the lower left would see the four curves superimposed as in Figure 6.10. Points of equal loudness are connected (as with the dashed curved line) to make contours, and these are projected down onto the horizontal plane to make a contour map (solid curves). An observer stationed above the top of the page and looking down upon this contour map sees the Fletcher–Munson diagram (Figure 6.12).

dependence of pitch upon intensity or waveform, and of loudness upon frequency.

The first case is a relatively minor effect. Large changes in intensity can produce shifts of as much as a semitone in pitch perception, even when the frequency remains constant. Low pitches tend to sound a bit lower, and high pitches a little higher, when made very loud; medium pitches are hardly affected. This effect occurs mainly for sine waves or other quite simple waveforms; I occasionally catch myself thinking I have hit the wrong low pedal note on the organ if playing only pipes with a very gentle timbre. But the complex waveforms most often used in music have stable pitch that does not depend on intensity, so we will concern ourselves with this effect no further.

The second case is extremely important; we already have met several pieces of evidence that sounds of the same intensity but different frequency may have vastly different loudness. We need some way of representing how loudness depends simultaneously on *both* intensity and frequency. Because there are two independent variables, a single graph will not do. We could make a family of many graphs; indeed, we showed a few members from such a family in Figure 6.10.

A clever way to understand these graphs more clearly is to imagine them spaced out along a third axis according to frequency, as in Figure 6.11. If we fill in enough of them, they merge together into a three-dimensional graph,

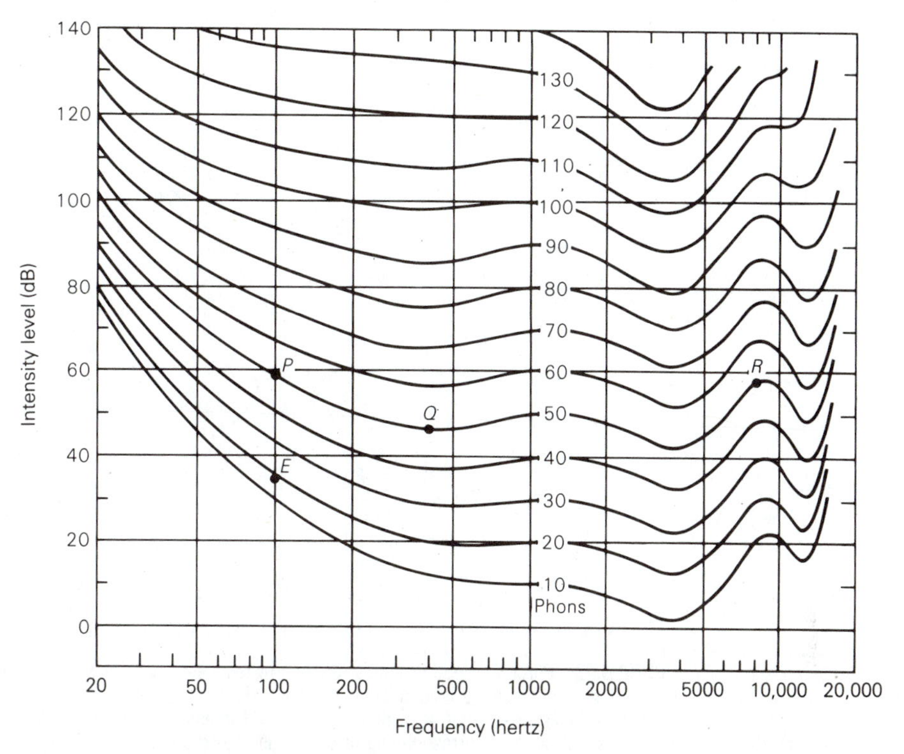

FIGURE 6.12 Contours of equal loudness level for sine waves, with the source straight ahead of a listener with two normal ears. Each contour is labeled with its loudness level in phons, which is by definition the same as the intensity level where that curve crosses $f = 1000$ Hz. The corresponding loudness in sones must be found on Figure 6.13. The crowding of contours on the left corresponds to a steeper drop from the highlands to sea level in the three-dimensional picture of Figure 6.11. Points *E*, *P*, *Q*, and *R* represent examples discussed in the text. (Original data by Robinson and Dadson, 1956.)

where location in a horizontal plane tells both intensity and frequency, and then height of some odd-shaped surface above this plane tells loudness. But because such a supergraph requires either a solid model (perhaps in plaster of Paris) or else difficult perspective drawing to show three-dimensional information on a flat sheet of paper, we might also consider showing the view from above. This is conveniently presented as a contour map; that is, a map showing where the shoreline would be if the supergraph were immersed in water to various depths. The dashed curve in Figure 6.11 is an example of such a shoreline.

In the present case such a contour map is called a *Fletcher–Munson diagram* (Figure 6.12), and its contours are *equal-loudness curves*. That is, each contour represents a family of sine waves, with such combinations of intensity and frequency that they all sound equally loud. The points *P*, *Q*, and *R* on the contour labeled "50 phons," for instance, tell us that a 100-Hz sine wave at 59 dB, 400 Hz at 46 dB, and 8000 Hz at 57 dB all have the same loudness.

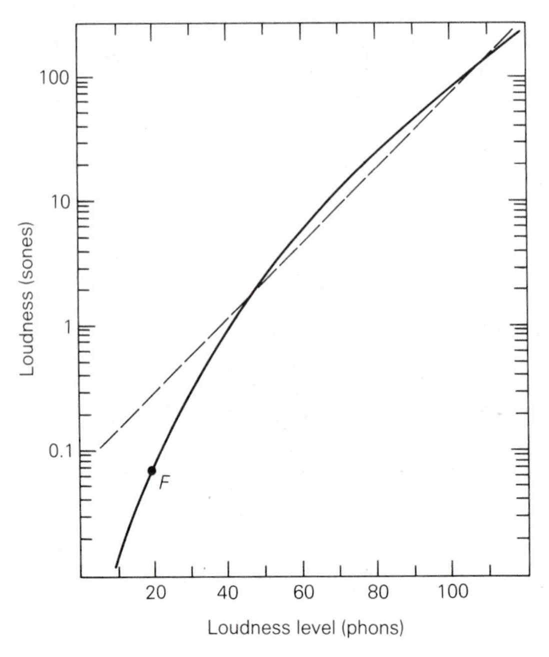

FIGURE 6.13 Perceived loudness as a function of loudness level. Now the same curve applies for sine waves of all frequencies; it is identical to the 1000-Hz curve in Figure 6.10. The dashed straight line represents the approximation that increases of 10-dB intensity produce doublings of loudness. Point *F* represents the example corresponding to point *E* in Figure 6.12.

This map can be used to find the loudness of any sine wave of given intensity and frequency simply by finding which contour the corresponding point falls on. (Only a few representative contours are shown; you are left to imagine many others in between.) Although each contour in the diagram could just as well be labeled with the corresponding loudness in sones, it is customary (and, unfortunately, confusing) to use an intermediate step called *loudness level.* Any two sounds of equal loudness are also of equal loudness level. The confusing part comes in assigning a number. The **loudness level** in **phons** is always the same number as the intensity level of the 1000-Hz tone of that same loudness. Sounds P, Q, and R in the example above all have loudness level $LL = 50$ phons, meaning they are just as loud as a 50-dB, 1000-Hz tone; that is, 2.3 sones.

Thus, we determine loudness in two steps. First, from the intensity level and frequency we get the loudness level on the Fletcher–Munson diagram and then go from the loudness level in phons to the corresponding actual loudness in sones. The second step is represented by a simple graph, shown in Figure 6.13. Let us work an example. Suppose we want to know how loud a 100-Hz sine wave would be at 35 dB. We locate point E on the Fletcher–Munson diagram, and find that it is just below the 20-phon contour. We estimate that this sound corresponds to 19 phons, then examine Figure 6.13 to see that this means 0.07 sones (point F).

The normal *threshold of hearing* is the intensity (depending on frequency) of sound that is barely audible under ideal conditions. It varies greatly from one person to another but corresponds to one or another equal-loudness contour, so it is easily seen on the Fletcher–Munson diagram. In a typical group of Americans, 1 percent or fewer have a threshold as low as the 0-phon contour; around half have a threshold below the 20-phon contour. The term *threshold of feeling* often is seen near the 120-phon contour, because sounds above that level are not only extremely loud but also accompanied by an uncomfortable tickling sensation.

An interesting application of the Fletcher–Munson diagram is in understanding volume controls in hi-fi systems. Suppose a live performance of some music is faithfully recorded in a good listening location, and that there are similar loudness levels of 90 phons both at 1000 Hz and at some extremely low bass frequency such as 50 Hz. Suppose now that the recording is played back in a small room with the volume set quite low to avoid disturbing the neighbors. A simple volume control will treat all frequencies equally; suppose it is set so that all intensities are reduced to 0.01 percent of their original values. At 1000 Hz, the original 90 phons or 90 dB is reduced to 50 dB and thus to 50 phons. But at 50 Hz, the original 90 phons was 102 dB; this is reduced to 62 dB, which is only 36 phons. Instead of the previous match in loudness (about 45 sones each), we now have 2.3 sones of treble and 0.6 sones of bass. It seems to the listener that the music is badly lacking in bass!

One solution to this problem is to allow the listener to boost (or reduce) the bass with a separate control whenever the overall level is low (or high). Another is to provide a slide switch labeled *loudness* or *presence* that inserts or removes a bass filter. Another, provided on some higher-quality equipment, is to have two separate volume-type control knobs called *level* and *loudness*. The first of these increases the intensity of all frequencies equally (appropriate for filling a larger room with the same sound), while the second changes the intensity for bass frequencies less than for the others (as needed to keep all loudness levels in proper relation for either loud or soft listening in the same room).

6.7 TIMBRE AND INSTRUMENT RECOGNITION

The third primary relation among sound attributes is the determination of timbre by waveform. It is easy to see that there is a close connection, either by using electronic oscillators (as suggested in Section 6.3) or by playing instruments such as flute, oboe, and violin into a microphone and oscilloscope (recall Figure 2.5, page 22). But the relation is much more difficult to describe than for loudness/intensity or pitch/frequency, because timbre and waveform are both complex quantities. In later chapters we will develop some concepts that make this problem more tractable, and in Chapter 17 we will answer our original question about whether timbre represents a finite amount of information.

We can, however, continue and consider the problem of instrument recognition. An experienced musician usually finds it quite easy to tell by sound alone which instrument he or she hears; some can even tell the difference between a violin and a viola. How? Because you can easily arrange for a trumpet, a violin, and a clarinet all to be heard at the same pitch and loudness, it may seem as if steady-tone timbre (and therefore waveform) is the basis on which instruments are recognized.

But that is not entirely true. If tape recordings are spliced so that you are allowed to hear only the steady middle part of each tone, judgments become much more difficult; the waveform alone is hardly sufficient for clear instrument identification after all. Instrument recognition depends a great deal on hearing the **transients**, the beginnings (attacks) and endings (decays) when the waveform is rapidly changing. This can include not only the initiation of the main vibrating mechanism (the initial scrape of the bow on a violin string, a tiny squawk from a clarinet reed before it settles down to steady oscillation, or the percussive sound of the first puff of air released by a trumpet player), but also incidental noise (clicking of keys or valves, rebounding hammers in a piano, sharp intakes of breath).

The durations of initial transients on any instrument generally vary between high and low notes. But typical attack times range from 20 ms or less for an oboe, through 30–40 ms for trumpet or clarinet, to 70–90 ms for flute or violin. Notes in the octave above middle C have periods of 2 to 4 ms. Thus, it may take anywhere from two or three up to several dozen vibration periods for the tone to be established and become fairly steady. This transient differs enough among various instruments to give vital clues to their identity.

The instrument designs and playing styles of the Baroque era may have emphasized further the differences in attack times between double-reed and string instruments. It has been suggested that Bach was aiming for greater precision and definition in the sound when he often had violin or cello parts doubled by oboe or bassoon.

SUMMARY

The human ear is a highly sensitive receptor. It achieves this through a lever system in the middle ear that transfers vibrations from the eardrum to the inner ear. Different frequencies preferentially produce motion of different parts of the basilar membrane, so that different fibers of the auditory nerve receive the greatest stimulation.

As a first approximation, it is useful to think of each psychological attribute of a steady sound (loudness, pitch, and timbre) as being produced mainly by one of its physical attributes (intensity, frequency, and waveform, respectively). But frequency actually has an important influence on loudness, too, leading us to represent a more complete picture with the Fletcher–Munson diagram. That diagram contains the same information as would a more complete family of curves (Figure 6.10).

Comparison of two loudness levels in phons can tell only whether one is louder than the other; only by finding the corresponding loudnesses in sones can we tell how much louder. The ear compresses the scale of loudness, so that an actual intensity range such as 100 million to 1 (for example, 1000 Hz between 20 and 100 dB) may be perceived as only a 1000-to-1 range of loudness.

Typical *JND*s for sounds of musical interest are 0.5 to 1 dB in intensity level and 1 Hz in frequency. These do *not* provide the basis for the sone and octave scales of loudness and pitch. Musical pitch is best measured in octaves, corresponding to factors of 2 in frequency. This scale of pitch does not have any zero.

The timbre produced by a steady waveform helps us recognize different musical instruments, but the transients when notes begin and end are also important for this purpose.

REFERENCES

This material is treated by Roederer in Chapters 2 and 3. Perhaps the best source of further details on this level is Denes and Pinson, Chapter 6. There is also interesting information on hearing loss from noise exposure in Section 17 of Strong and Plitnik. For a more detailed survey, see T. S. Littler, *The Physics of the Ear* (New York: Pergamon Press, 1965). There is an interesting article by A. J. Hudspeth about the workings of the hair cells of the inner ear in *Scientific American, 248,* 54 (January 1983), and some outstanding electron micrographs of cochlear structures in *Discover, 3,* 92 (November 1982).

Georg von Békésy won a Nobel Prize in physiology and medicine for his many difficult and ingenious experiments on the human ear, some of which had to be completed within an hour of dissection of a fresh corpse before the delicate tissues deteriorated. He is responsible for much of our present knowledge of how the inner ear works, and you can read about his original experiments in a collection of his papers titled *Experiments in Hearing* (New York: McGraw-Hill, 1960). A symposium summarizing more recent work on cochlear mechanics appears in *JASA, 67,* 1679–1728 (1980).

The interesting views of a psychologist who did a great deal of work on sound perception are found in *Psychophysics* by S. S. Stevens (New York: Wiley, 1975). They should be read with awareness that Stevens was a partisan in the ongoing debate over the "psychophysical law." For a good, clear presentation of differing opinions, see R. M. Warren, *Auditory Perception* (New York: Pergamon Press, 1982). B. Schneider in *JASA, 69,* 1208 (1981) also disagrees with Stevens—but in the opposite direction.

Studies of the importance of transients for instrument recognition are reported by M. Clark et al., *J. Audio Engr. Soc., 11,* 45 (1963), and by W. Strong and M. Clark, *JASA, 41, 39* and 277 (1967).

SYMBOLS, TERMS, AND RELATIONS

JND just noticeable difference
SIL sound level (dB)
LL loudness level
Fletcher–Munson diagram
equal-loudness curves
intensity
frequency
waveform
phons
basilar membrane
loudness
pitch
timbre
eardrum
sones
octaves
transient
cochlea

EXERCISES

1. Think through and write out your own feelings about the validity of quantifying human perceptions of sensory stimuli. You may wish to do reading on this basic question in an introductory book on experimental psychology.

2. Because of its odd shape, the pinna may collect sound better from some directions than others, so perhaps this gives us a way of judging which direction a sound comes from. Recall what was said about diffraction in Chapter 4, and explain whether this direction-finding mechanism works better at high or low frequencies.

3. Describe three different effects in the middle ear or eardrum that can decrease the amount of vibration reaching the inner ear.

*4. A sound pressure level of 160 dB means maximum pressure fluctuations of 2800 N/m^2 = .028 atm; according to Backus and Hundley, sound levels this high occur inside a trumpet mouthpiece. If such a sound were to fall on your eardrum, how large would be the pressure fluctuations in the inner ear if we ignore the protective mechanisms in the middle ear? How much displacement of the eardrum and of the oval window? Do these figures suggest why such a high level of sound is very damaging?

5. Think about the way fluid flows in the cochlea and the role of the round window for various frequencies. In your own words explain the differences in terms of physical reasons.

6. Think of the possibility of making a list of frequencies such that (a) the list is as long as possible, and yet (b) any two sounds with frequencies on this list are distinguishable in pitch. By using information on *JND*s in frequency, roughly estimate how many entries could be on such a list covering the entire audible range. Are there enough hair cells for this to be a reasonable answer? (Caution: Your positive answer will *not* justify jumping to the conclusion that only one hair cell or one small group responds to each frequency.)

7. If you go to low enough levels, the *JND* in sound level may become as large as 2 dB. Under those conditions, what is the minimum detectable change in intensity, expressed as a percentage?

8. What frequency is one octave below 300 Hz? Two octaves above 500 Hz?

9. Suppose you change a sine wave's frequency from 180 Hz to 400 Hz and then again from 400 to 700 Hz. Which change will be perceived as a greater change in pitch? (If possible, do the actual experiment rather than trusting claims in the text.)

10. From Figure 6.13 determine the loudness corresponding to loudness levels of 60 and 70 phons. Does the answer agree fairly well with the "doubling" rule of thumb? Do the same for 90 and 100 phons, then for 20 and 30 phons.

11. What loudness level in phons corresponds to a loudness of (a) 0.5 sone, (b) 4 sones, and (c) 25 sones?

12. If you hear three sounds with $f = 100$, 1000, and 3000 Hz, all at the same intensity level $SIL = 60$ dB, which will sound loudest and which least loud? If you hear three sounds (same three frequencies), each at the same loudness level $LL = 60$ phons, which actually has the greatest intensity and which the least?

13. Use Figures 6.12 and 6.13 to find the loudness in sones for the following sine waves: (a) 500 Hz at 30 dB, (b) 4000 Hz at 80 dB, and (c) 50 Hz at 70 dB.

14. What sound level in dB is required for a 10-sone loudness at (a) $f = 1000$ Hz, (b) $f = 4000$ Hz, and (c) $f = 50$ Hz?

15. Make a copy of the Fletcher–Munson diagram and relabel equal-loudness curves with their loudness in sones.

16. In Figure 6.5, how much average hearing loss is indicated for 55-year-old men at 2 KHz? At 4 KHz? How much at the same two frequencies for 55-year-old women?

17. Suppose your hi-fi set is adjusted to produce sound with a good balance of treble and bass, each at a loudness level of 40 phons. Then suppose a simple volume control is turned up so that all intensity levels are increased 40 dB. Explain, using specific numbers for illustration, how the sound will now be perceived and what complaints are likely to come from the neighbors.

18. About how much must you change the frequency of a sound to make it recognizably different if you start from (a) 1 KHz? (b) 4 KHz?

19. Starting at 1000 Hz and 95 dB, how much must you change the sound level to make a noticeable difference? What percentage change in intensity does this represent?

20. For all levels to sound equal in loudness to a 30-dB, 1000-Hz sine wave, what sound levels in dB would be required at (a) 50 Hz? (b) 300 Hz? and (c) 10 KHz?

PROJECTS

1. Devise experiments to illustrate bone conduction. You may be surprised how strong (and different) the sound can be just with such a simple object as a screwdriver handle held against the side of your head while its tip touches some vibrating object. You also can get interesting effects from a spoon or wire coat hanger struck while dangling from one end of a string, with a finger holding the other end in the ear. Do you find any evidence that bone conduction favors high or low frequencies in comparison to the normal air/eardrum/middle-ear path?

2. Use an oscilloscope together with either an electronic oscillator and speaker or a microphone and musical instruments. Look and listen, keep track of what happens when you change one variable at a time, and see whether you can verify the influence of each physical variable upon each psychological perception.

3. Conduct a *JND* experiment. This will require two sine wave generators and speakers, and either an electronic frequency counter or a good way of measuring amplitude or intensity (an oscilloscope may well be better for this purpose than a sound level meter). Try to be conscious of details in testing procedure that might lead to different results.

4. Conduct a ratio-estimation experiment to determine the relative loudness of 1000-Hz sine waves. How well do your results agree with the rule of thumb that every 10-dB increase in sound level doubles the loudness?

5. Record long, steady notes from several different instruments on tape. You can get rough information easily by playing it backward so that attacks and decays are reversed. But for more precise conclusions, edit the tape so that comparisons can be made both with and without transients. Ask other people to listen, and see how much difference the transients make in their ability to recognize instruments.

CHAPTER

7 Elemental Ingredients of Music

We have enlisted the aid of science in trying to understand sound in a fundamental way, and we have found that sound consists of extremely rapid vibrations. Certainly that is an important insight, but it leaves a considerable gap between that overly detailed picture and the way we consciously comprehend sound. Even when science measures our perceptions of pitch, loudness, and timbre of steady sounds, it still makes only partial contact with the world of real music.

How should we bridge that gap? We must study the various ways simple sounds are put together—the starting and stopping, the durations, the sensations of hearing several sounds at once. The central concept in such a study must be that "the whole is greater than the sum of its parts." That is, we account for only a fraction of the meaning in a piece of music by describing in complete detail how each successive note would be perceived all by itself.

There are patterns of organization in music, and our memories enable us to perceive these patterns as something beyond the mere succession of single tones. We understand very little of how our brains accomplish the higher-order processing of sound signals to recognize, judge, compare, and appreciate these patterns. So the language of art must work along with the language of science as we try to relate basic scientific understanding of sound to the realities and the beauties of music.

This chapter will begin with a discussion of how we perceive the organization of events in time, with little regard to which musical notes they might be or from which instrument. In Section 7.2 we take an initial look at how the combination of time structure and pitch arrangement gives us the musical ingredients of melody and harmony. Sections 7.3 and 7.4 introduce the special pitch relationships called musical intervals and relate them to the harmonic series. This series of intervals eventually turns out to be a key unifying concept for physical, psychological, and musical aspects of acoustics (sound production, timbre, and tonality).

For those readers who have not previously learned to read written music, this is a good time to turn to Appendix A.

7.1 ORGANIZING MUSICAL EVENTS IN TIME

It is no accident that we use the second as our basic unit of time measurement. If we had much smaller units (anything less than about a fifth of a second), we could not count them one by one with unaided human perceptions. But if our units were much larger (say, 5 seconds or more), we would soon find ourselves subdividing them. This is because we would perceive each one as a succession of several present moments during which several different things might occupy our attention.

It is natural to think that this has something to do with our heartbeats, which usually occur slightly less than a second apart. But several other physiological reasons are probably more important. One second is a comfortable time in which to perform various motions with our arms or legs, and this is simply because distances on the order of 1 meter are involved. A pendulum 1 meter long takes approximately 1 second to swing each way, and our muscles are suited to deal with such motions of our limbs. By this criterion, a natural time scale for a giraffe would be a little longer and for an insect much shorter than for us.

Our nervous systems also impose limitations. Suppose you rest your hand on a freshly painted surface, quickly notice that it feels sticky, and lift your hand. It actually remains in contact for at least a few tenths of a second, even though you think you remove it right away. This is because the warning nerve impulses take some time to reach your brain, and even if the decision to react is instantaneous it still takes more time for the "move" command to get back to your hand. After that, the motion itself takes time.

In music, too, we need to consider limitations on both movement and perception. As to movement, think first of playing a bass drum. If you swing your whole arm, you cannot play much faster than one beat per second. But now consider a good clarinet player, who can do scales and arpeggios at 20 or more notes per second. This is only possible, of course, because no arm motion is required. Does this mean that finger motion defines a shorter natural time scale for human actions? I believe not, because as a keyboard player I can testify that these finger motions are not commanded and carried out one by one. Indeed, when beginning to learn a passage and still thinking of it as a succession of single notes, I must go much slower. When I have learned the passage and brought it up to full speed, I treat groups of notes (that is, groups of motions by several fingers) as if they were a single (although complex) motion, and my fingers execute these complex commands at a rate of not more than three or four per second. It is much the same with a typewriter: to achieve 80 words per minute your mind and fingers must process entire words together, not one letter at a time.

This still leaves the question open as to what the natural time scale is for a listener. Again, I argue that it is around 1 second. We can listen to a passage with 20 notes per second, but we cannot perceive them each individually. We have instead a series of more complex percepts, each encompassing several notes, being comprehended at a rate of only a few per second. Close inspection of good music will reveal that the composer never attempts to communicate 20 really different ideas in a second; the fast passages are largely frill and figuration, with a great deal of redundancy. The progression of the major events of a piece (such as the *harmonic rhythm*, the sequence of significantly different chords, ignoring ornamentation), is generally in the neighborhood of one or two per second. Much slower than this will be boring, and any attempt by the composer to go much faster is doomed to esthetic failure.

Thus far we have spoken mainly of tempo, which should not be confused with meter and rhythm. **Tempo** refers to the "pulse rate" of the music. It is often left to the performer's discretion. But sometimes tempo is indicated by the composer at the beginning of a score with a marking such as ♩ = 120 or M.M. 120, either of which would indicate that the basic note values are to be played at the rate of 120 per minute, or two per second. Tempo also sometimes is indicated by code words such as those in Table 7.1.

A *metronome* (originally an adjustable spring-driven pendulum, but now available as an electronic device) can be set to produce audible clicks at any desired rate and thus set a uniform tempo for practicing. Changes in tempo as small as 10 percent can appreciably alter the mood conveyed. A good musician does not follow the absolutely constant tempo of a metronome in performance but uses carefully controlled flexibility in tempo as another artistic tool.

Meter indicates the way the basic time units are grouped between vertical bar lines in the written score; these groups of time units are called *measures.* The performer is notified at the beginning of a piece (and sometimes of changes in the middle) by signs such as $\frac{4}{4}$ (meaning four quarter-note beats in each measure) or $\frac{6}{8}$ (six eighth-note beats per measure, felt in two groups of three). The bar lines are intended as a convenience, making it easier for performers to keep track of their place; but if the structure of the music itself is

TABLE 7.1 Ranges of tempi, in beats per minute, corresponding to various directions found in the score. These ranges were inscribed on one particular metronome, and should not be taken very seriously. I find allegro at 132 on another metronome, for instance; they also disagree on whether adagio is faster or slower than larghetto. But that is not just the metronome makers' fault, for tempo words are not used uniformly by either composers or conductors. There is, in particular, a tendency for these words to imply greater contrasts and extremes now than they did a few hundred years ago.

Tempo Marking	Translation	Metronome Setting
Largo	Broad	40–70
Larghetto		70–100
Adagio	At ease	100–128
Andante	Going	128–156
Allegro	Gay	156–184
Presto	Fast	184–208

FIGURE 7.1 Some common rhythmic patterns with strong beats indicated by accent marks above the notes. Part (c) shows several closely related patterns that all fit the same meter.

irregular, bar lines can be more of a hindrance than a help. Many pieces of music (especially some written before 1400 and since 1900) use no regular meter or bar lines.

Rhythm denotes patterns of strong and weak beats or subbeats (Figure 7.1a,b). Do not confuse this use of the word *beat* (for one in a regularly spaced series of accented events) with the auditory beats (gradual fluctuations of intensity from the combination of tones with slightly different frequencies) of Chapter 4. Rhythmic patterns often are repeated (including minor variants of the main pattern) throughout a piece. They can be one of the strongest influences in determining what mood is conveyed by the music. Rhythm tends to be confused with meter because there is usually one particular meter that forms the most convenient framework for writing a given family of rhythmic patterns, such as $\frac{6}{8}$ meter for Figure 7.1c.

As with the overall tempo, do not suppose that it is normal to perform individual rhythmic patterns with the perfect mechanical precision that a literal reading of the written notes might suggest. There are normally systematic departures on the order of 10 to 60 ms from metronomic regularity, the most familiar example being the Viennese waltz rhythm with its persistently early second beat in each group of three.

Rhythmic patterns in much European classical and American popular music tend to be simple and straightforward. They are drawn from a single family of closely related patterns throughout a piece. They usually involve division and subdivision of time units into either two or three parts and so are written in what we call duple and triple meters. Rhythms that divide each of two main beats into three subbeats call for a *compound* meter such as $\frac{6}{8}$. Although $\frac{3}{4}$ meter also has six eighth-notes per measure, it arranges them into three groups of two and so is appropriate to an entirely different family of rhythmic patterns. A distinctive and interesting flavor can be added (as in Figure 7.2a) by a device called *hemiola.* In a piece with strong (S) and weak (w) subbeats normally occurring S w w S w w, a S w S w S w pattern is occasionally inserted (or vice versa) while the tempo of the subbeats is kept constant. This device is quite common in fifteenth-century music, in Flamenco, and in late-Renaissance and Baroque dance forms (see, for example, the final measures of each courante in the *English Suites* of J. S. Bach). It also shows up in nineteenth-century scherzos, such as the second movement of Brahms' *Piano Concerto no. 2* (B-flat Major) or the third movement of

FIGURE 7.2 Examples of rhythmic displacements. (a) Opening phrase of "God Save the King" (known in the United States as "America"), altered to show the effect of hemiola. (b) Syncopation in the solo line of a popular song. The normally strong beats (accent marks) still are reinforced by the accompaniment. (c) A familiar rhythmic question-and-answer, showing a silence that functions as a stronger beat than the sounds that precede and follow it.

Dvorak's *Symphony no. 7* (D Minor). A regular and persistent alternation, $2 \times 3 + 3 \times 2$ continuing measure after measure, accounts for the distinctive flavor of large sections of the musical play *Man of La Mancha*.

Another, more common, rhythmic device is *syncopation* (Figure 7.2b), a deliberate disturbance of the normal pulse of accented beats. At first glance this seems to mean a confusion of weak and strong beats by temporary shifting of a rhythmic pattern, usually in the direction of anticipating or making a strong beat too early. Yet the metrical framework usually stands intact, and we do not really lose track of where strong beats normally occur. (An excellent example is provided by the second movement of the Brahms *Piano Quintet*.) An extreme case that illuminates an essential aspect of syncopation is the "shave and a haircut" routine (Figure 7.2c); it shows that a weak sound (even a silence) can still function psychologically as a very strong beat if the context is properly prepared. So during syncopation, the truly strong beats continue to come at the same times (and may indeed be played normally in an accompanying part) but are merely emphasized in a different way.

Occasionally we find rhythms involving patterns of five beats (for example, Tchaikovsky's *Sixth Symphony*, second movement; *Daphne of the Dunes* by Harry Partch; or "Everything's Alright" in *Jesus Christ Superstar*) or seven beats (for example, Stravinsky's *Firebird Suite*, finale; or "The Temple" and "The Arrest" in *Jesus Christ Superstar*). These usually are perceived as alternating subgroups, $2 + 3$ or $3 + 4$; the human mind seldom perceives five or more things in a group without breaking it down into smaller units. Even groups of four are perceived largely as $2 + 2$.

Music from certain other cultures (Africa and Indonesia provide good examples) uses much more complex rhythmic patterns, especially in the way several rhythms played by different instruments are woven together. This may be because these cultures have concentrated their attention on this musical element and made it a primary feature; it may also be because they use largely percussion instruments. Our culture, having many continuously sounding wind and string instruments, has concentrated on developing a large and complex vocabulary of expression through sustained harmonies. Attempts to combine the rich harmonic structures of Western music with the most complex rhythmic intricacies from other cultures meet with considerable esthetic problems. The combination is challenging to effectively perform or comprehend, for it tends to contain too much musical information for most listeners to digest at once.

7.2 MELODY AND HARMONY

Each beat in a rhythmic pattern presents choices. There may be a new sound, a continuation of an old one, or a rest. There may be a single new note or several. We may choose a percussive sound or a sustained sound of definite pitch. This pitch may be higher or lower than its predecessor by various amounts. It takes all of these choices to make a tune.

Melody is a succession of different pitches in time that are perceived as a continuing line. It gives horizontal structure to music, a progression from one note to another and then another. *Harmony* is the combination of several pitches at one time; it gives a vertical structure, a cooperative effect from several notes sounding simultaneously. Although harmony is an optional feature specially characteristic of our culture, melody almost always is present. Simple vocal melodies seldom cover a range much more than an octave from lowest to highest pitch. One of the main reasons why "The Star-Spangled Banner" is notoriously difficult to sing is its range of more than $1\frac{1}{2}$ octaves.

Sometimes a single melody is accompanied by harmonies that exist purely for the support of that melody; this texture is very common in popular music. It is more difficult to write music that is simultaneously highly structured both vertically and horizontally—for example, to write pleasing four-part harmonies so that each of four players has an interesting melody of his or her own. *Counterpoint* is the art of writing such interwoven part-music; Giovanni da Palestrina (1525–1594) and Johann Sebastian Bach (1685–1750) are famed as great masters of this art.

An important aspect of the performance of any melody is **articulation**, the controlled manner in which successive notes are joined to one another by a performer. At one extreme a series of notes may be joined smoothly together in obedience to a *legato* indication (Figure 7.3a); indeed, a performer may even deliberately make them overlap slightly. Or we may, in response to a *staccato* marking (Figure 7.3b), release each note quickly and make no sound during the remainder of the time that would ordinarily be occupied by that note.

Professional musicians use many gradations in between, even when there is no direction to do so in the score. They realize that careful control of the spaces between notes can be just as important as the notes themselves in

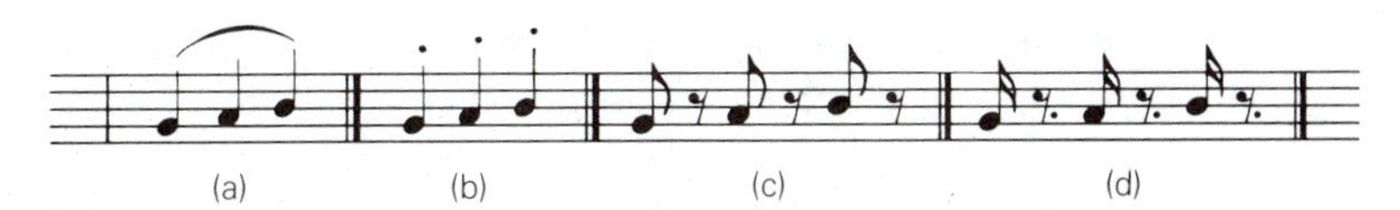

FIGURE 7.3 Either a *legato* direction or a slur mark (a) indicates that notes are to be smoothly joined together. Either a *staccato* direction or the equivalent dots over the notes (b) indicate that they should be detached; in that case the actual performance may be as in (c) or even (d).

FIGURE 7.4 Opening phrase of the Prelude from Bach's *English Suite* no. 4. Above the usual score are added marks to show approximately what articulation might be used in proper performance. A slur mark (⌢) shows where one note would merge directly into the next with no space in between. A minimal space, so short as to be hardly perceptible unless listened for carefully, is denoted by |. V indicates a slightly longer space, and U a space longer enough to be easily noticeable. This articulation will quickly establish which are the stressed notes and enable the listener to correctly recognize the duple meter.

conveying the desired mood. These brief gaps in the sound act as punctuation marks and can greatly help in showing where the strong and weak beats occur. Figure 7.4 shows an example where strict legato would leave a listener unable to understand the rhythmic structure until well into the second measure. It is possible on the piano and with many orchestral instruments to stress strong beats by playing them louder or with a more forceful attack. But on the harpsichord (for which this example was intended by the composer) and the organ, articulation is the only form of stress available and must be thoroughly understood by anyone who plays those instruments.

Stress by articulation has a solid acoustical foundation. It is not just a matter of perceiving the silent gaps as spaces; there is also a change in the apparent loudness of the notes. As shown in Figure 7.5, a note must be sustained several tenths of a second before we actually judge its loudness by

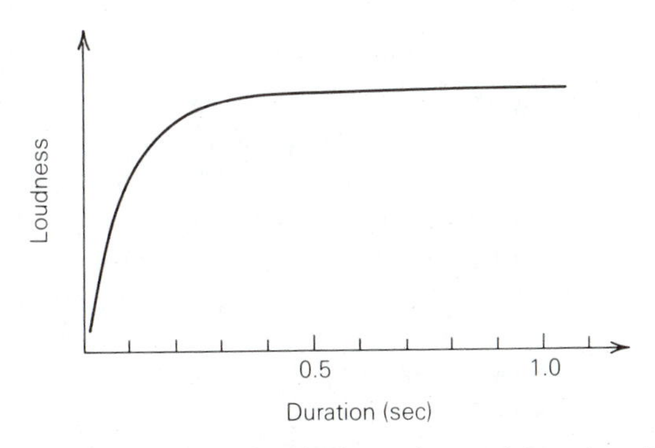

FIGURE 7.5 The curve shows how the perceived loudness of a sound depends on its duration. Notes lasting less than a few tenths of a second seem less loud than continuous sound of the same intensity.

its steady intensity. For shorter notes, we judge instead by the total energy received, which is proportional to the note's duration. Thus, for any tempo faster than two or three notes per second, those notes held longer actually seem louder, and those shortened by articulation seem softer.

The gaps are not really totally silent anyhow, because of room reverberation. The stretching out of each note by reverberation may be quite different in different rooms, so a good musician will adjust the degree of articulation accordingly. An extremely "live" room requires more brisk articulation to avoid the impression of muddiness caused by overlapping reverberation of successive notes.

7.3 SCALES AND INTERVALS

Of all possible pitches, which ones do we actually use to make a melody? If the *JND* in frequency is as little as 1 Hz, then there are a couple hundred recognizably different pitches in the octave above C_4 (from 262 to 523 Hz). In many other cultures (and to some extent in American jazz) performers are encouraged to make full use of these microtones. But with pianos, guitars, and standard orchestral instruments, we are heavily committed to using only 12 distinct pitches per octave; any deviation is considered erroneous ("out of tune"), or at best a subtle added coloration. Such a group of "allowed" pitches (arranged in ascending order) is called a **scale**.

* The term *scale* is notorious for vague and multiple meanings. Our use here is compatible with the *Harvard Dictionary of Music* and with common usage among musicians. We might point out, however, that some musicologists prefer to maintain a distinction between *gamut* (list of all available notes) and *scale* (subset of the gamut, those notes involved in a particular melody or musical passage).

There are many different scales, and we will consider some of them in Chapter 18. But for now let us look only at the *chromatic scale* (shown in Appendix A, Figure A.5), which consists of those 12 familiar pitches per octave found on a piano keyboard. The step from each note to the adjoining one is called a *semitone* (or half-step). Two adjoining white keys on the piano that have a black key between are a *tone* (or whole tone or whole step) apart.

Standard orchestral instruments are tuned to produce (at least approximately) the notes of an equal-tempered chromatic scale. The calculation of their frequencies is explained in Box 7.1. They are listed in Figure E (inside the back cover), which also gives the standard names of the notes (C_3, A_4, and so on).

The perceived spacing between two pitches is called an **interval**. We may listen to any interval in two different forms: One note following after another

BOX 7.1 THE CHROMATIC SCALE IN EQUAL TEMPERAMENT

There are many different ways to tune an instrument (defining its playing frequencies precisely) to play a chromatic scale. The version most often used as a standard of comparison in our culture today is called *equal temperament.* An especially strong influence is exerted by pianos, which are tuned approximately that way.

Suppose we take it as given that we want a scale with 12 equal semitones per octave, leaving for Chapter 18 such questions as why 12 rather than 11, or why we insist on making them equal. As a perceived interval in *pitch*, then, each semitone is to be precisely $\frac{1}{12}$ of an octave. But how much difference in *frequency* will it take to produce that perception?

We take a cue from the octaves, which have the property that every additional octave corresponds to a multiplication of frequency by 2. We hypothesize that every semitone should similarly correspond to a multiplication of frequency by a certain number, always the same one; we just don't happen to know which number yet, so we call it x. If we start from a frequency f_0 and go up one semitone, the new frequency is xf_0. A second semitone raises it to $x(xf_0) = x^2f_0$ and a third to $x(x^2f_0) = x^3f_0$. We keep going up step by step, and after 12 semitones we arrive at frequency $x^{12}f_0$. But 12 semitones up means precisely one octave up, which is $2f_0$, so we must have $x^{12} = 2$. That is an equation easily solved for x; trial and error, for instance, is a perfectly good method (see Exercise 9). The answer is approximately $x = 1.05946$, which is also called the twelfth root of 2.

Now we can construct a 12-tone equal-tempered scale by starting at any frequency (say $A_4 = 440$ Hz) and multiplying (or dividing) it repeatedly by 1.05946. That is how the numbers in Figure E were obtained.

forms a *melodic interval*; both notes sounding simultaneously form a *harmonic interval*. Three or more notes sounding together form a *chord*, which contains a combination of several intervals.

You should be familiar by now with the sound of one particular size interval, the octave. Well-trained musicians easily recognize each of several other intervals most often used in our music, for they have distinctive aural qualities. The intervals C_4–G_4 and A_4–E_5, for instance, are both given the same name (perfect fifth) even though they involve different pitches, because they sound alike in the same sense that any two octaves sound alike. For musical purposes it is the interval, the difference in pitches, that is much more important than the actual value of either individual pitch.

The common names for the intervals (third, fifth, octave, and so on) are just a way of telling how far apart they are written on the musical staff, or (usually) how many white keys they span on the piano. But to make precise distinctions about the actual sounds, this scheme adds adjectives (such as major, augmented), and the complete rules for their proper use are rather complicated. Fortunately, we need only the information in Table 7.2 for our purposes. Even if you have not used these terms before, you can easily translate as necessary by using the table along with Figure E.

If you have either the two note names of an interval or notes on the staff, you need only count how many semitones (every key, both black and white) to get from one to the other on the keyboard, and consult Table 7.2 to find the

TABLE 7.2 Naming of common intervals found in the chromatic scale. The reason any given interval may or may not require sharps or flats in written notation is that adjoining white keys on the piano do not always have a black key between. The last entry is an example of how the terminology can be extended beyond an octave; the twelfth is an octave plus another fifth. See also Figure E.

Interval Name and Abbreviation		Number of Semitones	Examples
Unison or prime	P1	0	
Minor second	m2	1	
Major second	M2	2	
Minor third	m3	3	
Major third	M3	4	
Perfect fourth	P4	5	
Tritone or Augmented fourth or Diminished fifth	TT A4 d5	6	
Perfect fifth	P5	7	
Minor sixth	m6	8	
Major sixth	M6	9	
Minor seventh	m7	10	
Major seventh	M7	11	
Octave	P8	12	
Twelfth	P12	12 + 7 = 19	

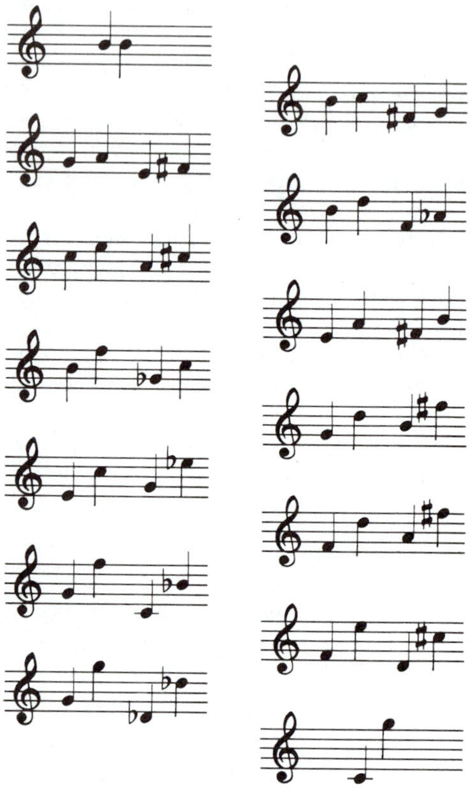

name of the interval. Or, given a named interval, you can start by getting the number of semitones from Table 7.2 and then count them off from any desired starting point on Figure E; if you have a keyboard accessible you then can strike those two keys and hear the actual sound. You may check your understanding of these procedures by working these examples: (a) B_3 and F_4 are six semitones apart on Figure E, and so form a tritone; (b) if a major sixth is to go upward from A_4, its nine-semitone width means the upper note must be $F_5^\sharp$.

Just as octaves always correspond to frequency ratios of 2 to 1, and equal-tempered semitones to frequency ratios of 1.05946 to 1, so each species of interval always corresponds closely to its own characteristic frequency ratio. But before we jump to any conclusions about the precise numerical values of these ratios, we must learn about the harmonic series.

7.4 THE HARMONIC SERIES

To become acquainted with the crucial concept of the harmonic series, let us consider the following experiment. Suppose we have a set of electronic oscillators that can be easily set to produce sine waves of any desired frequency. Out of the vast number of possible combinations, most of which would sound rather strange, let us investigate the special case in which all of the frequencies are simple multiples of a single frequency. Such a group is called a **harmonic series**. Let us take as an example the series $f_1 = 110$ Hz, $f_2 = 220$ Hz, $f_3 = 330$ Hz, $f_4 = 440$ Hz, $f_5 = 550$ Hz, . . . $f_{10} = 1100$ Hz, . . . $f_n = n \times 110$ Hz. Notice that these are equally spaced in frequency—each is 110 Hz above its predecessor.

Even before setting up the experiment, however, we know that these sounds are *not* equally spaced in pitch. We recognize that the ratios f_2/f_1, f_4/f_2, f_8/f_4, f_{16}/f_8, . . . are all octaves, so it is no surprise to see in Figure E that f_1, f_2, f_4, f_8 . . . correspond to A_2, A_3, A_4, A_5, ...

But the other notes in this series are also special, both aurally and mathematically. First, when you listen in succession to oscillators tuned to these frequencies, the notes sound to your ear as if most of them fit into our customary musical scales. More specifically, they remind you of the notes of a bugle call. Second, comparison with the numbers in Figure E shows that many, but not all, of these frequencies just happen to be extremely close to frequencies in the equal-tempered scale, enough so that we could justifiably assign them the same note names: $f_3 = 330$ Hz $\cong 329.6$ Hz $= E_4$, $f_5 = 550$ Hz $\cong 554.4$ Hz $= C_5^\#$, and so on.

A harmonic series can be constructed on *any* frequency f_1, which is called its **fundamental** frequency. Other members of the series are called the second harmonic, third harmonic, and so on. Figure 7.6 shows other examples in musical notation; you can construct the series for any fundamental with the help of Figure E. As you can see in Figure 7.6 (or hear in the experiment), some of the harmonics (such as the seventh and eleventh) simply do not fit our customary musical framework; they fall "in the cracks" of the piano keyboard. Yet these harmonics are legitimate members of the series; they are acoustically respectable. Their absence from the chromatic scale indicates that the European musical tradition has a deliberate self-imposed limitation on its melodic and harmonic resources. In Chapter 18 we will see some of the reasons for this.

And what about the other discrepancies? Even though the third and fifth harmonics are musically recognizable, they do not correspond precisely to notes in the equal-tempered scale. Which is the "true" frequency ratio for the interval of a major third—$(1.05946)^4 = 1.2599$ to 1 as found for four semitones in the equal-tempered scale, or 5 to 4 (that is, 1.2500 to 1) as suggested by the harmonic series? The answer is that it depends on what purpose you wish the interval to serve, or what criterion it must satisfy. Fortunately for musical purposes, our ears often will accept a range of values for these frequency

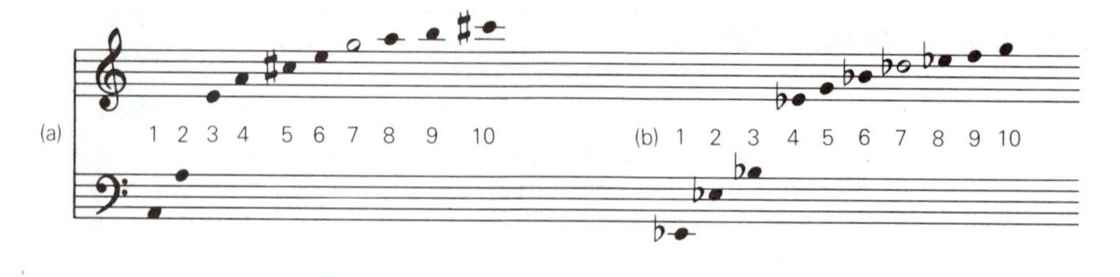

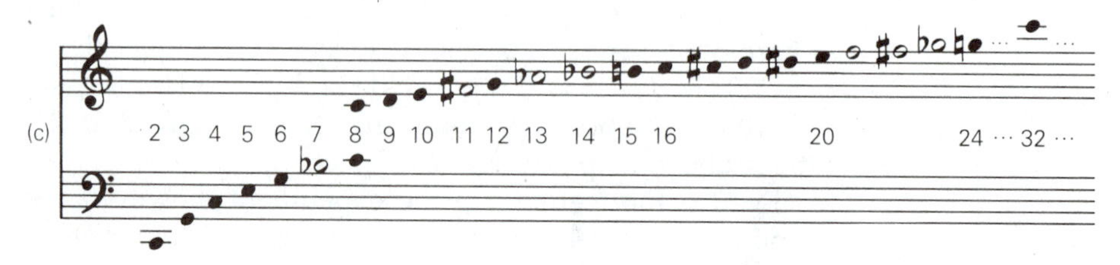

FIGURE 7.6 (a), (b) Ten lowest members of the harmonic series based on fundamentals 110 Hz (A_2) and 78 Hz ($E^{\flat}_2$). (c) More complete version of the harmonic series based on 32.7 Hz (C_1, fundamental not shown). Open note heads indicate harmonics (such as seventh, eleventh, and so on) whose pitches do not fall close to any note of the equal-tempered scale, and so cannot be accurately represented by the usual musical notation.

ratios so that either kind will serve. Again, a major portion of Chapter 18 will be devoted to showing how to deal with these discrepancies.

Our experiment also may give a clue to an even more important aspect of harmonic series. If you listen to the combined sound of many oscillators, they blend together so well when they belong to a harmonic series that you may perceive the combination as a single tone. Our task in the next chapter will be to explain in fundamental terms why the harmonic series has such an important role. It will be a central concept as we go on to contrast the nonharmonic nature of percussive sounds (Chapters 9 and 10) with the harmonic series that are naturally present in all instruments that produce steady tones (Chapters 11–14).

SUMMARY

The rate, or tempo, of musical events must be understood in terms of our ability to perform and comprehend them. Rhythm is the pattern of weak and strong beats that provides an organizing structure for music in time. Rhythmic patterns in melodies can be brought out clearly both by differences in intensity level of strong and weak notes and by careful articulation.

Musical intervals are differences or spacings in pitch between two notes. Melodic intervals together with rhythms make melodies, and harmonic intervals are the raw materials for chords, complex harmonies, and harmonic progressions in time.

Each species of interval corresponds to a certain frequency ratio. But the numerical values of these ratios in the equal-tempered chromatic scale correspond only approximately with those found in the harmonic series.

REFERENCES

For definitions, brief discussion, and examples of all musical terms, an excellent source is the *Harvard Dictionary of Music*, 2nd ed., edited by Willi Apel (Cambridge, Mass: Harvard University Press, 1969). Even for terms with which you think you are already familiar, it is good to check here because so many (such as rhythm and scale) often are used imprecisely or incorrectly. More extended discussions by experts on each topic appear in *The New Grove Dictionary of Music and Musicians*, 6th ed., edited by Stanley Sadie (London: Macmillan, 1979).

Research on rhythm and timing in music has been reviewed by Alf Gabrielsson in *Humanities Association Review*, *30*, 69 (1979), and *Music Perception*, *3*, 59 (1985); and by P. Fraisse in Chapter 6 and S. Sternberg, R. L. Knoll, and P. Zukofsky in Chapter 7 of *The Psychology of Music*, edited by Diana Deutsch (New York: Academic Press, 1982). For a musician's view of the time element in music, see *The Rhythmic Structure of Music* by G. W. Cooper and Leonard Meyer (Chicago: University of Chicago Press, 1969).

SYMBOLS, TERMS, AND RELATIONS

tempo
meter
rhythm
syncopation
$f_n = nf_1$: harmonic series
melody
harmony
articulation
fundamental
semitone
interval
scale
equal-tempered chromatic scale

EXERCISES

1. How much time is there between one beat and the next if the tempo indication is M.M. = 30? If M.M. = 300? Discuss the difficulties of performing, and especially of conducting, at these tempi.
2. Devise and carry out a simple experiment to measure roughly the shortest time in which you can repeat a motion using your entire arm. Do the same for a motion of the hand alone, pivoted at the wrist. Do the same for a tapping motion of a single finger. Finally, try to estimate how rapidly a pianist can hit successive notes in a continuous scale using all fingers. Discuss your results.
3. If successive beats follow each other 0.4 s apart, what metronome setting M.M. corresponds to this tempo?

4. In view of typical reaction times and their individual variation, how can an orchestra attack a chord in response to a conductor's signal in any other than a very ragged fashion? Answer the question first for the continuing beats in the midst of a piece. Hint: Think about ongoing sequences of signals from the conductor. You might enjoy an article in *Scientific American, 234* (May 1976), p. 74, that tells how some species of fireflies use a similar mechanism to flash in unison. Can you explain in similar terms how the conductor solves the more difficult problem of getting everyone to attack the very first chord of a piece together?

5. How long does it take sound to travel 35 m? Discuss the difficulties of directing antiphonal choirs located 35 m apart. To be specific, consider main beats at M.M. 120 divided into four subbeats each.

6. Study the score of some rapid passage in sixteenth notes, and mentally remove the frills and decorations. Then estimate how often the remaining essential musical gestures occur. Submit a marked photocopy to support your description.

7. Find two or three other examples of syncopation, and discuss whether they conform to the comments in Section 7.1. Illustrate your conclusion with marked copies of specific passages from the music.

8. Discuss the extent to which you believe complex rhythmic elements from other cultures have made their way into our music. Cite examples to show what types of music have been thus influenced.

9. Using a pocket calculator, show that $(1.10)^{12}$ is greater than 2, but $(1.05)^{12}$ is less than 2. Without looking at Box 7.1, guess better and better approximations to the solution of $x^{12} = 2$, and then see if your answer agrees with the number given there. Is that number an exact solution?

10. What note name corresponds to each of the following frequencies? (a) 2093 Hz, (b) 587 Hz, (c) 311 Hz.

11. What is the nearest note name in the equal-tempered chromatic scale to each of the following frequencies? (a) 256 Hz, (b) 1000 Hz, (c) 1710 Hz.

12. What frequency corresponds to each of the following notes in the equal-tempered chromatic scale? (a) E_1, (b) $F_3^\sharp$, (c) G_7, (d) $D_4^\flat$.

13. What intervals are formed by the following pairs of notes? (a) E_4/G_4, (b) G_4/E_5, (c) $D_4/C_5^\sharp$, (d) $D_3^\flat/F_3$.

14. What note is an octave above E_4? A perfect fifth above B_3? A minor seventh below $B_4^\flat$?

15. What three intervals are contained in the chord $C_4/E_4/G_4$ (called a major triad)? What three intervals in $D_4/F_4/A_4$ (a minor triad)? What is the difference?

16. Give the names of the nearest notes of the chromatic scale to each member of the harmonic series based on fundamental D_2.

17. Show the first 12 members of the harmonic series for fundamental F_2 as notes on the staff.

18. If A_5 is the fifth harmonic in a series, what is the fundamental? If $B_6^\flat$ is the ninth harmonic, what is the fundamental?

19. If you want to demonstrate a harmonic series by playing notes on a piano tuned in equal temperament, which harmonic numbers are readily and precisely available? Which ones available to a pretty good approximation? Which ones not available at all?

20. In Exercise 9, Chapter 3, you showed that reducing the temperature from 20°C to 8°C will lower the speaking frequency of organ flue pipes by approximately 2%. Are they speaking then sharp or flat from normal pitch? What percentage of change in frequency corresponds to a semitone? So how many semitones has the pitch been shifted?

PROJECTS

1. Interview several good keyboard or woodwind players or fast typists and find out whether they claim to control their finger motions individually or in clusters.
2. Is articulation important for string instruments? For woodwinds? Talk to people who perform on these instruments. Find recorded examples to illustrate the use of articulation.

CHAPTER

8 Sound Spectra and Electronic Synthesis

Details can be fascinating. Each petal of a blooming rose has an attractive individual shape and texture and color, not to mention the delicate veined structure revealed by a microscope or the story of the chemical changes that cause it to fade and curl.

But details also can be confusing. Will you ever understand the meaning of the whole rose by studying one petal at a time? No, much of that meaning lies in the pattern by which the detailed parts are organized. We would do well to learn first about the overall structure of the blossom and about those things that all petals have in common before we study differences in detail between one petal and another.

We would like to adopt a similar attitude now as we undertake a closer study of the processes by which musical instruments produce sounds. These sounds come from a great variety of sources, and if we start with any particular one—the trumpet, for instance—we run the risk of confusing details relevant only to that instrument with the general and essential properties of sound production.

It may help you to understand the organization of the chapters ahead if you think about the issue of energy supply for the creation of sound. In Chapters 9 and 10 we will study instruments that make transient sounds, with vibration energy simply delivered in a single lump at the beginning by striking or plucking. In a sense, all we have to do then is describe how this energy gradually leaks away in the form of sound waves. But steady sound sources require a continuing supply of energy. There are several different ways of supplying that energy; each (such as the trumpet player's blowing or the violinist's bowing) turns out to be somewhat complicated, and we will consider them in Chapters 11–14.

We might be wise to start if possible with perfectly steady sounds that do not entangle us in questions of starting or stopping, or of how they get their energy. The sound produced by electronic oscillators nicely fits this description. We can focus on properties essential to all steady sounds without being distracted by incidental acoustical or mechanical characteristics of a particular instrument such as the trumpet. And although electronic oscillators do use a continuing supply of energy, we will simply take this for granted with no attempt to explain the details because electronic circuit theory lies outside the usual realm of acoustics. This chapter thus is directed toward an understanding of the various idealized signals that can be generated electronically, and is supplemented with an optional section that introduces terminology and concepts used in practical applications of computers and synthesizers in music.

But we must begin by providing a proper framework for appreciating these issues in the first two sections. We shall find that we can make powerful statements about how seemingly complex sound waves can always be understood in terms of simple sine waves. The concept of the sound spectrum developed here will be used repeatedly throughout the rest of the book. In particular, we will see that there is an important connection between the steady or transient nature of the sound and the harmonic or nonharmonic nature of its spectrum.

8.1 PROTOTYPE STEADY TONES

The ideal source of steady tones is an electronic oscillator. We shall ignore nearly all the details of how these oscillators work, but we should at least know the difference between two words you often have heard—voltage and current. The *voltage* (or electrical potential) at each point in an electrical circuit is analogous to the pressure aspect of sound waves. An alternating voltage pushes electrons back and forth, and it is their motion that constitutes an alternating *current*. Although the two always go together, it is enough just to specify how the output voltage of an oscillator changes with time.

Several different waveforms can be made easily with electronic circuits, and these serve as the basic materials of electronic music. We shall assume for now that any electrical-signal waveform is going to be perfectly converted by an ideal speaker into a sound wave whose graph is precisely the same.

The **sine wave** (Figure 8.1a) results when electrons shuttle back and forth around a circuit in simple harmonic motion. This motion is closely analogous to the mechanical motion of a mass on a spring, and it has a natural frequency determined jointly by the masslike and springlike electric circuit elements called *inductance* and *capacitance*, respectively. In older vacuum-tube oscillators, turning a knob to change frequency simply meant adjusting a variable capacitor to change a resonant circuit's natural frequency. Generation and frequency control for sine waves is more complicated with modern integrated-circuit chips.

The **square wave** (Figure 8.1b) can be made by simply flipping a switch back and forth, connecting it first to the positive and then the negative terminals of a battery and repeating this process rapidly. This was formerly done with vibrating-reed switches, thin strips of metal that can switch back and forth hundreds or thousands of times per second. Now the switching can be done electronically with transistors, not only more reliably but also at frequencies well beyond a megahertz.

The **pulse wave** (Figure 8.1c) is a simple generalization of the square wave. Its *duty cycle* tells what fraction of the time the voltage stays at one of its

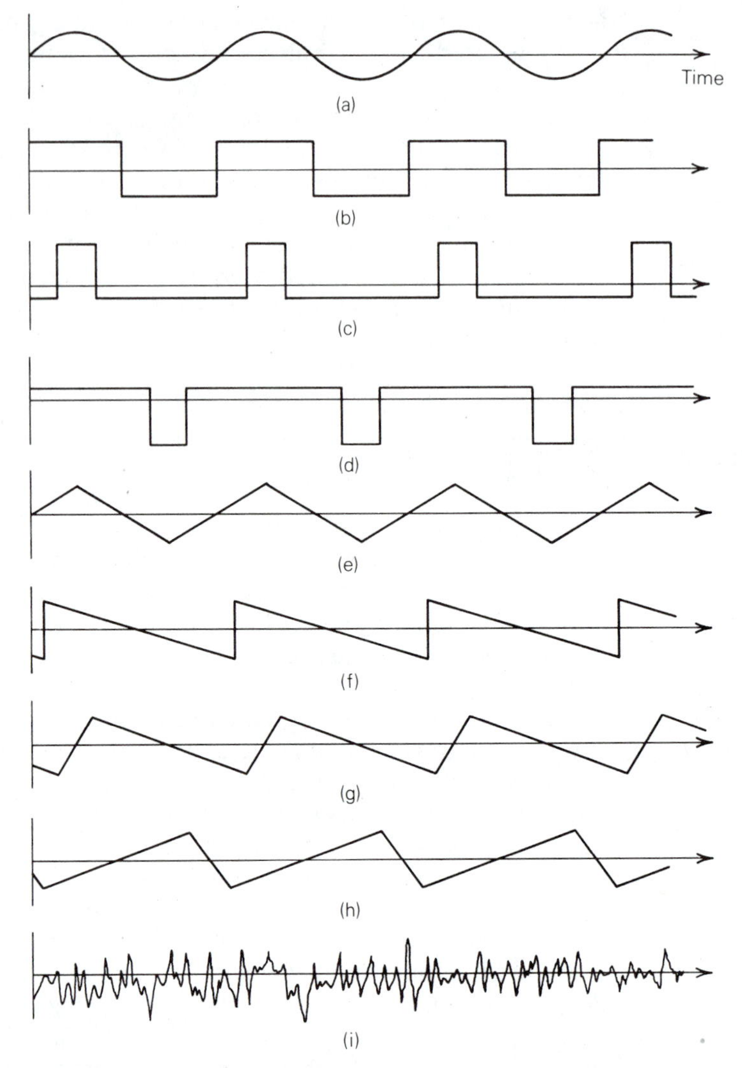

FIGURE 8.1 Waveforms commonly available from electronic oscillators. You may think of each graph as showing either the history of voltage output from the electrical circuit or the history of air pressure near an ideal loudspeaker driven by that voltage. (a) Sine wave. (b) Square wave. (c), (d) Pulse waves with 20% duty cycle. (e) Triangular wave. (f), (g), (h) Sawtooth waves. (i) Random noise.

two values. We can describe the square wave as a pulse wave with duty cycle one-half. When a pulse wave drives a loudspeaker, it makes no difference to your ear whether it is the compressions or the rarefactions that last longer (Figure 8.1c or d).

The **triangular** and simple **sawtooth waves** (Figure 8.1e,f) are extreme examples of another whole family of waves whose voltage (or resulting air pressure) increases for a while at a constant rate and then suddenly switches to a constant rate of decrease before returning to the minimum value and beginning another cycle (Figure 8.1g,h). Again, it makes no difference to your ear if the waveform is turned upside down.

Another kind of signal that is sometimes generated electronically is *random noise* (Figure 8.1i). This differs from all the others in that the waveform never repeats itself, no matter how long we wait. The only sense in which this is a steady, continuous sound is that its noisiness, its very randomness, always remains the same. We shall be able to describe this noisiness much more precisely in the following section.

But must we keep listing more and more kinds or is there some underlying unity among these waves? If we ask which is simplest of all, it may not seem entirely obvious whether we should pick sine or square waves. Here it is helpful to recall the simple harmonic oscillator: The mass on a spring, certainly one of the most elementary systems we can observe, shows smooth sinusoidal vibrations rather than the sudden jumps of a square wave. Smooth and gentle swells on the ocean are better candidates for use as simple descriptive tools than are breakers crashing on the beach. And one of the smoothest, purest sounds is that of a tuning fork, whose oscilloscope trace looks very similar to a sine wave. In the following chapter, we will spell out more detailed physical arguments for choosing sine waves. For now, let us see the consequences of combining sine waves in various ways.

8.2 PERIODIC WAVES AND FOURIER SPECTRA

A wave that repeats exactly the same pattern over and over is called a **periodic wave**; its period P is the length of time it takes to complete its basic pattern, and its frequency $f = 1/P$ is how many times per second that whole pattern repeats. We want to do three things in this section: first, to take simple periodic waves and put them together to make more complex waves; second, to take complex periodic waves and break them down into simple components; and, finally, to say what differences occur in the case of nonperiodic waves.

The first task is relatively easy. Take any two sine waves with different frequencies and let them both run at the same time. Will the total disturbance be periodic? Not in general; suppose, for instance, the frequencies happen to be $f_1 = 243.72\ldots$ Hz and $f_2 = 539.08\ldots$ Hz (periods $P_1 = 4.1031\ldots$ ms and $P_2 = 1.8550\ldots$ ms). Then, because $f_2/f_1 = 2.2119\ldots$, every time the first wave completes a full cycle the other completes two of its cycles plus another odd fraction. If we start them together at time zero, we can wait practically forever without having both start another cycle at precisely the same time. The combined disturbance is nonperiodic, as in Figure 9.6 (page 154).

But for carefully chosen frequencies, we can get a periodic combination. If, for instance, f_1 and f_2 were precisely 110.00 and 440.00 Hz, every cycle of the first wave would take the same time as exactly four cycles of the second, after which both would be doing the same thing at the same time again

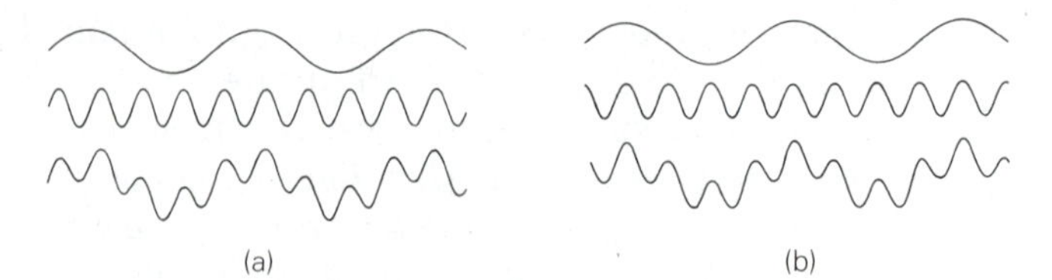

FIGURE 8.2 Addition of a fundamental and its fourth harmonic to form a complex periodic wave: (a) with both passing through zero at the same time; (b) with the harmonic shifted toward earlier time by approximately 25% of its period to give a different complex waveform.

(Figure 8.2a). It is not necessary that both start from zero at the same time. We can introduce a *phase shift* (moving one entire wave forward or backward in time with respect to the other) as long as we leave the frequencies the same; the resulting complex wave has a different shape but still the same periodicity (Figure 8.2b).

There is nothing special about the 4-to-1 ratio; what is important is only that both waves complete an integral number of cycles in the same time, such as three of one and two of the other. We can even add three or more sine waves and get a periodic result, as long as all of them complete some integral number of cycles in the same time. The most general case is where one wave completes one cycle, another completes two of its cycles, another completes three, another four, and so on, all in the same length of time. That is, if $f_2 = 2f_1, f_3 = 3f_1, f_4 = 4f_1, \ldots f_n = nf_1 \ldots$ then the combination of all of them is still periodic, with period $P = 1/f_1$ (Figure 8.3). But this set of frequencies is just the harmonic series!

A general statement helps us understand why the harmonic series is so important:

> ***Any set of sine waves whose frequencies belong to a harmonic series will combine to make a periodic complex wave, whose repetition frequency is that of the series fundamental. The individual components may have any amplitudes and any relative phases, and those determine the shape of the complex waveform.***

Because a periodic wave does exactly the same thing over and over as time goes by, the combined sound of a harmonic series is a steady sound. You still

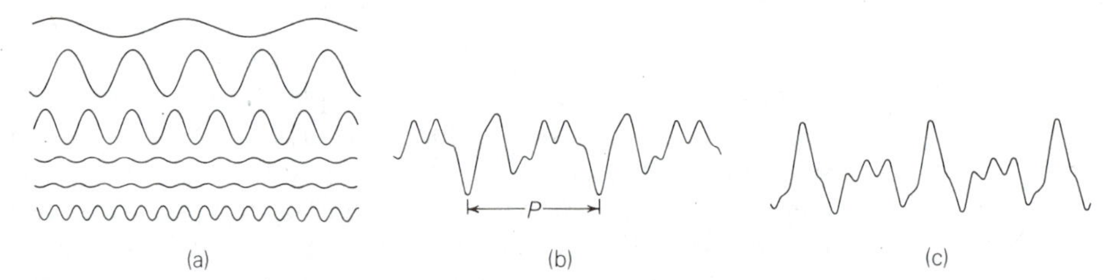

FIGURE 8.3 (a) Arbitrarily chosen amounts (and phases) of a fundamental and its second through sixth harmonics. (b) The complex wave with period *P* formed by combining them. (c) With all the same component amplitudes, 180° phase shifts of the third and sixth harmonics result in this waveform.

might wonder whether there is any periodic wave that could *not* be constructed in this way, but in fact there are no exceptions.

It is most remarkable that we *can* turn this whole concept around and make it work the other way. We can take any sound whatsoever that is steady in the sense of being periodic, and no matter which periodic waveform we choose or how complex it may appear, we always can break it down into sinusoidal components. The only sinusoids needed are those that form a harmonic series, and all have the same periodicity as the complex wave:

Any periodic waveform of period P may be built from a set of sine waves whose frequencies form a harmonic series with $f_1 = 1/P$. Each sine wave must have just the right amplitude and relative phase, and those can be determined from the shape of the complex waveform.

There is solid and beautiful proof to back up that claim; unfortunately the level of its mathematics makes it inappropriate for this book. But you may hear people refer to it sometimes as Fourier's theorem, after the young French mathematician who developed this method in the early nineteenth century. Putting sine waves together to make complex waves is called *Fourier synthesis;* taking complex waves apart into their sine wave components is called *Fourier analysis.* The recipe of sine wave amplitudes involved in a complex wave is called its **Fourier spectrum**, and each sine wave ingredient individually a *Fourier component*. From here on we will use the words *recipe* and *spectrum* interchangeably.

It also is possible to represent nonperiodic waves as sums of Fourier components. In place of the first rule, we have this statement about synthesis:

Any set of sine waves whose frequencies do not belong to a harmonic series will combine to make a complex wave that is nonperiodic, and will generally sound impure or unsteady in one way or another.

And the second rule about analysis also has its parallel:

Any nonperiodic waveform may be built from a set of sine waves, but their frequencies will not belong to a harmonic series. Each component must have the right amplitude and relative phase, which can be determined from the shape of the complex waveform.

The statement that the frequencies do not belong to a harmonic series actually includes two possibilities. One is that there could be only a few frequency components (such as 243.72, 539.08, 647.92, . . . Hz) that are not all multiples of any single number, and this is what we shall find is needed to describe the transient sounds of percussion instruments in Chapter 9. The other possibility is a continuous spectrum with components present at every frequency, and this is generally the case with continuous-noise sounds.

Perhaps you noticed the wording indicating that the amplitudes of the Fourier components "can be determined." That means there is a well-defined procedure (using calculus) that is always guaranteed to work: You show me a complex waveform, give me enough time to perform the computations, and I

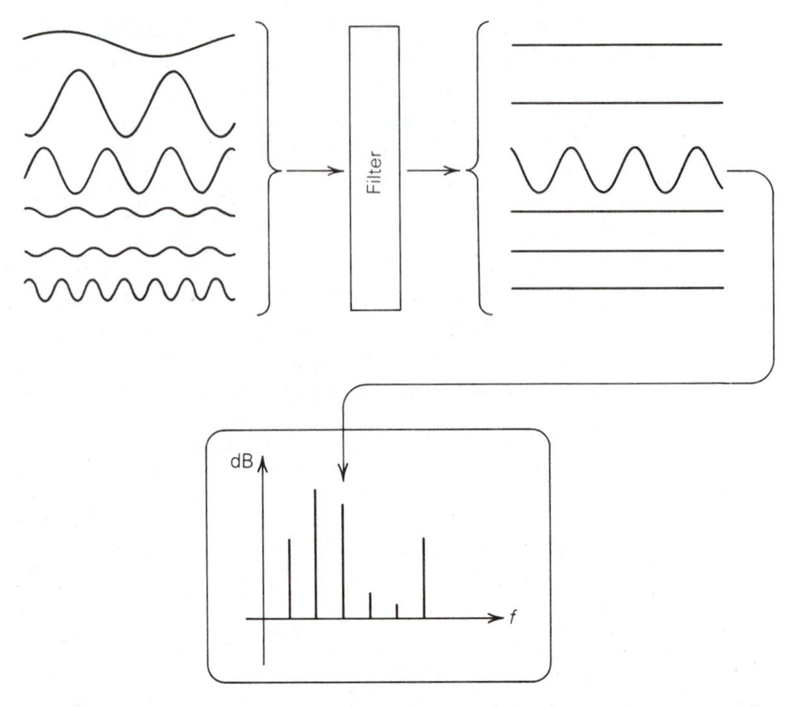

FIGURE 8.4 Sending the complex waveform of Figure 8.3 into a spectrum analyzer is equivalent to sending all its components through a filter that allows only one to pass. Its strength is used to move the dot on an oscilloscope screen (or the pen on a chart recorder). As the filter is tuned through a range of frequencies, each component, in turn, registers its strength to create a record of the spectrum of the original waveform.

can tell you all the right amplitudes and phases for its spectrum. Fortunately for the nonmathematician, it is enough for many purposes just to know that these numbers exist; we do not always actually need to compute them.

There is another way to determine spectra: You can buy (for a few thousand dollars) electronic gadgets called *Fourier analyzers*. You feed in any complex waveform and you get in return the spectrum, either as a list of numbers or in some kind of visual display. Some of these analyzers are simply special-purpose computers that automatically carry out the voluminous calculations just mentioned, relieving us of months of labor with paper and pencil. We may get a better physical feeling for what this means by describing an older type called *heterodyne analyzers*; leaving aside various details, these basically are just narrow-band filters that can be tuned. For each setting they block out most frequencies (that is, most of the Fourier components) but allow one narrow range of frequencies to survive. The decibel level of the signal that passes through the filter is displayed on an oscilloscope screen (Figure 8.4), and it represents the strength of whatever Fourier component has a frequency matching the filter's. As the filter setting is (electronically) changed, we get the strength of first one, then another and another component. You can think of your AM radio as a crude hand-operated

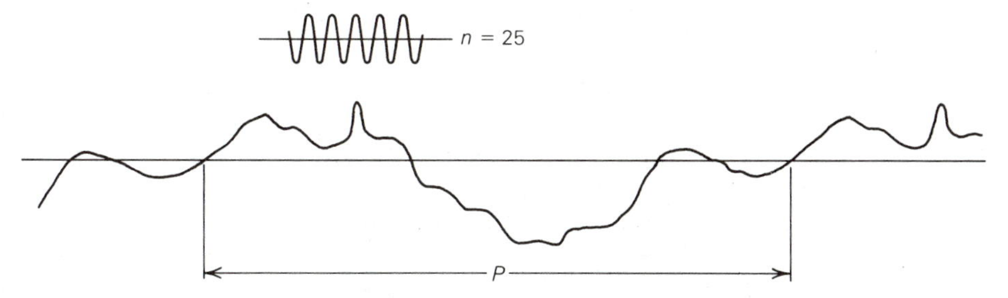

FIGURE 8.5 A complex periodic waveform including a narrow subpeak that requires the spectrum to include strong components for harmonic numbers near 25.

Fourier analyzer, which tells you as you twirl the dial what frequencies between 550 and 1600 KHz are being broadcast in strength by your local stations.

But even without the aid of gadgets or calculations, we can make useful qualitative statements about Fourier components. The strength of each component is a measure of the extent to which the complex wave does the same thing as the component. If our complex periodic waveform varies fairly slowly and smoothly (as did the resultant wave in Figure 8.2), it probably does not require more than the first few harmonics to be very strong. But if the waveform has extremely rapid changes, it must have strong components that themselves change that rapidly. Consider, for instance, a waveform whose repetition period is 10 ms, having a little subsidiary peak within it that rises and falls within 0.2 ms. Then we can estimate that there must be some harmonic components that complete half a cycle in about 0.2 ms or a complete cycle in about 0.4 ms. That is only $\frac{1}{25}$ of 10 ms, so the wave must contain harmonics with n around 25 that are quite strong (Figure 8.5). Sudden jumps, as in ideal square waves (Figure 8.1b), require that component strengths fall off only slowly with n, no matter how large an n we consider.

Each of the waves we described at the beginning of the chapter has a definite spectrum; these are shown in Figure 8.6. Several features are noteworthy. First, any wave whose second half merely repeats the first half but upside down requires only odd-numbered members of the harmonic series; any wave that lacks this symmetry requires at least some of the even-numbered components as well. Second, the smoother the waveform or the more gradual its changes, the more rapidly the spectrum drops off as we go to a higher n. Third, random noise, being nonperiodic, cannot be represented by a harmonic series; instead it requires the combination of sine waves of *all* frequencies.

This terminology makes it easier to differentiate among various kinds of random noise. *White noise* is defined as noise whose spectrum includes equal amplitudes at all frequencies (Figure 8.6i). This is what you hear between stations on your radio or TV. Because its hissing sound contains strong high-frequency components, white noise is quite bright and tiring to listen to. Electronic music synthesizers thus sometimes offer *pink noise* (Figure 8.6j) as

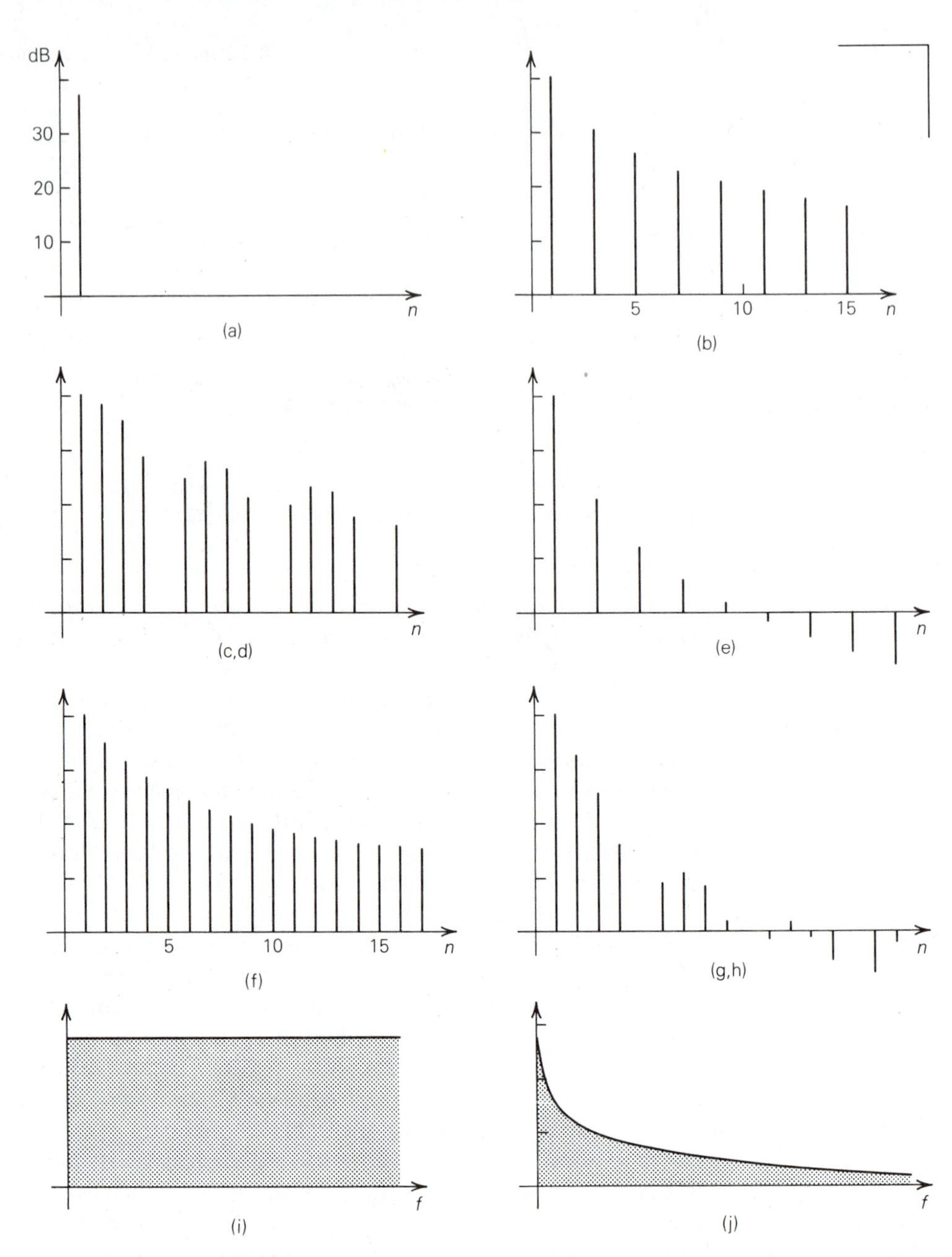

FIGURE 8.6 Spectra of the waveforms shown in Figure 8.1 in terms of component energies. The choice of zero point on the decibel scale is arbitrary and has no physical significance. (a) The sine wave involves only the fundamental. (b) The square wave contains only odd-numbered harmonics, whose amplitudes A_n are proportional to $1/n$ and energies to $1/n^2$; this can be described as an energy falloff at the rate of 6 dB per octave. (c), (d) The pulse wave includes all harmonics except the reciprocal of the duty cycle (here $1/0.2 = 5$) and its multiples. (e) The triangular wave again has only odd-numbered harmonics, but with amplitudes proportional to $1/n^2$ and energy falloff at 12 dB/octave, twice as fast as for the square wave. (f) The simple sawtooth wave requires both even and odd components with amplitudes proportional to $1/n$. (g), (h) The generalized sawtooth has both even and odd components; for this example every fifth component is missing because of a 20% duty cycle again. (i) White noise has equal strength at all frequencies. (j) Pink noise favors lower frequencies, dropping off 3 dB/octave for increasing frequency.

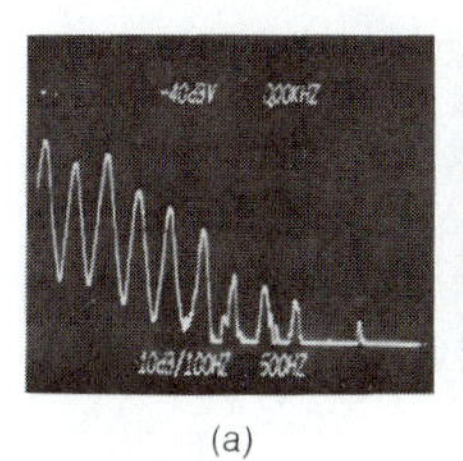
(a)

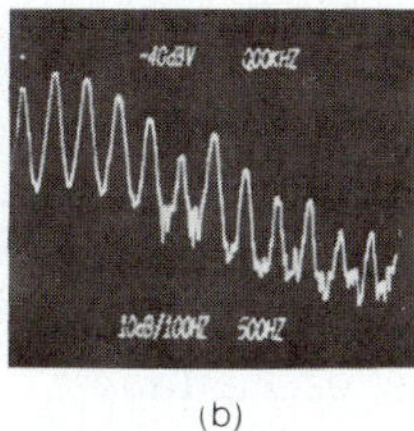
(b)

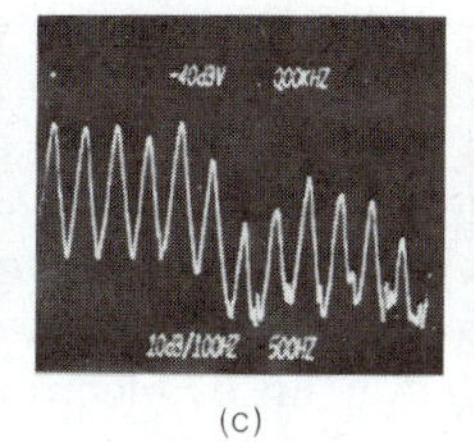
(c)

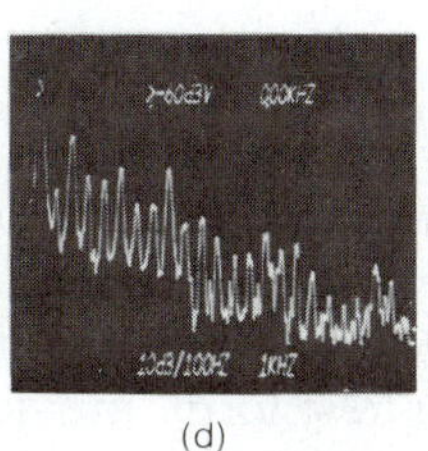
(d)

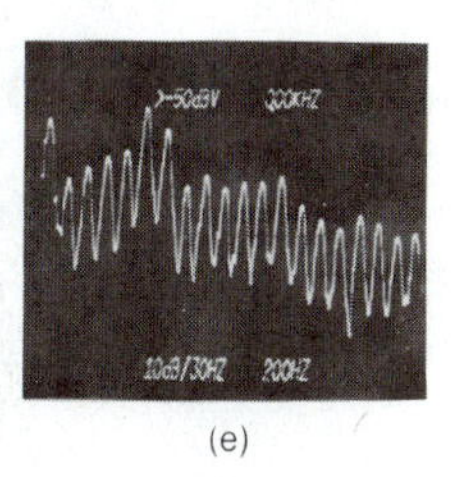
(e)

FIGURE 8.7 Measured spectra corresponding to Figure 2.5 (a) Flute, (b) trumpet, (c) saxophone, (d) violin, all playing A_4 with fundamental $f_1 = 440$ Hz. (e) Bassoon A_2, $f_1 = 110$ Hz. The first peak on the left is a zero marker, so it is, for instance, the fifth harmonic in the bassoon spectrum that is strongest. The spectral range shown extends up to 5 KHz (a, b, c), 10 KHz (d), and 2 KHz (e); in most cases additional harmonics also are present at higher frequencies. These peaks are not perfectly sharp spikes (as in Figure 8.6) because of a basic limitation of the measurement process (see Box 17.1).

an alternative; it has equal amounts of energy in each octave bandwidth. This may remind us of wave analogies again; each separate bit of a rainbow represents a light wave with spectrally pure color; that is, a sine wave of unique frequency. A mixture of all these colors (a continuous spectrum, *not* just a harmonic series) makes *white light*.

As further examples, consider the realistic musical waveforms shown in Figure 2.5. Their spectra, as measured with an electronic Fourier analyzer, are shown in Figure 8.7. We shall show similar recipes repeatedly in the next several chapters as we try to characterize what lies behind the timbre of various musical instruments.

You should wonder why we keep showing Fourier component amplitudes (or equivalent energies) but hardly ever worry about their phases. This is because for steady sounds our ears respond mainly to the amplitudes and are almost completely indifferent to the phases—a statement known as **Ohm's law**. That is, if you heard the two sound waves represented in Figure 8.2, you probably would not notice any difference. Similarly, the two waveforms of Figure 8.3b,c would be aurally indistinguishable. Ignoring phase information when giving sound recipes is somewhat like making vegetable stew in the kitchen: To a large extent it only matters how much of each ingredient goes into the pot, not in which order or at what precise times you add them.(We must state for the record that Ohm's "law" has a few minor loopholes, but for most practical purposes it is true. We defer until Chapter 17 the question of *why* our ears should behave that way.)

We now can make a little improvement on our rather vague statements in Section 6.3 about timbre being determined by waveform. We can say instead that each periodic waveform has its corresponding spectrum, and it is the spectrum that more directly determines the steady timbre, because, according to Ohm's law, the ear responds almost exclusively to that aspect of the wave shape. Many different waveforms all can have the same spectrum (by having different phases), so it is much more specific and useful to realize that only a different spectrum should produce a different timbre. (You might even want to cross out *waveform* in Figure 6.9 and replace it with *spectrum*.)

8.3 MODULATED TONES

How much of the preceding section really applies to ordinary musical instruments? After all, they do not create perfectly steady tones; they add complications such as transients and vibrato. Let us look first at these effects in their purest form, provided again by electronic oscillators.

The simplest kind of unsteadiness in a sound is a change in amplitude while the wave shape remains the same. If this takes the form of a repeated rise and fall, we may obtain waveforms such as those in Figure 8.8. We say that these waveforms show *amplitude modulation* (AM for short). If the modulating process is repetitive, the modulated tone still has a sort of steadiness in the continuing modulation.

Some oscillators have accompanying circuits that allow the amplitude of a waveform to be determined by a voltage instead of a knob setting. If that voltage comes from a second oscillator, we get a repetitive amplitude modulation of the first wave (called the carrier in radio applications). Both the carrier and the modulating wave can be either sine waves or any other waveforms.

Commercial AM radio uses carrier frequencies of approximately 1 MHz, modulated by a microphone signal at audio frequencies up to 5 KHz. The modulation constitutes a form of coding: Your radio receiver can tune in on the carrier wave, "read off" the modulation, and use that to reproduce a sound signal such as the one picked up by the microphone in the broadcasting studio.

Steady sound waves of audible frequency also can be carriers; amplitude modulation at subaudio frequencies (especially between about 1 and 10 Hz) then produces the effect musicians call **tremolo**, in which the loudness regularly fluctuates while the pitch remains unchanged. Note that tremolo is characterized by both a rate (the modulating frequency) and a strength (the depth of modulation or percent change in amplitude).

Another kind of repetitive change often applied to waveforms is *fre-*

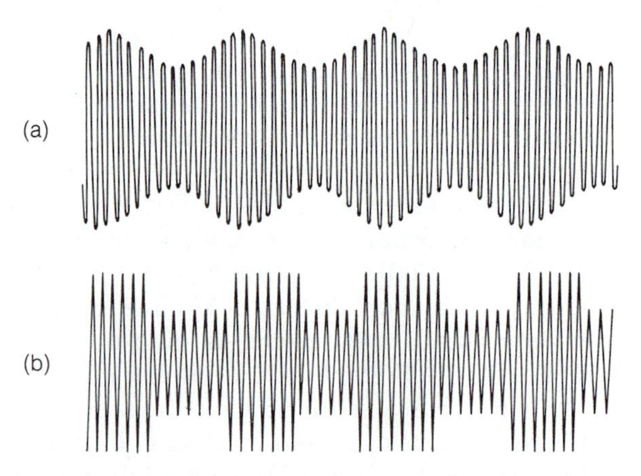

FIGURE 8.8 Amplitude modulation. (a) A sine wave carrier modulated by a sine wave. (b) A triangular wave carrier modulated by a square wave.

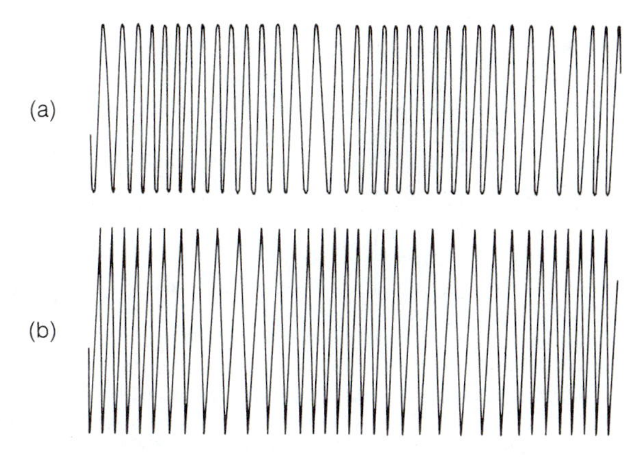

FIGURE 8.9 Frequency modulation. (a) A sine wave carrier modulated by a sawtooth wave. (b) A triangular wave modulated by a sine wave.

quency modulation (FM), illustrated in Figure 8.9. Again, some oscillators allow their frequency to be controlled by a voltage instead of a knob, so that FM is achieved simply by connecting a second oscillator to that control input.

Commercial FM radio uses carrier frequencies of approximately 100 MHz and modulates them at audio rates up to 15 KHz. This is just an alternate way of coding information about a sound wave onto a radio wave and requires a correspondingly different receiver to decode it and recreate the sound wave in your home.

Steady sound waves of audible frequency again can act as carriers; modulation of their frequencies with subaudio modulating frequencies (especially between about 1 and 10 Hz) produces the effect called **vibrato**. This, too, is characterized by both a rate (the modulating frequency, how often you notice the pitch wavering) and a strength (the amount the carrier frequency is altered, corresponding to how far up and down the pitch changes). Violinists, for example, typically use a vibrato rate of approximately 5 to 7 Hz, with an excursion of approximately 0.2 semitone either way from the average pitch.

---

*During frequency modulation of a sine wave, the frequency clearly varies a little above and below the original carrier frequency. The modulated wave in some sense must contain that entire range of frequencies; we then expect that its Fourier analysis must involve all of those frequencies. It is also true (but much less obvious) that amplitude modulation requires groups of Fourier components of slightly differing frequency. It is possible to give specific formulas, but it will be sufficient for our purposes to state that a small amount of *either* kind of modulation, applied to any waveform, has the effect of smearing out each original Fourier component over a range of frequencies whose width is about twice the modulating frequency (Figure 8.10). Throughout the following chapters, all information about spectra of different instruments should be understood in this light—insofar as the instrument is played with tremolo or vibrato, each Fourier component must be regarded as smeared out this way.

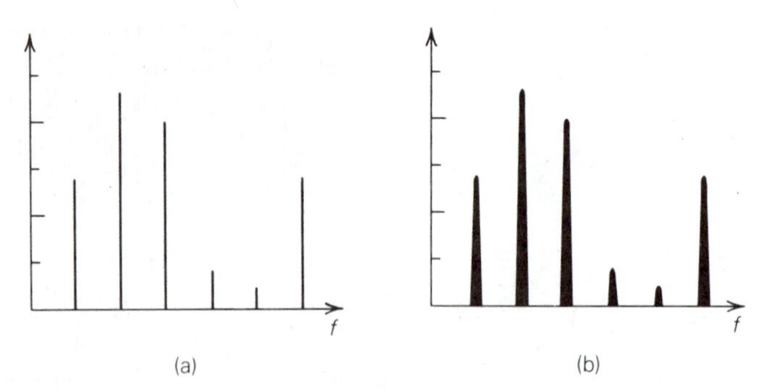

FIGURE 8.10 (a) The spectrum of the unmodulated complex wave of Figure 8.3. (b) If the wave is modulated, each component becomes a cluster of slightly differing frequencies, with the width of the cluster determined by the modulation frequency. The case shown is for modulation frequency of approximately one tenth of carrier frequency.

The ability to modulate tones is of crucial importance in electronic music. Without tremolo or vibrato, electronic tones seem too lifeless and artificial and do not hold the listener's interest well.

Another important modification of electronic tones to make them more lifelike is the addition of transients. A good electronic synthesizer allows you to shape each tone with an attack-decay-sustain-release (ADSR) unit. You can turn four knobs on the ADSR, each of which controls one aspect of the transient as shown in Figure 8.11.

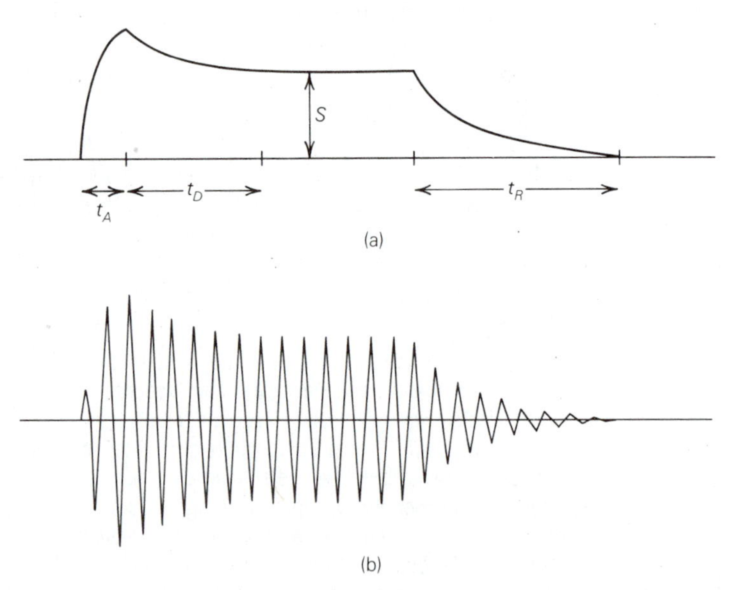

FIGURE 8.11 (a) Amplitude envelope provided by an ADSR module; t_A = attack time, t_D = initial decay time; S = sustained level; t_R = release or final decay time. (b) Triangular carrier wave after imposition of this envelope.

*8.4 ELECTRONIC AND COMPUTER MUSIC

The development of integrated-circuit electronics in the 1960s made electronic music synthesizers widely available in the 1970s. The further development of digital electronics that produced the personal-computer revolution also gave us a new generation of music synthesizers in the 1980s whose internal workings are quite different from those of the older ones. But the concepts of modularity and control signals remain important in understanding their overall organization.

Some of the modules that make up a synthesizer act as individual musical instruments, each of which can contribute one voice to the total sound produced; a sine or square wave oscillator is an example. Other modules act to combine or alter those sounds—for instance, by amplifying or filtering them. On older synthesizers, the modular organization often was obvious in the physical arrangement of a row of small boxes mounted in a large panel, with patch cords plugged in by hand to carry signals from one module to another. Many of the smaller commercial models avoided the tangled thicket of wires by providing sets of switches on the front panel that created such connections internally (Figure 8.12).

It is these interconnections that make interesting electronic music possible, for if you had to control every small detail of a sound sequence by connecting cords, turning knobs, or flipping switches, it would be hopelessly slow. With control signals we can let one module determine what another one will do. For example, a *voltage-controlled oscillator* (VCO) by itself would

FIGURE 8.12 A small portable electronic music synthesizer. (Photo courtesy of Moog Music Inc.)

produce only a perfectly steady tone. It can receive a signal from a second oscillator to control its frequency so that the output acquires a vibrato, or from an attack generator to give a short glissando into and away from each note. It also can receive successive steady voltages from a keyboard (each key providing a different voltage, adjusted to elicit the corresponding pitch from the oscillator) or from a sequencer (which is programmed to provide a series of voltages that plays a desired group of notes); or it may respond simultaneously to several such controls.

Similarly, a *voltage-controlled amplifier* (VCA) can be run by an oscillator (to provide tremolo), or by an ADSR unit (to provide transients), or by a sawtooth oscillator set to a very low frequency (to give gradual crescendo and decrescendo effects). *Voltage-controlled filters* (VCF) also are useful; by controlling the filter setting with an oscillator or an ADSR, you can cause either repetitive or transient changes in timbre of whatever signal enters the filter. Applying the same control signal to a VCA and VCF in tandem can produce electronic sounds that seem much more musical; we shall see in later chapters that most regular musical instruments have frequency spectra that change according to how loudly they are played. Lack of this feature is one of the main reasons for the "artificial" sound of the simplest synthesizers.

Another important concept is *nested control.* All of the controls mentioned above still leave rather narrow limits on the kinds of sound produced. But each oscillator is a module that can both control and be controlled. If you have several, you can let oscillator X control Y, and Y, in turn, control Z. A simple example would be setting Z to 440 Hz, Y to 8 Hz, and X to 0.2 Hz, and then interconnecting. The resulting sound is an A_4 with vibrato, the vibrato rate, in turn, becoming faster, slower, and faster again every 5 seconds.

You can appreciate how many possibilities are opened up by nested control by this crude estimate: If you can get, say, 10 different effects with direct single controls, you may expect to get 100 with controls nested two deep, 1000 for three deep, and so on. Even if three-fourths of the nested combinations are not useful, you still end up with a tremendous variety of possible sound outputs.

Another powerful aspect of nested control is the opportunity to create feedback loops. If you let X control Y, Y control Z, and Z, in turn, have some influence on X again, the results can be bizarre. Third-order control and feedback loops both are sufficiently complex that it can be quite difficult to predict what they will do; typically you just try some combinations and listen to what happens. In this sense making electronic music is still much more an art than a science. But some scientific understanding of the basic oscillator and control properties still can be a useful supplement to the musician's intuitive experimentation.

Until the recent advent of computer control, the complexity of large synthesizers has tended to discourage their use in live performance. Instead the musician spends hours and days in the studio, trying out various settings, listening to the results, and saving a few that appeal to him on a tape recording. Eventually, he selects some of these favorite sounds, edits, combines, and arranges them in some order. The final tape is played back for the audience, often as an accompaniment to a live instrumental or vocal group.

Electronic organs have been around for some sixty years and may be regarded as primitive forerunners to the modern synthesizer. The essential difference is the use of fixed-frequency oscillators and preset filter combinations in the electronic organ, because cheap and practical voltage-control modules are a recent development.

It is interesting to regard most electronic organs and pre-1980 synthesizers as small, special-purpose *analog computers*: They are machines that take some kind of input signals (settings of knobs, keys, and switches; signals from master oscillators) and combine or process them to produce a different output signal. The word *analog* here means specifically that the machine handles voltages that may vary continuously and take on many different values. Analog computers in general are little known among the general public, but they are sometimes useful in solving scientific and engineering problems. The name reflects the fact that these computers solve problems by analogy; that is, by connecting electrical circuits that happen to obey the same equations as other physical systems (for example, an airplane subject to wind gusts while under the control of an automatic pilot). Whatever the electrical circuits do, as evidenced by their voltage outputs, tells how the analogous system should behave.

What most people associate with the word *computer* is the *digital computer*, which handles numbers rather than continuously varying signals. The numbers are represented by electrical voltages, but they do not have many possible values; they are either "on" or "off," to represent ones and zeros, which in turn form a code for other numbers.

Digital computers have become increasingly useful to musicians. Besides the incidental uses listed in Box 8.1, they can be used to generate sounds more or less like those from analog synthesizers. For this application, sequences of numbers representing desired waveforms are stored in the computer memory. Repeated rapid readout of these numbers through a digital-to-analog converter (DAC), which produces corresponding voltages, replaces the old analog oscillators and allows a much wider choice of initial waveforms. Such digitally stored waveforms have been used by one major manufacturer of electronic organs since the mid-1970s, and underlie the sound production in the new generation of commercial music synthesizers (Figure 8.13).

But digital computers also can oversee the performance of an entire piece of music, eliminating any pushing of keys or buttons by a human player. The computer offers the advantage of exercising absolute and precise control over every feature of the sound—both the succession of notes and the exact microstructure of each note. In principle, you can produce any possible sound without limitation. The great disadvantage is that this is too much control; you never really want to consciously plan and specify the sound pressure for every tenth of a millisecond throughout an entire piece. For five minutes of

*BOX 8.1 MUSICAL USES OF COMPUTERS

Digital computers can perform music by generating sequences of numbers that are converted to voltages to drive a loudspeaker. But musicians also use them in several other ways.

1. Information storage and retrieval. Musicologists sometimes compile large catalogs of information about composers or about collections of musical scores in libraries. A computer can help collect and arrange this information and make any desired part of it quickly available.

2. Score writing. This could be characterized as an information-translating function. Several programs now are achieving modest success at producing a traditional written score from some other form of information about a piece of music. The other form may be a list of symbols entered on a typewriter keyboard, or in some cases it even can be the audible sound of live music reaching a microphone connected to the computer.

3. Composition. You can write a program that specifies a set of rules for acceptable melodies, or even counterpoint, and let the computer choose a series of notes that satisfies your rules. (See, for instance, the articles by Pinkerton and Hiller in the references for Chapter 19.)

4. Harmonic and structural analysis. In trying to understand how a piece of music is put together, you can store masses of information in a computer about what combinations and sequences of notes occur. The computer then can help sort, analyze, and summarize those features that may give the piece its distinctive character or be typical of its composer's style.

5. Associated mathematical or scientific problems. Many problems in acoustics (for example, multiple sound reflections in a concert hall or optimum design of a musical instrument) present difficult equations that are best solved with computers. There are also mathematical problems in the theory of musical scales and harmonies that can be treated with computers; we will discuss an example of this in Chapter 18.

FIGURE 8.13 A modern digital synthesizer. This model uses a frequency modulation process to provide a wide choice of waveforms. (Photo courtesy of Yamaha Music Corporation.)

music, that would mean providing a list of 3 million numbers! What you really want is for the computer to use a library of preprogrammed waveforms (corresponding to different "instruments" or to different types of analog oscillators) so that you need give only a few instructions about how these waveforms are to be combined and sequenced.

The techniques for controlling computer-generated sound have changed greatly since the 1960s; then you probably would have learned a specialized programming language such as MUSIC V. In this language you could give instructions to control sound parameters in ways that often corresponded, more or less, to control settings and connecting wires on an analog synthesizer. Even on the large mainframe computers of that time, this was quite laborious because each second of sound ultimately produced would require many seconds of computing time, which in turn represented the execution of a program that took minutes or hours to prepare. Thus, after you wrote your instructions and transferred them to punched cards, their execution would produce a tape recording of the sound. When you listened to the result and decided you would like something about it to be different, you changed some of your punched cards and put the stack back in the hopper to be run again.

Modern computers allow two important improvements on this process. First, they are much faster, so that sounds can be generated in real time during the execution of your instructions. Second, the user either has direct time-shared access through a video terminal to the central computer, or has exclusive control over a special-purpose microcomputer that is powerful enough to accomplish all these chores. Together, these make it possible for you to hear sounds respond immediately when you enter instructions for their properties to be changed. Thus, it is far easier to tinker with many details and shape a musical phrase in a way that pleases you within a reasonable amount of time. There are now many commercial products that combine some computing power of their own with the ability to exchange data and instructions with other synthesizers and with a master control computer, so that the old distinction between computer and synthesizer has been nearly obliterated.

SUMMARY

Every waveform can be described in terms of a spectrum of simple sine waves. This spectrum either is or is not a harmonic series, depending on whether the waveform is or is not periodic. The spectrum is especially important in understanding musical tones, because it provides the most direct link between our understanding of the vibrations naturally occurring in each musical instrument and our perception of its tone color. For this reason, we will use the language of spectra in discussing various instruments in Chapters 9–14 and in considering the ear's response in Chapter 17.

Electronic music synthesizers have changed a great deal from their origin in the interconnection of a few simple oscillators. Important modular functions such as voltage control of oscillator frequency, amplifier output, and filter settings now can be implemented on digital computers. Modern computer-controlled synthesis has the power and versatility to produce a wide variety of interesting music.

REFERENCES

For an analysis of various experiments showing the limited validity of Ohm's law, see Chapter 3 in *Aspects of Tone Sensation* by R. Plomp (New York: Academic Press, 1976). R. D. Patterson reports in *JASA, 82,* 1560 (1987) that human ears can detect considerable perceptual differences when phase changes occur in the harmonic components of tones with fundamental frequency below 200 Hz, but that we are essentially phase-deaf above 400 Hz.

Good practical introductions to electronic music of the 1980s are given by Barry Schrader, *Introduction to Electro-Acoustic Music* (Englewood Cliffs, NJ: Prentice-Hall, 1982) and R. T. Adam, *Electronic Music Composition (for Beginners)* (Ames, IA: William C. Brown, 1986).

For my taste, some of the more worthwhile original electronic music has been produced by Morton Subotnick and Wendy Carlos. Outstanding and creative electronic presentations of traditional music have been made by Carlos and by Isao Tomita. For a good article on the computer as a musical instrument by two pioneers in that field, see M. V. Mathews and J. R. Pierce, *Scientific American* (February 1987), p. 126.

SYMBOLS, TERMS, AND RELATIONS

voltage and current
sine, square, pulse, triangular, and sawtooth waves
random noise
Ohm's law
tremolo = amplitude modulation
vibrato = frequency modulation
periodic wave
harmonic series
Fourier synthesis
Fourier analysis
Fourier component
Fourier spectrum = vibration recipe
amplitude and phase

EXERCISES

1. (a) If the components being added in Figure 8.2 have frequencies 110 and 440 Hz, what is the pattern repetition frequency of the resultant complex wave? (b) If in a separate experiment you combine two components with frequencies 220 and 330 Hz, what is the frequency of the complex wave? (Hint: To what harmonic series do they belong?)

2. If sine waves of frequency 200, 300, 500, 800, and 900 Hz are added together, what are the repetition frequency and period of the resulting complex wave?

3. Consider three sine waves belonging to a harmonic series, with frequencies f_1, $f_2 = 2f_1$, and $f_8 = 8f_1$. Sketch these above each other as in Figure 8.4, with amplitudes $A_1 = 5$ cm, $A_2 = 2$ cm, and $A_8 = 1$ cm on your paper; let a full cycle at f_1 nearly fill the width of the page. Let the relative phases be anything you like. Sketch the approximate waveform made by combining these three harmonics. Would your complex waveform (if done exactly) necessarily match those of everyone else? Would they all *sound* the same?

4. The note G_3 is produced with a complex waveform. What are the frequencies of its Fourier components?

5. A complex waveform repeats every 4 ms, What are the frequencies of its Fourier components?

6. Discuss the relation between the harmonic spectra shown in Figure 8.7 and the corresponding waveforms of Figure 2.5 (page 22). Explain specific features of the connection, such as the relative smoothness of waveform (a), prominence of the fifth peak in spectrum (e), and so on.

7. Listen to a good singer and estimate his or her vibrato frequency. Estimate the fastest and slowest rates you think would be practical to perform or pleasing to hear.

8. In view of the rules in Section 8.2, in which of the following cases is the vibration recipe required to be precisely a harmonic series, and in which is it permitted to be nonharmonic? (a) Flute continuously sounding a steady note; (b) xylophone bar; (c) steadily bowed violin; (d) snare drum; (e) steadily blown oboe; (f) struck piano string.

PROJECT

1. Obtain access to an electronic synthesizer and learn how to operate it. Demonstrate for the class various effects described in this chapter.

CHAPTER 9 Percussion Instruments and Natural Modes

Bong . . . thud . . . pop . . . crack . . . tinkle. . . . Percussion instruments produce a great variety of sounds. Why are they different? How can we describe the differences accurately? Is there some unifying concept that can provide a framework for understanding such diverse sounds?

It is time to take a closer look at sound production. In the coming chapters we will develop answers to these questions for all types of instruments. The percussion instruments are a good place to start, for in studying them we will gain more physical insight into the meaning of the Fourier spectrum as a tool for describing sound.

The vibrations of percussion instruments are in some cases extremely complex, and to understand them it is wise to seek first the simplest examples. Some astute observations with a tuning fork suggest that we examine an even simpler system, the coupled pendulums of Section 9.2. This will lead us to the idea of natural modes of vibration, which we then apply in Sections 9.4 and 9.5 to xylophones and drums. Then we will discuss how different sounds can come from the same instrument when it is hit in different ways, and end the chapter by considering how the sounds change while they are dying away.

Our emphasis here will be on how to accurately and efficiently describe the physical properties of these sounds; we will defer certain questions of how our ears perceive their pitch and timbre until Chapter 17.

9.1 SEARCHING FOR SIMPLICITY

Look back now at the complex waveform shown in Figure 3.1 (page 39). The irregular pattern of air disturbances as time goes by suggests that the surface of the wooden stick itself must have moved in a jerky, irregular way. That is, the stick was apparently not vibrating in smooth, always-the-same, simple harmonic motion. The task of understanding the complex motion of the stick (and thus explaining the sound it makes) appears quite discouraging.

But this is always one of the fundamental challenges of science: to cut through the details of seemingly complex phenomena, to identify a few essential ingredients, and to explain how the combined action of these few simple and basic elements leads to the complex result originally observed. Such insights are likely to come from broad experience with many related examples rather than a narrow head-on attack on a single case.

Thus, before investing too much effort in the "clack" of one particular wooden stick, let us look around more. Suppose we connect a microphone to an oscilloscope and record traces of the sounds made by as many different percussion instruments as we can find. This experiment produces a great variety of pictures (see Figure 9.1); many of them are just as complex as Figure 3.1 and equally discouraging. But a pattern begins to emerge, for the instruments that produce a more musical tone (one of definite pitch) apparently undergo a more regular and orderly motion.

The next logical step is to search out the simplest among all these pictures. If we have included a tuning fork in our experiment, we find that it is a prime candidate for further study (Figure 9.2). If we look especially at the signal from the fork 10 or 20 seconds after it was struck, we see a smooth, regular, and repetitive trace. Although it is difficult for our eyes to judge exactly, it at least closely resembles simple harmonic motion (*SHM*). And the corresponding sound in our ears is the very essence of a pure, smooth tone of well-defined pitch—the *fundamental tone* of that fork.

So here we have a case of simple motion at a unique, well-defined frequency. How do we work back from this to the other cases of complex motion? A helpful hint comes from remembering that even the tuning fork, when first struck, shows some complexity—and that is just when our ears notice a second, much higher distinct pitch. This is called the *clang tone*, and you may mistake it easily for the lower "intended" pitch of the fundamental tone if you do not hold the fork close to your ear. Is it possible that the clang tone also is a simple harmonic vibration and that the total sound seems complex only because the clang tone and the fundamental tone are both going at the same time?

Even though we cannot answer the question with certainty yet, let us speculate about where that train of thought might lead. If the initial sound of the tuning fork proves to be two simple motions mixed together, might all the other more complex cases also turn out to be just two (or more) simple motions happening simultaneously? Bold as it may seem, that is exactly what we will claim before we are through.

To make a convincing case for that claim, we must devote a few pages to the important scientific activity of abstraction. Even the tuning fork, being a real object, is a little bit complicated. To isolate our physical ideas in their purest form, we will study some highly idealized systems, imagining that all masses are concentrated at a few points and that they are connected together with perfect, linear springs. After using this abstraction to develop concepts and terminology, we will return to the case of the tuning fork and then to other percussion instruments.

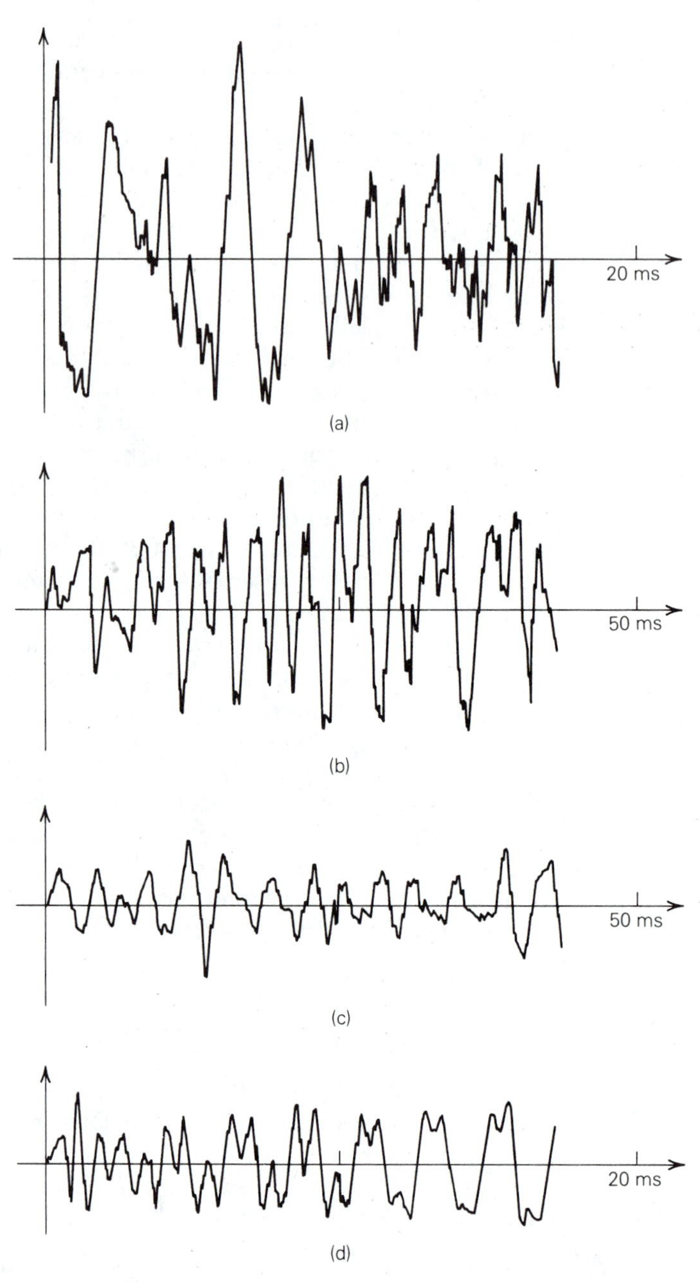

FIGURE 9.1 Oscilloscope traces for several percussion instruments: (a) snare drum, (b) tympani, (c) large brass gong, (d) xylophone. Trace (c) was taken approximately 10 seconds after the gong was struck; all others begin at the initial impact. Note the shorter time scales for (a) and (d); ms = millisecond.

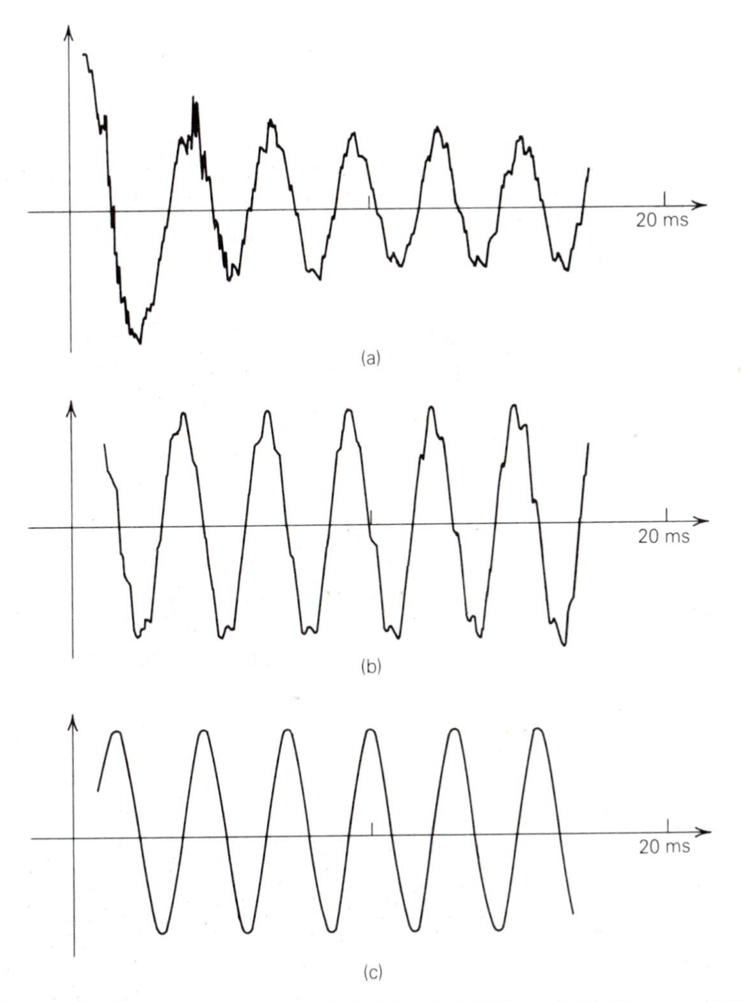

FIGURE 9.2 Oscilloscope traces from a tuning fork: (a) at the initial impact, (b) after approximately 1 second, and (c) after approximately 10 seconds. The amplitude slowly and continually decreased, but the oscilloscope was reset to increasingly sensitive scales for (b) and (c) to make the traces visible.

9.2 COUPLED PENDULUMS

Suppose we set up a pair of pendulums as in Figure 9.3 and study their combined motions. Before the spring is attached, each pendulum swings independently; if both are the same length, they swing with the same frequency. In that case the overall motion of the system is quite simple: the same pattern repeats, continuously, and each part of the system executes simple harmonic motion.

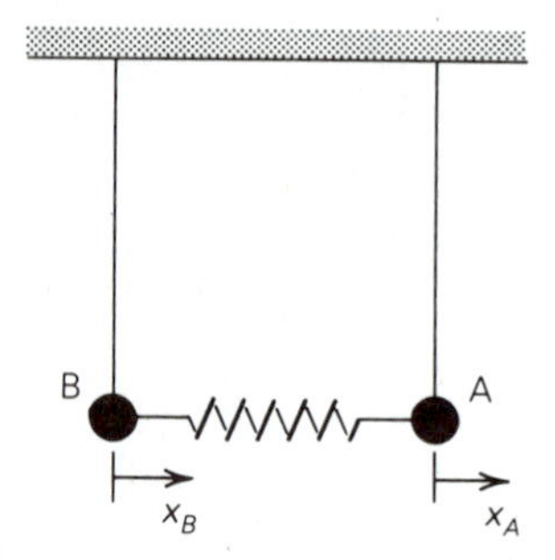

FIGURE 9.3 Two coupled pendulums. For the discussion of Section 9.2, they are allowed to move only left and right in the plane of the page. Arrows x_A and x_B indicate direction of positive displacement for graphs in succeeding figures. For simplicity, we choose a spring whose relaxed length matches the distance between the pendulum rest positions. We also suppose the spring is stiff enough to support its own weight and thus not sag when the pendulums move closer together. Then, if the pendulums separate, the spring is stretched and pulls inward; if they come together, it is compressed and pushes outward on both.

What happens after the spring is attached? If we pull one pendulum aside and release it, we see a complex motion. Each pendulum moves in an irregular way. Not only is it not *SHM*, it is not even periodic (exactly repetitive) at all. Figure 9.4 shows graphs for a typical motion of this sort.

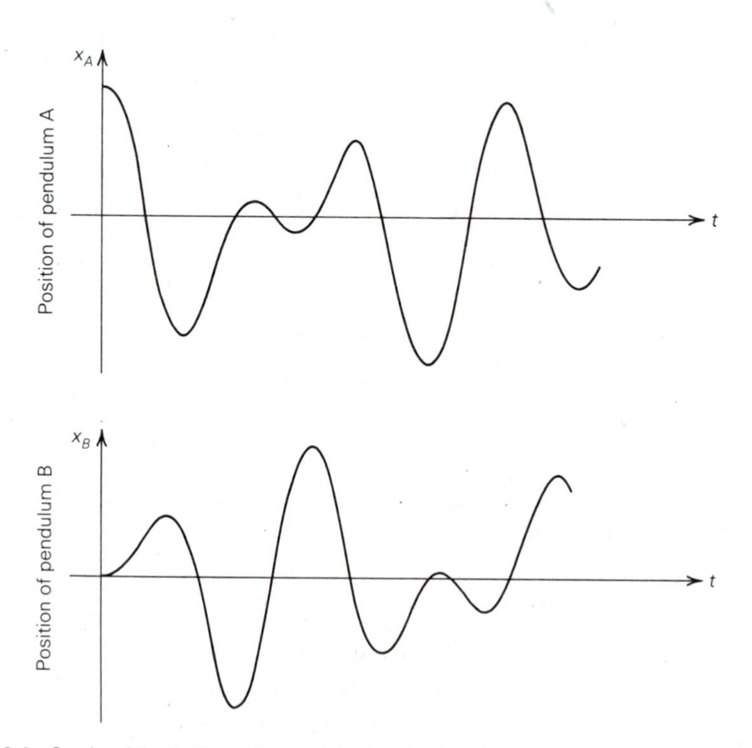

FIGURE 9.4 Graphs of the motion of the pendulums in Figure 9.3 for the case where pendulum A is pulled some distance to the right and released from rest at time $t = 0$.

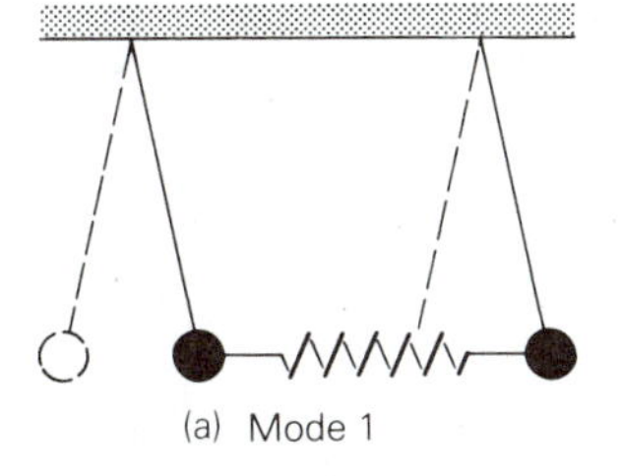

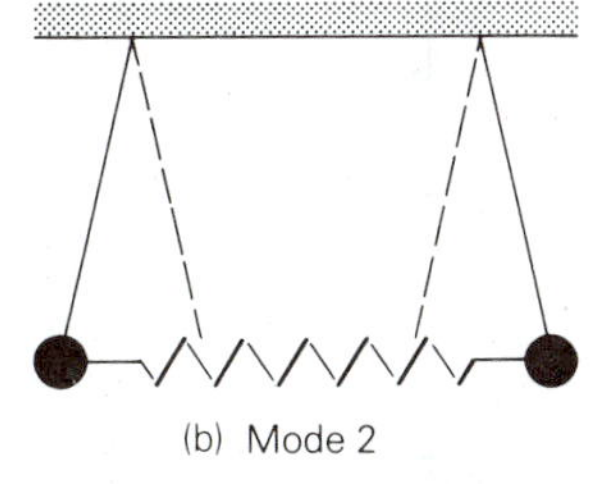

FIGURE 9.5 Natural modes of two coupled pendulums, showing the system at maximum displacement. Dashed lines indicate displacement a half-cycle later. Because of restoring force from the spring, mode 2 has a higher frequency than mode 1.

But this complexity means we have not approached our study of the system in the right way. When we pull one pendulum aside without touching the other, it implies a subconscious hope that the resulting motion will have the simplicity of a lone pendulum, with the second one never becoming involved. But, of course, the spring makes it quite impossible for one to move without the other also moving.

Let us be more clever and ask, "Is there some way we could start them so that the resulting motion would be very simple?" Either contemplation or experiment soon provides an affirmative answer: If we pull both pendulums the same distance to the same side and release them simultaneously, the subsequent motion is exceedingly simple (Figure 9.5a). It is a periodic motion—the two pendulums stay in step with each other and each executes *SHM*. And because during this motion the spring length never changes, it might just as well be absent, and so the frequency f_1 of this *SHM* is the same as for either pendulum alone.

That is not the only way to obtain a simple motion. It also is easy to see that releasing the pendulums from equal displacements to *opposite* sides (Figure 9.5b) results in an *SHM* in which they move toward and away from each other. But this time the stretching and squeezing of the spring provide an additional restoring force to aid gravity in returning the pendulums to the vertical position. So the frequency f_2 for this second type of motion must be greater than f_1; how much greater depends on the stiffness of the spring.

Each of these special patterns that gives *SHM* is called a **natural mode**, and each has its own characteristic *natural mode frequency*. Natural modes and their frequencies are properties of the system as a whole, not of either pendulum individually.

Try as you may, you cannot come up with any third natural mode for these pendulums (as long as they are restricted to motion in the plane of the page). But that is not really disturbing; it seems reasonable that because there are only two pendulums, there are in some sense only two things they can do, and we already have found both. Now comes the great leap of speculative insight: If in some sense the two natural modes represent all possibilities, then perhaps we should be able to understand *any* motion of the pendulums as being a combination of these two special motions.

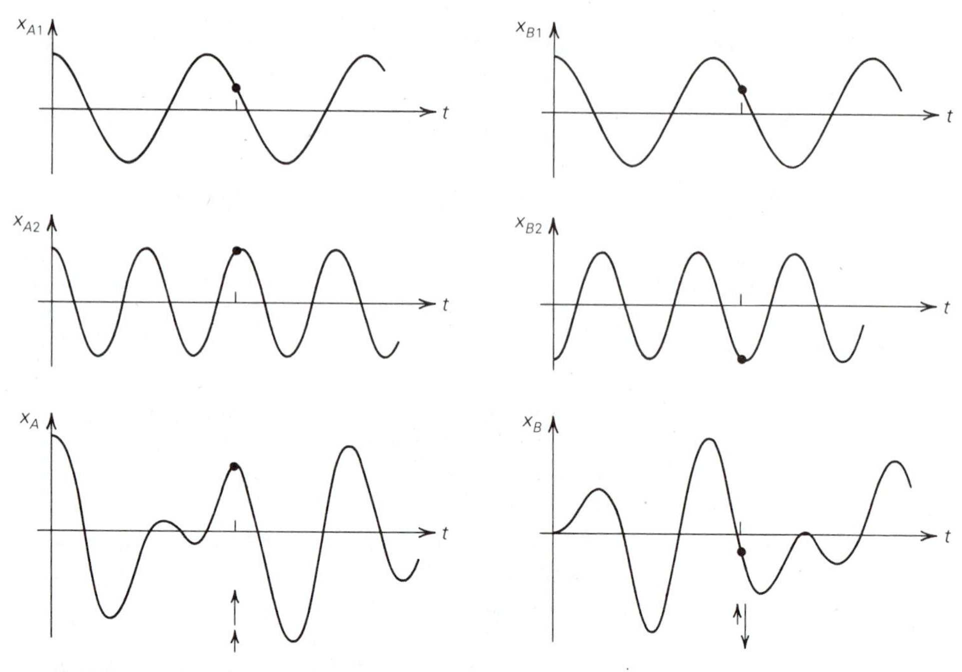

FIGURE 9.6 At left, graphs of motion of pendulum A because of mode 1 alone (top), motion of pendulum A because of mode 2 alone (middle), and total motion of pendulum A with both modes present (bottom). At right, the same information for pendulum B. Each graph at the bottom is the sum of the two above it, in the sense illustrated for one particular time by the dots and arrows. See also Figure 9.7.

For example, what would it look like if both modes were going at the same time with equal amplitudes? Figures 9.6 and 9.7 show how we can add them together to find out what each pendulum does individually. The result accounts precisely for the seemingly complex motion of Figure 9.4.

No matter how you start the pendulums in motion—pulling them aside unequal distances, even giving unequal shoves as they are released—there is always some combination of appropriate amounts of the two natural modes that exactly accounts for the seemingly complex motion. In principle, we need

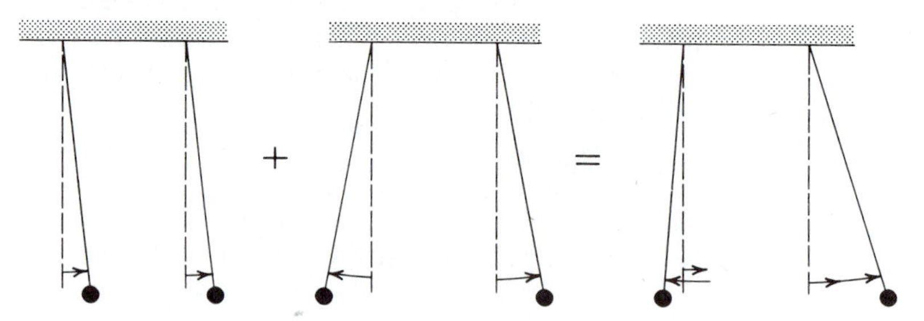

FIGURE 9.7 Further illustration of what is meant by Figure 9.6. Pendulum configurations corresponding to the dots in Figure 9.6, showing how appropriate amounts of the two natural modes together can account for any pendulum displacements.

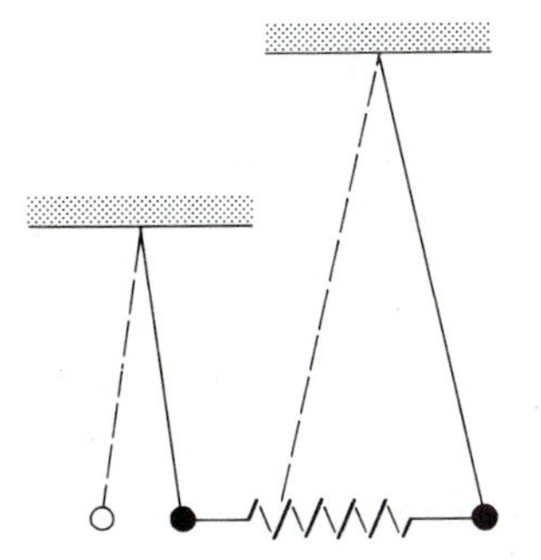

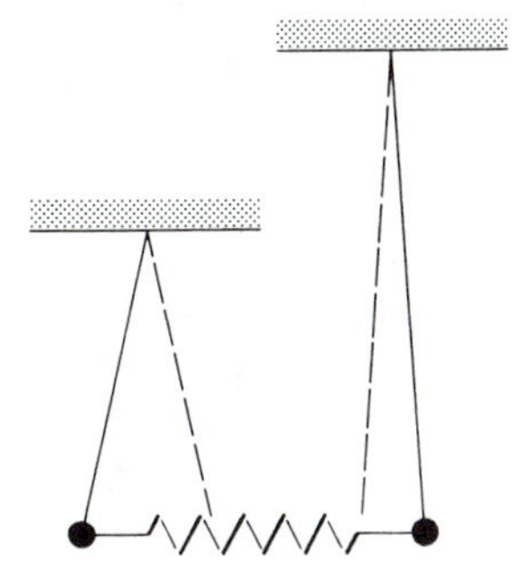

FIGURE 9.8 Natural modes for pendulums of unequal length. Now the lower-frequency mode is predominantly a motion of the longer pendulum, and the higher-frequency mode involves more motion of the shorter pendulum.

not rely on guessing what combination; there exists a definite mathematical procedure that can always tell how much of each mode is involved. We will not use that exact procedure here; it is standard fare in senior-level texts on mechanics for physics majors. But we will be able to deduce many important qualitative features without using such mathematics.

Before describing the general theory of natural modes, we should remark on one detail. It is not actually essential that the two pendulums have equal length; it is merely easier to see how our ideas should develop. If the lengths are unequal, there are still two natural modes (Figure 9.8), although it is no longer easy to guess what they look like just on the basis of symmetry. But again there are mathematical procedures for finding the natural modes, and they still have the same property that an appropriate combination of these two modes can account for every possible motion of the system.

9.3 NATURAL MODES AND THEIR FREQUENCIES

We want to apply the concept of natural modes to the tuning fork and eventually to all percussion instruments. To do this, we must generalize what we learned about the twin pendulums so that it will apply to any complex vibrating system. You should be aware that we are only presenting arguments to make it seem reasonable that natural modes have all the properties claimed, but we are not proving those claims. Such proofs exist but require more mathematics than is appropriate for this book.

One way to proceed is to consider three pendulums, then four, five, and so on. It is still fairly easy with three to guess, and to verify experimentally, that there are three natural modes (shown in Figure 9.9a). Although the guessing and the demonstrating rapidly become harder, we also can show four natural modes for four pendulums, five for five, and so on. It is helpful to agree that we always will list these modes in order from least to greatest involvement of the springs, so that f_1, f_2, f_3, ... will be a list of mode

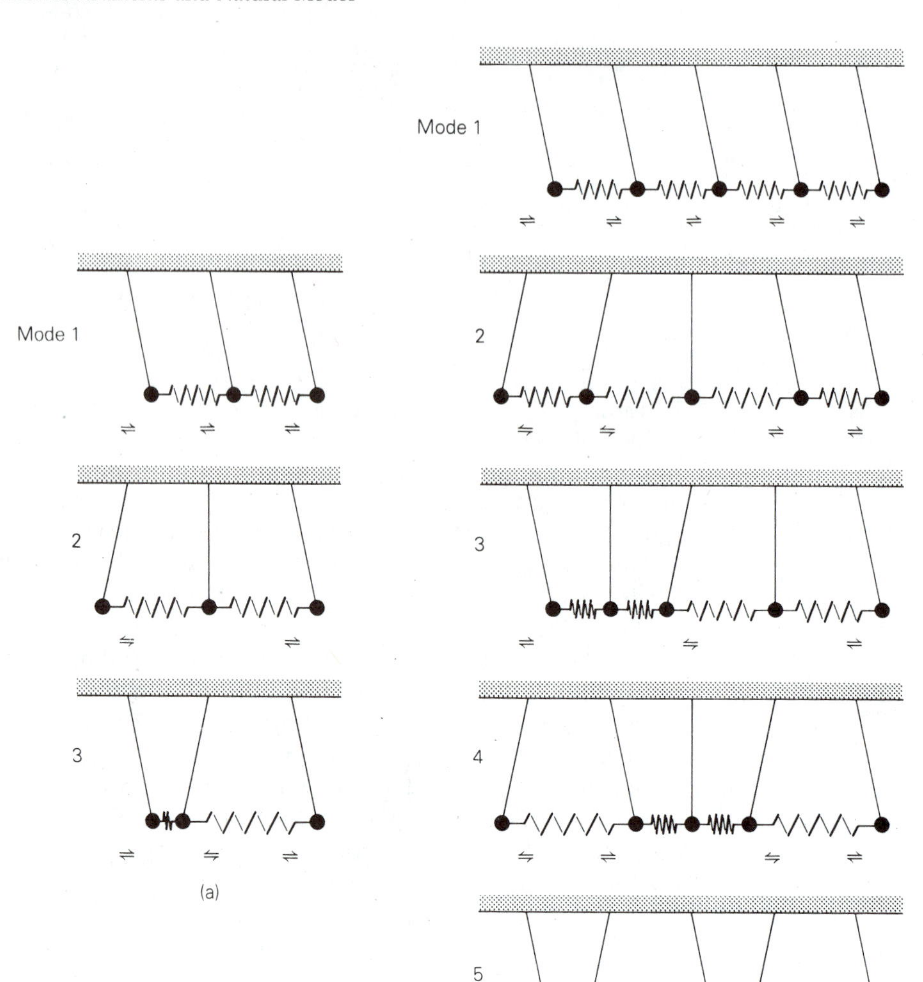

FIGURE 9.9 Natural modes for (left) three and (right) five pendulums. Note the progression in each case from no spring involvement in the first mode (top) to greatest involvement for the highest-frequency mode (bottom).

frequencies in order from lowest to highest. These modes will serve as models for the longitudinal vibrations of air columns in wind instruments in Chapter 12.

A different sort of generalization is achieved by imagining that the pendulums are constrained (say, by hanging in some kind of grooves or tracks) to vibrate in and out of the page when viewed as in all the preceding figures, and are not allowed to move right or left. Similar statements still apply: The

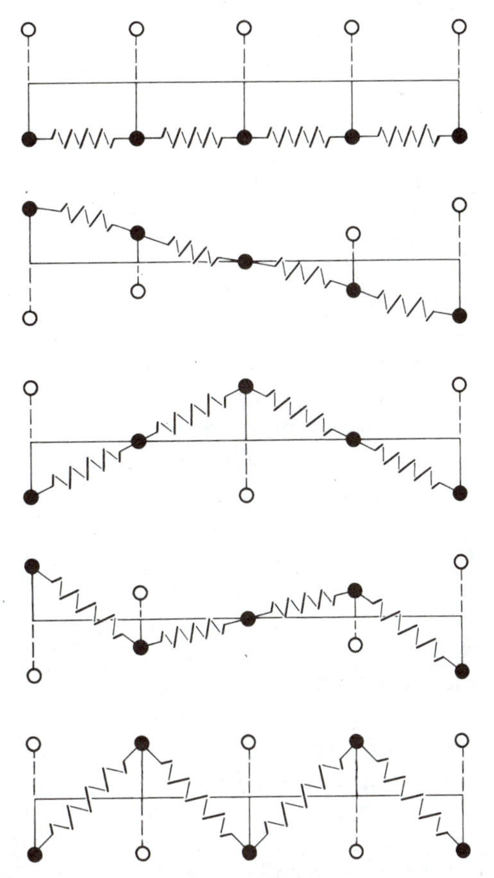

FIGURE 9.10 Looking down from above at five coupled pendulums allowed to move only transverse to the line along which they are mounted, showing the five natural modes in order from lowest to highest frequency. Open circles show configurations a half-cycle later.

general combined motion may seem complex, but it can be accounted for with an appropriate mixture of natural modes. These natural modes are shown for the example of five pendulums in Figure 9.10, where we have changed to a view from above to make the displacements visible. These modes will help us understand transverse vibrations of percussion instruments later in this chapter.

What happens now if we remove the constraints and allow the pendulums to swing in any direction? Each pendulum can do two independent things (right/left or in/out); motion in any other diagonal direction is merely a combination of these two. Then five pendulums (for instance) are capable altogether of 10 independent motions, and there prove to be precisely 10 natural modes (the five shown in Figure 9.10 together with those in Figure 9.9b). Any seemingly chaotic motion of these pendulums can, in principle, be described as a mixture of 10 simple periodic motions, each repeating at its own characteristic frequency.

Before we return from the abstract world of point masses and ideal springs, let us state the properties of these vibrating systems in the most general way.

1. Every system of point masses has some number of *degrees of freedom,* which we call N. This means the number of independent motions possible for the system; there are no more than N ways the system can move that are truly different from one another. (Our original pair of pendulums, for instance, had two degrees of freedom.) An equivalent statement is that it requires precisely N pieces of information to tell where every part of the system is located or N functions of time to completely describe the history of its vibration (for example, the two graphs of Figure 9.4 for a system with $N = 2$).

2. A system with N degrees of freedom is guaranteed to have precisely N *natural modes*—no more and no less. To repeat our definition, each of these modes constitutes *a pattern of motion in which every part of the system undergoes simple harmonic motion.* They do this in unison, with all parts reaching their greatest displacement from their respective equilibrium positions at the same time. But this motion generally has different amplitudes for different parts of the system; in particular, any point that remains at rest during a modal vibration is called a *node* of that mode. Each natural mode is described by telling or showing the directions of motion and the relative amplitudes of the individual masses.

3. Each natural mode has its own *characteristic frequency,* and usually all N frequencies are different. Each mode frequency is determined jointly by some effective average inertia of all the masses that participate in the motion of that mode and by some effective average springiness of all the restoring forces involved in that mode. This is a generalization of the formula $f = (1/2\pi)\sqrt{K/M}$ that we gave in Section 2.4 for a single mass M on a spring of stiffness K. Each mode also has its own characteristic decay rate at which the vibrations die away; we will discuss this further in Section 9.7.

4. Any vibration whatsoever in a system with N degrees of freedom, no matter how complex it may seem, can be accounted for entirely as a *superposition* of the N natural modes. There exists a precise mathematical procedure that can always determine just what the recipe is for how much of each mode is involved. Technically, this claim holds only for vibration amplitudes sufficiently small that nonlinear effects are unimportant; in most of our acoustical applications that condition is quite well satisfied.

The meaning of these general statements gradually will become much clearer as we illustrate them with numerous examples during the next several chapters.

---

*As you read other books, you will sometimes find different terminology. For one thing physicists often refer to natural modes as *normal modes.* The natural frequencies and the modes themselves often are referred to jointly (especially by musicians) by terms such as *partial*, *overtone*, or *harmonic*. *Partial* simply means one part of the total vibration, but I dislike using an adjective as a noun. *Overtone* refers to any mode other than the one of lowest frequency (the fundamental) and generates unnecessary confusion in numbering (the third overtone means the fourth mode, and so on). *Harmonic* really should be limited to those special cases in which all mode frequencies happen to be exact integral multiples of the fundamental (that is, a harmonic series, as described in Section 7.4). All the cases we study in this chapter have *inharmonic* mode frequencies, and thus do not satisfy that condition.

9.4 TUNING FORKS AND XYLOPHONE BARS

If you have watched a tuning fork with strobe lighting, you know that it can vibrate as shown in Figure 9.11. But does this correspond to the fundamental or the clang tone? Are there other vibrations that might deform it in different ways?

To see how we can apply all the rules for natural modes, consider the idealized model of a tuning fork prong in Figure 9.12a. This is a cousin to the simple pendulum; it is a minor detail that the restoring force is provided by the stiffness of a springy metal strip rather than by gravity or by a coil spring. This mass can move in only one way; the system has one degree of freedom and one natural mode. Because this mode occurs alone, the vibrations are easily recognizable as *SHM* and their frequency is simply $(1/2\pi)\sqrt{K/M}$, as in Section 2.4.

Now the mass of a real tuning fork prong is not all concentrated in one lump at the end, so consider the improved model of Figure 9.12b. This has two degrees of freedom and two natural modes. For a given amount of motion of the masses, it is clearly mode 2 that requires more bending of the metal strips; thus it has greater restoring forces and the higher of the two frequencies. We may speculate at this point that mode 1 is analogous to a tuning fork vibration that produces the fundamental tone, and mode 2 to a vibration that produces the clang tone.

Think now of a series of models in which three, four, and ever greater

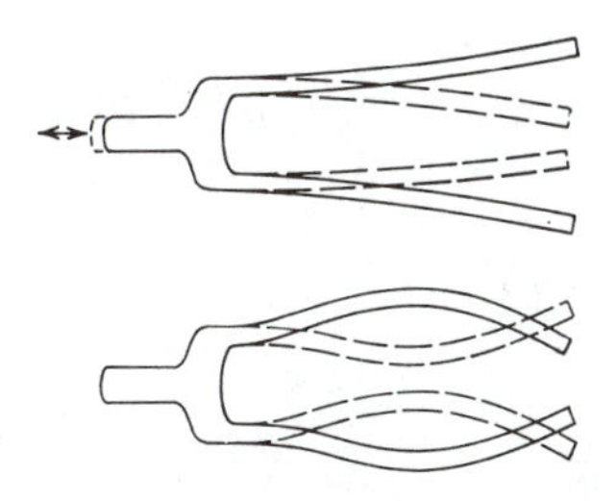

FIGURE 9.11 Motion of a tuning fork in its first two modes. The top configuration corresponds to the fundamental tone, and the bottom to the high-pitched clang tone. The slight motion of the handle indicated by the arrows is what transfers vibrational energy from the fork to any surface against which it is held, such as a table top.

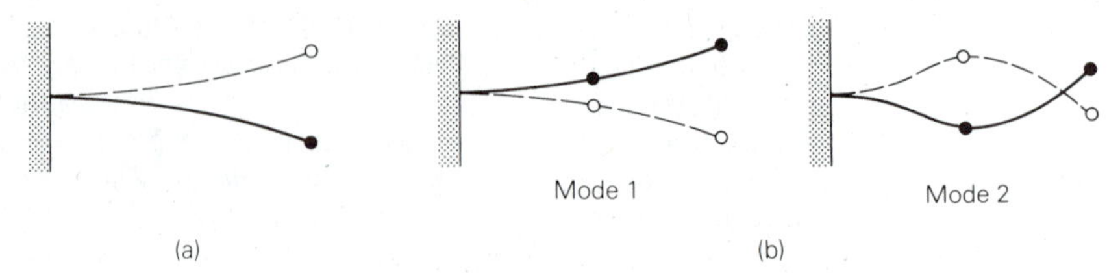

FIGURE 9.12 Idealized models for a tuning fork prong. Edge-on view of thin springy metal strips supporting small massive balls. The mass of the strips is considered negligible compared to the balls (a) A single ball has only one natural mode. (b) For two balls there are two natural modes. Increasing the number of balls without limit leads to the modes shown in Figure 9.13.

numbers of smaller masses are connected by more and more springy strips that are made shorter and shorter so that the total length of the system remains the same. These models become increasingly like the real tuning fork prong, which has the mass and stiffness distributed smoothly all along its length. Because the models have three, then four, and ever greater numbers of natural modes, it appears that the real tuning fork has an enormous number of natural modes. (Sometimes this is indicated by words such as "*N* is infinite.") Calculations with these models predict the mode shapes and frequencies shown in Figure 9.13 for a thin uniform bar firmly clamped in a rigid mounting at one end.

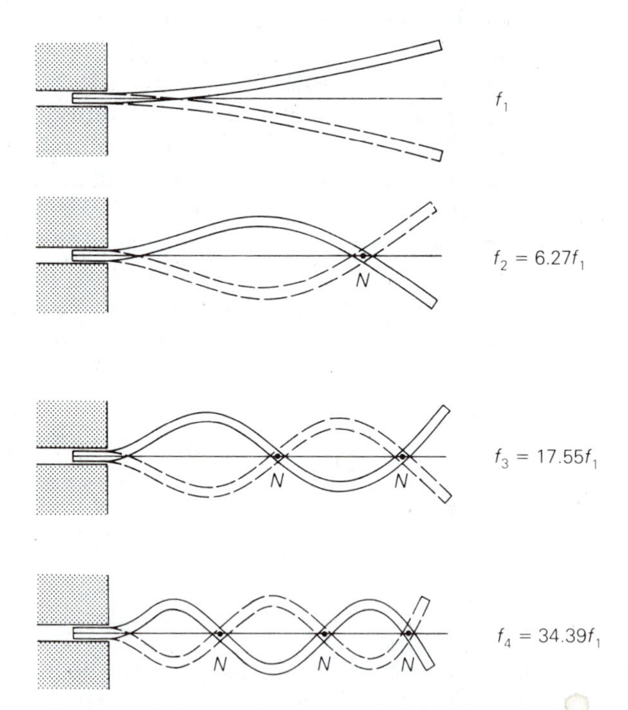

FIGURE 9.13 The first four natural modes of a continuous bar of uniform cross section, clamped at one end. Notice the nodal points *N*, where the displacement always remains zero.

Because a tuning fork prong merges gradually into the base, its frequencies are slightly different from those given in Figure 9.13. The fork used for Figure 9.2 has $f_2 \cong 6.1 f_1$. In fact, by proper design of the base, the prongs are (sometimes) deliberately made to have $f_2 = 6.00 f_1$, so that the clang tone forms a musical interval (a nineteenth; that is, two octaves plus a fifth) with the fundamental. The third and higher modes are of little musical interest, for two different reasons. First, they are at extremely high frequencies (the fifth or sixth mode already will be beyond the audible range); second, they die away very rapidly (as we will explain further in Section 9.7). The third and fourth modes help make the brief metallic ping when the fork is struck, but only the first and second give a long-sustained tone. You should be able now to go back to Figure 9.2 and explain all the major features shown there.

---

*You may suspect that there should be another low-frequency mode in which both prongs move (as in Figure 9.13) to the *same* side together and similarly close cousins to the second, third, and higher modes described above. There are such modes in principle, but their energy is lost almost immediately through motion of the handle when the fork is held in a soft hand. Only if the handle were clamped in a rigid and tight mounting would these modes be of any importance, so they are customarily not even listed or assigned mode numbers. The reason the usual modes do survive is that the counterbalancing opposite motions of the two prongs require practically no motion of the handle.

Also possible physically but not interesting musically are natural modes in which the prongs twist and others in which they vibrate longitudinally.

Now let us go on to other wooden and metal bars, such as those on xylophones and vibraharps. How do these differ from a tuning fork prong? First, they are free at both ends rather than being attached to a handle; second, they do not have constant thickness. We will look first at bars that do have uniform cross sections to see why they are not very satisfactory musically.

Just as for clamped bars, mathematical theory can predict exactly what the mode shapes and frequencies should be for the simple case of uniform bars free at both ends. The first few transverse modes are shown in Figure 9.14a.

---

*For a free thin bar of length L and rectangular cross section of thinkness a (the width does not matter), the fundamental frequency is $f_1 = (1.028\ a/L^2)\sqrt{Y/D}$. Here D is the mass density of the material and Y is a measure of its stiffness called Young's modulus; for example, aluminum has $D = 2700\ \text{kg/m}^3$ and $Y = 7 \times 10^{10}\ \text{N/m}^2$. The higher-mode frequencies are given by $f_n \cong 0.441\,(n + \frac{1}{2})^2 f_1$. For a bar clamped at one end, the corresponding formulas are $f_1 = (0.162\ a/L^2)\sqrt{Y/D}$ and $f_n \cong 2.81(n - \frac{1}{2})^2 f_1$. In both cases, note especially that (a) doubling the length cuts all frequencies to one-fourth, not one-half, of their former values; and (b) the frequency ratios for the higher modes are not small-integer ratios.

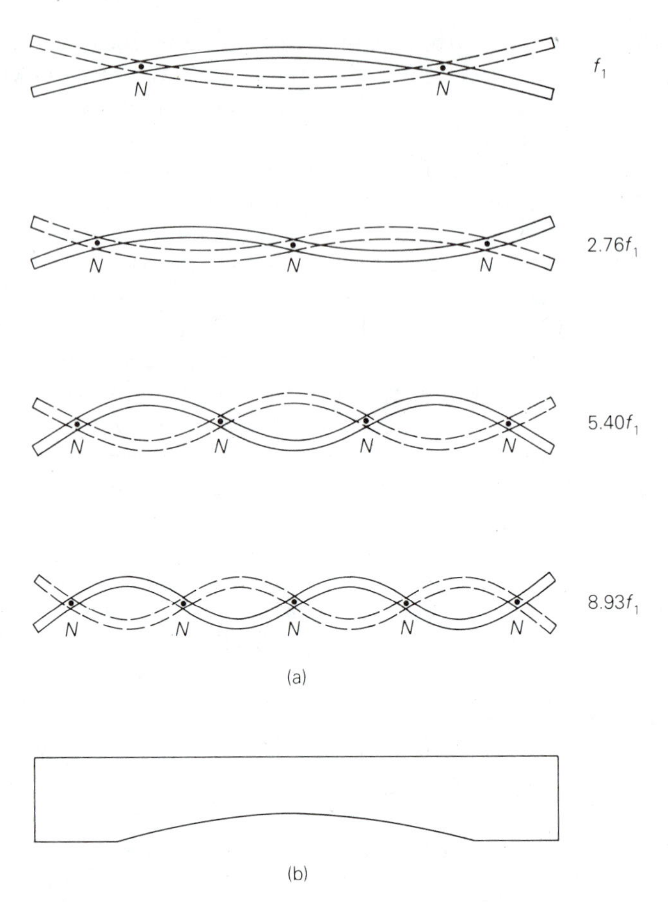

FIGURE 9.14 (a) First four natural modes of a uniform bar free at both ends. (b) Modification of the bar to produce musically helpful frequency ratios. Marimba bars may be thinned even more than shown here.

The rules for the natural modes tell us that when we hit such a bar it must be possible to describe the resulting vibration as a combination of these modes. So the sound we hear should be a combination of pure tones at these mode frequencies. Will that combination have a pleasing musical sound? Will we hear it as a note of definite pitch? No, because these frequencies do not belong to a harmonic series; the corresponding nonperiodic waveform does not give a clear impression of a single definite pitch. At best you might think you heard two distinct pitches approximately $17\frac{1}{2}$ semitones apart, corresponding to the ratio 2.76 to 1 for the first two mode frequencies.

If only we could make, say, $f_2 = 4f_1$, the two lowest modes would reinforce each other by producing tones exactly two octaves apart. We can obtain such special frequency ratios by using nonuniform bars. In particular, thinning a bar at the center reduces the restoring forces that act when it is bent. The mass also is reduced, but the reduction in stiffness turns out to be more important, and the frequency of the first mode is greatly lowered. But the higher modes are much less affected, because they still require bending the thick parts of the bar.

The right amount of thinning can lower f_1 enough to change the frequency ratio f_2/f_1 from 2.76 to 4.00, and this (theoretically) is how marimba bars are made. (In real life the manufacturers do not bother to achieve this condition on the dozen or more highest notes of the instrument, whose sounds die away quickly anyhow.) Xylophone bars have far less thinning, so that the ratio is raised only to 3.00; this accounts for their distinctly different tone quality, which involves a strong component of the musical interval of a fifth above the fundamental (remember the harmonic series).

As with the tuning fork, the third and higher modes do not last long enough to contribute to the pitch, so we need not worry much about their frequencies.

9.5 DRUMS, CYMBALS, AND BELLS

It is logical to proceed now with a program of showing the natural modes and their frequencies for each percussion instrument, in turn, and then accounting for all its acoustical properties in those terms. That program would, of course, result in a book in itself. But we can get a good idea of what it would contain by looking at one or two more representative cases that illustrate the important concepts.

Consider first a drumhead attached to an isolated hoop. It is the tension (the force per unit length with which the membrane is pulled outward in every direction by the hoop) that plays the role of restoring force and so helps determine the vibration frequencies. The mass of the membrane is spread out, and various parts can vibrate with different amplitudes. Figure 9.15 shows the sort of idealized model whose natural modes a physicist would study in order to arrive at a theory of the vibrations of the continuous membrane. The predictions of such a theory are shown in Figure 9.16. Because the vibrating surface is two-dimensional, we must adopt a different way of representing the mode shapes. During half of each cycle all the "+" regions are above the equilibrium plane and the "−" regions below; during the other half-cycle it is the other way around. In between there is a moment when the membrane is flat but adjoining regions are moving in opposite directions. There are now *nodal lines* where the membrane never moves up or down. These natural mode patterns can be confirmed experimentally, as shown in Figure 9.17.

---

*The fundamental frequency for the membrane alone (with no air-loading effects) is $f_1 = (0.765/d)\sqrt{T/\sigma}$, where d is the diameter, T the tension, and σ the mass per unit area. Typical values for a Mylar tympani head (thickness 0.2 mm) are $d \cong 0.6$ m, $\sigma = 0.26$ kg/m^2 and $T \cong 2 \times 10^3$ N/m. For large values of n, a rough approximation of f_n is provided by the expression $1.3\sqrt{n}\,f_1$. Notice that higher modes of a drumhead crowd closer and closer together in frequency, unlike solid-bar modes, which spread farther and farther apart.

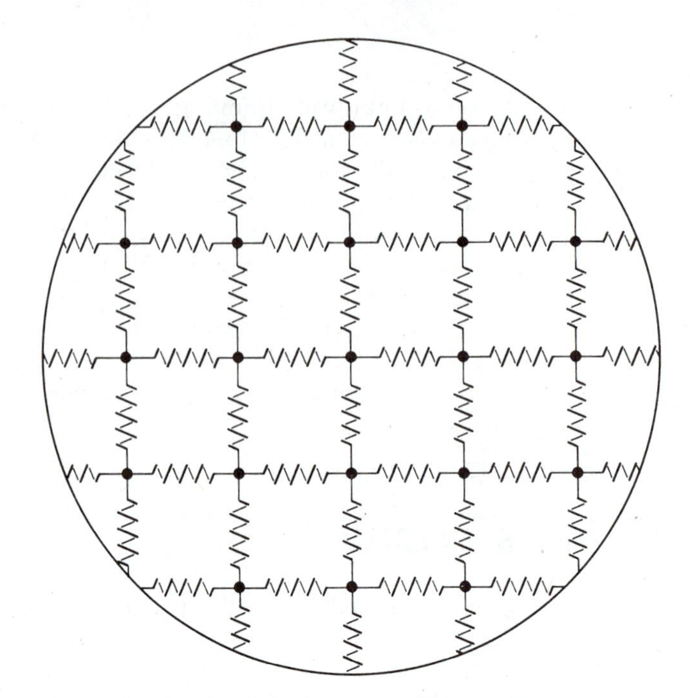

FIGURE 9.15 An array of ideal springs and point masses, which serves as a model for a stretched membrane. Motions in and out of the page are the only modes of interest; there are 21 such modes here. As the number of masses and springs is made very large, this model becomes a good approximation to the continuous membrane.

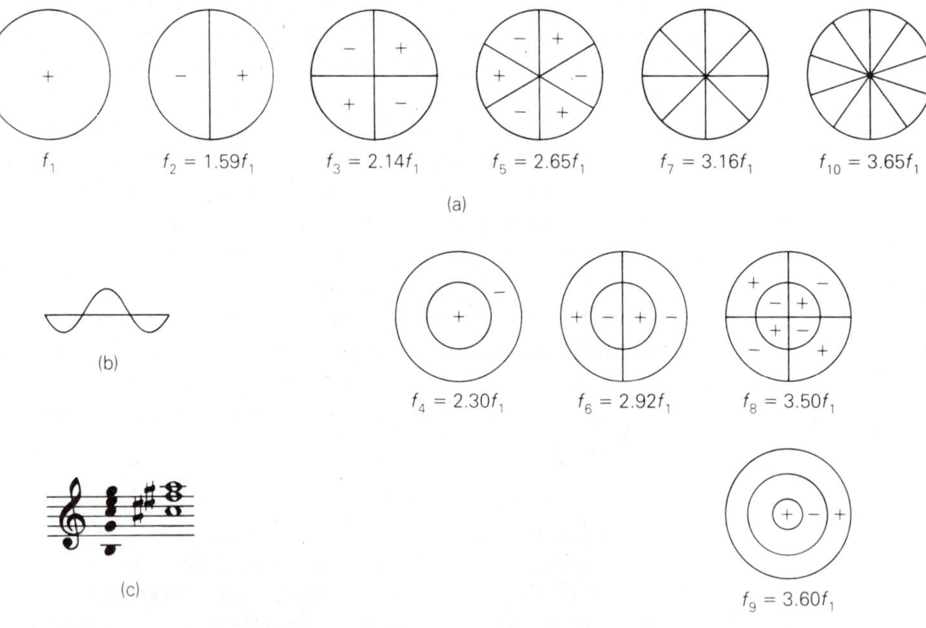

FIGURE 9.16 (a) The first 10 natural modes of an ideal stretched membrane, arranged in families according to how many circular nodal lines are involved. (b) Side view of the shape of mode 4 along its diameter. (c) Corresponding pitches when mode 1 is B_3. Open note heads do not correspond closely to the equal-tempered scale. Modes 8, 9, and 10 all are within a semitone of the A_5 shown. Higher modes contribute many additional pitches, continuing upward from these and crowded close together.

FIGURE 9.17 Photographs of a rubber membrane vibrating in several of its natural modes. Each pair represents the same mode at two times a half-cycle apart. (Courtesy of National Film Board of Canada and BFA Educational Media)

To understand why drumhead vibrations do not generally sound very musical, suppose we translate the frequency ratios (with the help of Figure E and the Harmonic Series Slider) into musical notation. The results (Figure 9.16c) indicate a clamorous mixture for which it is difficult to assign any definite pitch. Addition of a body to the drum creates a resonant cavity below the membrane; this alters both the mode frequencies and the efficiency with which they radiate so that the tone color changes. A judiciously shaped body even can make certain mode frequencies nearly harmonic so that a sound of definite pitch is produced. For further discussion of how this is done with tympani, see Box 9.1.

Think now about the vibrations of a flat, uniform disc of sheet metal. The restoring force is provided by the stiffness of the metal itself (as with the tuning fork) rather than externally applied tension. Although the exact frequencies differ, the natural modes of vibration are qualitatively similar for the disc and the stretched membrane. The greatest difference comes from allowing the edge of the disc to move freely (like the end of a xylophone bar) instead of being fastened down.

Now let the disc be permanently deformed a bit at a time—first to the shallow-dish shape of a cymbal or gong, then more deeply like a church bell, and finally stretched out into a long hollow tube like an orchestral chime. At each stage we can list natural modes and identify each one with its ancestor on the drumhead (Figure 9.18). It is impractical to predict all the mode

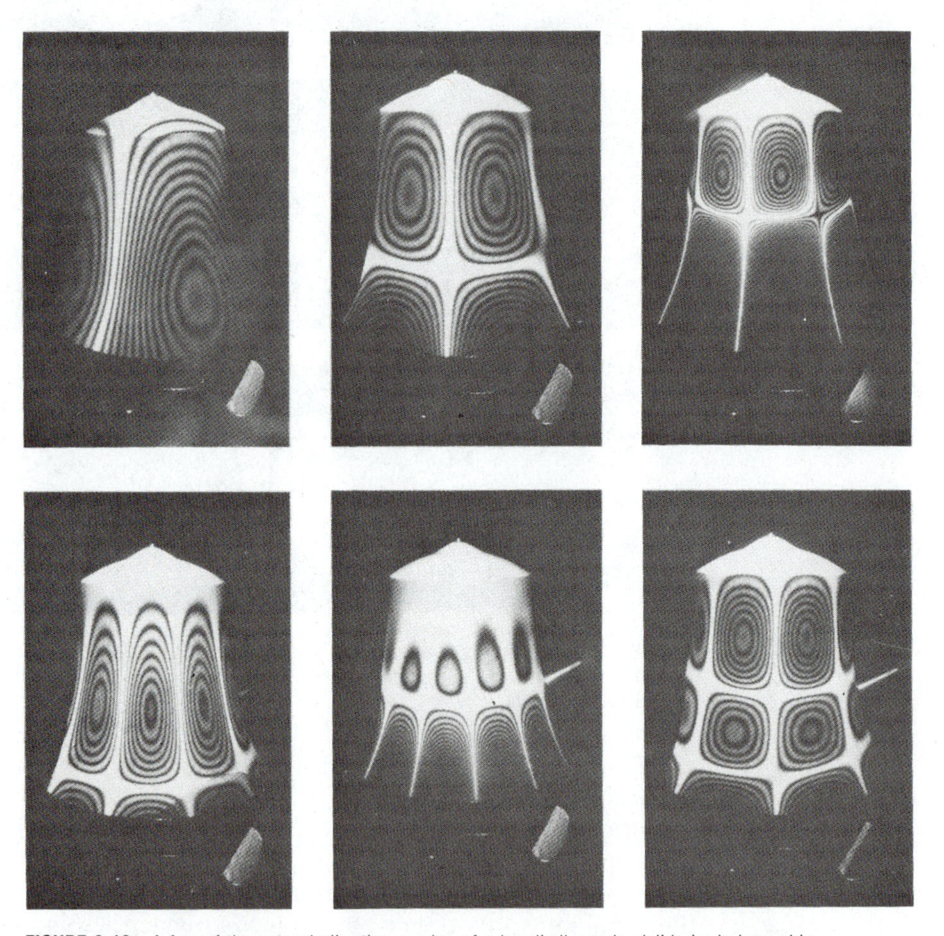

FIGURE 9.18 A few of the natural vibration modes of a handbell, made visible by holographic interferometry. The brightest parts show the nodal lines, and the centers of the bullseye patterns are the points of maximum amplitude. (Photos by George Tarbag, courtesy of T. D. Rossing)

BOX 9.1 TYMPANI

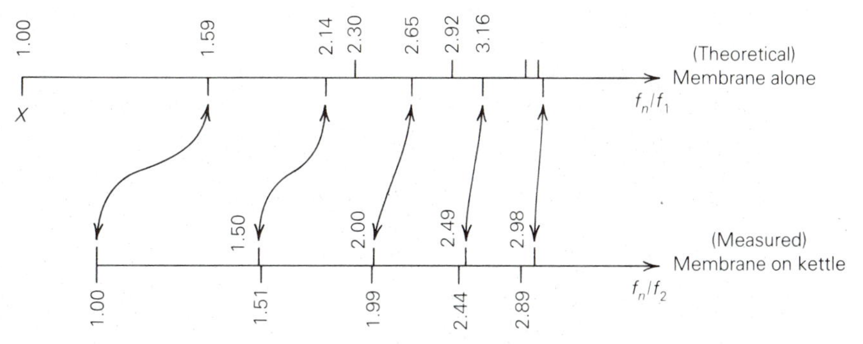

FIGURE 9.19 Drumhead frequencies plotted on the same pitch scale as Figure E. The top line shows frequencies (as in Figure 9.16) predicted by idealized theory for membrane alone. Markers above the line are for modes with circular nodal lines. Markers below the line include those modes important in establishing pitch for tympani. *X* designates fundamental mode killed by radiation loss. Arrows indicate lowering of frequency by air mass. The lower line shows frequencies as multiples of f_2. Markers above the line show exceptionally good fit to desired ratios, measured by Benade for one particular kettledrum tuned to one particular note. Markers below the line show measurements of a different drum (by Craig Anderson, reported by Rossing) for which harmonic relations are achieved only very approximately; these probably are more typical of average tympani.

Most drums do not produce a tone of definite pitch, and we blame that on the lack of clear musical relationships among the many natural mode frequencies shown in Figure 9.16. How is it, then, that in the special case of tympani we can obtain definite pitch? It turns out that this is the combined result of several effects.

First, the fundamental mode pushes all the adjacent air in the same direction, unlike all the others that produce mainly a sloshing of nearby air back and forth between the "+" and "−" regions of the membrane. When the drumhead is mounted over a closed bowl, this results in mode 1 radiating its sound to the outside world far more efficiently than any of the others. Because of this strong radiation, mode 1 loses its energy rapidly in the strong initial thump, and thus is practically missing from the later persistent sound. That is good, because its frequency does not fit well into the desired pattern.

Second, the usual striking point, half to three-fourths of the way out from the center, is near the circular nodal lines of many modes (such as modes 4, 6, 8, 9 in Figure 9.16). Thus the vibration recipe contains only small traces of these modes; they contribute a little pungency to the flavor, but the burden of producing a tone of definite pitch is left almost entirely to those modes that do not have circular nodal lines (numbers 2, 3, 5, 7, 10, and so on).

Third, the frequency calculations must be corrected because those given in Figure 9.16 apply only to a membrane in vacuum. In real life, motion of the membrane requires motion of the adjoining air as well. Thus the air contributes additional mass, just as if the membrane were somewhat heavier, and this lowers the frequencies. The air may reduce the frequency of mode 2 by 25 percent or more; higher modes are less affected, because they do not efficiently move the air as far above and below the membrane. These adjustments depend somewhat on the exact shape of the bowl, so that choosing the right shape can bring the frequencies into desirable patterns.

Because the fundamental dies out quickly, mode 2 usurps its role as the lowest-frequency mode. Suppose we express the other frequencies as multiples of f_2 instead of f_1, taking into account the corrections for air mass, and plot them (Figure 9.19) on the same musical scale as Figure E. You now can verify with the harmonic series slider that the adjusted frequencies of the important modes form important musical intervals with one another; more specifically, they approximately match numbers 2 through 6 of the harmonic series. This is why they reinforce one another in producing an impression of definite pitch.

frequencies mathematically, but we can measure them on any real cymbal or bell.

As in the simpler case of xylophone bars, judicious variations in thickness can adjust the frequency ratios so that at least some of the modes cooperate in producing a definite musical pitch. We will not pursue the details further here but refer the interested reader to Rossing's article. (See also Section 17.3 for a discussion of how this mode cooperation produces a clear pitch sensation.)

9.6 STRIKING POINTS AND VIBRATION RECIPES

In the last two sections we have described the natural modes of bars and membranes and claimed that these account for every possible vibration of these objects. But out of all the possible mixtures of various amounts of each natural mode, which *vibration recipe* actually occurs when you strike the bar or the drum? That mixture must somehow be determined by how or where you strike. To see why, let us consider the simplest possibility.

How could you make a drumhead vibrate in a single, pure, natural mode with no admixture of any others? There are only two ways to do that. One is to push steadily on all parts of the membrane to deform it into the shape characteristic of that mode and then quickly release it; the other is to strike all parts simultaneously with just the right strength to produce the motions characteristic of that mode when it passes through the equilibrium position. These cases require applying different amounts of force (steadily in the first, impulsively in the second) to different parts of the drumhead; that is, strongest in the center of each region, gradually decreasing toward the boundaries, and exactly zero on the nodal lines. And with the exception of the first mode, you must simultaneously apply upward forces in some regions and downward in others. That is extremely difficult to do, so for practical purposes we never even attempt to produce one mode all by itself. Nor would it have any real use even if we could, for it would only sound like a tuning fork instead of like a drum.

So when we strike a drum in the ordinary way, what is the vibration recipe that tells how much of each mode is mixed into the total? The last example suggests that the answer is something like this: If the force pattern is very much like one of the mode patterns, then that mode is efficiently set in motion; but there is little motion of a mode whose pattern is dissimilar to the force distribution. That general rule, like the others, has a precise mathematical form, but this qualitative version serves our purposes well:

Each mode is excited more or less, according to how closely the pattern of the disturbing force matches the pattern of the mode itself.

The special case where the disturbing force is localized to a very small region allows this restatement, which may be easier to understand:

Striking an object at any given point excites each natural mode in proportion to how much that mode involves motion of that particular point.

For instance, the first mode of a drumhead is most efficiently excited by striking it in the center, and as you strike closer and closer to the edge you will get vibrations involving smaller and smaller proportions of that mode.

It is worth spelling out a more specific rule that is hidden in the last one above:

If the striking point is on a nodal point or line of some particular mode, then that mode is completely left out of the recipe.

If you hit a drumhead in the center, you get only a dull thump. This is because the center is on the nodal lines of the great majority of natural modes (Figure 9.16, modes 2, 3, 5, 6, 7, 8, 10, etc.); only those few circularly symmetric modes (numbers 1, 4, 9, etc.) are excited, and the sound lacks richness and brightness because so many of the drum's natural frequencies are missing from the recipe. To make the piece-of-pie modes strong, you should hit approximately half to three-fourths of the way out to the edge. (For further application of this rule, see Box 9.1.)

Here is another important aspect of the basic rule:

Striking an entire region at once instead of just a point produces the same recipe as you get by adding together the recipes for striking at each of the points contained in that region.

If a large, soft object hits a drumhead entirely within one "+" or "−" region of some mode, all of the points hit convey the same message—that mode is being strongly excited (Figure 9.20a,b). But if the area struck includes both "+" and "−" regions, the messages are contradictory; the effects of different points trying to excite the mode a half-cycle out of step with one another cancel out, and that mode has very little share in the vibration recipe (Figure 9.20c).

Even more important in many cases may be the analogous effect in the time domain: If an applied force persists for a half-cycle or more of any mode,

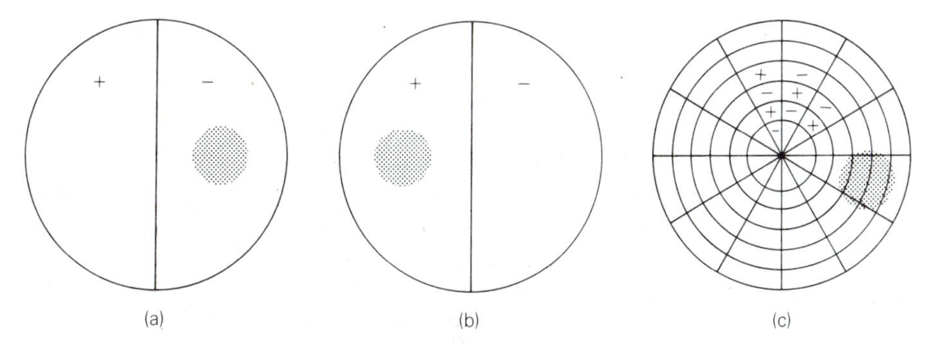

FIGURE 9.20 (a) A mallet impact well within one region of a natural mode excites that mode efficiently. (b) This is true regardless of whether the region is labeled "+" or "−"; the only difference is in shifting the vibration a half-cycle forward in time. (c) An impact covering adjoining regions will produce practically no motion of the mode, as shown here for mode 88 ($f_{88} = 11.1 f_1$) of an ideal drumhead. The vibration recipe for an impact region as large as this will include practically nothing for frequencies more than three octaves above f_1.

it will not continue to feed more energy into that mode. Rather, once the membrane begins moving up again, it will give energy back rather than accept more. Applying this argument to all modes in turn, we can say:

If a striking force has finite duration T in time, then only modes whose frequencies are less than approximately 2/T are efficiently excited.

Now we can explain what happens when a drum or xylophone player changes from hard to soft mallets. The former exert a large pressure on a tiny area and excite many modes (all those involving overall motion of that area), making a very bright sound. Softer mallets exert smaller pressure over a larger area of contact and so excite only the lower modes. Any modes whose nodal lines are closer together than the diameter of the contact region cannot be strongly excited, and this effectively imposes a cutoff frequency above which practically nothing happens. The softer the mallet and the larger the contact area, the lower that cutoff frequency will be and the duller the tone color. To whatever extent the softer mallet is slower in bouncing back, the discrimination against higher frequency modes will be all the greater because of the longer contact time.

9.7 DAMPED VIBRATIONS

It has been useful to concentrate on the initiation of vibrations, just as if, once started, they would continue forever. But in real life they die away. As we pointed out in Chapter 3, this results from loss of energy both through radiation of sound into the air and through friction. These are called **damping** processes, and the vibrations of gradually decreasing amplitude are called *damped vibrations*. How can we describe damping more precisely, and what effect does it have on the perceived tone?

The simplest type of damping, called **exponential decay**, occurs when the rate of continuing loss of energy is always in direct proportion to the remaining amount of energy. To see what exponential damping means, suppose after a length of time $t_{1/2}$ (the halving time) the amplitude of some vibration mode decreases to half its original value, and thus the vibrational energy to a fourth of its original value. Then the amplitude of the sound waves being produced also is halved at $t_{1/2}$, and the rate at which they drain energy away is cut to one-fourth the original rate; this also holds true for some kinds of frictional damping. If you wait a further time $t_{1/2}$, the vibration does not vanish (Figure 9.21); its amplitude just becomes half of what it was at the beginning of that time, which means one-fourth of the original. Additional waits of $t_{1/2}$ reduce the relative amplitude to one-eighth, one-sixteenth, and so on, but never quite to zero. This description is often true to a good approximation in acoustics. (Decay of radioactive substances follows a similar rule.)

For musical purposes, it is good to define the *damping time* τ (tau) of any mode as the length of time necessary for the vibration amplitude to fall to $\frac{1}{1000}$

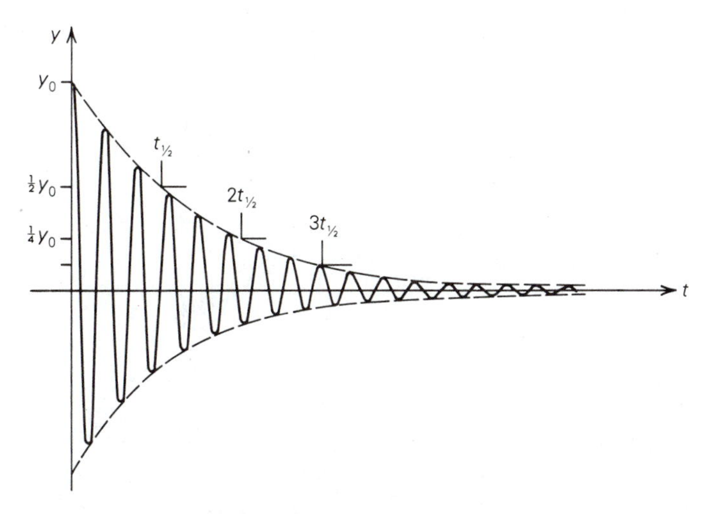

FIGURE 9.21 An exponentially damped vibration. The dashed lines (often called an *envelope*) represent the gradually changing amplitude. Each additional time $t_{1/2}$ reduces the amplitude to half its previous value. The curved envelope on this graph corresponds to a straight line in Figure 9.22.

of its original value; it is related to the halving time by $\tau \cong 10t_{1/2}$. The sound being emitted after a time τ then falls to $\frac{1}{1{,}000{,}000}$ of its original intensity; that is, to a sound level 60 dB lower. That corresponds approximately to your psychological impression of fading into inaudibility.

In general, damping times are different for different modes. Any mode for which the sound radiation is more efficient has its damping time reduced, and any mode for which frictional energy losses are more severe has shorter damping time. The higher-frequency modes of bars and membranes generally have shorter damping times, because for a given amplitude of displacement the higher modes involve more severe (and more frequent) bending and thus heavier frictional losses in the repeated flexing of the material.

Frictional losses sometimes come from contact between the vibrating object and something soft and yielding, such as a finger or a piece of felt. There is a useful rule governing this, which runs closely parallel to our rules for mode excitation.

> ***Localized frictional damping will affect each mode in proportion to how much motion that mode causes at the point of application of the damping; in particular, it leaves undisturbed any mode that has a node at the point of application.***

Lightly touching a drumhead at its center, for instance, will quickly damp out modes 1, 4, and 9 while leaving most of the others vibrating normally.

Because of the variation in damping time from one mode to another, the effect of damping on the perceived tone is much more interesting than a simple dying away. The tone color also changes during the decay. Suppose, for example, some object has a recipe of mode strengths when first struck,

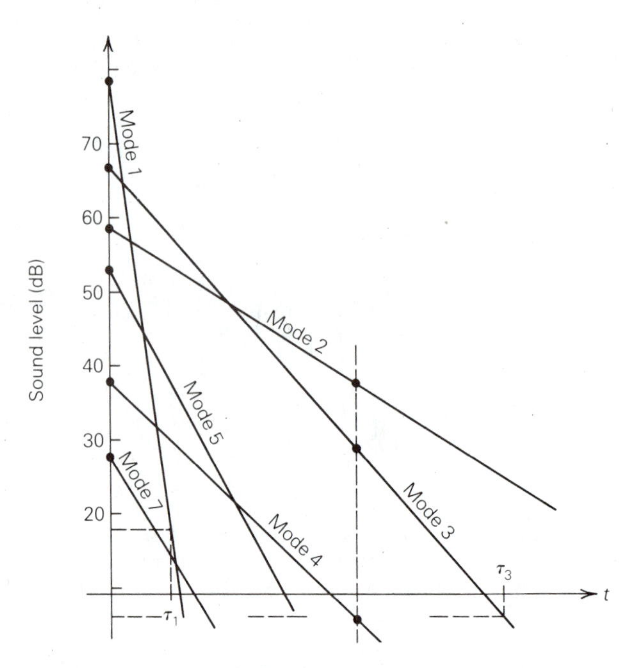

FIGURE 9.22 A hypothetical example of decay of different natural modes at different rates. Each mode begins with a contribution to the sound intensity indicated by a dot on the vertical axis. Exponential decay produces equal reduction in sound level for equal amounts of time, indicated by straight lines on this type of graph. At time zero many modes are present, including some of high frequency. At a certain later time (vertical dashed line) only a few of the lower modes (three dots) retain any appreciable strength, so the timbre is much simpler. The damping time of each mode is the time required for the line to drop 60 dB. Damping times are indicated for mode 1 (very short) and mode 3 (longer).

as shown at the left edge of Figure 9.22. Because there are large amounts of higher modes, the tone is bright and complex. But after awhile, practically nothing is left at an audible level except a few of the lowest modes, so the tone color becomes simpler and more subdued. As you look at the graph, think of the initial short "clack" from a xylophone bar or "thunk" from a kettledrum, each followed by a more slowly decaying sound carrying definite pitch and involving only a few modes of vibration.

SUMMARY

The vibrations of percussion instruments are complex and nonperiodic, but they can be neatly explained with the concept of natural modes. Any solid body has many such modes, but the first few of them account for most of the prominent acoustic characteristics.

A natural mode is a periodic vibration in which every part of the body is in simple harmonic motion; all parts remain in step, reaching maximum displacement (either positive or negative) together. Each natural mode has its own characteristic frequency, which is determined by the inertia and restoring forces involved in its motion. Any seemingly complicated motion of the body may be explained as a combined motion of its natural modes.

We have shown the natural modes of tuning forks, xylophone bars, and drumheads. These examples illustrate how the natural mode idea can be used for many other percussion instruments.

The initial forces that set an object in motion (usually by striking) are especially important, for they determine the recipe telling how much of each natural mode is present in the total vibration. We have stated several definite rules for how this recipe is affected by our choice of striking point. This recipe can be expected to change as time goes on, because different modes generally die away at different rates. Other things being equal, a high-frequency mode usually is damped out more quickly than one of low frequency. In cases of exponential decay, the decibel level from any one mode decreases at a uniform rate, and the time required for a 60-dB drop is called the damping time for that mode.

REFERENCES

Extensive research on many percussion instruments has been done by Thomas Rossing and his collaborators at Northern Illinois University. A general survey on the acoustics of percussion is given in the two-part article by Rossing in *The Physics Teacher, 14,* 546 (1976), and *15,* 278 (1977). Rossing also is the author of specific articles on tympani in *Scientific American* (November 1982), p. 172; on bells in *American Scientist, 72,* 440 (1984); and (with R. W. Peterson) on gongs and cymbals in *Percussion Notes, 19*(3), 31 (1982).

An interesting series of articles on carillon bells can be found in *Music Perception, 4,* 241–280 (1987). It includes both scientific redesign of the bells to give them new acoustic properties, and evaluation of how the new sound was received by listeners.

Benade develops the normal mode concept at greater length and with more detailed application to percussion in Chapters 4, 6, and 9. Derivations of the mathematical formulas for bar and membrane modes can be found in Morse and Ingard, Chapter 5.

SYMBOLS, TERMS, AND RELATIONS

SHM	simple harmonic motion	natural mode
N	node	nodal line
N	total degrees of freedom	vibration recipe
n	mode index label	exponential decay
f_n	frequency of nth mode	
$t_{1/2}$	amplitude-halving time	$\tau \cong 10t_{1/2}$
τ	damping time (60 dB)	

EXERCISES

1. If the same spring were attached higher on the pendulum shafts in Figure 9.3, how would that change the natural mode frequencies f_1 and f_2? If the spring were kept in the original position and made stiffer instead, how would that affect the mode frequencies?

2. What would happen to the modes and their frequencies if the spring in Figure 9.3 were very weak? Explain in physical terms why f_2 would be only slightly greater than f_1. Describe the mode behavior in words and with a sketch analogous to Figure 9.6. Use the graph to argue that vibration energy will seem to shift slowly from one pendulum to the other and back again.

3. Sketch the natural modes for in/out pendulum motion, as in Figure 9.10, for the case $N = 4$; assign mode numbers in order from lowest to highest frequency.

*4. Sketch the natural modes for right/left pendulum motion, as in Figure 9.9, for the case $N = 4$; assign mode numbers in order from lowest to highest frequency.

5. Two (equal) masses are connected with springs (Figure 9.23) and are free to move in three dimensions. How many degrees of freedom does this system have? Sketch and describe its natural modes. (Hint: Simply take one dimension at a time.) Are there any pairs of modes in which both have the same frequency? (These are called *degenerate modes*.)

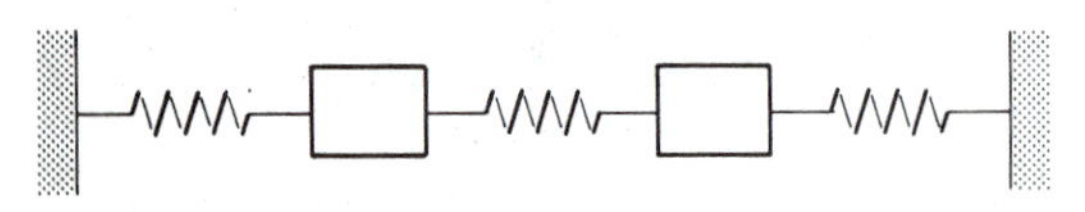

FIGURE 9.23

6. Explain in physical terms *why* mode 5 in Figure 9.10 has higher frequency than mode 2.

7. Consider the transverse motions of the two masses in Figure 9.23 as a crude model of guitar string vibration. (a) Sketch displacements of +1 unit for both masses to represent one natural mode, and displacements −1 and +1 (left and right masses, respectively) for the other. (b) Sketch the system for displacements +2 and +4, representing a string being plucked nearer one end than the other. Show that three units of one natural mode mixed with one unit of the other can make this shape and therefore would be the recipe for the plucked vibration.

8. Make a series of sketches like Figure 9.7 for each of five or six evenly spaced times in Figure 9.6 to show further what it means.

*9. Compare the xylophone bar modes of Figure 9.14 with the transverse pendulum modes of Figure 9.10. Explain why the analogs of the first two pendulum modes were omitted from the discussion of bars. (Hint: Consider what kind of restoring forces would act in such a motion of a bar and what the resulting frequency would be.)

10. Use Figure E and Figure 9.14 to find the nearest note names for the first four modes of a free uniform bar whose first mode is at A_2 (110 Hz), and show these on the staff as in Figure 9.16c.

11. Account for the various features of Figure 9.1d with ideas drawn from Section 9.4. Especially explain how often the subsidiary peaks occur toward the right side of the trace. (Hint: Is this a xylophone or a marimba?) Measure the dura-

tion of the nearly periodic waveform toward the right, and infer which note in the scale was played.

12. Explain Figure 9.2 in detail. Tell why parts (a), (b), and (c) must involve different magnifications of the vertical scale.

13. Guess and sketch the first few natural modes of the orchestral triangle, a metal rod bent into the shape of a triangle open at one corner. Describe how to get different timbres by striking the triangle at different points.

*14. Measure the length and thickness of an aluminum tuning fork prong and check how closely the formula given predicts its fundamental frequency f_1.

15. If you strike a xylophone bar at its center, how much mode 2 will be in the recipe? Where would you strike it to get a strong component of mode 2? Where would you strike to banish mode 1 from the recipe and thus have a better chance to hear and recognize mode 2 separately? Where could you touch the bar with your finger after striking to damp out mode 1 but leave mode 2 undisturbed? If possible, use these techniques to verify that you can actually hear a tone a twelfth above the fundamental on the xylophone but a fifteenth above on the marimba.

16. A good xylophonist knows how to obtain varying timbre by striking either at the center of the keys or at various distances out from the center. Explain these variations.

17. Consider bars supported as shown in Figure 9.24 by soft material (a) at the extreme ends and (b) at the middle. In each case, which modes involve bar motion at the supports so that they are strongly damped? Which modes have nodes at the supports so that they are not damped? Where would you place the supports to let mode 1 ring freely? Does this explain why xylophone bars are mounted as they are? Is it possible to mount the bars so that modes 1, 2, 3,. . ., all can ring freely at the same time?

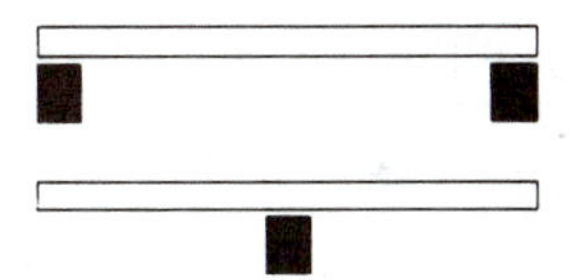

FIGURE 9.24

18. Which four of the drumhead modes of Figure 9.16 are shown in Figure 9.17?

19. Sketch what you expect to be the first few natural modes of a cymbal, and guess qualitatively as to their relative frequencies. How would the entire set of frequencies be shifted if the diameter of the cymbal were increased? Or if it were made of thicker metal?

20. Although it would actually be quite difficult, suppose you somehow managed to strike a drumhead simultaneously at two symmetrically located points, as shown in Figure 9.25. Which modes will be strong and which absent in the vibration recipe?

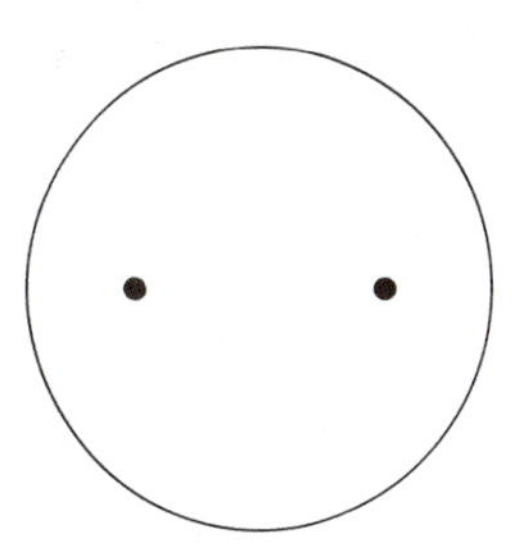

FIGURE 9.25

21. Start with 1 and multiply repeatedly by $\frac{1}{2}$. Use the result of your tenth multiplication to verify that the relation $\tau \cong 10\, t_{1/2}$ is valid within 3 percent.

22. If the amplitude of some mode dies to half its original value in 0.2 s, what is its 60-dB damping time?

23. Describe how you could use your finger to damp out mode 1 of a drum with minimum disturbance to the other modes. How could you use this technique to verify our claim that mode 1 is unimportant for tympani? If possible, find some tympani and try it; see what happens when you touch various points during the sustained vibration.

24. Describe the vibration recipe immediately after striking a bar (a) precisely at its center, (b) nearly at the center, and (c) well off-center. How will these spectra differ after one or two seconds have elapsed?

25. Suppose a drumstick remains in contact with a wood block for 2 ms. At roughly what frequency is the dividing line between efficient and inefficient mode excitation?

26. Can you extend the reasoning in Section 9.4 to suggest what shape you could make a bar to give the first two modes a 2.00 to 1 frequency ratio?

PROJECTS

1. Sketch your prediction of what a marimba waveform will look like if recorded in the manner of Figure 9.1d. Use a storage oscilloscope to get a picture and check out your prediction.

2. Study the bars on a vibraharp and a glockenspiel. Are they undercut? If so, how much? Can you guess and then verify with the techniques of Exercise 15 how much higher mode 2 is in pitch than mode 1? Explain as many similarities and differences as you can in comparison with xylophone or marimba bars.

3. Use your ears and wristwatch to obtain crude estimates of damping time τ for a variety of percussion instruments.

CHAPTER

10 Piano and Guitar Strings

The piano, with its struck strings, is closely allied with the percussion instruments. Yet it sounds so different from drums and cymbals, perhaps even from xylophones, that it seems inadequate or even misleading to classify them all together. Piano sounds have clear, sustained pitch and provide all the necessary melodic, harmonic, and rhythmic elements to make a complete piece of music.

But what underlying acoustical characteristics of the piano set it apart in this way? Our first job in this chapter is to show that long, thin strings have a special property: Their natural mode frequencies form a harmonic series. This is what makes their tone more musical than that of other percussion instruments.

As in the last chapter, we will go on to consider how the recipe of these vibrations is determined by the place and manner in which we first excite the string. We will describe the spectra produced by the plucking process on the guitar and contrast the results with the striking process in pianos. Finally, we will consider how the choice of string size, tuning, and multiple stringing contribute further details to what we recognize as piano tone.

10.1 NATURAL MODES OF A THIN STRING

Several popular instruments involve the sound from a string vibrating freely after it is plucked or struck. These include harps, guitars, mandolins, harpsichords, and pianos. We would like to understand those string vibrations as completely as possible. Judging by what we learned in the last chapter, a good way to do this is to find the natural modes of a string and then consider the recipes for various mixtures of those modes.

The vibrating strings of all these instruments are stretched tightly between two (nearly) rigid supports, as pictured in Figure 3.4 (page 43). One, called the *bridge*, is mounted on a soundboard; the other may be either the *nut* (another bridge mounted on the heavy frame of a piano or at the end of a guitar neck) or one of the frets on a guitar fingerboard. Although the last bit of string at the end cannot move up or down, it can tilt because it is merely supported and not clamped.

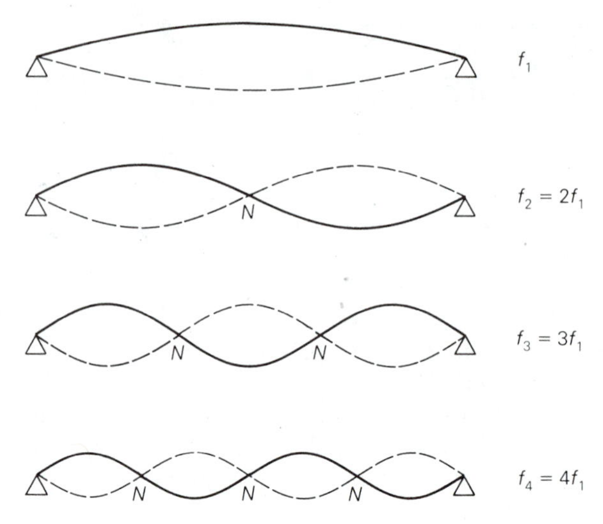

FIGURE 10.1 The first four natural modes of an ideal string. The solid and dashed lines show maximum displacements at two times a half-cycle apart. There is never any displacement at the nodes *N* for one mode vibrating alone.

You can imagine how we might start again with a few point masses connected by springs all in a row. We could find their natural modes for transverse vibrations (similar to those of Figure 9.10, page 157) and repeat this process for longer and longer chains. Then as the number of masses becomes extremely large, this serves as a model for the vibrations of a continuous string. But by now we should be able to omit the details. You probably are not surprised at all when we claim that the natural modes of a string are shown in Figure 10.1.

We should state clearly that these are the modes for a highly idealized string. It is not only perfectly uniform but also perfectly flexible; it presents no resistance whatsoever to bending. That is equivalent to saying it has zero stiffness, so that the restoring forces must come entirely from the externally applied tension. Later in the chapter we will point out how the stiffness of real strings modifies some of our claims about their modes.

These natural mode vibrations are sometimes called **standing waves**. They have a wavelike shape, yet they do not seem to be going anywhere. You may read more about the important relation between standing and traveling waves in Box 10.1.

The most remarkable thing about these modes is that their frequencies form a harmonic series. Thus *all* these modes have intimate musical relationships and cooperate in establishing a sense of pitch. The pitch corresponds to the fundamental frequency; that is, you judge it to be the same note as if you heard the pure sine wave from the first mode alone. A word of caution: Even though we recognize this is happening, we must not pretend our explanation is complete until we attempt (in Chapter 17) to understand why our ears like the harmonic series so well.

BOX 10.1 STANDING AND TRAVELING WAVES

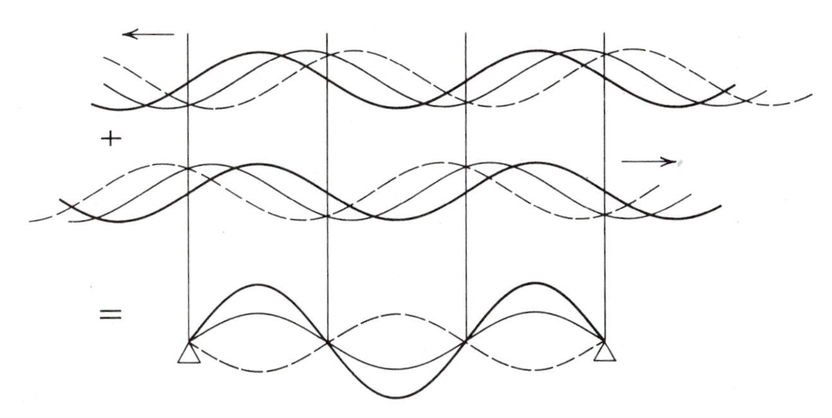

FIGURE 10.2 Multiple-exposure snapshots of string shape for two sinusoidal traveling waves and the standing wave that results when both are present together. Dashed, light solid, and dark solid lines indicate successively later times. The graph for the standing wave at each instant is the sum of the two corresponding graphs for the traveling waves, with the addition being performed the same way as in Figure 9.6 (page 154).

The standing waves introduced in this chapter are not as different as they may seem from the traveling waves described in Chapters 1 and 4. They are just a different way of looking at the same thing. It is really a matter of convenience, taste, or intuitive appeal when we choose to describe any general wave in terms of standing or traveling waves.

First, let us illustrate how *a standing wave can be described as the net result of two traveling waves.* Figure 10.2 shows two sine waves whose wavelength happens to be two-thirds the length of a certain string, one traveling to the left and the other to the right. Adding these two waves together produces the standing wave of the third natural mode of the string. It is a useful mathematical fiction to imagine the traveling waves moving freely along an infinite string; each is cleverly constructed so that as it enters the region occupied by the real string it produces exactly that motion which in reality is the reflection of the other wave from that support. Notice how at each node of the standing wave (especially at the ends of the string), any positive displacement of one traveling wave always is canceled precisely by negative displacement of the other. Thus, that part of the combined traveling waves lying between the bridges is identical in every respect to the standing wave. We will show a generalization of this approach for nonsinusoidal shapes (mixtures of standing waves) in Figures 10.4 and 10.8.

Second, *any traveling wave,* together with the reflections it generates at the ends of the string, *can be expressed instead as a combination of standing waves;* that is, as a natural mode recipe. This is illustrated in Figure 10.3. The truth of the claim is much less obvious, but its justification lies in the general theorems about natural modes, which we described in Chapter 9.

These statements apply not only to the easily visualized waves on a string but also to all other waves. In particular, we will use the standing/traveling wave equivalence for sound waves in Chapters 12, 13, and 15.

(continued)

BOX 10.1 *(continued)*

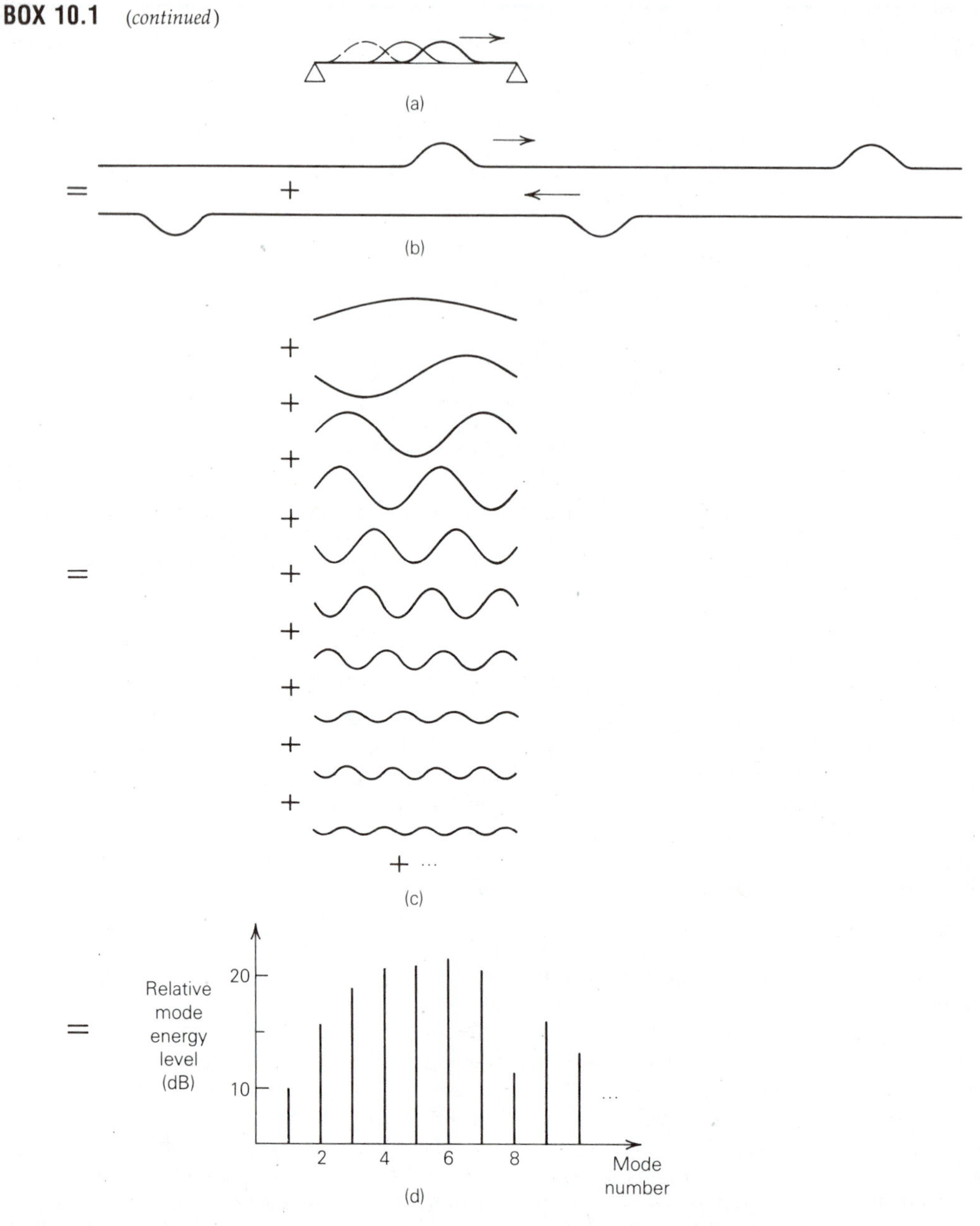

FIGURE 10.3 (a) Successive images in time of a traveling wave pulse. (b) The two continuous traveling wave trains that together can account for the original pulse and all its reflections. (c) Appropriate amounts of every possible standing wave on the original string, so that when all are added together the pulse and its reflections are reproduced. (d) A more convenient representation for the recipe telling how much of each standing wave mode must be included, given in terms of energy instead of amplitude. The zero point of the vertical scale is arbitrary; what is significant is the differences in heights of the bars, signifying the relative amounts of energy in each mode. In this case, modes 4 through 7 all have nearly equal energy, while modes 1 and 8 have approximately only a tenth as much. Caution: In part (c) the graphical addition is more complicated than in Figures 10.2 or 9.6, because different modes do not all reach their maximum displacements at the same times (they are not all in phase).

It also is remarkable that the shapes of these modes are perfect sine functions, with the nodes evenly spaced between the ends and every peak of the same height. (That was not true for the stiff bar modes we saw in the last chapter). This makes it easy to remember these modes: the nth natural mode simply fits n "loops" (vibrating sections between adjacent nodes) into the available string length L. Each loop is just half a wavelength, so we have

$$n(\tfrac{1}{2}\lambda_n) = L.$$

(We write λ_n as a reminder that different modes have different wavelengths). Because the wavelength is related to the frequency by $\lambda = v_t/f$ (remember Section 2.1), we can write this in terms of frequency instead as $n(v_t/2f_n) = L$. This in turn can be rearranged to say that the natural frequency of mode number n is

$$f_n = n(v_t/2L).$$

The subscript on v_t is a reminder that this means the velocity of transverse waves on the string and *not* that of sound in air.

The reason this equation is so useful for the ideal string is that v_t is the same for all wavelengths. That was not true for the stiff bar; there a change from $n = 1$ to $n = 2$ meant a change in v_t also, and thus f_2 was not just twice f_1. But the perfectly flexible string has the special property that v_t depends only on the string's mass per unit length μ (Greek mu) and tension T (strength of the force that is stretching it):

$$v_t = \sqrt{T/\mu}.$$

Typical values for a guitar (D string) are $T \cong 150$ N, $\mu \cong .005$ kg/m, and $v_t \cong 170$ m/s; and for a piano (C_4) $T \cong 650$ N, $\mu \cong .006$ kg/m, and $v_t \cong 330$ m/s.

Because $v_t/2L$ remains the same for all modes, the equation for f_n tells us we have a harmonic series. By putting the value for v_t into the equation in the case $n = 1$, we can write a useful expression for the fundamental frequency:

$$f_1 = v_t/2L = (1/2L)\ \sqrt{T/\mu}.$$

This makes it clear that doubling the length of a string (while keeping other things the same) lowers its pitch one octave. Increasing the tension raises the pitch and choosing a heavier gauge string lowers it, but because of the square root it takes a quadrupling of either of those factors to change the pitch by one octave.

10.2 VIBRATION RECIPES FOR PLUCKED STRINGS

Now that we know the natural modes of a string, we are ready to consider the proportions in which they may be mixed. The plucked string of a guitar or harpsichord is an easy place to start, for it is clear how the motion begins and all the rules about mode excitation developed in the last chapter should

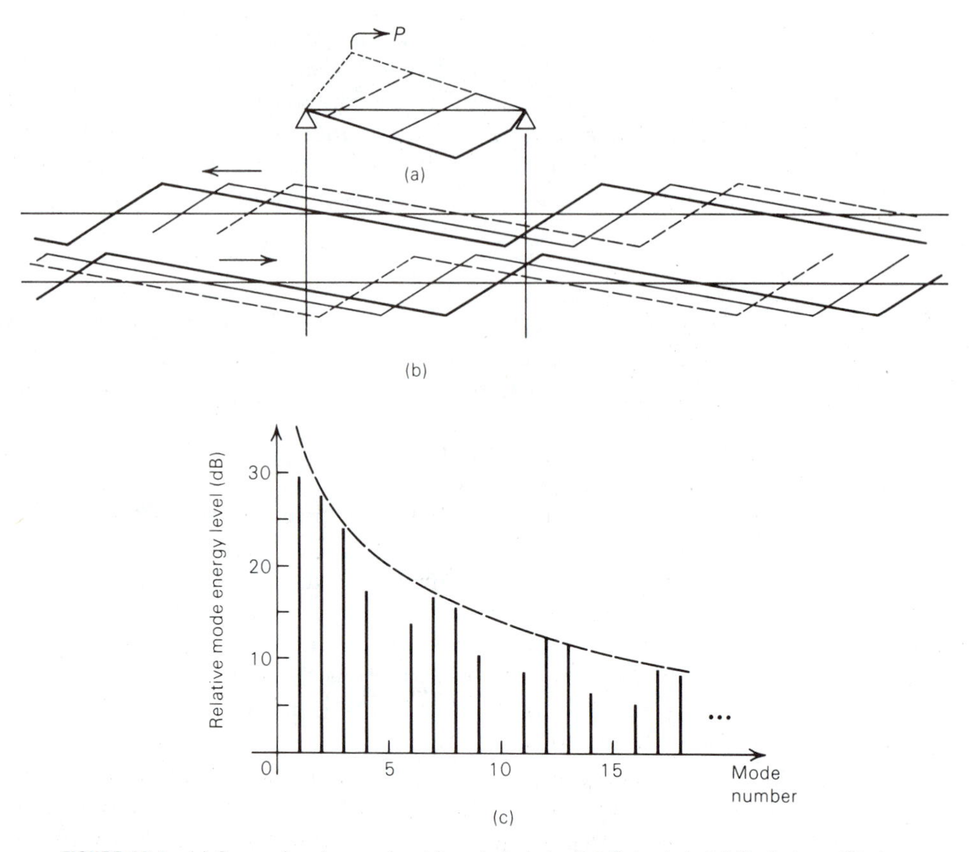

FIGURE 10.4 (a) Successive shapes of a string plucked at point P, located at $L/5$, during a little less than half of the first cycle of vibration. Two sharp kinks move away from P in opposite directions, then reflect back and forth repeatedly. (b) The equivalent pair of infinite traveling wave trains. (c) The initial recipe of mode energies included in this vibration, with its envelope decreasing at 6 dB per octave. Modes 5, 10, 15 . . . are absent because they have nodes at the plucking point.

apply. After being pulled aside by a finger or plectrum, the string is practically at rest at the moment it is released. One way to understand the string motion is to imagine the initial triangular shape split into two waves of similar shape that then travel along the string, as shown in Figure 10.4a,b. At any subsequent moment, one can add these two waves back together to obtain a new string shape, which in general consists of three straight segments. The sequence of shapes repeats itself with the fundamental mode frequency f_1.

But the frequency spectrum corresponding to this motion would be more musically informative. Fourier theory tells us that the mere presence of a kink in the initial string shape, regardless of its exact location, requires a distribution of energy among the natural modes whose general trend is to fall off at 6 dB per octave as you go toward higher frequencies. This is shown by the dashed line in Figure 10.4c. But superimposed upon this trend is the specific property that the excitation of any mode is proportional to how much motion that mode has at the plucking point, meaning in particular that any mode with a node at the plucking point is omitted from the recipe. Plucking at the midpoint of the string, for instance, would produce a strong fundamental component in the sound. Again consider Figure 10.1 to see why the third, fifth, and other odd-numbered modes also would be favored, while modes 2, 4, 6, and so on (which all have a node at the middle) would be completely absent. If you pluck a guitar this way, you will hear a much different, less colorful sound than usual. Plucking one-fourth of the way along the string would favor modes 2, 6, 10, . . . (which have their greatest motion there) but still omit modes 4, 8, 12, . . . (which have nodes there). Plucking close to the bridge makes the fundamental and other low-numbered modes relatively weak, so that the resulting sound has more of a biting edge.

The recipe of string mode energies does not yet correspond to what you hear, for the vibration must be transmitted to the air first. The electric guitar offers the easier case for analysis, for its amplified signal comes from pickup detectors that are mounted close underneath the strings and respond directly to the amount of string motion where they are located. (We wait until Section 16.2 to comment about the actual mechanism of those pickups.) We can state a general rule about this detection process that is quite analogous to our previous ones about excitation:

Each natural mode excites the pickup in proportion to how much motion that mode produces at the pickup location. In particular, any mode that has a node at the pickup is absent from the amplified sound, even if it is present in the string vibration.

Figure 10.5 illustrates how this rule can be applied to determine the spectrum delivered to the guitar amplifier. Good electric guitars give the

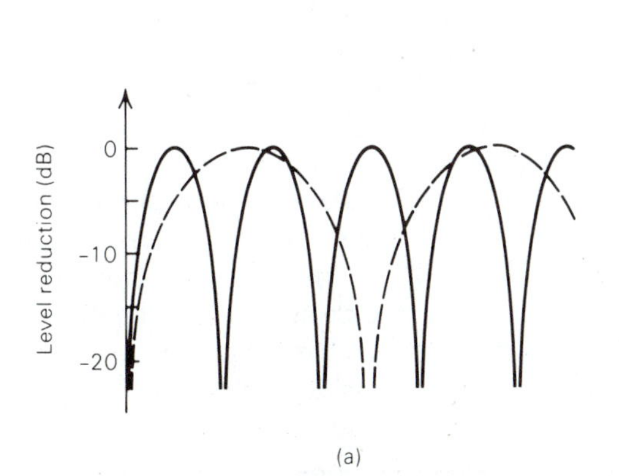

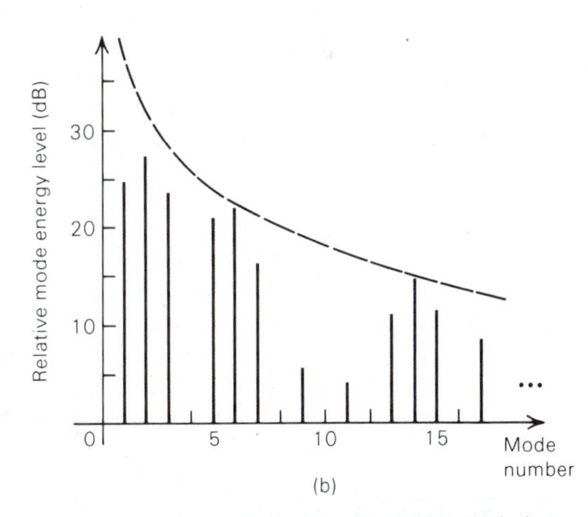

FIGURE 10.5 Spectrum of mode energies delivered to an electric guitar amplifier for a string plucked at $L/10$ with pickup detector at $L/4$. Modes 4, 8, 12, . . . are suppressed (node at detector), as are 10, 20, . . . (node at plucking point). The two corrections shown in (a) for inefficiency of excitation (dashed) and of detection (solid) must both be subtracted from the 6-dB/octave dashed envelope in (b) to obtain the final spectrum.

player a choice of at least two pickup points, one close to the bridge ($L/18$, for instance) and another farther away (maybe $L/4$ or $L/5$). The pickup close to the bridge responds to many modes and can produce a bright, piercing sound that is suitable for solos. The other pickup favors the lower modes for mellower tones and would be appropriate for a background rhythm part; it also imposes its own unique signature by discriminating against a few modes that have nodes near the pickup location, producing a sound recipe unlike that of any acoustic guitar. The ultimate electric guitar would enable you to quickly and easily change the pickup location to any point along the string, giving a wide choice of tone colors all extracted from the same string vibration. The electronic amplifier also offers further opportunities to modify the tone, for instance, by sending it through filter circuits that emphasize the bass or by some deliberate distortion such as clipping (see Chapter 16).

An acoustic guitar differs in several important ways from its electric cousin. First, with no electrical assistance, it cannot compete in sheer volume of sound; it is limited to what can be radiated by the vibrations of its own body. That body (especially the upper surface on which the bridge is mounted) must be quite light to maximize its response to the string motion, and with a hollow cavity whose resonance assists the lower pitches. The body of an electric guitar, on the other hand, is merely a solid platform for holding the strings, with no acoustical function at all. A second difference is the natural decay of sound from the acoustic guitar, as the strings lose their energy through the soundboard into the air. The bridge of an electric guitar is a much more rigid mounting, so that its string vibrations and their resulting sound both decay much more slowly. Finally, we cannot reasonably expect more than a vague resemblance in tone color, for the efficiency with which the body of an acoustic guitar converts various frequencies from string motion into air motion will never vary in the same way from mode to mode as does the detection efficiency of electric pickups. We will wait until our discussion of the violin in Chapter 11 to look more closely at this radiation-efficiency problem.

The harp and harpsichord also have their strings mounted on a soundboard that is mainly responsible for radiating their sound into the air. Effective shapes for these soundboards evolved long ago through trial and error by Renaissance instrument makers, but their action is too complex for us to make blanket statements about the resulting spectra. The string motion, however, must be as we have described above. Like a guitar player, a harpist can change the instrument's timbre at will by plucking closer to or farther from the string ends. Larger harpsichords also offer some choice, rather like that of multiple pickups on an electric guitar, by having two or more complete sets of jacks on the key backs to pluck at different fixed positions along the strings.

10.3 VIBRATION RECIPES FOR THE PIANO

Much of what we have said about guitar strings applies equally well to pianos; but there are also several differences. Most obvious is that the piano

BOX 10.2 THE PIANO ACTION

The mechanical action of a piano is extremely complicated, involving many moving parts for each key. It accomplishes several things: (1) It is a lever system, so that the hammer can be given a greater velocity than that of the key itself. (2) It includes an escapement mechanism so that the hammer is no longer in contact with the key when it strikes the strings; thus it bounces free rather than being held against the string and stifling the vibrations. (3) Each key also raises a felt damper off the string and lets this damper down again when the key is released.

The key action is supplemented by three pedals. On the right is the *damper pedal,* often incorrectly called the "loud pedal." It merely holds the entire set of felt dampers off the strings, so that (1) a string can continue to vibrate even after its key is released and (2) other strings whose keys were not struck can resonate in sympathy. The pedal in the middle (often omitted on spinet models) is the *sostenuto pedal,* which enables you to sustain one chord while going on to play other notes. It holds up the dampers for whichever keys are down when the pedal goes down but lets all the remaining dampers operate normally. This effect is rarely used.

On the left is the *una corda pedal,* also called the *soft pedal.* On upright models this pedal merely moves all the hammers closer to the strings so they will not hit as hard. (That is, even if you push the key in the same way, your force acts through a smaller distance and does less work, so the hammer receives less kinetic energy.) But on a grand piano this pedal shifts the entire action mechanism to the right, so that the hammers strike only two strings in each triplet and miss the third. (The name *una corda,* meaning "one string," is left over from a time when pianos were built to allow enough shift to strike only one string.)

string is struck rather than plucked, and this generates a somewhat different vibration spectrum. Let us see what we can say about that spectrum, putting aside some of the details about how the hammer is controlled in Box 10.2.

First, the softer the hammer, the more gentle and gradual its contact with the string, the less the excitation of higher modes, and the duller the sound. A piano technician can prick and loosen the compressed felt with special needles as an antidote for too sharply percussive sound made by hammers packed and hardened by years of use. Of course, you can go the other way too; for certain types of entertainment, thumbtacks stuck in the hammers make for more abrupt string contact, putting more energy into high modes and producing a very bright, tinny sound. Modern felt hammers are made so that they act quite soft if they hit the strings at low speed; but faster impacts bring the harder inner layers of the hammer into play. Thus, when a pianist changes from *piano* to *forte,* the tone is not only louder (as if it had been amplified) but also brighter in color.

Second, the high-frequency vibrations die away more quickly; 60-dB decay times vary from about 15 seconds or more in the extreme bass, through 2 or 3 seconds in the midrange, to a half-second or less in the extreme treble. This is true not only of the overall strength from different notes, but also of the different components within the spectrum of a single note. The initial transient is relatively complex and bright, but the sustained tone is milder because the high-frequency components drop out more quickly.

Third, any mode that has a node precisely at the impact point will be discriminated against. You may encounter folklore to the effect that piano designers place the hammers one-seventh of each string's length from one end, with the intention that modes 7, 14, 21 ... will be absent from the spectrum. This is superficially attractive, because the seventh harmonic does not appear to fit in the chromatic scale; in fact, the seventh is acoustically as respectable as any other. If you measure hammer positions in various pianos, you will find that $L/7$ is only at one extreme end of the range of positions actually used. You are more likely to find values around $L/8$ to $L/9$ in bass and midrange, decreasing to as little as $L/10$ or $L/12$ at the treble end. These positions, found long ago by trial and error, produce a reasonable compromise in tone color and in efficiency of energy transfer from hammer to string.

We cannot make such a simple generalization about piano spectra as we did for plucked strings in Figure 10.4. Figure 10.6 shows samples of measured string motion. In explaining these, we must understand that the hammer

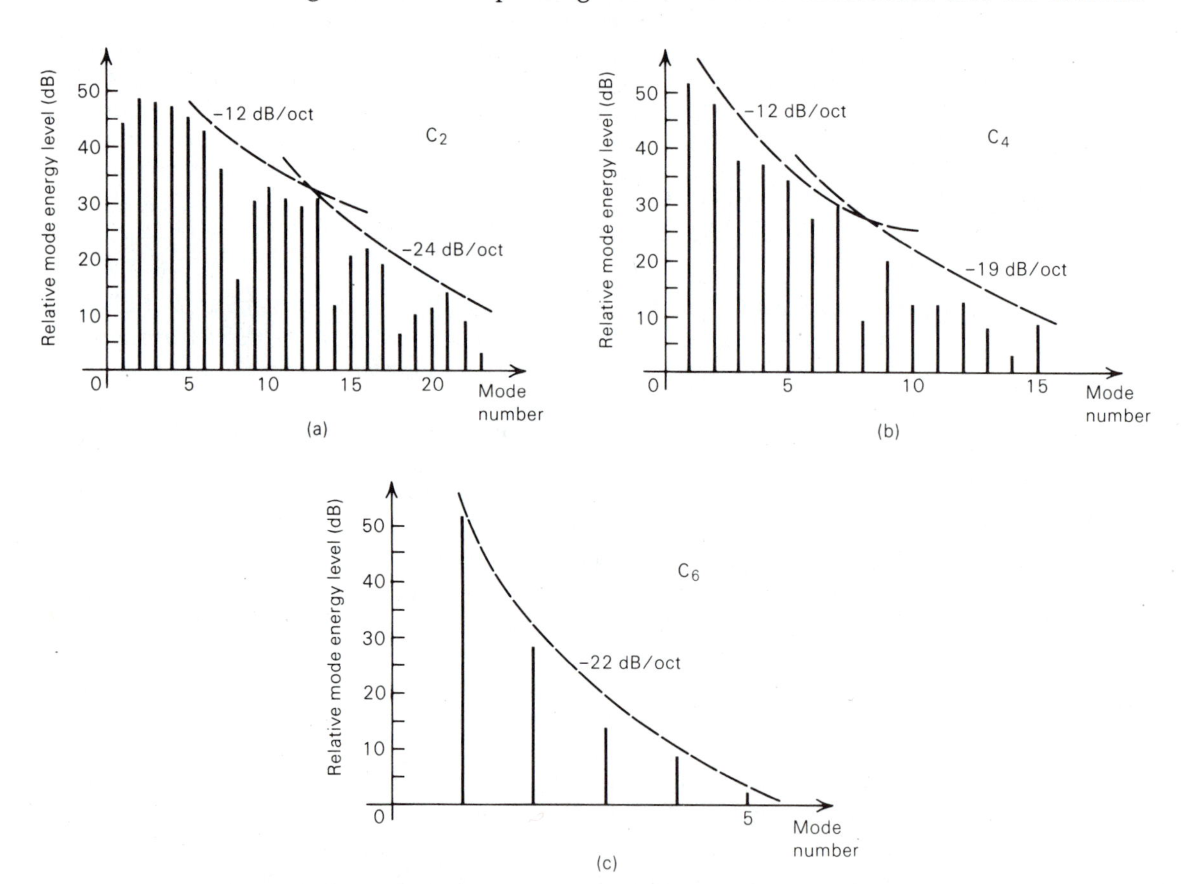

FIGURE 10.6 Sample piano string motion spectra at *forte* dynamic level as measured by Hall and Askenfelt. The dashed envelopes shown for comparison all fall off much faster at high frequencies than the 6-dB/octave shown in Figure 10.4 for plucking. The highest frequency shown is approximately (a) 1.5 KHz, (b) 4.0 KHz, and (c) 5.3 KHz.

does not rebound instantly of its own accord; only its reaction to reflected waves returning from the near end of the string can throw it free. The duration of contact ranges from approximately 5 ms at the bass end to 0.5 ms at the treble end, and is somewhat longer for soft than for loud notes. Only in the extreme bass are the hammers enough lighter than the strings to bounce free within a fraction of the string's fundamental vibration period; in that case the first several modes may have comparable strength before a declining trend sets in at higher frequencies. In the extreme treble, the hammers are much heavier than the strings, and so remain in contact long enough for transverse waves on the string to make several round trips. This sharply reduces the efficiency of excitation of the higher modes (as we discussed in Section 9.6), and the spectra fall off steeply from the beginning. Comparing these with the 6-dB/octave envelope for plucking, we can explain how the piano's sound is less bright and tinkly than the harpsichord's. As with the acoustic guitar and harpsichord, transmission through the bridge and soundboard makes the sound spectra that finally reach your ears somewhat different from the string vibration spectra.

The decay of piano tones actually is more complicated than the idealized picture given in Chapter 9. These tones begin to decay quite rapidly, but then (somewhat to your ears' surprise) the decay shifts to a slower rate and the tones last a long time after all. Near C_4, for instance, the characteristic decay time scale might be approximately 3 to 5 seconds at the beginning but then change later to 10 or 20 seconds.

There are several possible explanations for this. First, each string can vibrate not only up and down but also sideways. (Visualize here a grand piano, not an upright.) That is, for each natural frequency there is really a pair of natural modes, not just one; we may call them the *vertically* and *horizontally polarized modes*. Now we expect that a vertical hammer blow should excite only the vertical mode. But what if the striking direction is not precisely vertical? Then we start with a small admixture of horizontal string motion. The vertical string motion is efficiently coupled by the bridge to the soundboard, and that energy is rapidly radiated away. But the bridge and soundboard respond poorly to the horizontal mode, so what little energy it has takes a long time to leak away. The total radiation from both modes could then be as in Figure 10.7a. Measurements indicate, however, that this mechanism is less important than the following one.

A similar effect also can arise from multiple stringing. Most piano keys strike a set of three strings, not just one. For example, the piano in my classroom is triple-strung from F_2 on up. What happens when three strings are set vibrating in unison on the same bridge? They produce a force on the bridge three times as great as one string alone, and in response to this force the bridge and soundboard should vibrate with three times greater amplitude. But this means each of the three strings must exert its force through three times the distance on each cycle that it would if alone and so loses its energy three times faster by doing work upon the soundboard. This accounts for the initial rapid decay.

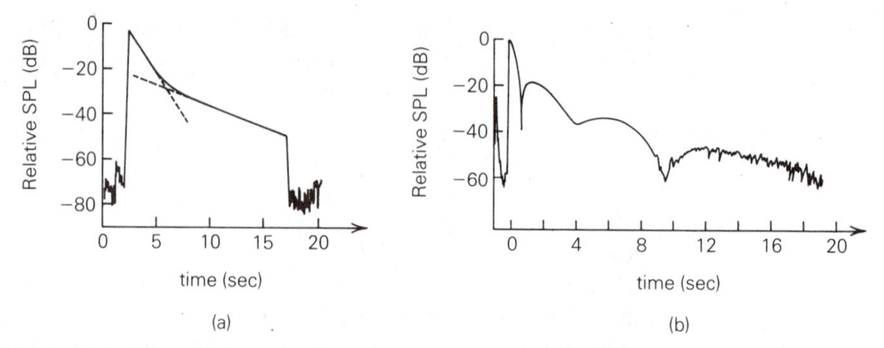

FIGURE 10.7 (a) Decay of sound from a single piano string as measured by Weinreich, showing contributions from vertical and horizontal string motion (fast and slow decays, respectively). (At $t = 17$ sec, the key was released and the damper came down on the string.) (b) Beats during the decay of sound from a pair of strings.

Suppose, however, that the strings are not tuned in perfect unison—439.5, 440.0, and 440.3 Hz, for instance. Then after approximately 1 second they become out of step with one another and no longer drive the bridge cooperatively. From time to time they may be nearly in step again briefly, so we hear beating, corresponding to the fluctuations shown in Figure 10.7b. But the long-term net effect of the beating is that each string radiates its energy at an average rate the same as it would alone, giving the slower late decay. Slight mistuning of the unisons actually is desirable for the greater warmth and interest of the slow beating it produces as well as because it prevents the initial fast decay from proceeding too far and making a "dead" tone rather than a "singing" tone.

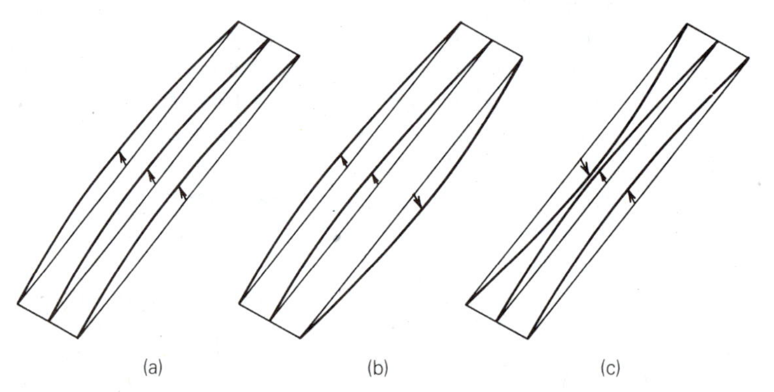

FIGURE 10.8 Coupled fundamental modes of a triplet of strings. (a) The symmetric mode in which all strings cooperate in their action upon the bridge. (b) An unsymmetrical mode that twists the bridge without producing an overall up-and-down motion; whenever the two strings on the left pull upward, the third string pulls downward just as hard so that there is zero net force. (c) Another unsymmetrical mode; this mode is preferentially excited if the hammer hits only the two strings on the right. Each mode shown is the first member of a whole family of modes, made in a similar way by combining higher-numbered modes of the single strings.

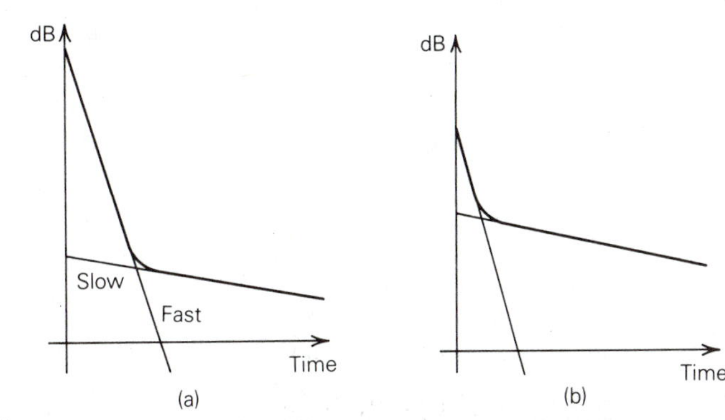

FIGURE 10.9 Change in tone decay with the *una corda* pedal. (a) With the pedal up, all three strings are struck and most of the initial energy goes into the fast-decaying mode; the initial sound is strong, but the long-lasting component is weak. (b) With the pedal down, only two strings are struck, and a major portion of the initial energy goes into the slow-decaying mode; the sound is smoother and less percussive.

It is instructive to explain this process in more sophisticated terms. Because the three strings are mounted on the same bridge, they are not really independent; the bridge couples them together, just as the springs did the pendulums in Chapter 9. So the natural modes of the triplet of strings are like those shown in Figure 10.8. The first of these modes moves the bridge quite efficiently and thus has a short decay time; ideally it would be the only one excited by a hammer striking all three strings together. But if the hammer is slightly lopsided and does not hit all three strings in precisely the same way, the other two modes also will show up in the recipe. They produce mainly a twisting motion of the bridge; but it cannot transmit such a motion to the soundboard as well, so these modes have a long decay time.

We now have some insight into the operation of the *una corda* pedal, which makes the hammers strike only two strings in each triplet and miss the third. This means deliberately putting a large share of the original energy into the softer and slower-decaying modes of family (c) in Figure 10.8. So using this pedal is not merely a matter of reducing the intensity of the sound; it alters the characteristic piano timbre by changing the relative balance between fast- and slow-decaying components of the tone, as indicated schematically in Figure 10.9.

10.4 PIANO SCALING AND TUNING

The preceding section has described the main features of sound production for any single note of the piano. We want to understand more now about the overall design of the instrument, especially how best to choose string sizes. The way in which string lengths and diameters change from note to note and from octave to octave is called the *scaling* of the instrument.

Certainly the central feature of any plan of scaling must be the equation $f_1 = (1/2L)\sqrt{T/\mu}$ for the fundamental frequency of each string. But this is not enough. It appears to allow us to pick any values we please for two out of the three quantities on the right (linear mass density and length, for instance); then we can use the equation to find the value of the third quantity (tension, in this instance) that will produce the desired frequency. So there must be some other grounds for pinning down the choices. Cost, convenience, and tone quality all enter in.

Let us consider each string property in turn, beginning with the length L. The string cannot be too short without creating problems with tone quality, as we shall see below. But if it is too long, the piano becomes bulky and expensive. Nevertheless, there seems to be a wide range still feasible, anywhere from a few centimeters up to one or two meters.

Next, the tension must not be allowed to exceed the breaking strength of the wire. This depends both on its cross-sectional area and on its composition: $T < (\pi/4)\, d^2H$, where d is the diameter and the tensile strength H is characteristic of the material from which the wire is made (steel, for example, has $H \cong 1.5 \times 10^9\ \mathrm{N/m^2}$). There is no definite lower limit to T, but smaller tension decreases the ability of the string to feed its energy to the soundboard. This is because of the difference in their impedances, as explained in Box 10.3. Thus, if we wish to get as much sound as possible out of the instrument, we have strong motivation to put all strings under the maximum tension they can stand without breaking.

Third, small and large diameters both create problems. Too small a diameter worsens the impedance mismatch; that is, a lightweight string is unable to make the heavy soundboard move much, and its sound is weak. But large diameters mean wires that are not fully flexible. The stiffness provides an additional restoring force along with the tension, and this greater restoring force raises the frequencies of vibration. But the frequencies of the higher modes are affected more by the stiffness than the fundamental, so the harmonic-series relationship is destroyed; we say the stiffness causes **inharmonicity**. For modest stiffness, the frequencies remain nearly enough harmonic that the tone still is pleasing and musical. But excessive stiffness makes the string behave more like a rod; the frequencies become more and more inharmonic, producing a more metallic sound.

What is the best compromise to minimize all these difficulties? This compromise was developed long ago through trial and error. It involves (1) putting all strings under tension close to the maximum they can withstand to get a bright and long-lasting tone, (2) getting high frequencies for the treble mainly by decreasing the string length but also by decreasing the diameter slightly to avoid too much stiffness, (3) getting low frequencies for the bass largely by making strings heavier rather than longer, and (4) triple stringing in the treble to deliver adequate energy to the soundboard. The following six optional paragraphs show in some detail why piano string lengths and diameters are chosen as they are, why the middle and high ranges are triple-strung, and why the bass strings are overwound.

BOX 10.3 IMPEDANCE MATCHING

We often are interested in how well vibrational energy can be transferred from one object to another. The energy traveling along a piano string, for instance, does not all pass on to the soundboard the first time it reaches the bridge; most of it is reflected and bounces back and forth many times. Only a small fraction is transmitted to the soundboard on each bounce.

This is one example of a common and general situation in physics. The concept of **impedance** helps us see what many different examples have in common. The definition $Z = F/V$ says that impedance is a measure of how much alternating force F must be applied to produce a vibrational velocity V (*not* the wave speed v) in a string, in the air, in an electrical circuit, or in any other wave-carrying medium.

The wave impedance of a thin flexible string is given by $Z_{st} = \sqrt{\mu T}$. This says that greater inertia and greater tension both make for greater impedance: A thick string has larger impedance than a thin one.

The aspect of impedance that most concerns us here is *impedance matching*. At a boundary between two media of similar impedance, a wave will travel from one into the other with little reflection; it hardly notices the difference. But any boundary where the impedance changes greatly will reflect most of the energy back, regardless of whether that change is an increase or a decrease. If we wish for efficient energy transfer, we should match impedances. If we wish to keep energy bottled up in one region, we should mismatch impedances at the boundaries.

We will meet examples of the first kind in Chapter 16; if we want efficient transfer of energy from a hi-fi amplifier to a speaker, the electrical impedances should be matched (both 8 ohms, for instance). Camera lenses provide another example: The impedance for light waves is different in glass than in air, so an ordinary lens surface (no matter how carefully ground, smoothly polished, and cleaned) always reflects a small percentage of the light falling on it. In a good camera there may be six or eight lens surfaces, so that half or less of the incident light intensity finally is transmitted all the way through to the film. Coated lenses cut down on the light loss by providing a layer with impedance intermediate between that of air and glass, allowing the light to get across in two stages, both with smaller impedance change and correspondingly less reflection. The use of a soundboard to mediate between a thin string and the surrounding air is analogous.

But the string and soundboard also illustrate the other case, where we do *not* want a perfect match. If the string and soundboard had about the same impedance, the string would deliver most of its energy on the first vibration; the damping time would be a small fraction of a second, there would be no persistent standing wave, and we would hear only a loud but dull and pitchless thump. On the other hand, if the soundboard impedance were millions of times greater than the string's, nearly all the energy would be reflected back from the bridge. The string would keep vibrating a long time (an hour, for instance if it did not lose the energy through bending friction or direct radiation instead), during which the slow leakage of energy to the soundboard would produce only a very faint sound. What we need is a soundboard impedance only a few thousand times greater than the string impedance. Then the string retains much of its energy for thousands of vibrations (several seconds) but also gives up the energy fast enough that the soundboard can produce a loud sound.

---

*Let us start with the assumption that we will put all strings under near-maximum tension, say $T = k(\pi/4)d^2 H$ with safety margin $k \cong 0.5$. Notice that the linear mass density μ depends in the same way on the diameter, $\mu = (\pi/4)d^2 D$. (D is the mass density of the material from which the wire is made; steel, for instance, has $D = 7.8 \times 10^3$ kg/m^3.) Then the diameter cancels out of the equation for the frequency: $f_1 = (1/2L)\sqrt{T/\mu} = (1/2L)\sqrt{kH/D}$. Thus, for a given material, the length is practically determined by the frequency, leaving the diameter to be chosen independently on other grounds.

Let us follow through on that. It suggests that a steel string for C_4 on the piano should have $L = (1/2f)\sqrt{kH/D} \cong (1/523\text{ Hz})\ (8 \times 10^8\ \text{N/m}^2/8 \times 10^3\ \text{kg/m}^3)^{1/2} \cong 0.6$ m. And indeed, this is about what you will find for middle C on any piano. But it also suggests that, as long as we continue to use steel, we should cut the length in half for every doubling of frequency. If this were actually done, it would give a C_8 string less than 4 cm long, which makes things a bit crowded, as well as creating problems with stiffness. Thus, in practice, lengths only are divided by some smaller scaling factor such as 1.88 for each octave to give a C_8 string approximately 5 cm long. But this means the only way to get the frequency doubled is to increase k (by a factor of 1.13 for each octave), skirting closer and closer to disaster, with k rising to approximately 0.8 for C_8.

Going toward the bass, doubling the string length for each octave lower would result in $L \cong 5$ m for C_1 and $L \cong 6$ m for the lowest note, A_0. Even multiplying by only 1.88 for each octave still gives $L \cong 4.7$ m for A_0. Hardly anyone could afford such a piano, even if they had room for it. If we merely use shorter strings with less tension, the tone suffers, becoming thin and dead. The solution is to wrap a copper winding around an ordinary-size wire core; this increases the mass per unit length without appreciably changing the stiffness. (The increased impedance is taken care of by mounting these strings on a separate, heavier bridge.) It is quite a satisfactory solution in large grand pianos, bringing the lowest string lengths down to approximately 1.5 to 2 meters. But in small spinet models with A_0 string lengths of 1 meter or even less, the bass octave is doomed to low tension and muddy failure. (That must, of course, also be partly blamed on the soundboard.)

Now with the lengths fairly well fixed, how do we choose diameters? For a more powerful sound, we should like to make the diameters large, allowing a correspondingly larger tension and better coupling to the soundboard. The limiting factor is stiffness, and its effect is best expressed in terms of the quantity $J = \pi^3 Yd^4/128\ TL^2 = \pi^2 Yd^2/32\ kL^2 H$. ($Y$, as in Chapter 9, means the Young's modulus of the material, which is approximately 1.9×10^{11} N/m^2 for steel.) For larger diameter d, the greater stiffness is represented by a larger J; but greater tension T or length L both reduce the relative importance of stiffness, as represented by a smaller J. For steel with $k = 0.5$, $J \cong 80\ (d/L)^2$.

The inharmonicity of the string is conveniently given in terms of J by the formula

$$f_n \cong nf_1[1 + (n^2 - 1)J].$$

This says that $f_2 \cong 2f_1(1 + 3J), f_3 \cong 3f_1(1 + 8J)$, and so forth, which would not be a harmonic series unless J were zero. For instance, $J = 0.02$ will raise the second mode frequency an entire semitone (6 percent) away from the desired exact octave above the fundamental, with the departures becoming rapidly worse as we go to higher-numbered modes.

An acceptable J for C_4 is approximately 0.0002, which keeps all modes up through the sixth within an eighth of a semitone of the harmonic series. This limits the string diameter to approximately 1 mm. Because a single string this size cannot drive the soundboard well enough, we use three strings. As we go to the top of the keyboard, dividing the string length by 1.88 for each octave while keeping the same diameter will multiply J by

$(1.88)^2/1.13 = 3.12$ per octave, bringing it up to .02 for C_8. That is a little too much, so we sacrifice some loudness for better harmonicity by gradually changing to slightly smaller diameter (and correspondingly less tension); a typical C_8 string has $d = 0.8$ mm, keeping J down to approximately .015.

The tuning of the piano is affected by the inharmonicity of its strings. In the middle range the vibrations are nearly harmonic, and careful measurements show that a skilled piano tuner sets the octaves quite accurately at 2-to-1 frequency ratios. But in the top octave, where the strings are relatively stiff, neither the tuner nor most listeners will be satisfied unless the octaves (such as C_7–C_8) are stretched slightly wider to ratios as high as 2.025 to 1. There is a similar effect in the lowest bass notes.

One explanation for these *stretched octaves* is physical: Perhaps the criterion for tuning is that the fundamental of the higher note be matched precisely in frequency with the second mode of the lower note so that they produce no beats when sounding together. Because stiffness gives the second mode of the lower note a frequency slightly more than twice its fundamental, this would cause the tuner to set stretched octaves, especially in the high treble range.

Another possible explanation is psychological: In extremely high and low ranges where pitch perception becomes difficult, perhaps the ear demands an exaggeration of the usual 2-to-1 ratio in order to be sure the two tones are at least an octave apart. It is possible to test which reason is more important by accurately measuring the work of a skilled tuner in the bass register on a variety of pianos. If the psychological mechanism is dominant, the tuner should produce similarly stretched octaves on any instrument. But if the physical mechanism is more important, she should produce relatively little stretching of the bass octaves on a large concert grand (for which the inharmonicity of the bass strings is hardly any greater than in the midrange) but much more stretching on a small spinet (whose relatively short, stiff bass strings are quite inharmonic). The results of such an experiment (as reported by Backus) indicate that the physical mechanism of string inharmonicity accounts for most or all of the stretching.

SUMMARY

The natural modes of a thin string have the shape of sinusoidal standing waves, and the mode frequencies have the very special property of forming a harmonic series. These frequencies are related to string length, tension, and density by $f_n = (n/2L)\sqrt{T/\mu}$. Any disturbance of such a string can be described with a natural mode vibration recipe or equally well in terms of traveling waves reflecting back and forth between the ends of the string.

Some acoustical properties of the piano can be explained with thin strings: the presence of definite pitch; the dependence of the vibration recipe on the striking point, hardness, and breadth of the hammer; and the complex manner in which the tones decay, especially for multiple stringing. Other acoustical properties require consideration of the stiffness of thicker strings: the stretched octaves in extreme bass and treble, the efficiency of transfer of energy from string to soundboard, and the variation of string sizes in a well-designed piano.

The proportions in which wave energy is transmitted or reflected at any boundary (such as between string and soundboard) are determined by the relative impedances. Optimum design requires trade-offs between strong transmission for louder sound (requiring impedance matching) and strong reflection for longer-lasting vibrations (requiring impedance mismatching).

Guitar-string vibration recipes are somewhat different from those of pianos because the vibration is initiated by plucking, but they depend in a similar way on whether the plucking or striking point is close to the bridge. Electric guitars modify the vibration recipe by preferentially amplifying those natural modes whose motion is greatest close to the pickup point.

REFERENCES

There is an interesting article on the piano by E. D. Blackham in *Scientific American, 213* (December 1965). It is reprinted in the Hutchins collection, as well as in *Musical Acoustics: Piano and Wind Instruments*, an anthology of original research papers edited by Earle L. Kent (Halsted Press, 1977). Backus treats the piano in his Chapter 13. You will especially find many more details given by Benade in his Chapters 7, 8, and 17. A nice account of the historical development of piano design is given by Edwin M. Good in *Giraffes, Black Dragons, and Other Pianos* (Palo Alto, CA: Stanford University Press, 1982); I consider it worthwhile reading even though I disagree with a few of Good's comments on acoustics.

A classic study of piano tones was made by Harvey Fletcher, E. D. Blackham, and R. Stratton, *JASA, 34,* 749 (1962). They synthesized many artificial tones and succeeded in making some that were practically indistinguishable from real piano tones. Multiple stringing and double decay curves were originally explored by T. C. Hundley, H. Benioff, and D. W. Martin in the 1950s, although their work was not published until much later in *JASA, 64,* 1303 (1978). Additional studies by Gabriel Weinreich appear in *JASA, 62,* 1474 (1977), and Weinreich presents a good popular account of this topic in *Scientific American* (January 1979). Measurements of piano-string vibration spectra and a discussion of attempts to construct a theory that can predict them are given by D. E. Hall and A. Askenfelt in *JASA, 83,* 1627 (1988).

For more information on guitars, see the introductory article by T. D. Rossing in the *Quarterly* of the Guild of American Luthiers, *11,* 12 (1983) and *12,* 20 (1984), and numerous articles in the *Journal of Guitar Acoustics*, beginning in 1980.

A detailed analysis of harpsichord scaling is worked out by N. H. Fletcher in *Acustica, 37,* 139 (1977). (This is Neville Fletcher from Australia, not to be confused with Harvey Fletcher of Bell Labs and Brigham Young University. Harvey Fletcher was one of the great pioneers of musical acoustics in the 1920s and 1930s, and it is his name that goes with the Fletcher–Munson diagram.) A preliminary analysis of the clavichord is given by Suzanne Thwaites and N. H. Fletcher in *JASA, 69,* 1476 (1981).

SYMBOLS, TERMS, AND RELATIONS

n	mode index
λ_n	standing wavelength
f_n	mode frequency
v_t	string transverse wave velocity
T	tension (a force)
μ	string mass per unit length
L	string length
d	string diameter
D	density (mass per unit volume)
H	tensile strength
Z	impedance

$f_n = n(v_t/2L)$

$v_t = \sqrt{T/\mu}$

$T_{max} = (\pi/4)d^2H$

natural modes
standing waves
traveling waves
nodes and loops
damper pedal
una corda pedal
impedance
impedance matching
harmonic series
inharmonicity
stretched octaves
mode recipes
missing modes

EXERCISES

1. What are the wavelengths of the first three natural modes on a string of length $L = 0.6$ m? (Caution: This means wavelength of the transverse string vibration, *not* of the resulting sound in air.)

2. If the velocity of transverse waves on a certain string of length $L = 0.3$ m is $v_t = 240$ m/s, what is the natural frequency f_1 of its fundamental mode?

3. Suppose you have a string of length $L = 0.5$ m on which waves travel at speed $v_t = 150$ m/s. What is the frequency f_5 of the fifth mode of this string? Regarding the mode as a standing wave, what is its wavelength λ_5? (Caution: same as for Exercise 1.)

4. A string vibrates in mode 6. Sketch its appearance at several successive moments during a cycle, also indicating direction of motion.

5. If you double the tension applied to a string, how much is its fundamental frequency increased? Show how many semitones this raises the pitch.

6. If a string with linear mass density $\mu = 0.002$ kg/m is placed under tension $T = 180$ N, with what speed v_t will it carry transverse waves?

7. If a string of linear mass density $\mu = 0.006$ kg/m and length $L = 0.75$ m is subjected to tension $T = 540$ N, what is the frequency f_1 of its first mode?

*8. Suppose you have a piano bass string of length $L = 2$ m and linear mass density $\mu = 0.06$ kg/m, and you want it to vibrate with fundamental frequency $f_1 = 30$ Hz. How much tension must you apply?

9. For a midrange string on a piano, look inside and estimate the effective length of string contacted by the hammer (the indentations in the felt are a good clue). Express this as a fraction of the total length of the string. Use this to estimate the harmonic number (and its frequency) above which it would be a poor approximation to pretend that the hammer makes a simple point contact.

10. Compare the richness of vibration recipes of a bass and a treble piano string a second or two after they are struck.

11. Explain in terms of soundboard flexibility why horizontal string motion is poorly communicated to a grand piano soundboard.

12. Explain physically why, for two strings of the same diameter under the same tension, the shorter one shows more inharmonicity in its vibrations.

13. What specific faults will listeners find with a piano tuned precisely to the equal-tempered scale as given in Figure E?

*14. Verify that $J = 0.0002$ gives f_6 within one-eighth of a semitone of $6f_1$.

*15. If a steel string and a brass string have the same length and diameter and both are stressed up to the same safety factor k, how much higher pitch will come from the steel? Show that under those circumstances the steel string will have lower J than the brass. (See Project 3 for data on brass.) Now suppose instead that the tensions are chosen to give both strings the same pitch (still with same lengths and diameters), and show that under these circumstances the steel will have higher J than the brass. Use these conclusions to discuss the tendency toward higher absolute pitch or different tone color when modern harpsichord builders faithfully copy old designs but string their instruments with modern high-tensile-strength wire.

16. Why are the damper felts in a grand piano located immediately above the hammers instead of somewhere else along the strings?

17. Does the soft pedal of an upright piano produce the same acoustic effects as that of a grand? Explain.

18. Sketch initial vibration recipes for a string *plucked* at points (a) $L/3$ and (b) $L/8$ from the bridge. Explain the difference in perceived timbre in terms of these recipes.

19. (a) State which modes would be missing if you plucked a string at $L/4$ or $L/5$ from one end. Think of those points now as being 2/8 or 2/10 of the way down the string, and admit that there is nothing to prevent you from plucking in between, say at 2/9 of L. Make a sketch like Figure 10.4 to indicate what the spectrum would be like in that case. (b) Is there some discrimination against modes 4 and 5?

20. Savage reports measuring the vibration recipe shown in Figure 10.10 from a harpsichord string. Describe where this string was plucked.

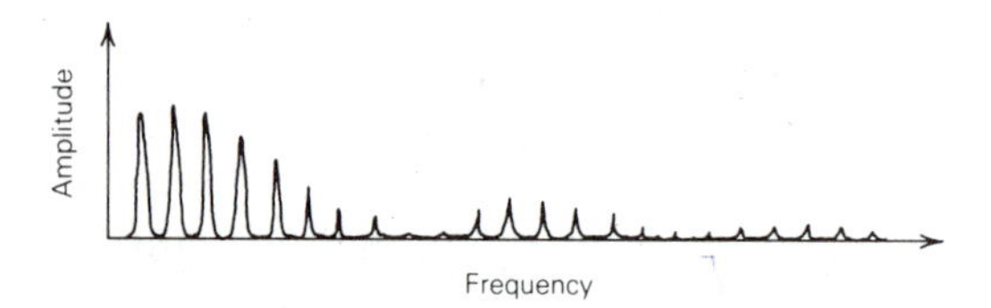

FIGURE 10.10 (From *Problems for Musical Acoustics* by William R. Savage. Copyright © 1977 by Oxford University Press, Inc. Reprinted by permission.)

21. Where would you pluck a guitar string to most effectively excite mode 3? Where would you touch it lightly with your finger to kill modes 1 and 2 but leave 3

vibrating? If you do this, how much higher is the pitch of this harmonic than the pitch you ordinarily hear from that string? Hint: Look at the harmonic series slider.

22. Suppose you pluck a guitar string that sounds the note G_2. Now you touch your finger lightly on the string at $L/4$ down from the nut (that is, at the fifth fret). Which modes of vibration survive? Why? What pitch do you hear now?

23. Suppose a string is plucked at $L/5$ as in Figure 10.4, and an electric pickup is located at $L/7$. Sketch the vibration recipe that will be broadcast by the amplifier.

24. Consider the example of Figure 10.5. Suppose the roles were reversed, so that you plucked the string at $L/4$ and the pickup was at $L/10$. Explain what the resulting spectrum would be.

25. Will a guitar bridge present different impedance to the string depending on whether the string vibrates perpendicular or parallel to the top plate? Why? Explain how a guitar then might have a rapidly decaying initial transient followed by a more sustained sound. How could you vary your plucking technique to change the relative amounts of fast- and slow-decaying sound?

26. Consider the transmission of sound from the air outside through a window to the air inside your house. What is the relative efficiency of transmission if the window material is thin plastic film, ordinary window glass, or inch-thick plate glass? Explain why, in terms of impedance.

PROJECTS

1. Measure how far pickups on an electric guitar are located from the bridge, and express this as a fraction of the string length. Discuss how this affects the vibration recipe, and relate your explanation to the actual tone color you hear. If possible, check several different guitars to see whether they vary in their pickup locations.

2. Learn how a harpsichord buff stop operates, listen to its sound, and explain its effect on timbre in terms of how it modifies the vibration recipe.

*3. My harpsichord is strung partly with steel wire, ranging gradually from 0.36 mm diameter at $G_2^\sharp$ to 0.22 mm at G_6. But the bass is strung with simple brass wire (no overwinding), ranging from 0.36 mm diameter at G_2 to 0.56 mm at F_1. Approximate string lengths are 14 cm for G_6, 24 cm G_5, 47 cm G_4, 88 cm G_3, 136 cm G_2, 171 cm G_1, and 174 cm F_1 (where the bass end of the bridge is recurved). The density of brass is slightly greater than steel, approximately 8.4×10^3 kg/m^3; its yield strength is only approximately 4×10^8 N/m^2 and Young's modulus approximately 9×10^{10} N/m^2. Discuss as completely as you can the acoustical considerations in this scaling.

CHAPTER 11 The Bowed String

The violin family holds an honored place in our musical culture. For more than 300 years it has served as the foundation for chamber and symphony orchestras. Eminent composers have chosen to express themselves through music for string quartet and numerous solo works for violin and for cello. This must have something to do with both the beauty and the power of the tone that these instruments produce and with their wide range and versatility.

We should like to understand the physical principles of the operation of these instruments and if possible to see some of the reasons why the violin design in particular has been such a lasting success. Our goals for this chapter thus are to explain (1) the mechanism by which the bow excites the string, with accompanying insight into bowing techniques; (2) the nature of the resulting string vibrations; (3) how the body of the instrument transmits these vibrations to the air; and (4) the resulting characteristic timbre of this family. In doing so, we will find that we need to develop and use two important physical concepts—dynamic instability and resonance. We must begin by considering briefly the construction of these instruments and the proper terminology for their parts.

11.1 VIOLIN CONSTRUCTION

The design of the violin has remained fairly stable since the time of such great Italian master craftsmen as Niccolo Amati (1596–1684), Antonio Stradivari (1644–1737), and Giuseppe Guarneri (1698–1744), who each came from a famous family of instrument makers. The main features of their design are shown in Figure 11.1.

Note first the neck, whose strength withstands the string tension and supports the fingerboard. Recent work has indicated that vibration of the neck and fingerboard may be important to the sound as well as to the player's feeling of responsiveness in the instrument.

The *body* construction is even more critical. The *top and back plates* must be carved with great care to achieve the proper curvature and thickness (as little as 2 mm near the edges). The top plate usually is made of spruce and the back plate of curly maple; the wood must be seasoned carefully without kiln-

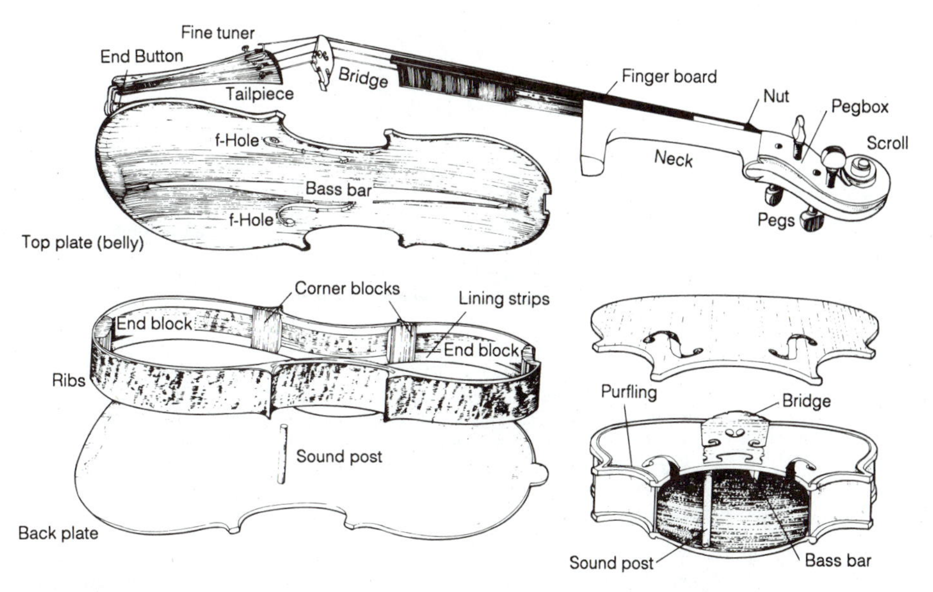

FIGURE 11.1 Anatomy of a violin.

drying. The grain of the wood runs parallel to the strings to make the plate stiffer in that direction; it is relatively flexible across the grain.

The wood is finished with a thin protective layer of varnish. An otherwise fine violin may be ruined by applying a thick coat of poorly chosen varnish. Many musicians still believe that some secret formula for varnish is the key to building a great instrument, despite the generally skeptical opinion of scientists. Although it seems reasonable that the right varnish could help slightly, by far the greatest share of an instrument's quality must be determined before the varnish is ever applied. According to John Schelleng, one of the few people who have ever made real scientific tests of this question, what may be most important about a good varnish is that you can apply it in an extremely thin coat.

To allow easy vibration, the plates are deliberately weakened all along the edge that is glued to the ribs. A small channel or groove is cut out and inlaid with separate thin strips of wood called *purfling*. These allow the plates to vibrate more as if they were hinged along their edges instead of clamped. Cracking of the glue that holds the purfling in the groove may make it even easier for the plate-rib joint to bend, and this is one way age may enhance an instrument's responsiveness.

Violin strings are supported on the body of the instrument by a *bridge*. If this were a simple low bridge rigidly mounted on the top plate (Figure 11.2a), the strings would find it hard to pass on their energy to the body. That is, such a bridge would present extremely high impedance to the horizontal motion of a bowed string and reflect nearly all arriving vibrational energy back toward the nut. (Recall the similar behavior of a piano soundboard for horizontal string motion described in Chapter 10.)

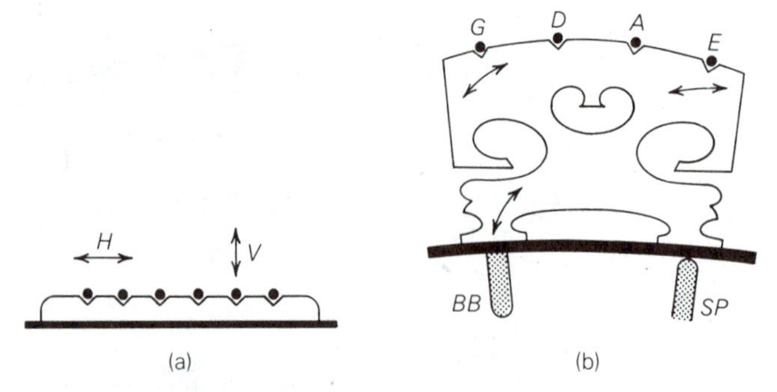

FIGURE 11.2 (a) For a low, flat bridge, horizontal string motion (*H*) is far less efficient than vertical motion (*V*) in moving the plate on which it is mounted. This is satisfactory for a guitar because the player can include at least some vertical component in plucking. (b) Bowed string motion is limited to the bowing direction, so the violin bridge (viewed here from the tailpiece) acts as a lever pivoted on the sound post *SP* to impart vertical motion to the rest of the top plate (and bass bar *BB*). This picture applies only for frequencies up to approximately 2 KHz; above that the bridge itself becomes distorted, although its net effect is still to aid in transferring string motion to the plate.

So the violin instead uses the delicate, flexible high bridge shown in Figure 11.2b. The slightly asymmetrical and carved-out shape is not just decorative; it also encourages a rolling motion of the bridge, which, in turn, can push in and pull out on the top plate. The plate is much more yielding to this type of motion (has a smaller impedance) and so accepts a larger share of the string's energy on each cycle. It is this in-and-out motion that most effectively moves the nearby air to produce useful sound.

The total tension in the four strings of a violin is typically on the order of 220 N (50 lb). This results in a downward force on the bridge of approximately 90 N, called the down-bearing (Figure 11.3). The *sound post* provides a connection that allows the back plate to help bear this steady load, and supports the treble foot of the bridge like the fulcrum of a rocking lever when the strings vibrate. If the sound post is missing or improperly placed, the instrument's tone suffers drastically.

The *bass bar*, too, strengthens the top plate and helps to distribute vibrational energy from the bridge to all parts of the plate. The presence of the bass bar is not really surprising if you are aware that guitar top plates and piano and harpsichord soundboards also are stiffened by numerous ribs to give their best sound. (Note that *ribs* means "reinforcing bars" on piano soundboards but means "side walls" to a violinist.)

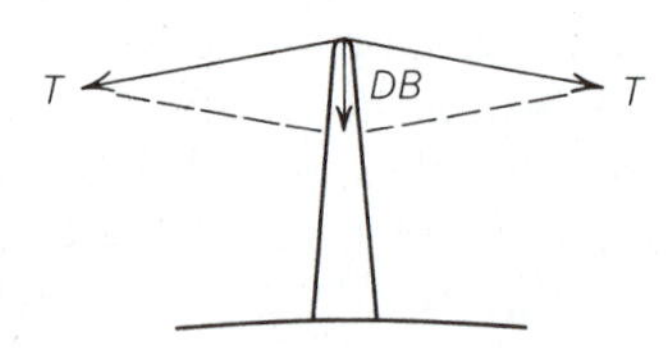

FIGURE 11.3 Side view of a violin string passing over the bridge. Arrows show how the combined tension *T* from each end of the string produces a net downward force on the bridge, the down-bearing *DB*.

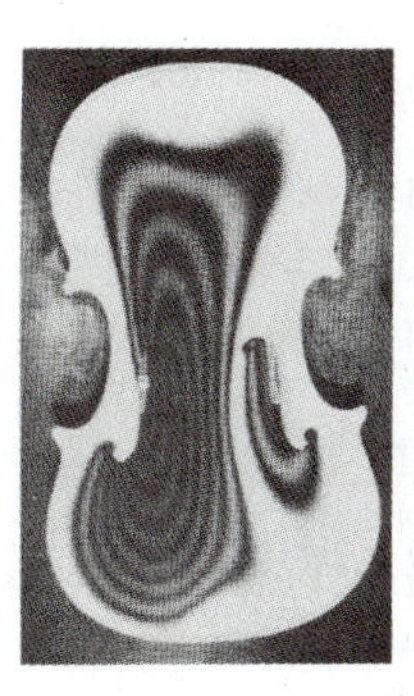 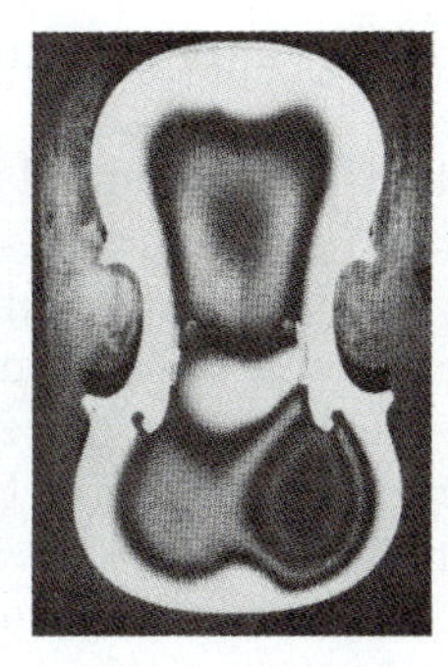 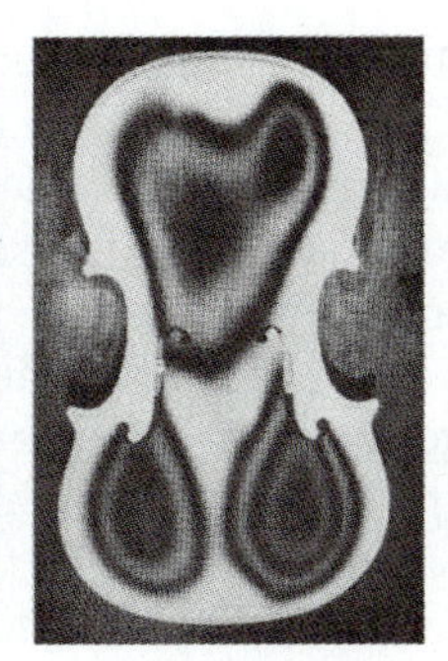 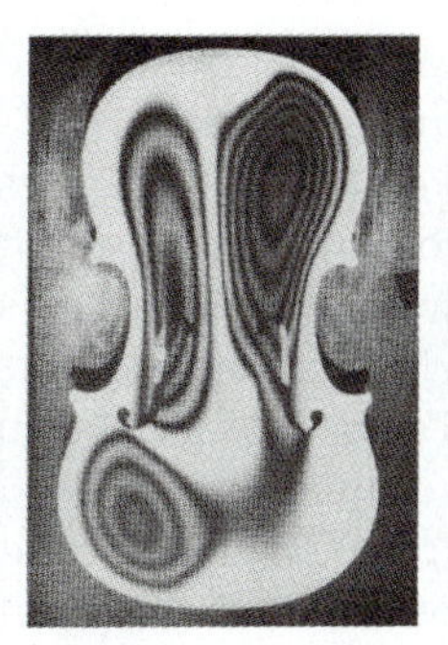 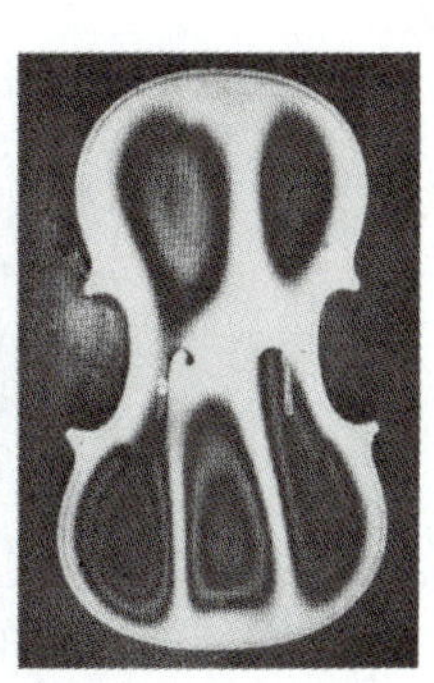

FIGURE 11.4 Vibration of a violin top plate (including bass bar and artificial sound post), studied with the technique of hologram interferometry. The light and dark bands form contour maps of vibration amplitudes. These vibration modes occurred at frequencies 540, 775, 800, 980, and 1110 Hz. Both the modes and their frequencies are changed considerably in the assembled instrument. (Photos provided by E. Jansson, N. Molin, and H. Sundin)

The violin body has many natural modes of vibration, just like any other solid object. In recent years new optical techniques have made it possible to produce photographs showing vibrational motions of less than a thousandth of a millimeter. These have been used to map out the shapes of some of these modes, as illustrated in Figure 11.4.

---

*There are several mutually reinforcing aspects of stringing practice that make even old violins different in modern use from what they were 300 years ago. These are the replacement of gut strings by metal-wound or steel strings, the gradual rise of standard pitch for A_4 from approximately 420 Hz (or even less) toward 440 Hz, lengthening of the neck and use of greater string tension, and greater bridge height (which increases the down-bearing force). The result is a louder and more brilliant tone, which does not always represent an improvement; it can make modern performances of baroque music more strident than the composer intended. Another result is that many violins of the old Italian masters have had thicker bass bars installed to sustain the increased tension and down-bearing, so they are no longer entirely authentic representations of their makers' art. It is quite a tribute to that art that the instruments survive the modernization process at all!

The general features of the viola and cello are similar, but each has some unique character of its own. They do not merely sound like recorded violins played back at slower speed, because their sizes are not in direct proportion to the wavelengths of sound they must produce. The cello range begins at C_2, the viola at C_3, and the violin at G_3. But violas are generally only some 15–20 percent larger than violins; cellos are a little more than twice violin size. Thus the viola is not just a half-size cello even though its playing range is one octave higher. The viola is not even close to being a violin scaled up to precisely 50 percent larger in every dimension, even though that is what you might expect from its range, which goes a perfect fifth lower. As we shall see, this tends to make the viola quite weak in its lower register. Unfortunately, it cannot be made much larger without becoming impractical to play when held under the chin.

The general question of just how each instrument should be proportioned to make a cohesive family with similar sound characteristics is called *scaling*. I will not enter into scaling theory here, but you should know that Carleen Hutchins and others learned a great deal about violin-family scaling in the 1960s, both theoretically and experimentally. They designed and built a complete family of eight instruments, ranging from a large bass up through a treble, which is tuned an octave above the violin. Their oversize viola is held between the player's knees, cello-style. You may read more about these remarkable instruments in the references given at the end of the chapter.

11.2 BOWING AND STRING VIBRATIONS

How does the steady motion of the bow maintain vibration of the string? Why is it that the string cannot remain in one position pulled slightly aside, while the bow slips past (as in Figure 11.5)? Actually, if you do not push down hard enough on the bow, or if it is not well-rosined, you *can* make it slide or skitter across the string and produce very little vibration.

But when properly used, the bow catches the string and carries it farther and farther to one side. That is, as the bow starts to move, the string sticks to the bow hairs until the force (from tension) pulling it back becomes great enough to dislodge it from the sticky rosin. It flies back fast enough that the bow cannot recapture the string until it has gone clear to the other side. The string sticks to the bow and moves once more in the original direction, until finally it must break loose from the bow again. If this sequence can be repeated periodically, we have the desired string vibration.

This *stick-slip mechanism* is an example of **dynamic instability**. That means a situation in which we might imagine a perfectly steady equilibrium, but any slight disturbance causes the system to abandon that equilibrium state and change to an oscillatory state. In this case, consider how the frictional force exerted by the bow upon the string changes with their relative velocity

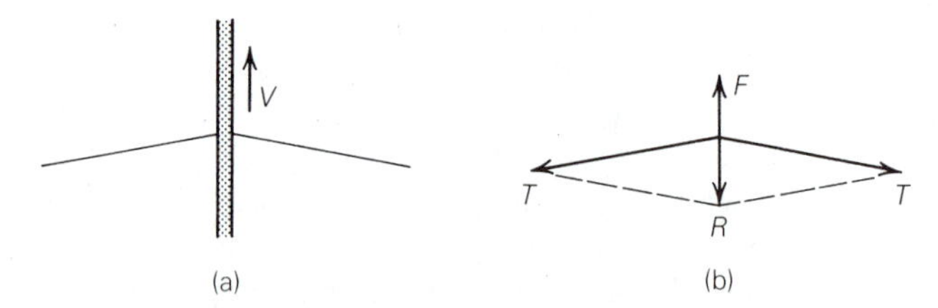

FIGURE 11.5 (a) A string pulled aside by a bow sliding across it moving at speed V. (Compared to Figure 11.3 this is a top view.) (b) Forces on the piece of string in contact with the bow. The tension T from each end of the string gives a total restoring force R, which is opposed by the bow's frictional force F. The hypothetical equilibrium with F exactly balancing R is unstable, so a sliding rosined bow cannot maintain the string motionless in this shape.

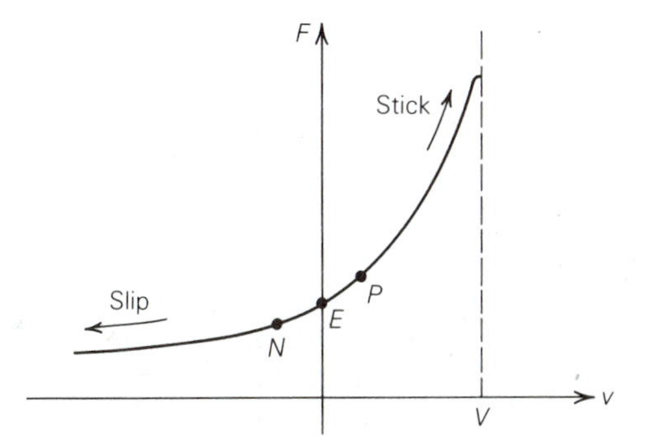

FIGURE 11.6 Schematic dependence of friction force F on string motion. The string velocity v takes on various values while the bow velocity V remains constant. When v matches V, string and bow stick together and the force of static friction can be quite large. The greater the difference in velocities, the smaller the force of sliding friction becomes. Point E represents hypothetical dynamic equilibrium with string not moving.

(Figure 11.6). When the bow is sliding past the string, the frictional force is quite small, but when they are traveling at about the same speed, stuck together or nearly so, the force becomes much larger.

Now consider the hypothetical equilibrium state with the string held aside by the frictional force but not moving (Figure 11.5 and E in Figure 11.6). Suppose for some reason the string momentarily slips back a bit toward its normal position; that is, it acquires a slight negative velocity. Then the frictional force is reduced (N), which allows it to slip even farther back. Again, if for some reason the string momentarily moves forward a bit with the bow, the slight positive velocity (P) means that it feels an increased frictional force tending to carry it even farther forward. So E represents an *unstable equilibrium*, just like a ball balanced at the top of a hill so that the slightest disturbance will make it roll down one side or the other.

Our picture so far has the bow capturing the string so that both travel together (the stick point in Figure 11.6) until something happens to break them apart. Once the string slips a little bit, it is easier for it to slip more, and it flies back almost entirely under control of the tension, like a plucked string. As soon as its natural vibrational motion starts to bring it forward again, the $E \rightarrow P$ instability works to make it stick to the bow again.

If you think of the string as vibrating smoothly and sinusoidally in one of its natural modes, it is hard to understand how the sticking and slipping could happen always at precisely the same point in every cycle to maintain a steady tone (a periodic vibration). But here is where it is more enlightening to think in terms of traveling instead of standing waves. The sudden release of the string, when slipping begins, sends a little kink traveling along the string. That kink is reflected from the bridge, and when it arrives back at the bowing point it gives that part of the string a sudden jerk that ensures its becoming restuck to the bow. And after reflecting from the nut, the kink gives another jerk in the opposite direction as it completes its round trip, and this helps break the string loose at just the right time to start another cycle.

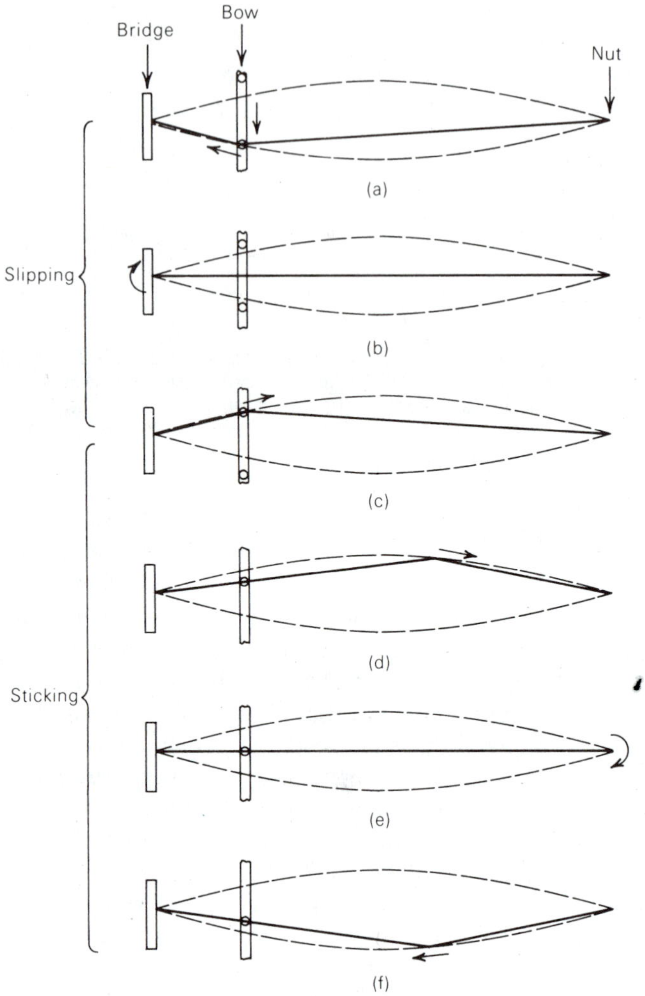

Bridge
Bow
Nut
Slipping
Sticking
(a)
(b)
(c)
(d)
(e)
(f)

FIGURE 11.7 Shape of the ideal bowed-string vibration at six times during its cycle. Strobe lighting makes these patterns easily visible. The string always is divided into two straight segments, with the kink traveling at uniform speed around the lens-shaped path. The motion shown is "down-bow"; for "up-bow" the kink travels around the loop counterclockwise instead. (a) The moment when the string breaks loose from the bow and the kink begins moving toward the bridge. (b) The kink has just arrived at the bridge. (c) As the kink passes the bow it changes upward to downward motion for that piece of the string so that it is recaptured by the bow. (d), (e), (f) The kink travels to the nut and reflects back; when it arrives at the bow it helps dislodge the string and start the cycle again. ○ denotes points on the bow where the string sticks.

When this self-regulating reinforcement mechanism is working properly, the string vibration is as shown in Figures 11.7 and 11.8. The corresponding mode energy recipe is shown in Figure 11.9. The general decreasing trend from low to high modes is the same as for plucking; for bowing, however, all modes participate, even those with nodes at or near the bowing point.

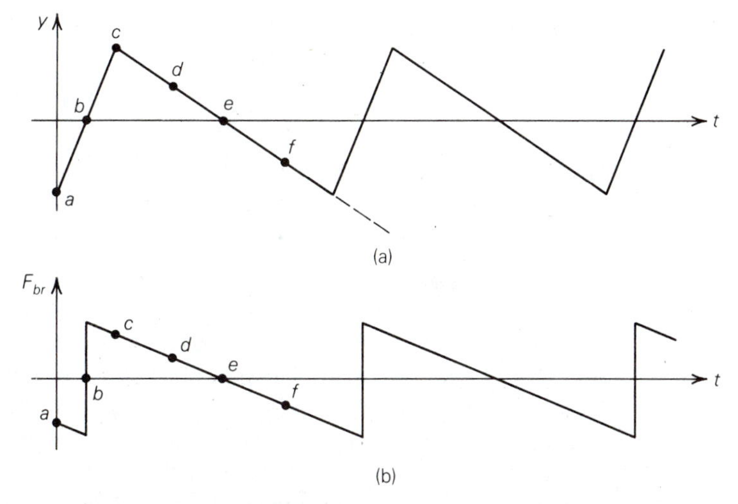

FIGURE 11.8 (a) Sideways displacement of the string at the bowing point as time goes by, for down-bowing. Dashed line shows how the bow continues to move after the string breaks loose. The closer the bowing point is to the bridge, the shorter the fraction of each cycle (a-b-c) during which slipping occurs. (b) Sideways force exerted by the string against an unyielding bridge; this is only an approximation to what happens at a real bridge. Letters refer to corresponding times in Figure 11.7, where the slope of the string as it meets the bridge determines the sideways force as a fraction of the (nearly) constant tension in the string.

How long does it take for the traveling kink to complete a round trip? The distance traveled is twice the string length, and the speed is $v_t = \sqrt{T/\mu}$ (as in Chapter 10), so the time period is $P = 2\,L/v_t$. The frequency of kink-triggered vibrations is $f = 1/P = (1/2\,L)\,\sqrt{T/\mu}$, which is simply the natural frequency of the string's first mode. Thus, the stick-slip mechanism automatically drives the string at just the frequency needed to get a strong response.

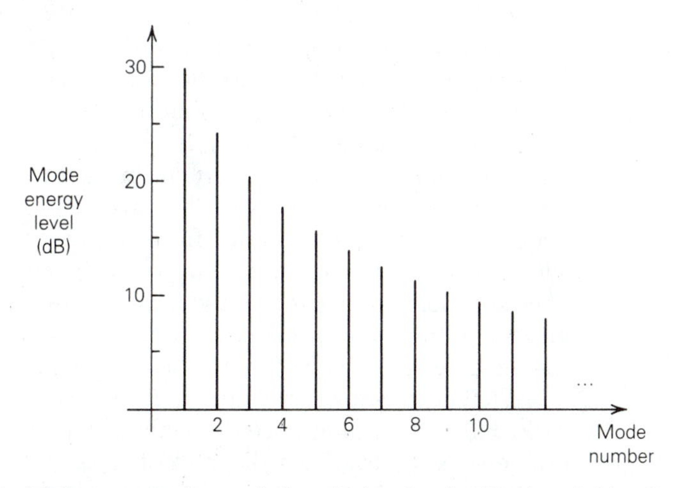

FIGURE 11.9 Relative amounts of energy in the natural modes of an ideal bowed string. Compare with Figure 10.4 for a plucked string, which has the same 6-dB/octave envelope.

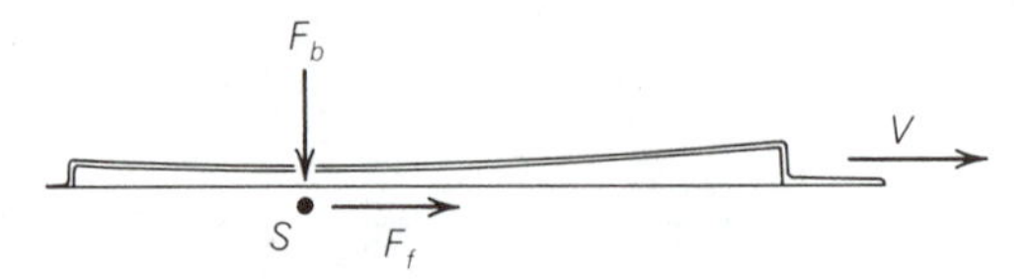

FIGURE 11.10 The bow moving with speed V exerts on the string S both a sideways frictional force F_f (Figures 11.5 and 11.6) and a downward bowing force F_b (Figure 11.11).

A remarkable feature of this mechanism is that the string vibration shapes, the spectrum of natural mode energies, and the force exerted on the bridge by the string are predicted to be always similar. Although the amplitude may be larger or smaller, the shapes should always be about the same, regardless of bowing force, speed, or location. There are, in fact, musically significant differences in spectrum and timbre when the bowing point is changed, which means this idealized picture has missed some details (for which you should read Schelleng). But at least bowing timbre depends upon bowing point significantly less than plucking timbre does on plucking point. So violinists move their bows toward or away from the bridge more for the sake of loudness and bowing speed control than for timbre changes.

String players often speak of "bow pressure," but this is more accurately called *bowing force*. This means the downward force holding the bow against the string, *not* the frictional force parallel to the bow (Figure 11.10). If the bowing force is too light, the slipping tends to take place too soon, before the kink returns. You may fail to get any steady tone, or you may even set up two or more traveling kinks and maintain oscillation at one of the higher natural frequencies of the string. We can get a similar effect deliberately but more reliably by resting a finger lightly on one of the nodes of the desired harmonic mode, leaving it free to vibrate while damping out all lower modes. This technique is used when the music calls for playing harmonics on open strings.

There is also an upper limit on the bowing force; if you push too hard, the kink will be unable to break the string loose from the bow at the right moment. The regularity of vibration is no longer self-maintained, and the sound becomes an erratic squawk. Figure 11.11 shows how the bowing force must be maintained within certain limits for normal playing and how that range becomes narrower as the bowing point is moved closer to the bridge. Beginners do well to bow far from the bridge, where the wider range of allowable force is more forgiving of their irregularities. Professionals who have sufficient control to use the narrower range near the bridge can achieve greater volume and brilliance, because the much greater bowing force they use increases the friction force, which then delivers more energy to the string.

What is not shown in Figure 11.11 is how the bowing speed enters in. For each different bowing speed a similar diagram could be made; or for each bridge-to-bow distance a diagram of force versus speed could be made. In summary, this is what those diagrams would show: *Louder playing requires*

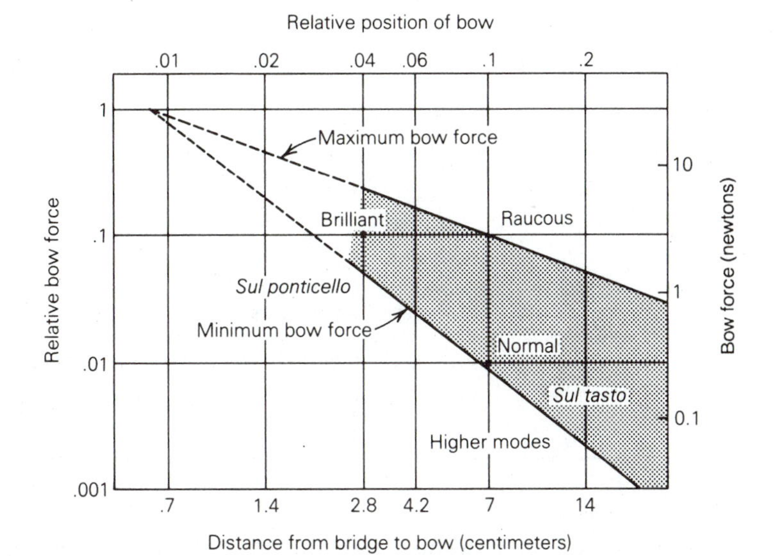

FIGURE 11.11 Diagram after Schelleng (1974) showing different bowing regimes for a cello A string with bow speed $V = 20$ cm/sec. Only in the shaded region between the two slanting lines can normal string vibrations be maintained.

greater bow speed or bowing closer to the bridge; these in turn tend to demand greater bowing force to maintain proper oscillations.

---

*Specifically, the total vibration energy E is fixed by the requirement that the bow speed V match the string speed v at the bowing point x_b (distance measured from the bridge) during the sticking part of the cycle. It can be shown that $E = (\mu L^3/6)(V/x_b)^2$, where μ and L are the linear mass density and length of the string. The required bowing force is not uniquely determined but varies in a qualitatively similar way with V and x_b. (See Exercises 11 and 12.)

This section has tacitly assumed a perfectly flexible string. As for the piano, string stiffness affects some details, such as reducing the strength of the very high harmonics. Stiffness also rounds off the sharp corner of the kink, which reduces the precision of its frequency-controlling feedback. These effects are discussed by Schelleng (1974).

Let us emphasize that most of the information in this section has been directed toward (a) satisfying our curiosity about just what the string does and (b) understanding the limits on how the bow is used. As we proceed to describe how the body of the violin radiates sound into the air, we need use only the information about the vibrational force exerted by the string on the bridge (Figure 11.8b) or, equivalently, the spectrum of mode energies (Figure 11.9).

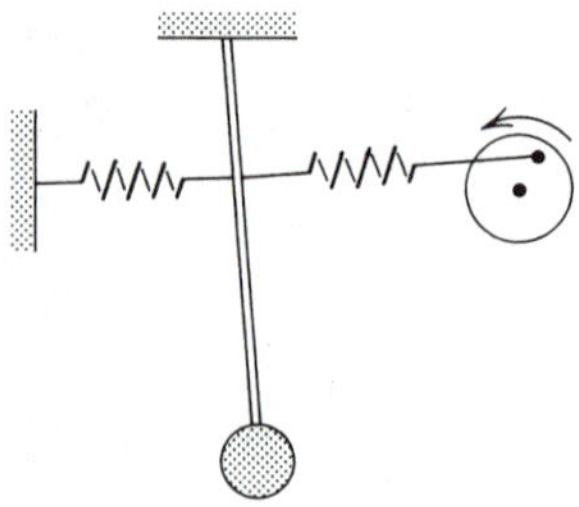

FIGURE 11.12 A driven harmonic oscillator. This pendulum has a solid shaft and is allowed to swing only in the plane of the page. Two rather weak springs are attached near the top; one is anchored on a solid wall at the left and the other attached eccentrically to a wheel. The springs apply a force to the pendulum that varies sinusoidally at the frequency of the wheel's steady rotation.

11.3 RESONANCE

To understand how a violin body responds to the vibrations of its strings, we must learn more about the physical phenomenon of resonance. While sound provides some of the most familiar cases, resonance can occur in vibrating systems of any kind, such as the simple example of Figure 11.12. This pendulum is a system with one degree of freedom and so has one natural mode frequency f_1 that depends on its length. The eccentric-mounted spring provides a way of exerting a sinusoidally alternating force on the pendulum. Let f stand for the frequency at which this driving force F repeats its cycle of pulling and pushing.

How does the pendulum respond? Imagine an experiment beginning with a very slow alternation in F, then gradually increasing f to higher and higher frequencies, always keeping the same amplitude of F. At each frequency we measure the amplitude of the resulting pendulum motion, and then graph its dependence on the driving frequency. The outcome (Figure 11.13) is that the motion is especially large when f is close to f_1. Such

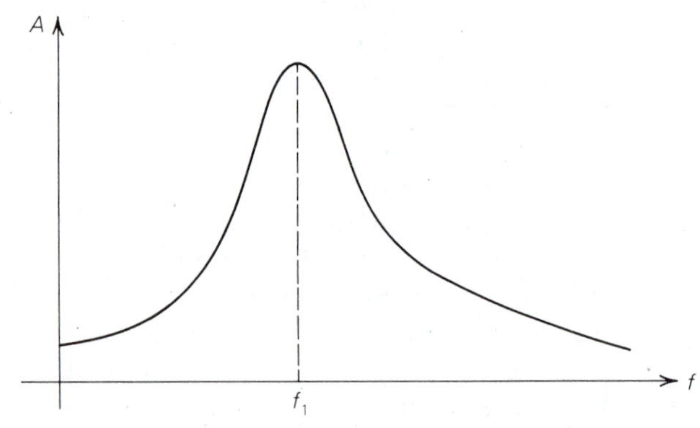

FIGURE 11.13 Amplitude A of motion of a driven harmonic oscillator, as a function of the driving frequency f. Peak response occurs when the driving frequency matches the natural frequency f_1.

experiments lead to the following definition:

> ***Resonance means the comparatively large-amplitude vibration that results whenever the frequency of some driving force closely matches a natural oscillation frequency of the system on which it acts.***

This is somewhat similar to pushing a child in a swing. For a given maximum force (say, the amount you can exert with one finger), changing the strength or direction of the force once per minute is too slow and once per second is too fast; neither produces much motion. But if you choose the driving frequency to be the same as the natural frequency of the swing (probably around once per 2 or 3 seconds), even a very small force persisting for many cycles eventually builds up a motion of large amplitude. (See also Box 11.1.)

An example of resonance in acoustics is the reputed shattering of wine glasses by opera singers (Figure 11.14). Here sound pressure variations provide the driving force causing vibration of the glass. It does no good to sing loudly if you do it at the wrong frequency. The trick is to tap the glass lightly, let it vibrate at its own natural frequency, and listen to the pitch of the emitted

FIGURE 11.14

BOX 11.1 ENERGY FLOW AND RESONANCE

Here is a way to understand why resonance occurs. If the driving frequency f is much less than the natural frequency f_1, the pendulum in Figure 11.12 has plenty of time to respond to the slow changes in the force F (Figure 11.15a); its position at each moment is practically the same as if the force at that moment were constant and had never been any different. The amplitude of oscillation for these low frequencies is *stiffness-limited*. During quarter-cycle 1, the force is in the same direction as the motion, so positive work is done and energy is delivered to the system. But during quarter-cycle 2, the force remains positive while the displacement is negative (moving back to the equilibrium position); multiplying these gives a negative product for the work done. That means the system returns the energy to its surroundings.

If f is much larger than f_1, the pendulum has trouble responding fast enough (Figure 11.15b); every time it gets turned around, say, moving toward the left, the force already is going on into the next positive part of its cycle. The amplitude for these high frequencies is *inertia-limited*. Again, the pendulum receives energy on one quarter-cycle but returns it on the next.

At resonance, $f = f_1$, the stiffness and mass are played off against each other and the amplitude is *dissipation-limited*. The pendulum position lags a quarter-cycle behind the force (Figure 11.15c). This is just enough so that positive force always occurs while the motion is toward the right and negative force for motion toward the left. The external force always does positive work on the system, the energy flows in the same direction on every quarter-cycle, and the vibration energy builds up steadily over many cycles to a large value. Only for large amplitude is there enough dissipation to use up the continuing energy input and keep the amplitude from growing any further.

sound. Then if you can sing a pure and steady tone at that same frequency, you can build up a strong resonant response, a vibration with amplitude large enough that it may exceed the breaking strength of the glass.

One detail we have left out so far is energy dissipation. This means anything that allows vibration energy to escape from the oscillator, so it includes both friction (conversion to heat energy) and radiation (conversion to sound energy). Friction in the wood, friction in the varnish, and radiation of sound all make important contributions to the total dissipation of energy in a violin body.

Dissipation is especially important near resonance, because large-amplitude motions give more opportunity for these forces to divert oscillation energy into other forms. If the dissipative forces are strong, they prevent the occurrence of very large vibrations in the first place (Figure 11.16c). If the dissipative forces are weak, the vibration amplitude must become large at resonance before the forces can remove as much energy during each cycle as is being steadily supplied by the driving force (Figure 11.16a).

Both extremes have important acoustical applications. The sharp, high peak for small dissipation indicates a spectacularly strong and finely tuned resonance. That is exactly what will be needed in the next chapter to explain how flutes and organ pipes build up such large air vibrations inside and only at certain strongly preferred frequencies. The broad, low peak for large

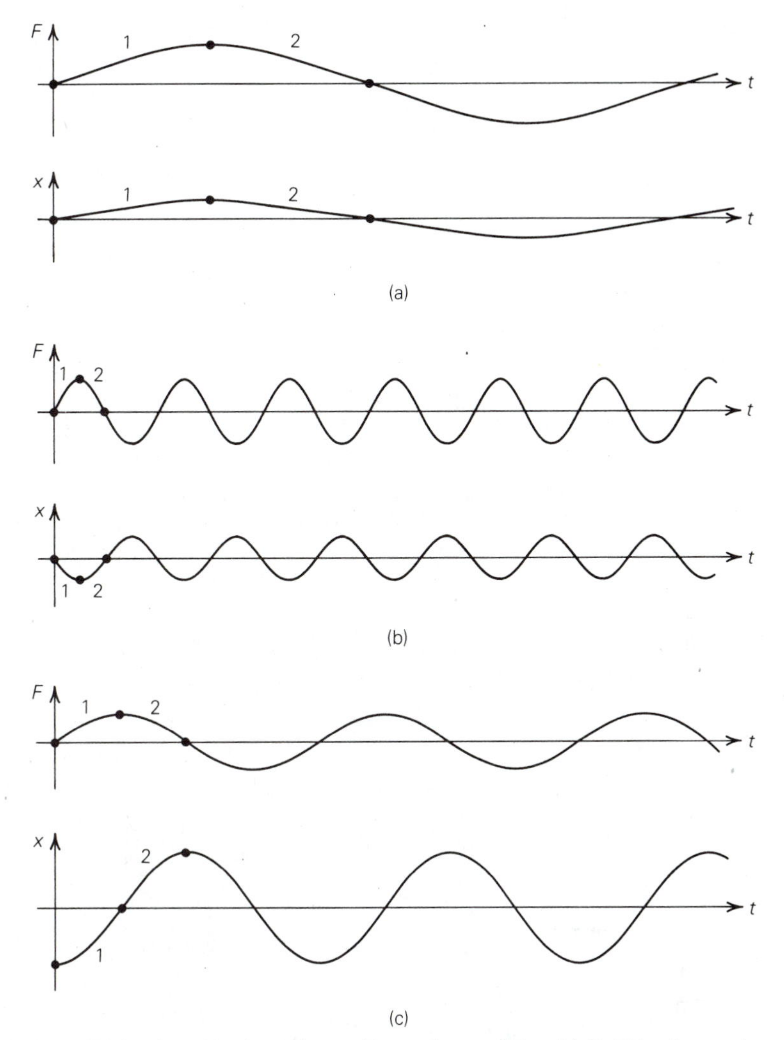

FIGURE 11.15 Driving force F and resulting position x of an oscillator. (a) At driving frequencies well below resonance, they are in phase. The force remains positive while the mass moves out on quarter-cycle 1 and back on quarter-cycle 2, resulting in no net work done. (b) Displacement from equilibrium lags a half-cycle behind force for driving frequencies well above resonance; again, negative work in quarter-cycle 1 cancels positive work in quarter-cycle 2. (c) At resonance, force and motion are always in the same direction and always doing positive work on the oscillator.

dissipation represents a much weaker resonance but one that is not especially choosy about frequency. In Chapter 14 we will see that this type of broad resonance is what enables you to sing a recognizable vowel sound at any of a wide range of pitches.

These ideas are easily extended to systems with many degrees of freedom. Each natural mode simply has its own resonant response, peaked at its own characteristic frequency. Thus, the total response of the system to a

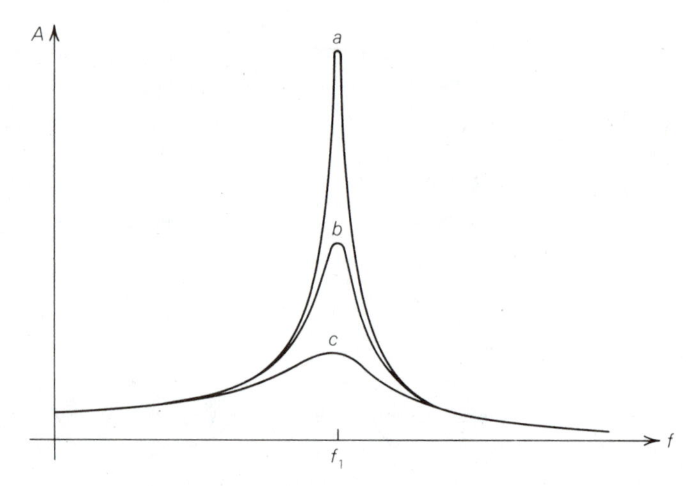

FIGURE 11.16 Influence of dissipation on resonance. (a) Weak dissipation processes allow amplitudes to become very large. (b) Moderate dissipation. (c) Strong frictional or radiative dissipation prevents large motions.

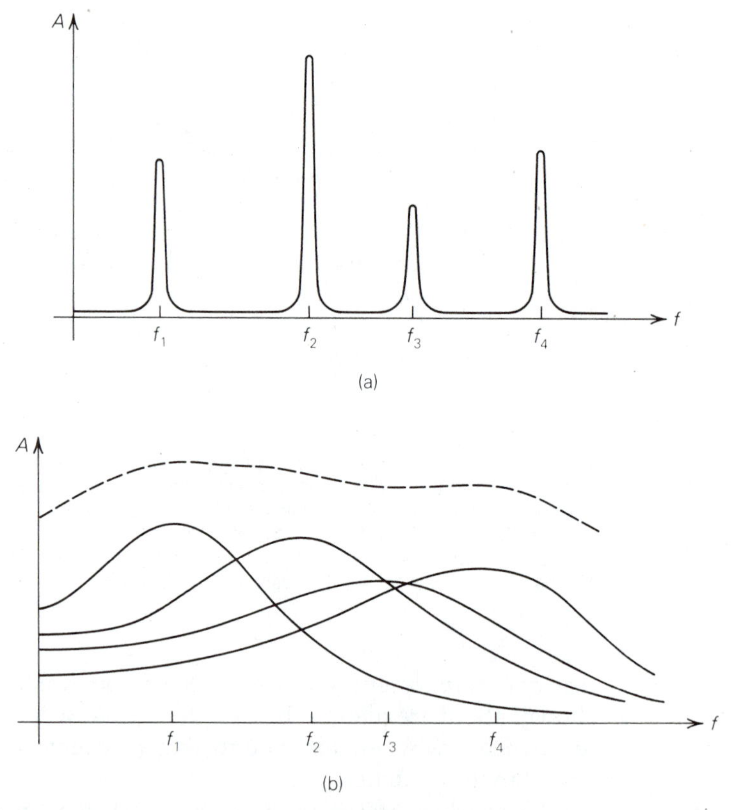

FIGURE 11.17 Resonance with several degrees of freedom. (a) Weak dissipation, well-separated resonances for each natural mode. (b) Strong dissipation, simultaneous response of several modes; total response of all modes (dashed line) may remain fairly uniform at different driving frequencies.

sinusoidal driving force, which is the combined motion of all these modes, may look like Figure 11.17. Case (a) is what we want for air columns in wind instruments—strong resonance on a few well-defined pitches but inability to vibrate much at "wrong" notes in between. Case (b) is what a piano soundboard needs—enough overlapping that sound of all frequencies will be similarly enhanced, so that the soundboard responds well at any frequency to the driving force it receives from the strings.

The violin is an in-between case. As we shall see in the next section, a good violin body may actually have a peaky, nonuniform response to pure sine waves, and this can give it a pleasant individual character. But some overall average of the total response to an entire harmonic series needs to stay fairly constant from one note to another.

11.4 SOUND RADIATION FROM STRING INSTRUMENTS

How can the concepts of resonance be applied to the violin body? Let us ignore at first the real nature of the bowed-string vibrations and imagine that a sinusoidally varying force is applied to the bridge. This can be done in the laboratory by attaching a small electrically driven shaker. We want to see how the response of the instrument to that driving force changes with frequency.

We must stop to ask here what will be the measure of the response. If we mean "amplitude of vibration of the violin body," that will be difficult to measure. Furthermore, we will have to separately determine the radiation efficiency; that is, how much vibration of the surrounding air is created by a given amount of motion of the body. That radiation efficiency can be expected to change in a complicated way with increasing frequency, and it also will be somewhat difficult to measure. We can bypass those difficulties (although missing out on some interesting physics) by defining *response* to mean "sound level produced in the air a few meters away for a given driving force on the bridge." For practical purposes we are interested mainly in this net effect of the two-step process.

A typical *response curve* for a violin is shown in Figure 11.18. The important features are (1) practically no response below the first resonant frequency at approximately 273 Hz; (2) another prominent resonance at approximately 460 Hz; (3) rather uneven response up to approximately 900 Hz, with a significant dip around 600 to 700 Hz; (4) better mode overlapping and more even response (with some exceptions) above 900 Hz; and (5) gradual decrease in response toward high frequencies. That decrease is not as rapid as would be predicted for the body alone, because resonance in the bridge strengthens the vibrations of about 2 to 3 KHz. You would get the impression from this response curve that different notes differ vastly in their loudness, and that the response is exceedingly poor and uneven between roughly 200 and 1000 Hz, corresponding to the notes G_3 through C_6 that are most often played. But that

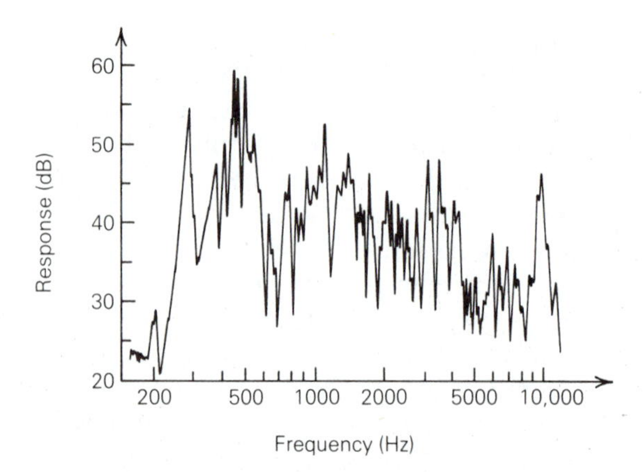

FIGURE 11.18 Response curve of one particular violin, showing how the sound level produced at one particular location nearby depends on the frequency of a sinusoidal force applied to the bridge. (From Hutchins, 1973. Courtesy of Journal of the Audio Engineering Society.)

impression is misleading, because the bowed string does *not* exert a sinusoidal force on the bridge; we must consider all harmonics, not just the fundamental.

This is where the power and beauty of Fourier analysis come into play. Rather than trying to figure out directly what would result from a sawtooth force upon the bridge, we can ask what each of its components would do separately. Figure 11.19 shows how the string energy spectrum is modified by the body's response characteristics to form the spectrum of the sound wave heard by a listener. The bar representing each harmonic of the input (a) is raised by the number representing the response at that frequency (b) to obtain the strength of that harmonic in the output recipe (c), (d), (e).

The output sound wave is not even approximately a sawtooth (see Figure 2.5d, for instance), and it has a tone quality determined largely by whatever harmonics fall between 800 and 4000 Hz. The presence of many high harmonics of appreciable strength is one of the main features that makes a "stringy" sound distinguishable from others. The unevenness of the response curve below 900 Hz is not disastrous after all, because those low components are only a part of the recipe for the sound you hear. In a much more subtle way the nonuniform response to low fundamentals also contributes to our mental image of a pleasing string sound—the constantly changing color from note to note helps give violin music its interesting and individual character.

---

*Because different frequencies radiate with different efficiencies in various directions, we must expect certain fine details of response curves to change with microphone position. All the spectra of Figure 11.19c–e must be taken only as typical of the possibilities, not as absolutely characteristic.

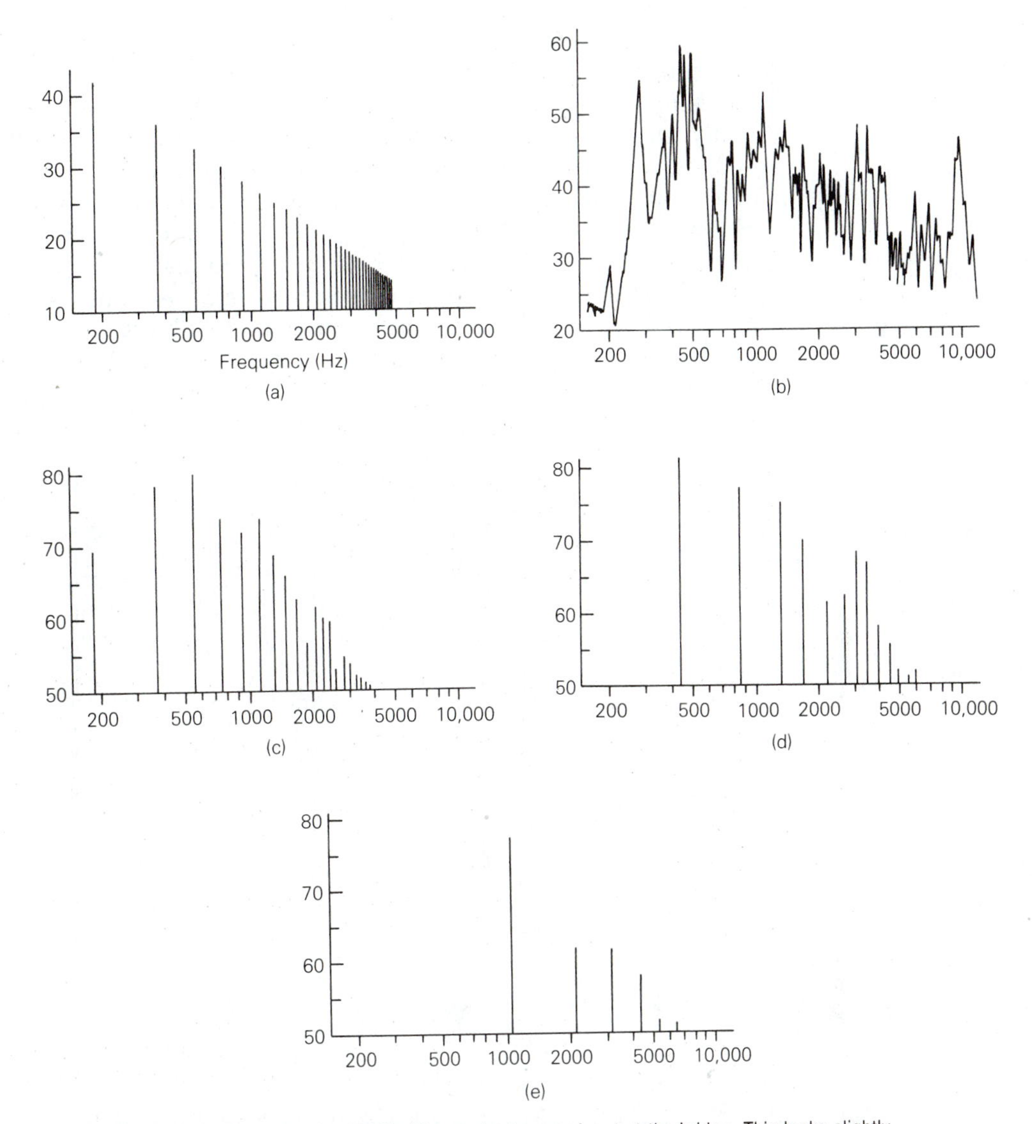

FIGURE 11.19 (a) Spectrum of a 196 Hz (G_3) sawtooth wave input at the bridge. This looks slightly different from Figure 11.9 because the horizontal axis is modified to represent octaves as equal distances. (b) Response curve determining how much each component is enhanced through radiation from the body. (c) Resulting sound spectrum. Note that use of decibel scales takes a multiplication (of string component energy by a response or amplification factor) and turns it into addition, so that each bar in (a) is merely lengthened or shortened according to the different heights in (b). (d), (e) Similar radiated spectra for 440 Hz (A_4) and 1047 Hz (C_6). These must be regarded only as typical; spectra from another instrument would differ in detail.

Determination of response curves requires considerable time and equipment, and the results differ a lot from one instrument to another. So another, much simpler test often is performed to give what is called a *loudness curve.* (Strictly speaking, that is a misnomer, because it is sound level that is measured, not loudness.) For this, a simple sound level meter measures the violin's sound output as it is bowed vigorously, as loudly as it will speak, on each note in its range. This test gives no detailed information about separate harmonic components, but it can be a very good diagnostic tool.

Figure 11.20 shows loudness curves for several instruments. The first case is a violin with weak response between D_4 and A_4 but B's that stick out like sore thumbs. The second curve is for a good instrument by Stradivarius and

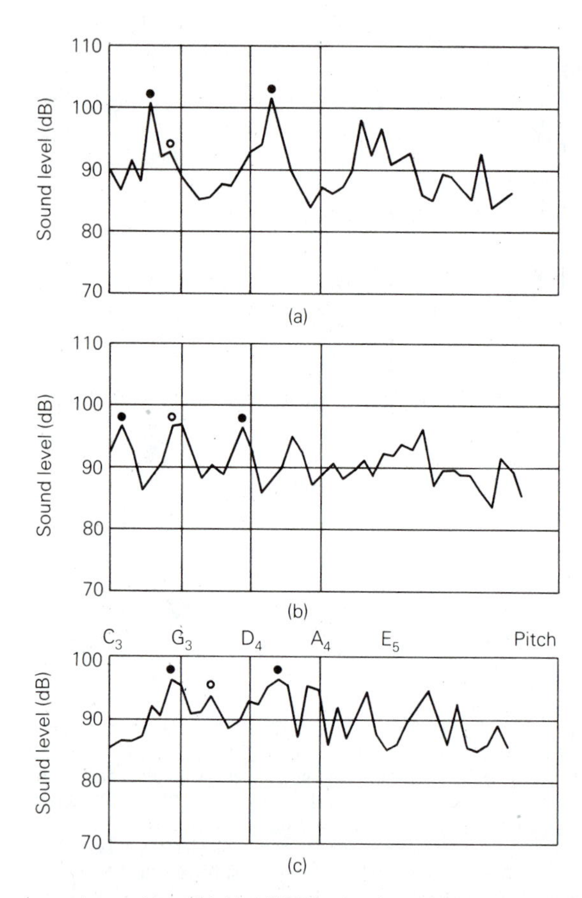

FIGURE 11.20 Loudness curves, from Hutchins (1962), showing sound levels produced by playing each note as loudly as possible. Vertical lines mark open-string pitches. (a) A poor violin from the early eighteenth century. (b) A good Stradivarius of similar age. (c) A good viola. Open circles indicate air resonance; solid dots indicate wood resonances. (By permission of *Scientific American.* All rights reserved.)

the third for a good viola. There are three resonances that often can be identified on these curves and seem to be crucial in determining the quality of the instrument.

The open circle above each curve indicates the main *air resonance,* which is determined primarily by the size of the f-holes and the volume of the body cavity, rather than any properties of the wood. At this frequency the air moves strongly in and out through the f-holes on each cycle, just as for the Helmholtz resonator action of an empty jug (Section 2.5 and Exercise 17 in Chapter 2). If the main air resonance is tuned to occur near the second open string frequency (D_4 for violin), it helps make a strong, solid sound for that note and others nearby. Higher cavity modes allow the contained air to slosh back and forth within the body but with rather little motion through the f-holes. This would imply that all air resonances except the lowest are relatively unimportant in sound radiation. Measurements by Jansson, however, indicate that their effect is not entirely negligible.

The solid dot toward the right above each curve identifies the *wood resonance,* which corresponds to the lowest natural mode of the body and is determined by the mass and stiffness of the wood (see Exercise 2). This seems to help most if it occurs near the third string frequency (A_4 for violin). The poor quality of violin (a) is largely due to both the weakness of the air resonance and too much separation between the air and wood resonances.

The solid dot toward the left denotes a resonance called *wood prime.* It is somewhat surprising to see a peak at such a low frequency, because Figure 11.18 shows essentially no response there; but that gives a clue to its meaning. Another clue is that it always occurs just one octave below the main wood resonance. Wood prime represents notes whose radiated fundamental is extremely weak but whose *second* harmonic is strongly reinforced by the wood resonance. This, too, gives a higher sound level and more solid tone, so that violin (b) produces its lowest notes with good strength. Violin (a) is weak on its three lowest notes, because its wood resonance frequency is too high, and an expert might use this information to do a little judicious carving to lower it.

The viola (c) has quite even response, and its resonances are spaced about right; but it also is weak at the bottom of its range. This is a serious problem for most violas and cellos, because they are too small (in comparison to the violin) for the notes they are supposed to play; the smaller body size makes it difficult to get the wood resonance low enough. The new Hutchins instruments are designed especially to deal with this problem.

String instruments are sometimes beset with an entirely different problem called the *wolf tone.* This is a very raucous and unsteady tone, which may seem to want to jump to a slightly different pitch or sometimes up an entire octave. It occurs when a string is fingered to produce a note too close to a main wood resonance that is too strong. If the body vibrates too easily at the resonant frequency, the bridge yields a great deal and does not provide a

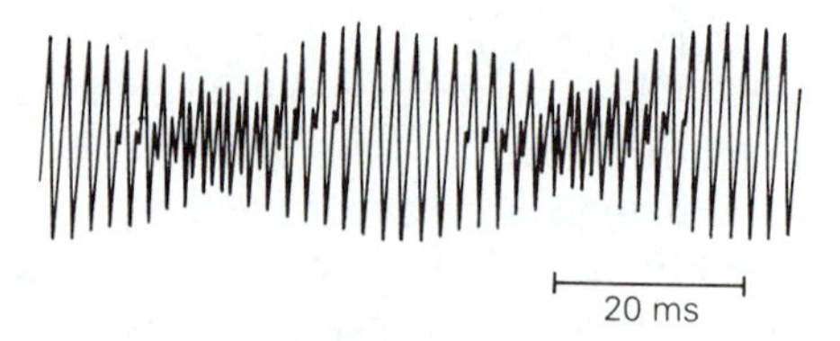

FIGURE 11.21 Waveform of the force exerted on the bridge by a violin string undergoing wolf-tone vibrations, measured by McIntyre and Woodhouse.

sufficiently rigid support for the string. That is, the bridge impedance is no longer enough larger than the string's for the string to resonate well; its energy is drained away too quickly. More specifically, reflections from the bridge are not strong enough at the body's resonant frequency to properly maintain the stick-slip mechanism. The string and the body may adopt the strange compromise of vibrating simultaneously at two slightly different frequencies, one above and one below the wood resonance. The combination of these two vibrations produces the curious beating effect of the wolf note (Figure 11.21).

Cellos are particularly prone to wolf tones, and they commonly occur in otherwise good instruments. The problem arises because the bridge impedance is not enough larger than the string's; the simplest solution is to change to a lighter string, which has less impedance. Wolf tones also can be suppressed by attaching a small mass near the middle of the short section of the offending string between bridge and tailpiece. This mass is chosen to make that short end alone resonate at the wolf frequency; it turns out to have an overall antiresonant effect that brings the bridge impedance back up. Then the string can work with the bow to vibrate normally and regain its rightful control of the whole instrument's vibrations.

The impedance seen by the string at the bridge is also the key to understanding the use of violin *mutes*. These are small pieces of wood or plastic that can be clamped on or against the top of the bridge without touching the active portion of the strings. The additional mass increases the bridge impedance, increasing the impedance mismatch between strings and bridge (see Box 9.3) and decreasing the transmission of energy from strings to body, especially for high frequencies. A practice mute with even larger mass can greatly reduce the audible sound output, so that a string player can practice long and vigorously without disturbing the neighbors.

In closing, let us briefly describe one other interesting property of sound radiation from string instruments. The shape of their bodies makes it seem quite unlikely that they are equally effective in pushing the air outward in every direction. The radiation patterns have been studied in some detail by Meyer, with results such as those shown in Figure 11.22. The patterns are quite complex and differ from one frequency to another. But there is a general tendency toward uniform radiation in all directions at the lowest frequencies and toward a distinct beam of radiation moving outward from the top plate at

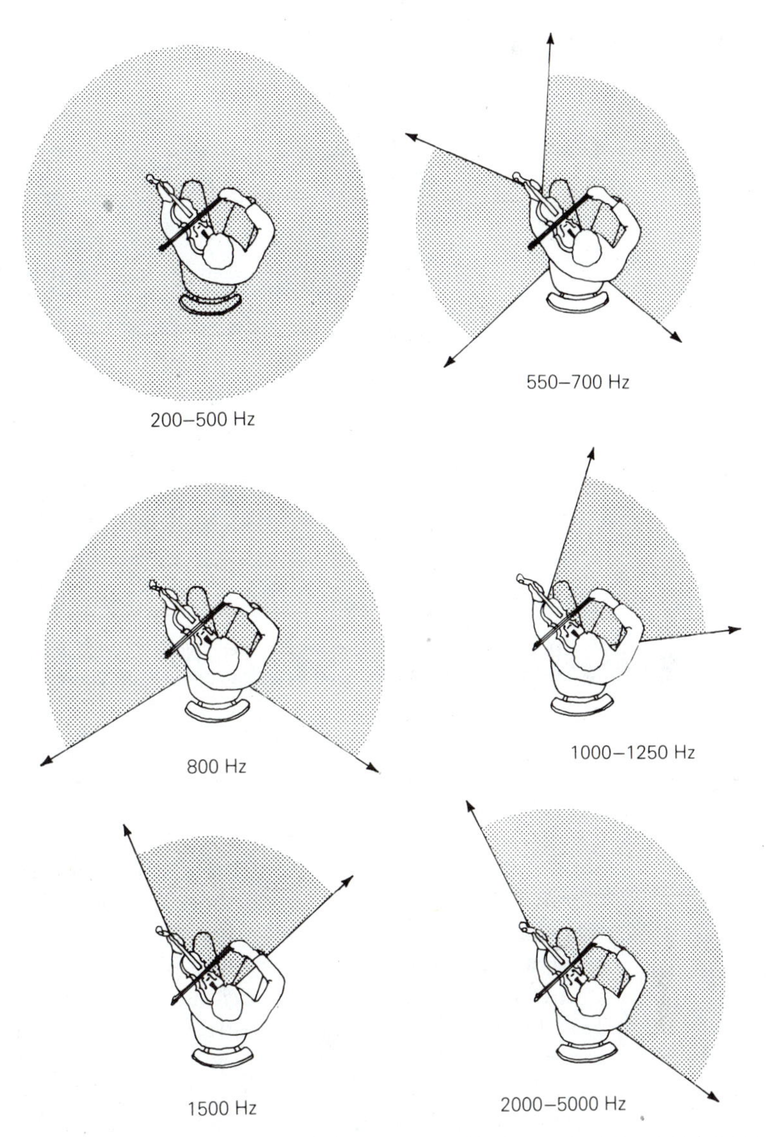

FIGURE 11.22 Directional radiation from a violin, as measured by Meyer. Shaded regions have sound level within 3 dB of maximum. Outside the shaded regions the intensity is merely weaker, not zero. (By permission of *JASA*)

high frequencies. Meyer uses his measurements to argue that the European orchestra seating system (Figure 11.23a) is often preferable to the American (Figure 11.23b), both for good tone quality toward the audience from the cellos and for contrast between first and second violin parts. This is not a universal rule, however; the best arrangement depends also on the auditorium, as well as on the nature of the music played.

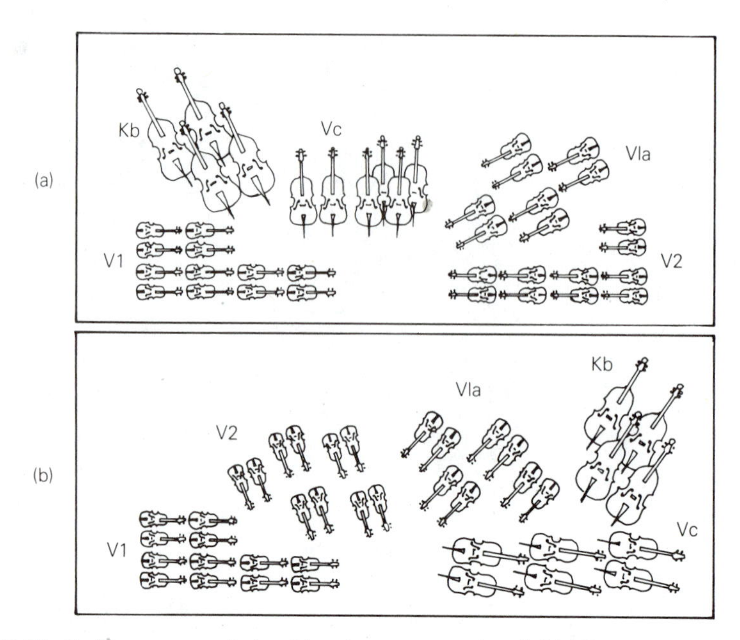

FIGURE 11.23 Seating arrangements for string players, commonly called (a) European and (b) American. The audience is toward the bottom of the page. (By permission of *JASA*)

SUMMARY

Sound production by bowed string instruments requires first that a repetitive vibration be maintained by the bow on the string, then that this vibration be passed on in turn to the bridge, the body, and the surrounding air. The reflection of wave pulses traveling along the string provides a stabilizing control on the alternation of sticking and slipping of the bow on the string. The resulting vibration of the string includes all members of the harmonic series, each of which acts nearly independently through the bridge upon the body.

The body and enclosed air each resonate at several of their own natural frequencies, strengthening some components of string vibration more than others and determining the characteristic tone color of the instrument. The combined effect of body response and radiation efficiency is summarized in the response curve (response to sinusoidal driving force on the bridge). Some of the musical consequences of that information are obtained more easily as a loudness curve (response to actual bowing of the strings). The best instruments have strong and relatively uniform loudness, with air and wood resonances near the frequencies of the second and third open strings.

Resonance in general means large-amplitude vibrations in response to an alternating driving force and may be characterized by (1) the frequency at which it occurs, (2) the maximum response, and (3) the width of the band of frequencies for which the response is large. The resonant frequency may be any natural mode frequency of the driven object, and the height and width of each resonance peak are determined by the dissipative forces involved. Stronger frictional or radiative energy losses tend to make resonance peaks lower and wider.

REFERENCES

The history and recent progress of research on the violin, including several topics not treated in this book, have been reviewed in highly readable style by Carleen Hutchins, *JASA, 73,* 1421 (1983), and by M. E. McIntyre and J. Woodhouse, *Interdisciplinary Science Reviews, 3,* 157 (1978). Both articles provide many additional references.

Every string player also should read the *Scientific American* articles by Carleen Hutchins ("The Physics of Violins," November 1962, p. 79), and John Schelleng ("The Physics of the Bowed String," January 1974, p. 87). Both are reproduced in the Hutchins collection. Those interested in violin construction and testing will find another article by Hutchins ("The Acoustics of Violin Plates") in the October 1981 issue.

For further understanding of the discussion of how modern violins differ from earlier instruments, it is very informative to listen to recordings made in the 1980s on authentic Baroque instruments by such groups as Concentus Musicus or the Academy of Ancient Music. The changes in instruments are tied closely to changes in performance practice.

Informative measurements on bow force and motion are described by Anders Askenfelt, *JASA, 80,* 1007 (1986) and *86,* 503 (1989). A more detailed account of Schelleng's work on bowing is found in "The Bowed String and the Player," *JASA, 53,* 26 (1973); this article is reproduced in the Rossing collection. Schelleng's article on the acoustical effects of varnish is in *JASA, 44,* 1175 (1968).

Those interested in string instruments of unusual size and modified design will find another interesting article by Carleen Hutchins, "Founding a Family of Fiddles," in *Physics Today, 20* (February 1967), 23.

Many original research papers are reprinted in two volumes edited by Hutchins titled *Musical Acoustics, Part I, Violin Family Components,* and *Part II, Violin Family Functions* (Halsted Press, 1975 and 1976). Early reports of ongoing research often appear first in the *Journal of the Catgut Acoustical Society.*

Detailed discussion on mathematical modeling of the violin is contained in the excellent book by L. Cremer, *The Physics of the Violin* (Cambridge, MA: MIT Press, 1983). Early results using two important new techniques for measuring violins and their radiation are reported by E. B. Arnold and G. Weinreich, *JASA, 72,* 1739 (1982), and *77,* 710 (1985); and by K. D. Marshall, *JASA, 77,* 695 (1985).

Studies of wolf tones have been reported in *Acustica* by Ian Firth (*39,* 252, 1978), by Colin Gough (*44,* 113, 1980), and by W. Guth (*41,* 163, 1978 and *63,* 35, 1987). Detailed investigations of bridge resonance have been made by M. Hacklinger, *Acustica, 39,* 323 (1978), and by W. J. Trott, *JASA, 81,* 1948 (1987). For more on directional radiation properties of all sizes of string instruments, see J. Meyer, *JASA, 51,* 1994 (1972).

SYMBOLS, TERMS, AND RELATIONS

P	period
f	frequency (variable)
f_n	natural mode frequencies
F	force
A	amplitude
v_t	wave speed along string
v	speed of transverse string motion at bow
V	speed of bow
T	tension
μ	mass per unit length
L	string length

bridge
bass bar
stick-slip mechanism
dynamic instability
bowing point
resonance
response curve
loudness curve
wolf tone

soundpost
scaling
bowing force
dissipation
mute

$$f_1 = \left(\frac{1}{2L}\right)\sqrt{T/\mu}$$

EXERCISES

1. (a) A coat of varnish adds mass to a violin body. What effect does that have on its resonant frequencies? (b) The fully dried varnish also adds stiffness. How does that affect the resonant frequencies? (c) The viscosity of the varnish when it is only partly dried provides a frictional energy-loss mechanism. How does that affect the resonance peaks? (d) If response curves were made several times during the drying of a new coat of varnish, how would they tend to evolve?

2. Explain how scraping bits of wood off a top or back plate may either lower or raise its natural frequencies by considering removal of material (a) near the middle of the plate and (b) very near the joint with the ribs.

*3. Sketch a diagram like Figure 11.3 for a bridge only half as high. Explain why this reduces the down-bearing, and explain its possible consequences on the sound radiated.

4. Our study of drumheads may suggest that the first mode of a violin top plate would be symmetrical, because the plate itself looks symmetrical. Yet the first mode of the plate shown in Figure 11.4a is clearly nonsymmetrical, moving mainly on the left side. Why do you think this happens?

5. The string bass normally has a range beginning at E_1, two octaves plus a minor third below the lowest note on a violin. What is the frequency ratio for those two notes? Normal total length for a violin is approximately 60 cm, of which the body alone is some 36 cm, and the active string length approximately 33 cm. The body is some 21 cm across at its widest point. If the double bass were just a scaled-up violin, how big would it be? What difficulty might you have in playing it?

6. When you hold a piece of chalk wrong as you move it across a blackboard (pushing instead of pulling, with the writing end leading instead of trailing), you may hear a terrible screech and see a series of dots

instead of a smooth line. Explain this in terms of dynamic instability, drawing a close parallel to violin bowing.

*7. If a violin A_4 string of length 35 cm is to be under 60 N tension, what linear mass density should it have?

*8. To see how important it is that violin strings be uniform, consider the effects of an extra lump of mass attached to the midpoint of a string. Using the traveling-wave picture, explain what happens as the kink travels down the string and why this may upset the stick-slip mechanism. Using the standing-wave picture, state how the added mass affects the natural frequencies of the odd-and even-numbered modes. Do they still form a harmonic series? Does that affect their ability to cooperate in producing a steady complex waveform?

9. In reality, a violin string has some stiffness, so like the piano strings in Chapter 10 its natural mode frequencies depart slightly from a harmonic series. If, for instance, the fifteenth mode of a certain string had $f_{15} = 15.5f_1$, explain why that mode could not participate in a complex waveform of frequency f_1. What effect does this suggest on the spectra of Figures 11.9 and 11.19?

10. Why is the production of "harmonics" (such as B_6 from an open E_5 string) encouraged by resting a finger lightly at one point on the string? Where would that point be for this example?

*11. Explain qualitatively how changes in bowing point and bowing speed are related to changes in overall vibration amplitude. Think carefully about how the bow and string must move at the same speed when sticking at the bowing point. It may help to sketch how these changes would alter Figures 11.7 and 11.8.

12. Suppose for a given bowing point you double the bow speed and so double the amplitude of string motion. Suppose also that you exert on average twice as much frictional force to sustain the motion (which is at least reasonable, although not entirely obvious). How many times greater is the work you are doing on the string? Show that this should give a sound level 6 dB higher.

13. From everyday life cite one acoustical and one nonacoustical example of resonance different from those mentioned in the text.

14. Spell out the application of the argument in Box 11.1 to the case of pushing a child in a swing.

15. How do you think the chances of shattering a wine glass by resonance are affected by whether it is empty or full? Explain.

16. If the fundamental frequency in Figure 11.19a changes by 1–2 percent because of vibrato, do the strengths of all radiated harmonics (Figure 11.19c) change in the same way? Explain in terms of Figure 11.19b. (This is another identifying characteristic for string tone, distinguishing it from the woodwinds, whose spectra tend to stay about the same during vibrato.)

17. For comparison with Figure 11.19c, sketch the approximate radiated sound spectrum for A_3 (220 Hz) played on the same instrument. You will get this by starting with a spectrum like Figure 11.19a (except shifted a little to the right for the higher frequency) and adjusting each line upward according to the response indicated for its frequency in Figure 11.19b.

18. Explain what difference it should make to a violin's tone if it had no f-holes.

19. Suppose you made a new violin, copying very precisely in every respect (including thickness everywhere of the top and back plates) the exact dimensions of a prized Stradivarius. Discuss as many physical reasons as you can why the result may still be a quite different instrument.

20. The strings on a double bass normally are tuned to E_1, A_1, D_2, and G_2. What is the interval between each pair? Why do you think they are not tuned in perfect fifths, as for the cello or violin? Justify your answer by discussing how far apart the finger positions are for adjacent notes in each case. (Hints: Refer to Box 7.1. Normal active string length is approximately 33 cm for the violin and 104 cm for the bass.)

21. Occasionally you may see a double bass with an extension on the fingerboard so that the lowest string can go beyond the scroll at the end of the neck. With the extension, the lowest possible note becomes C_1 instead of E_1. If the normal string length is approximately 104 cm, how much additional length must the extension provide?

22. "Playing harmonics" is done easily on open strings, which was the situation implied in Exercise 10 above, as well as in Exercises 21 and 22 in Chapter 10. But an important extension of the technique involves pressing the string firmly against the fingerboard (or guitar fret) with the index finger while resting the little finger lightly on the string farther down. If the desired note (say, for illustration, B_5) is two octaves above the one that would normally be heard for that position of the index finger, where must the little finger be? Express the answer in terms of what note you would hear if you pressed the finger down firmly at that point.

23. Consider further the nineteenth-century modification of old violins. Other things being equal, what percentage increase in tension would be required by a change in standard pitch for A_4 from 415 to 440 Hz? What increase in tension to maintain the same pitch if extension of the neck increases the active string length from 32 to 33 cm? What increase in tension if gut strings are replaced by 10 percent heavier wound strings? What overall increase in tension for all these changes together?

PROJECTS

1. If you are a string player, volunteer to give the class a demonstration of bowing techniques, illustrating as many points from this chapter as you can.

2. Measure all major dimensions of a violin and its strings. What would they be for a 50 percent larger viola? Discuss problems this causes in playing technique, and compare with actual viola dimensions.

3. Measure and interpret the loudness curve of some instrument available to you.

4. Cover an f-hole with masking tape and observe the difference in tone quality of various notes; perhaps even compare loudness curves with and without the f-hole covered. Are some notes affected more than others? Explain. Another possibility: Stick a lump of modeling clay at various places on a violin body, such as the mode 1 antinode seen in Figure 11.4a, and observe its effect on the tone.

CHAPTER

12 Blown Pipes and Flutes

The transverse flute, familiar from its use in concert bands and orchestras, reached its present form under the hands of Theobald Boehm in 1847. It is the most highly developed representative of another family of instruments that dates back several thousand years. Because its design involves several subtle features, we will save our discussion of the transverse flute for the last section of this chapter. To appreciate each of those features more clearly, we will spend most of the chapter on other examples that provide more straightforward illustrations.

Organ pipes, in particular, allow us to focus our attention on the mechanism of sound production by an airstream and resonating pipe and on the resulting tone color. After looking at idealized simple pipes in the first section and airstream oscillations in the second, we shall apply this information in Section 12.3 to understand how variations in pipe design can lead to widely differing types of tone. Before leaving organ pipes behind, we pause to consider in Section 12.4 how the sound from hundreds and thousands of pipes is combined to produce the majestic results that make the pipe organ the "king of instruments."

The recorder (German *blockflöte*) is described in Section 12.5. Its basic mechanism is clearly the same as that of organ flue pipes, but the added complication of fingerholes makes it possible to produce many different notes from the same pipe. The recorder provides a convenient step between organ pipes and transverse flute. The flute's key mechanism is a natural further development of the recorder fingerholes, but there are important differences in its mouthpiece design.

12.1 AIR COLUMN VIBRATIONS

Before we examined the bowing of violins, we learned first about the natural modes of stretched strings. Thus, before attempting to explain the operation of the driving airstream, we would like to know the natural modes of vibration of the air confined inside a long, thin tube. (We will use *tube* to mean the simplest possible idealization and *pipe* for real musical devices with mouth holes that are more complicated than a simple tube end.)

Let us consider for now only tubes of uniform cross section, leaving the possibility of tapers or bulges for later. In pipe organs you will find round metal pipes and square wooden ones, but those merely represent the easiest way to use the materials. It is of no acoustical importance whether the cross section is round or square, and we will use the term *cylindrical pipe* to include either case.

The uniformity of a cylindrical tube makes it easy to guess its natural modes, which must be standing waves of some sort. Suppose you create a sinusoidal disturbance at some point in the tube as time goes by. This launches a sinusoidal traveling wave, which propagates undistorted along the tube at constant speed. Reflection at either end of the tube provides a similar wave traveling in the opposite direction, and the two together make a sinusoidal standing wave (Box 12.1; see also Box 10.1). In the body of the

BOX 12.1 ACOUSTIC STANDING WAVES

Standing waves in an enclosed air column are analogous to standing waves on a string, as described in Chapter 10. But there are two additional complications: These waves are now longitudinal instead of transverse and so are a little more difficult to visualize; and the waves have two major aspects—pressure and motion—that must both be understood.

Consider Figure 12.1a. At the bottom you see uniform shading, representing uniformly dense air confined in a cylindrical tube. The arrows indicate air velocities at the moment when this "snapshot" is taken. Immediately above are three graphs—one for velocity v, one for displacement y, and one for pressure p, all as functions of location x along the axis of the tube. Remember that y represents a displacement along the x direction of an air parcel from its home location. This first snapshot catches every parcel moving through its equilibrium position, so that all the y values are identically zero. Because they all are evenly spaced, there are no compressions or rarefactions, and the acoustic pressure (the difference from atmospheric) is zero everywhere. Note that at this moment the vibration energy is entirely in kinetic form.

Another snapshot a quarter of a cycle later (Figure 12.1b) catches every part of the system at its maximum displacement from equilibrium. Each bit of air is momentarily at rest before traveling in the opposite direction, so there are no velocity arrows and the graph of v shows identically zero. In the vicinity of point C, all the air parcels have moved along together, so their density is unchanged and there is no pressure difference. But around point D, air has squeezed toward this point from both sides, so the strongest compression is created right at the point where there is no motion. Note that at this moment the vibration energy is entirely in the form of potential energy.

After another quarter-cycle we have a situation such as in Figure 12.1a again, except the velocity arrows are reversed and the velocity graph is turned upside down. Still another quarter-cycle gives a picture such as Figure 12.1b again, but with compressions and rarefactions interchanged and the y and p graphs turned upside down. The final quarter-cycle brings us back to the starting point and the entire sequence repeats.

We also can ask how things look at any one point as time goes by. When this is done for point C, we get Figure 12.1c, which presents the sequence of events described above. Because the pressure never changes at C, this point is called a **pressure node**; but the displacement and velocity amplitudes here are greater than anywhere else, and so it is called a **displacement antinode**. A quarter-wavelength away, at point D, we get the graphs of Figure 12.1d. This point has maximum pressure fluctuations and is called a **pressure antinode**; but the air at D never moves, so it is a **displacement node**.

(continued)

BOX 12.1 *(continued)*

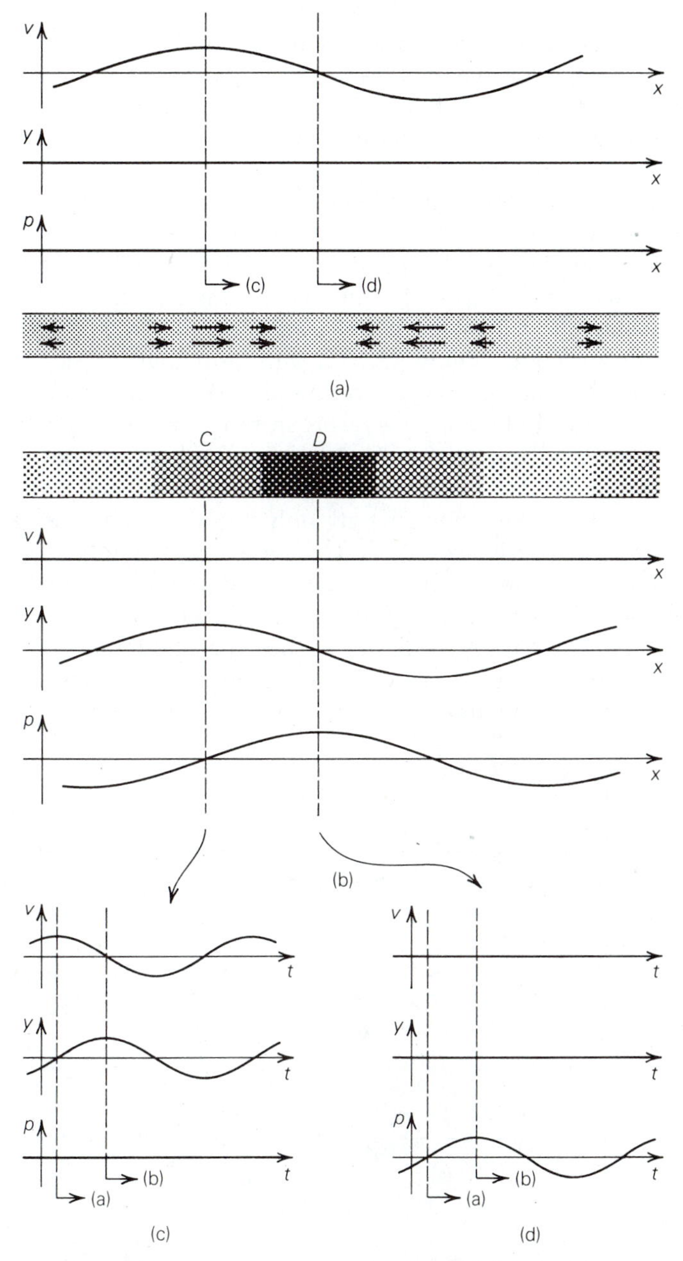

FIGURE 12.1 Standing waves in the interior of a cylindrical pipe.

These relations are critically important. You should study Figure 12.1 at length to be sure you see how parts (c) and (d) present an alternative form of the same information in parts (a) and (b). Pay special attention to the difference between position and time as independent variables (horizontal axis labels).

tube, then, sinusoidal standing waves are possible; the only question is whether they come out right at both ends.

What limitations are placed on a sound wave at the end of a tube? Consider first a closed end. Clearly the air molecules at the extreme end cannot go anywhere, because the end wall is in their way, so the displacement and velocity of the air at a closed end must always remain zero. But a displacement node must correspond to a pressure antinode (Figure 12.1b), and that is fine, because the end wall gives something to push against to build extra pressure.

At an open end the pressure must remain nearly atmospheric; only where the air is confined by the tube walls is it easy to build up appreciable extra pressure. The pressure node at an open end must correspond to a displacement antinode (Figure 12.1b again), and that is fine, because the opening makes it easier for the air to flow there than anywhere else inside the tube. The mouth hole of a flute or organ pipe always functions as an open end.

By imposing these boundary conditions we see that for most wavelengths a standing wave would come out wrong at one end or the other. A standing wave fits a tube of finite length only if it has one of a few special wavelengths that come out just right. These few allowed standing waves are the natural vibration modes of the air column and are closely analogous to the natural modes of a long row of pendulums (Figure 9.9, page 156).

What are the frequencies of these natural modes? Consider first an open tube, meaning open at both ends. Figure 12.2 shows how its natural modes must have some integral number of half-wavelengths of the sound match the

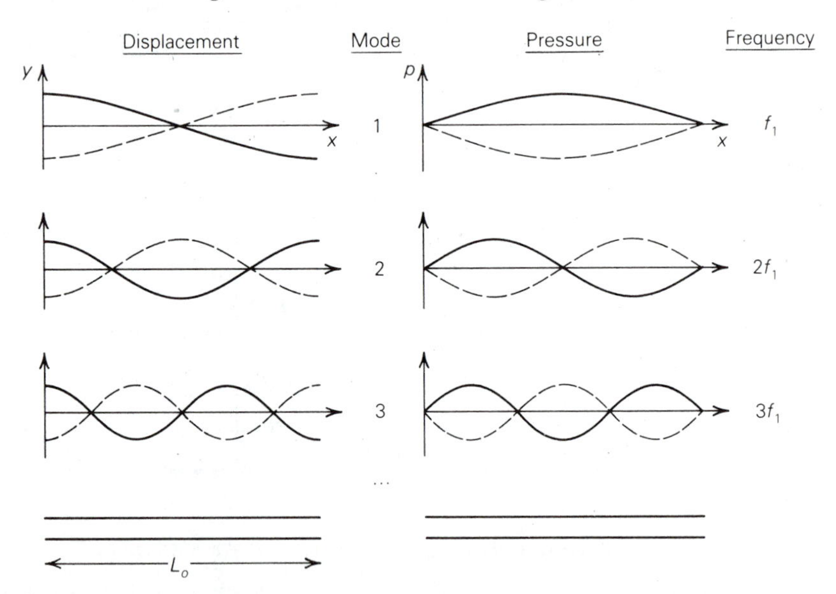

FIGURE 12.2 The first few natural modes of air confined in an open tube. At left, the curves are graphs of maximum displacements at two times a half-cycle apart; they must have antinodes at the ends of the tube. At right are corresponding graphs of acoustic pressure, which must have nodes at the ends. Fundamental frequency is $f_1 = v/2L_o$.

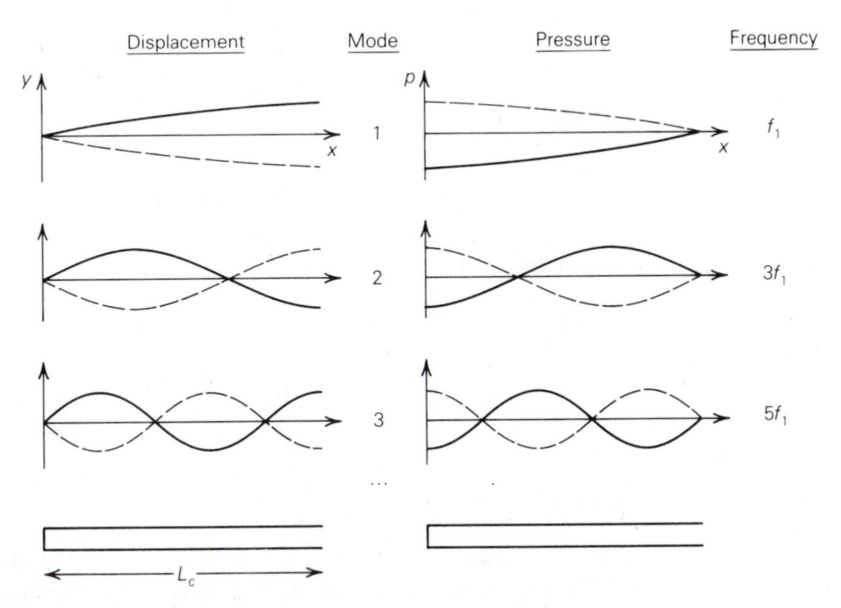

FIGURE 12.3 The first few natural modes of air confined in a closed tube (closed on one end). At left are displacement graphs; at right are pressure graphs for two times a half-cycle apart. Fundamental frequency is $f_1 = v/4L_c$.

tube length. For the nth natural mode, with wavelength λ_n, that means $n(\lambda_n/2) = L_o$, so that $\lambda_n = 2L_o/n$. But then the nth natural mode frequency is $f_n = v/\lambda_n = nv/2L_o$, where v is the speed of sound in air. Notice that these mode frequencies form a harmonic series, with $f_1 = v/2L_o$, suggesting that such pipes should be capable of producing highly musical sounds. This entire argument runs closely parallel to that for stretched string modes in Section 10.1.

The case of a tube closed at both ends produces similar formulas, but it is not musically interesting, because there is nowhere for the sound to come out. The term *closed pipe* (or, among organists, *stopped pipe*) normally is used to mean a pipe open at one end but closed at the other. As shown in Figure 12.3, the natural modes of such a tube must fit $\frac{1}{4}$, $\frac{3}{4}$, $\frac{5}{4}$, or in general some odd number of quarter-wavelengths into the length of the tube. This can be expressed mathematically as $(2n - 1)(\lambda_n/4) = L_c$, or $\lambda_n = 4L_c/(2n - 1)$. Then the natural mode frequencies are $f_n = v/\lambda_n = (2n - 1)v/4L_c$. Notice that f_2, f_3, f_4 . . . are merely 3, 5, 7 . . . times $f_1 = v/4L_c$; these, too, are members of a harmonic series, but all the even-numbered members are missing. This helps give stopped pipes a characteristically different sound from open pipes. Notice that if you have one open and one closed tube, both of the same length, the fundamental frequency for the closed tube is an octave lower.

These formulas can be turned around the other way, showing that the length needed to achieve a given fundamental frequency is $L_o = v/2f_1$ for an open tube but only $L_c = v/4f_1$ for a closed tube. For example, C_4 ($f = 262$ Hz) requires an open tube with $L_o = (344 \text{ m/s})/(524 \text{ Hz}) \cong 0.66$ m or a closed tube with $L_c \cong 0.33$ m. When organ builders are severely limited by space or cost, they have strong incentive to choose closed pipes for the low pedal notes, because the same note (although not the same tone quality) can be generated with a pipe only half as long as if it were open.

* The requirement that there be a pressure node precisely at the open end of a tube is an idealization that is not quite true. The effective location of the pressure node is really a little bit outside the end. Or, in terms of velocity, the airflow tends to continue a bit beyond the end of the tube as if still confined before it finally spreads out. The tube acts as if it were a little longer than it really is. This effect can be approximately allowed for by replacing every L in the formulas above by $L + l$, where the additional term is called an *end correction*. For low frequencies, $l \cong 0.3\, d$ for each open end, where d is the pipe diameter. In the example above, a diameter $d = 3$ cm would mean a correction of approximately 1 cm for each open end, and so actual pipe lengths $L_0 \cong 0.64$ m and $L_c \cong 0.32$ m.

Before proceeding with organ pipes, we pause to make a connection with ear sensitivity. We now can explain why the Fletcher–Munson curves (Figure 6.12) dip approximately 5 or 10 dB to indicate lower hearing thresholds at approximately 3 to 4 KHz. This is merely a resonance effect in the outer ear. The canal leading to the eardrum is roughly 2.5 to 3 cm long, open at the outer end and closed by the eardrum on the other. We see now that the first natural mode of this "closed pipe" should have a wavelength $4L_c$ of some 10 cm, corresponding to a frequency in the vicinity of $344/0.10 = 3.4$ KHz. The second natural mode produces another dip in the threshold curves above 10 KHz.

12.2 FLUID JETS AND EDGETONES

At the heart of every wind instrument must be a mechanism for converting energy of a steady flow of air into energy of air vibrations. Recall that this purpose was served for string instruments by the stick-slip mechanism, a form of dynamical instability. Now we consider *fluid-flow instabilities.* These also are examples of dynamical instability, because they represent situations in which small departures from a steady motion tend to become larger and larger.

Instabilities are a common feature of fluid flows. Remembering that the term *fluid* includes both gases and liquids, we can cite several familiar examples involving air and water (see Box 12.2). The following general properties are important:

1. In each case we can imagine a perfectly smooth and steady flow, as in Figure 12.4a, just as we imagined a steady slipping of a violin bow across the string. These sometimes are called *laminar* flows, which means

BOX 12.2 EXAMPLES OF FLOW INSTABILITY

1. Winter winds "sing" when blowing past a power line or the corner of a house.
2. Air escapes with a hissing sound through the heating vents in my house when their louvers are nearly closed.
3. When you drag a stick through a still pool, or a spoon through a cup of coffee, a turbulent wake is left behind.
4. Wind blowing past a flag does not just make it stand out straight from the pole; small ripples on the flag are amplified and it waves back and forth. There is a similar effect on leaves or even entire branches of a tree.
5. When I blow my nose during allergy season, there is a critical flow rate above which strong vibrations set in that help clear the nasal passages.
6. Wind blowing over a lake causes slight irregularities on the water surface to be amplified (as in Figure 12.4), thus generating the waves that rock your boat.
7. Smoke may rise a little way above a burning candle in a narrow plume, but then begins to waver and swirl.
8. The airstream from my hair dryer becomes turbulent by the time it has traveled 20 or 30 cm.

The last several examples are modified somewhat by the Bernoulli effect, which we will consider in Chapter 14.

"layered." But such steady flows are not observed unless they are quite slow.

2. Once a critical speed is exceeded, small perturbations tend to grow larger, taking energy away from the steady flow (Figure 12.4b). If the critical speed is only slightly exceeded, the new flow still may be steady in the more general sense of having regular, orderly *oscillations* superposed on it.
3. If the flow speed is well above the critical value, oscillatory disturbances of many different wavelengths and frequencies all grow, and the resulting chaotic state is called *turbulent* flow. Such flows are familiar; think of a towering thundercloud, whose lumpy shape shows the swirling motions of the upward-rushing air inside it.

For another example, a river 1 meter deep becomes turbulent for flow speeds around 1 m/s or more. (Even though the flow may look deceptively calm and smooth, notice that the pattern formed by leaves floating here and there on the surface continually shifts in a random, irregular way.)

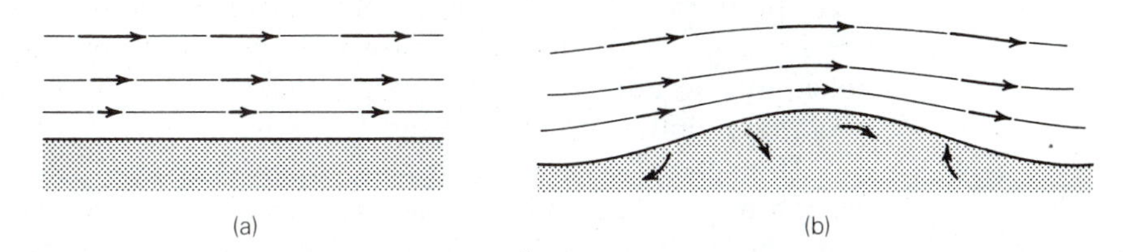

FIGURE 12.4 Flow of air above the surface of a lake. (a) The simple airflow that leaves the surface undisturbed occurs only for low speeds. (b) Instability sets in when the wind speed is high enough; any little ripple on the water surface alters the airflow in such a way that the air pushes the ripple along, does work upon it, and thereby supplies energy for the ripple to grow larger.

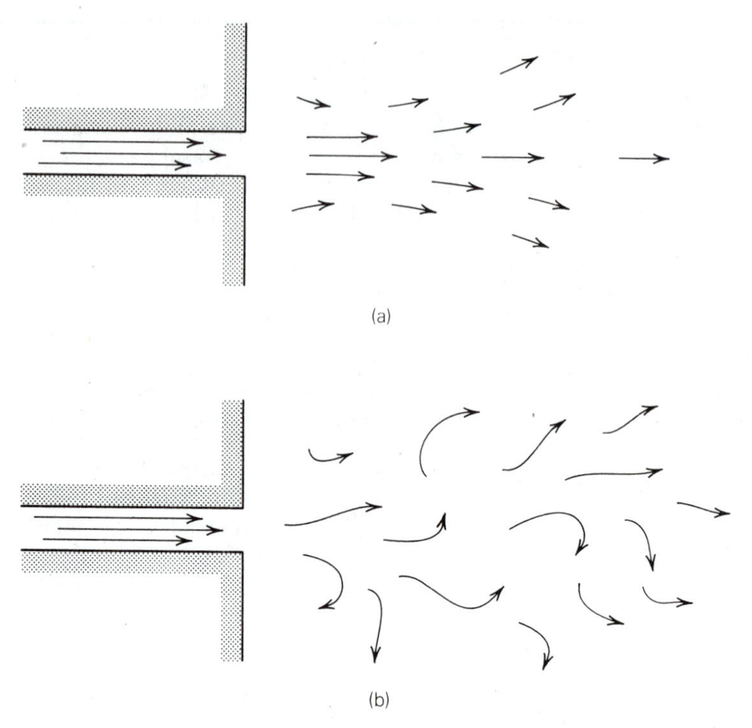

FIGURE 12.5 A fluid jet emerging from an opening. (a) Laminar flow at low speed. (b) Turbulent flow at higher speed.

Let us consider three classes of unstable fluid flows that give rise to sound: first a simple fluid jet, then a jet hitting an obstacle, and finally a jet controlled by a resonator. A fast stream of fluid emerging into a region filled with similar fluid at rest will not just smoothly and gradually diffuse and slow down (Figure 12.5a). It becomes turbulent (Figure 12.5b), as you know if you have stood in front of a swimming pool water inlet and felt the swirling, pulsating action of the emerging jet. You can make a similar jet of air by blowing through your mouth (not whistling). When you blow hard enough you hear noise, which is direct proof that the airflow is not smooth and steady. This broadband noise (the sound of *h* or *wh*) is not a periodic disturbance, and so has no definite pitch. A wide range of frequencies and eddy sizes participate. Small amounts of this jet noise still are present in flutes and organ pipes even when they produce notes of definite pitch. (This must be taken into account by anyone trying to use electronic synthesis to produce an accurate imitation of flute sound.)

The second case is much more interesting. Let a thin sheet of air issue from a narrow slit, called a *flue*, and place a sharp-edged obstacle directly in the way of this flat jet (Figure 12.6). There now may be much stronger vibrations, and they may be concentrated mainly about one preferred frequency. That is, you hear a louder sound with a more definite sense of high or low pitch. As we mentioned in Chapter 3, you can easily produce such a

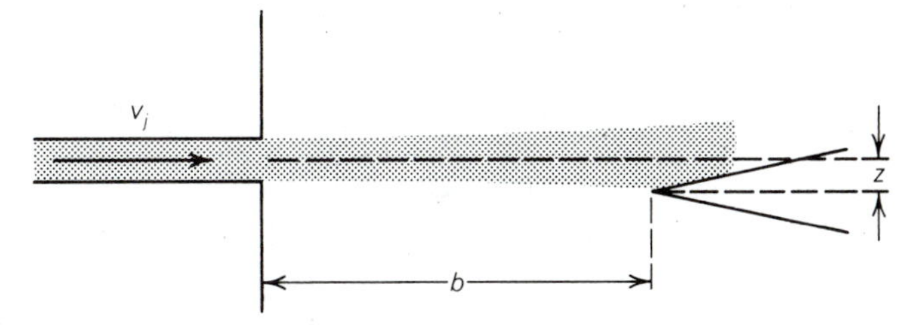

FIGURE 12.6 A fluid jet emerges from a flue with average speed v_j, crosses a gap of length b, and strikes a wedge to produce an edgetone. In general, the wedge need not lie precisely on the center line of the jet, but may have an offset z, which affects the tone quality.

sound, which is called an *edgetone*, by blowing as before while holding a small card or sheet of stiff paper in front of your mouth.

You quickly notice that the tone is sensitive to the placement of the edge in the airstream, and that its pitch depends on how hard you blow and on how far the jet travels before hitting the edge. Careful laboratory measurements reveal the features shown in Figure 12.7. Any explanation of how an edgetone works must account for rises in frequency when jet speed increases or edge distance decreases as well as for the possibility of several stages of operation.

Edgetone jet behavior is an especially difficult problem in fluid mechanics, one on which experts still are not fully agreed. Sometimes the jet has been pictured as forming vortices alternately on one side and then the other (Figure 12.8a) before reaching the edge. But those concerned with edgetones in musical acoustics favor the view that the instability is not so fully developed; rather, all that is necessary is for the jet to attain a sinuous form that will cause it to deliver air alternately above and below the edge (Figure 12.8b).

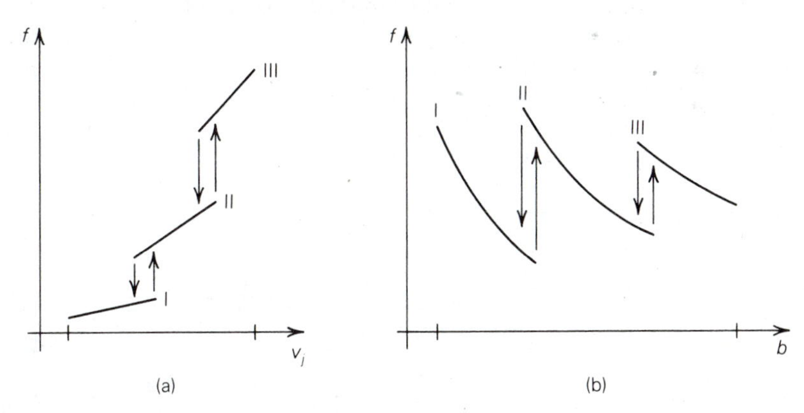

FIGURE 12.7 (a) For a given gap size, increasing the jet speed v_j increases the frequency of an edgetone. There are some speeds for which the edgetone may operate in either of two stages with different frequencies. Arrows indicate sudden jumps from one stage to another. Outside a certain range of speeds the edgetone will not speak at all. (b) For a given jet speed, increasing the gap distance b tends to lower the edgetone frequency. At certain points, however, the frequency may jump suddenly to a different value as the edgetone shifts to a more favorable stage of operation. It is only the first of these stages that corresponds to normal musical applications.

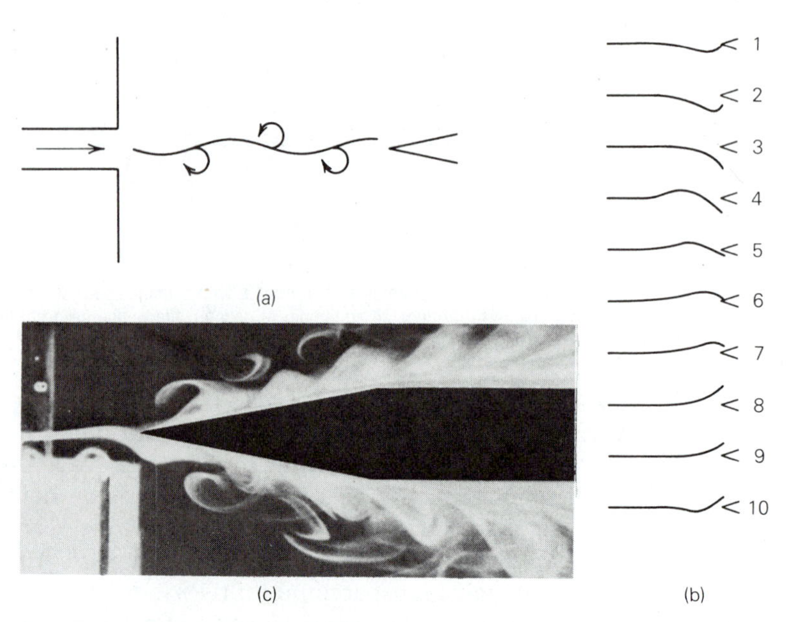

FIGURE 12.8 (a) According to some sources, the interaction of a series of vortices with the edge produces a disturbance that ensures continued formation of new vortices to maintain an edgetone. (b) But experiments show that it is only necessary for the jet to wave back and forth slightly. (c) A photograph suggesting that vortex formation does not occur until after the stream already has passed the edge. (Part (b) after Bouyoucos and Nyborg, *JASA*, *26*, 511, 1954; photo courtesy of J. V. Bouyoucos)

The edge serves to divide one region of space from another, so that when the jet begins to flow below the edge, for instance, it causes higher fluid pressure there than above the edge. (If the divider were removed, that pressure difference would be relieved quickly by a small amount of upward flow.) But this higher pressure forces other fluid to move out and make room for that added by the jet. In particular, part of this flow goes up through the gap between flue and edge (Figure 12.9). It pushes upward on the fluid that is just emerging from the slit, ensuring that a short time later the jet will flow above the edge instead of below.

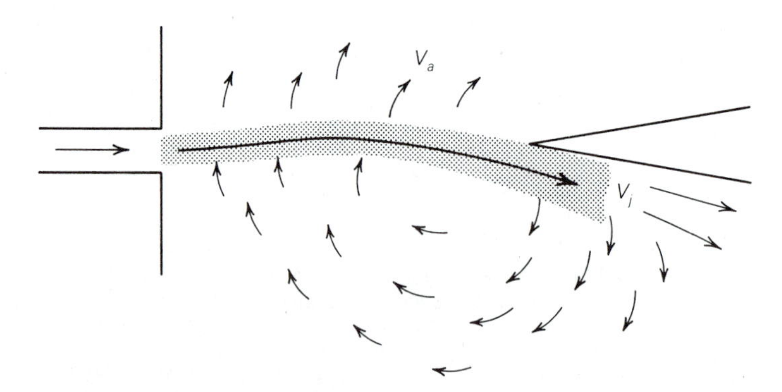

FIGURE 12.9 As the jet flows (v_j) to one side of the wedge, it forces other fluid out of the way; some of this flows through the gap to the other side. Notice that the resulting acoustic flow (v_a) tends to divert the jet to the opposite side of the wedge.

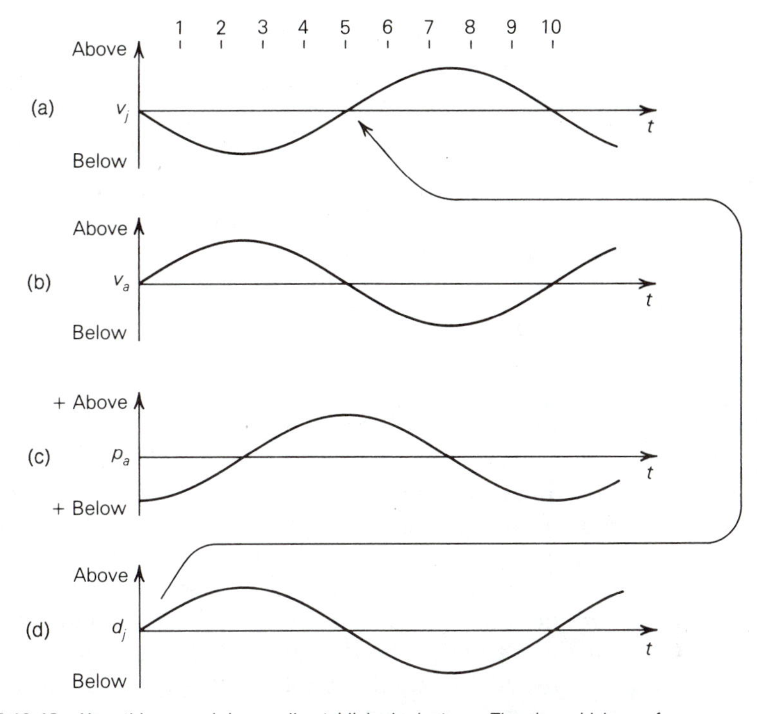

FIGURE 12.10 How things work in a well-established edgetone. The sinusoidal waveforms are an oversimplification. Numbered times correspond to jet pictures of Figure 12.8b. (a) Fluid flow provided by the jet; (b) resulting oscillating vertical flow through the gap (as shown by Figure 12.9); (c) resulting pressure differences; (d) displacement of the emerging jet by the acoustic flow. The arrow indicates how this disturbance is responsible for switching the jet at the edge a half-cycle later.

We have here an example of a *positive feedback* mechanism, which is important because it can (a) maintain an oscillation once it is started and (b) determine a preferred frequency of oscillation. Consider the graphs in Figure 12.10. Each period of jet flow below the edge creates an upward disturbance at the flue; when this disturbance travels across the gap it makes the jet flow above the edge. This, in turn, causes downward flow at the flue, which some time later switches the jet below the edge, and the whole cycle starts over again. The frequency of oscillation is controlled by the speed with which the disturbance travels across the gap. Surprisingly, that is approximately only 0.4 times the speed of the jet itself. The unstable wave, in accommodating its motion to the relatively still air on either side, actually travels upstream on the jet while the jet carries it to the edge. So half the period of oscillation for Stage I operation must be the time it takes to travel a distance b at speed $0.4v_j$: $P/2 \cong b/(0.4v_j)$, or $P \cong b/(0.2v_j)$. Then the preferred frequency of oscillation is $f_1 = 1/P \cong 0.2v_j/b$. For instance, a jet velocity $v_j = 10$ m/s and a gap $b = 1$ cm give an edgetone at approximately 200 Hz.

This formula accounts for the dependence of f_1 on v_j and b as shown in Figure 12.7. The switch to higher stages also can be explained by considering the possibility that the jet disturbance could take $1\frac{1}{2}$ or $2\frac{1}{2}$ periods to cross the gap instead of only $\frac{1}{2}$. Each of these also would generate positive feedback and maintain oscillations at frequencies $f_{II} \cong 3f_I$ or $f_{III} \cong 5f_I$. The mechanism does not work for other frequencies in between; at $2f_I$, for instance, there is negative feedback. That is, jet flow below the edge causes a disturbance to arrive one full period later, and that moves the jet above the edge instead of below again. Oscillation is effectively suppressed for any frequency that has negative feedback.

The third class of unstable flows, the jet in the presence of a resonator, brings us to organ pipes.

12.3 ORGAN FLUE PIPES

The mouth configuration of an organ pipe (Figure 3.6) directs a sheet of moving air against an edge. But we must not jump to the conclusion that it merely produces an edgetone that just happens to be reinforced by resonance in the adjoining pipe. That resonance builds up much higher pressures and stronger acoustic flows than the isolated edgetone would, and thus the pipe exercises strong control over the edgetone. Specifically, the presence of the pipe usually makes the whole system speak at a lower frequency than would the isolated edgetone (Figure 12.11). The pipe achieves this dominance by providing a strong positive feedback of its own in the form of a wave that

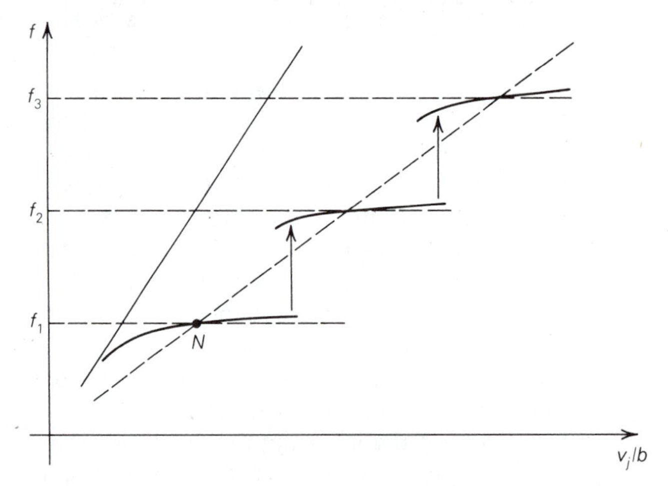

FIGURE 12.11 Modification of an edgetone by a resonator. The solid sloping line shows how Stage I of the isolated edgetone would behave (lower portion of Figure 12.7a). The dashed sloping line shows the effect of requiring quarter-cycle instead of half-cycle time for the disturbance to cross the gap (see Figure 12.13). Horizontal dashed lines show natural mode frequencies of the pipe alone. Solid curves show how a blown pipe actually behaves: Blowing harder slightly and gradually increases the pitch until it breaks up to the next higher mode, indicated by arrows. Resonance is strongest and most reliable where the dashed lines intersect, and the point *N* represents normal operation for a well-designed pipe. (After Coltman, 1976)

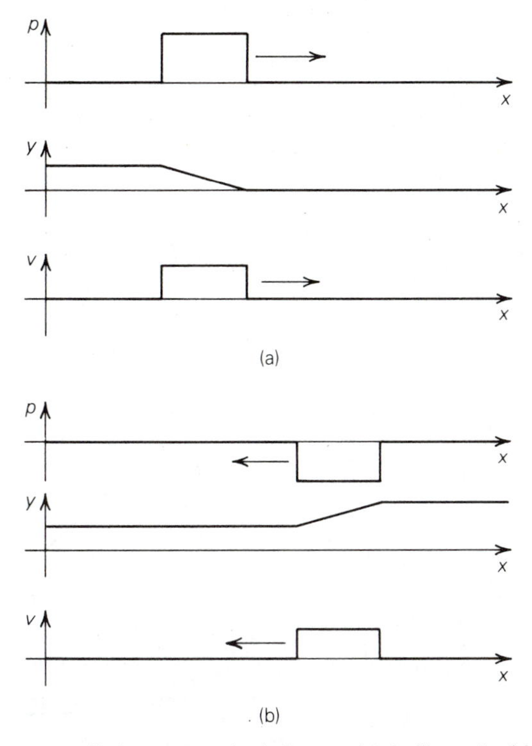

FIGURE 12.12 The pressure, displacement, and velocity associated with a pulse (a) traveling down a tube and (b) after a reflection from an open end. Notice that the positive values of v in (b) mean that each bit of air moves to the right again in order for the rarefaction to move to the left.

travels to the far end of the pipe, reflects back, and disturbs the jet upon its return.

Consider first a closed pipe. Suppose we follow a positive pressure pulse initiated by the switching of the jet from outside to inside. This compression travels to the closed end, reflects, and tries to push the jet back outside when it returns to the mouth. So we have self-sustaining oscillation if the round-trip pulse travel time is half a period of jet oscillation: $2L_c/v = \frac{1}{2}P$, or $P = 4L_c/v$. This is just the period of the first natural mode of the pipe!

A similar positive pressure pulse sent down an open pipe is changed to a negative pulse upon reflection (Figure 12.12). When this rarefaction arrives back at the mouth it will help suck the jet inside again, so we have self-sustained oscillation if the round-trip travel time matches a *full* period: $2L_o/v = P$. Again this turns out to be just the period of the first natural mode of the pipe. In both cases we see not only (from the standing-wave picture) that the fundamental natural frequency f_1 is one at which the pipe *could* respond strongly to an alternating driving force, but also (from the traveling-wave picture) that the pipe *will* use its feedback signal to persuade the jet to drive it at that particular frequency. If the jet velocity is too large or the gap distance too small, so that the edgetone's natural preference would be for a

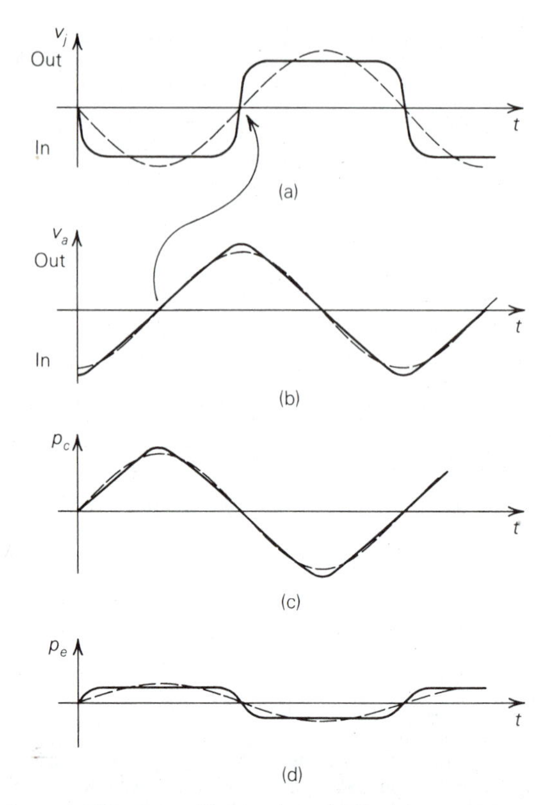

FIGURE 12.13 Edgetone quantities as modified by strong feedback in an open organ pipe: (a) jet flow; (b) resulting acoustic flow and disturbance of emerging jet; (c) pressure at the center of the pipe; (d) pressure just inside the end. Dashed lines show the fundamental components of each for comparison to Figure 12.10. If you compare (d) with Figure 12.10c, remember that positive pressure "inside" here corresponds to positive pressure "below" there.

much higher frequency, it may be the second or third pipe mode that captures the jet with its feedback. Then the jet switches above and below the lip more often, and a higher-pitched sound is produced; this is called *overblowing* the pipe.

Even when the jet oscillates at the pipe's fundamental frequency, it also may receive positive feedback from the second, third, and higher modes. So the jet motion, and the resulting sound, are not just sinusoidal; they tend toward waveforms such as those in Figure 12.13. When pipe resonance is strongest and most fully in control of the jet motion, the pressure and velocity phases are related approximately as in Figure 12.13. Note that the jet current is now only about a quarter-cycle behind the acoustic current instead of a half-cycle; thus jet disturbances should travel from flue to edge in about a quarter-cycle to obtain positive feedback: $P/4 \cong b/(0.4v_j)$, or $f \cong 0.1v_j/b$. The mouth configuration and air supply should be designed so that its pure edgetone would be an octave or more above the note to be produced when the resonator pipe is attached.

---

*For a given pipe length and pitch, the sound quality is strongly dependent on the gap b from flue to edge (upper lip), called the *cutup* by organists. A small cutup and low air pressure produce a soft and relatively stringy sound, while a large cutup and correspondingly higher pressure produce a louder note with poorer harmonic content.

Organ pipes are extremely sensitive to small adjustments (called *voicing*) of their flues. One of the most important is the *jet offset* (z in Figure 12.6). With no offset, a symmetrical motion is favored in which the jet injects similar squirts of air inside and outside the pipe; this gives a symmetrical waveform containing mainly odd-numbered harmonics (for open pipes as well as closed). Larger offset favors nonsymmetrical jet motion and nonsymmetrical waveforms, including strong even and odd harmonics.

Another important adjustment is *nicking*, or small indentations made in the soft metal along the edges of the flue opening. The roughness makes the jet thicker and more stable, resulting in smoother initiation of the pipe speech as well as gentler timbre in its continuing sound. This was considered desirable for better blending of voices in Romantic music, so organ pipes from approximately 50 to 100 years ago often are heavily nicked. Recently the pendulum has swung back toward "baroque voicing," with only slight nicking so that the pipe speaks with a more prominent transient. This *chiff* serves as a "consonant" preceding each steady "vowel" sound, and makes it easier to follow the individual parts in contrapuntal music. Heavily nicked pipes and high wind pressures are two among several reasons why old theater organs are entirely unsatisfactory for playing Bach.

In normal operation the recipe of air vibrations inside a flue pipe usually is dominated by the lowest natural mode. We must not assume, however, that the same is true of the sound outside the pipe, because the pipe radiates more efficiently at higher frequencies. Hence in the examples of Figure 12.14 the higher harmonics often rival the fundamental in strength.

The differences in these spectra are partly due to differences in pipe proportions. Flue pipes often are classified as *flute, principal,* or *string*; flute pipes tend to be fat and string pipes skinny, with principals in between. Remember, however, that voicing adjustments at the mouth are also important in determining whether the tone quality will tend toward flutelike or stringlike characteristics.

There are two reasons why wide pipes tend to be poor in upper harmonics: The higher pipe modes occur at the wrong frequencies to form part of the harmonic series (the formulas in Section 12.1 are accurate only for thin pipes), and they do not resonate strongly. Both of these effects stem from the behavior of traveling waves reaching the end of the pipe. The lower pipe modes have wavelengths that are large compared to the pipe diameter, and such waves diffract strongly at the end; they "feel" the end and are reflected by it quite effectively. Large amounts of energy stay inside the pipe, building up a strong resonance. But modes with wavelengths less than approximately twice the pipe diameter do not feel such a strong discontinuity; much of the energy in such a wave keeps going right on out the end, little stays inside, and the resonance is weak. Beyond a certain critical frequency, higher harmonics will be virtually absent from the recipe. The dividing line for an open

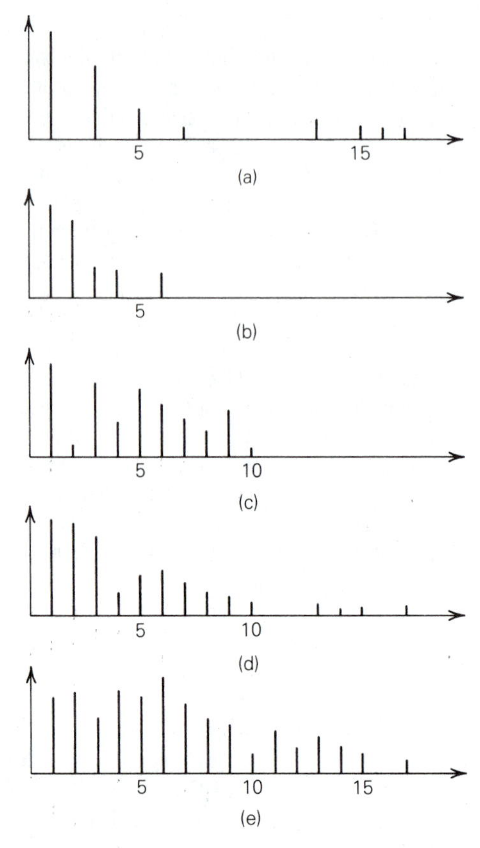

FIGURE 12.14 Typical spectra from several organ flue pipes, all speaking C_4. (a) Gedeckt, a closed flute pipe; (b) spitz flute, an open tapered pipe; (c) rohr flute, capped, with a chimney; (d) principal, open; (e) viola d'orchestre, an open string pipe. (After Strong and Plitnik)

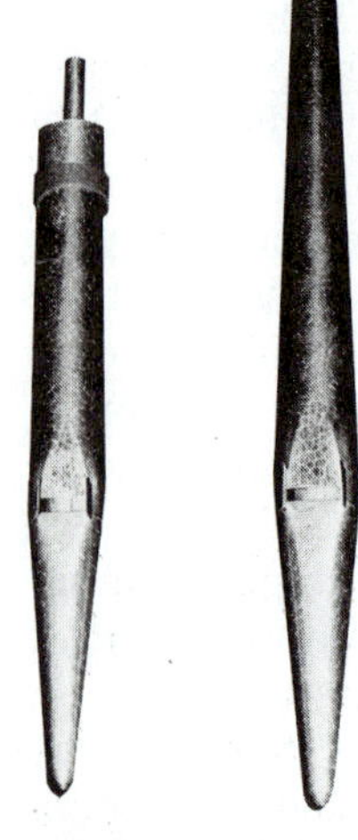

FIGURE 12.15 Two noncylindrical organ pipes. Note the small chimney on top of the rohr flute (left) and the taper of the spitz flute (German for "pointed flute"). (Photo by Stephen Hamilton)

pipe is at a mode number in the vicinity of $N = L/D$; for a given length of pipe, this occurs at smaller N when the diameter D is increased.

Open and closed cylindrical pipes of different diameters do not exhaust all the possibilities; tapered and chimneyed pipes also are fairly common (Figure 12.15). The tapered pipe does not differ essentially in its operation from a cylindrical pipe, because it happens that the natural mode frequencies for a partial cone open at both ends also form a harmonic series, even though the mode shapes are not sinusoidal. The taper does, however, tend to strengthen the second harmonic relative to the fundamental. The chimneyed pipe is to a first approximation a closed pipe; but the narrow chimney's secondary resonance can help emphasize one or more of the higher modes, as well as weaken the prohibition against even harmonics.

A complete set of pipes producing similar tone color is called a *rank*. Normal modern keyboard span for an organ is five octaves (C_2 to C_7), so that most ranks have 61 pipes, with the longest being $2^5 = 32$ times the length of the shortest. To make a usable rank, an organ builder must face the problem

of *scaling*, or deciding how the diameters and mouth proportions should vary on pipes of different length. One possibility is to use the same diameter regardless of length. But such a rank would have a very inhomogeneous sound; 4 cm diameter, for instance, results in a critical N of approximately 2 at the treble end (practically no harmonics, very fluty) but N of approximately 65 at the bass end (many harmonics, very stringy).

A more logical plan would seem to be to keep all proportions the same, with each small pipe an exact miniature replica of the largest one in the rank. That would keep the critical N the same for all pipes, thus seemingly allowing them to have similar spectra and tone color. Unfortunately, that does not work; the smaller pipes simply cannot produce as much sound, so the rank is unbalanced, weak in the treble and too strong in the bass. Actually, maintaining similar spectra is not the right way to get similar tone colors anyhow; the first 10 harmonics of C_2, for instance, span the range from 65 to 650 Hz (predominantly low frequencies), while the first 10 harmonics of C_7 extend from 2100 Hz to 21 KHz (beyond the audible range). Using the same recipe of relative harmonic strengths in these two ranges would result in quite different timbres.

Good loudness and tonal balance is achieved when the bass pipe spectra are somewhat richer in harmonics than the treble. A reasonable compromise long ago established by trial and error is to let the diameters change a little more slowly than the lengths—just enough so that you must go some 16 or 17 semitones up the rank before you find the diameter halved, whereas the length is halved after 12 semitones. Then the ratio of diameters of the largest and smallest pipes is 12 or 13 instead of 32.

*12.4 ORGAN REGISTRATION AND DESIGN

A complete pipe organ may have several dozen ranks of pipes, each offering a different tone color; some of the largest have more than 100 ranks. The organist needs effective ways of choosing and controlling which of these pipes shall sound at any given time. We begin with brief comments about the artistic choices, after which we will describe certain mechanical controls.

The organist's choice of *stops* (meaning in most cases one rank of pipes for each stop or control knob) is much like that of the orchestrator or arranger who decides which instruments to assign to play each line in the music. Some stops are best suited to solo work, while others blend well so that a chorus of several stops can be used to make a richer sound. Reed pipes often fall in the former category and principals in the latter.

One way to make a fuller sound is simple doubling, letting two or three pipes all sound the same pitch. A more significant strategy is to have every key activate not only one pipe at normal pitch but also another an octave higher. Doubling at the octave is successful (both for organists and orchestrators) only if it is perceived by the listener as a new tone color for that succession of notes, *not* as two separate musical parts moving always in parallel octaves. Hence the usual preference for achieving a good blend is by combining two members of the same family (such as two principal stops) rather than two dissimilar stops (such as one flute and one reed).

In organists' terminology, the basic pitch is that of the 8′ stops. The name comes from the fact that the largest pipe (C_2) in an open rank is approximately 8 feet long; unfortunately, this terminology is sure to survive on pipe organs long after all other vestiges of English units are gone. Good organs have 4′ and 2′ stops to sound one and two octaves higher, and sometimes 1′ as well. Pedal divisions normally offer 16′ stops for doubling at the lower octave, and occasionally 32′ as well. It also is useful sometimes to add pitches corresponding to the third and fifth harmonics of the 8′ pipes, hence the *mutation* stops designated $2\frac{2}{3}'$ and $1\frac{3}{5}'$. These stops sound G_4 and E_5, respectively, when the C_3 key is pressed, for instance.

We must emphasize that organ registration is not just Fourier synthesis. The 8′ pipes already produce complex periodic waves that involve many members of the harmonic series. Addition of a 4′ stop merely adds further strength to harmonic numbers 2, 4, 6, . . . ; a $2\frac{2}{3}'$ stop reinforces numbers 3, 6, 9, . . . ; and so on. Only on the Hammond electronic organ can the player adjust the strength of individual harmonics independently by using the drawbars.

It becomes impractical to provide stops for each individual harmonic beyond the sixth or eighth, so good organs also offer *mixture* stops that add a festive and penetrating brightness to the sound. Such a stop controls several ranks of pipes at once, playing pitches corresponding to several higher members of the harmonic series. When the C_3 key is pressed, for example, a three-rank grave mixture might sound G_5, C_6, and G_6; an acute mixture would sound even higher notes such as C_7, G_7, and C_8. The profile of a mixture rank (Figure 12.16) illustrates an important point about tone quality. The desired brightness is determined not so much by which harmonic numbers are involved but by whether those harmonics fall at frequencies between about 1 KHz and 10 KHz. This requires higher harmonic numbers for the low notes (sometimes as high as 48 or even more) and lower harmonic numbers for the high notes (such as 2, 3, or 4), with breaks where the harmonic numbers change at several pitches between C_2 and C_7.

There is much interesting physics and technology in the devices that control organ pipes. The pipes must stand on a wind chest supplying air at slightly above atmospheric pressure. This pressure is traditionally measured in "inches of water," meaning how far it will push water around a U-shaped tube open to the atmosphere at the other end (Figure 12.17). Centimeters of water are more convenient, for each one is very nearly 0.001 atmosphere. Appropriate pressure in an organ is generally between 5 and 10 cm of water; higher pressures sometimes have been used in the past but accomplish little other than a loud, bombastic sound.

There must be a set of valves and channels that allows each pipe to speak only when its corresponding key on the keyboard is pressed *and* the stop-

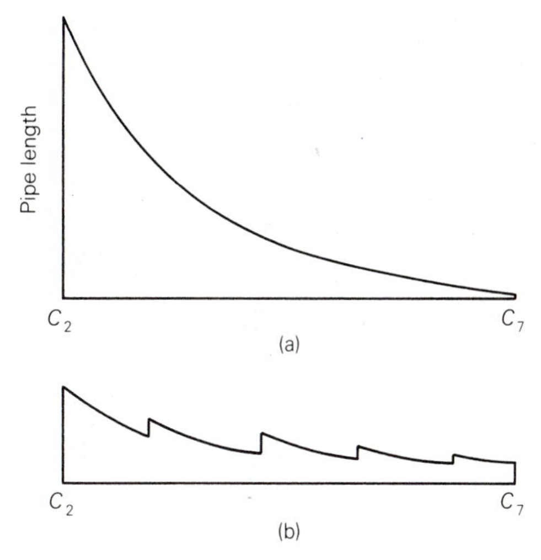

FIGURE 12.16 Profiles of most organ pipe ranks (a) show a smooth decrease in length from bass (left) to treble (right). But mixture ranks (b) exhibit breaks to keep their pitches within the range that best adds the desired brightness to the total tone color.

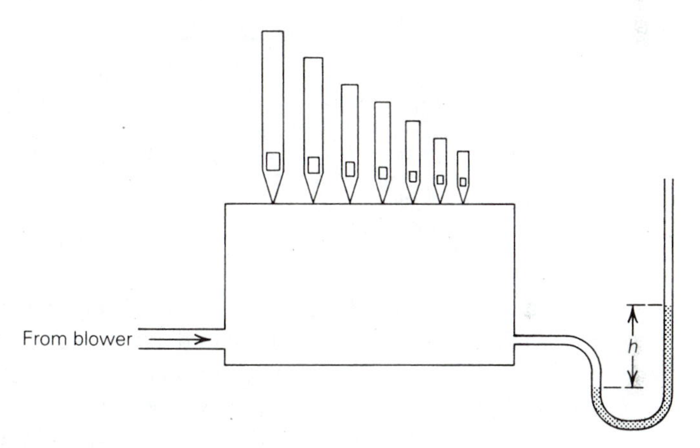

FIGURE 12.17 The height h of one water level above the other is a measure of the excess pressure above atmospheric inside the wind chest.

knob for that rank is "on." This can be accomplished in several ways (Figure 12.18). The *box chest* stands always ready to supply air to any pipe, and each pipe has its own separate valve; the logic decisions (which valves to open) must be accomplished somewhere else, as is now easy to do with electric actions. The *stop-channel chest* admits air only to those channels whose stops are "on"; again every pipe has its own valve, but now pressing a key always opens the valves for every pipe corresponding to that note, although only those on pressurized channels speak. The *tone-channel chest* is divided crosswise instead of lengthwise; there is only one valve for each key, which admits air to that tone channel when the key is depressed. Stop control can be

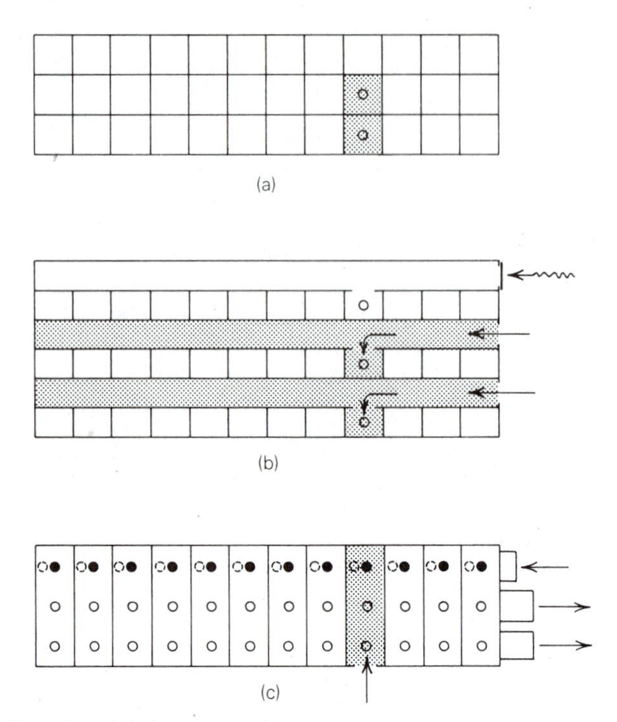

FIGURE 12.18 Alternative wind chest designs (top view) for a miniature organ with only three stops and 12 keys. Shaded portions are pressurized, in this example with two stops on and one key down. Pipe feet rest in holes on chest top to receive air. (a) The box chest has an individual air compartment for each pipe. (b) The stop-channel chest has three pipe valves open but air only reaches two pipes. (c) The tone-channel chest still admits air to only two pipes, because holes in the top slider (in the "off" position) do not match up with holes in the chest itself.

accomplished by sliding templates in the top of the chest that block the airways of all pipes in a rank when in the "off" position. Tone-channel chests with sliders have several mechanical and acoustical advantages over the other two types.

The *action* of an organ is the scheme used to translate motion of a key on the console into motion of the appropriate valve on the wind chest, which may be located some distance away. *Mechanical action* (or *tracker action*) held a monopoly for centuries; it simply has a series of wooden strips and levers, where A pulls B, B pulls C, and so on, until finally H pulls the valve open. *Pneumatic action* became popular in the nineteenth century. Here air is supplied to the console, where pushing a key admits this air into one of a large bundle of long flexible pieces of tubing. This tubing leads to the wind chest, where the arriving air opens the pipe valve. *Electropneumatic action* allows even greater flexibility in placing the console some distance away from the pipes; here the key makes electrical contacts and it is the wires that go over to the wind chest, where they operate electromagnets to open small secondary valves, which then open the primary valves by pneumatic-lever action. *Direct electric action* uses magnets to open the primary valves for the pipes.

It would be natural to suppose that electric action is most modern and has replaced all the other types, but that is not so. In fact it has been difficult to develop an electric action as quiet and reliable as either the mechanical or pneumatic actions. There has been a resurgence of interest in mechanical actions since 1960, and tracker design is now the choice for many new instruments. This is largely because direct mechanical contact gives the player a feeling of control over the precise time and manner in which the valve is opened. After playing trackers for a while, it seems as if keys that only make electrical contacts are "mushy" and hard to control; it is much the same as the feeling of being in contact with the road when driving a car with rack-and-pinion steering but isolated from it by power steering. Sometimes we discover that the old ways were pretty good after all!

12.5 FINGERHOLES AND RECORDERS

Pipe organs are neither cheap nor easily portable, so it is worthwhile inventing ways to get several different notes out of the same pipe. This can be done even on a simple pipe by overblowing; but only a few notes of very uneven tone quality are available this way, and in any case they are spaced too far apart to play a good melody. A continuous range of notes can be obtained by having a movable stopper to change the air column length; most of us have played with such slide whistles as children. But fingerholes make it much easier to jump rapidly and accurately from one note to any other, so this is what we find on a great variety of wind instruments, both ancient and modern.

What effect should a single large hole in the side of a pipe have? Consider Figure 12.19a. With the hole closed, the fundamental mode of the pipe involves a large fluctuating pressure at that point. But with the finger removed, there is no longer a solid wall confining the air; such pressure can no longer build up to the previous value in the presence of the open hole. Or, in terms of oscillating airflow, little air will flow toward the pipe end when it can so easily go in and out through the hole instead. So the open hole, instead of the pipe end, becomes the pressure node (or displacement antinode) that determines the allowed wavelengths and thus the natural mode frequencies. The farther the hole is moved toward the top of the instrument (as the blowing end is customarily called), the shorter the allowed wavelengths, the higher all the mode frequencies, and the higher the pitch of the note heard.

But that is an oversimplification that is not really true unless the hole is so large that the air moves through it just as freely as if the tube were completely chopped off at that point. The smaller the fingerhole, the more acoustic pressure can build up inside the tube in spite of it, and the less the vibratory airflow is diverted through it. Thus, a small hole higher on the instrument can produce a pressure distribution in the upper part of the tube with the same wavelength as would a large hole farther down (Figure 12.19b,c). It is p

(a)

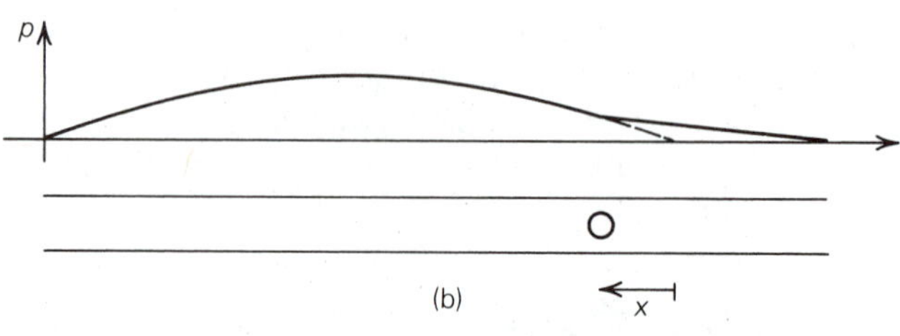

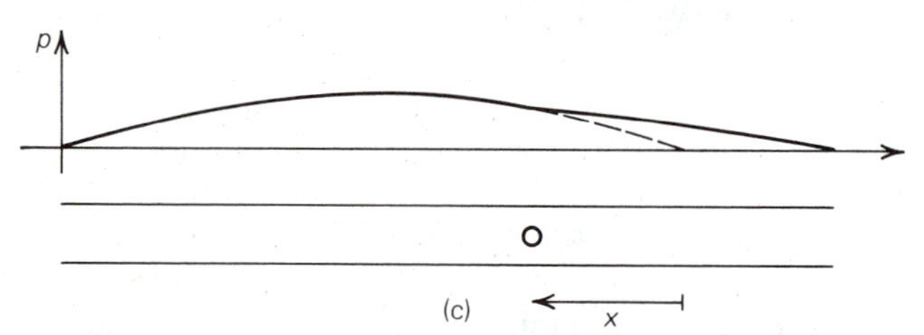

FIGURE 12.19 (a) The pressure standing wave of the first mode inside a pipe with a large open hole; the dashed line shows the first mode with the hole closed for comparison. (b), (c) Pressure patterns for smaller holes at distances x higher on the pipe, with dashed lines indicating how the internal wavelength and therefore the frequency can be made the same as in (a).

fortunate that the same note can be obtained either way, for this allows some leeway in placing holes where the fingers of a normal person can reach them.

---

*One can get rough ideas about placement of a single hole of diameter D_h on a long pipe of diameter D_p from the formula $x \approx t_e(D_p/D_h)^2 - 0.3\,D_p$. Here x is how far the hole is placed above the point where the tube would have to be cut off completely to produce the same note. The effective length of the hole, $t_e \approx t + 0.8\,D_h - 0.5\,D_h^2/D_p$, includes both the actual wall thickness t and an end correction. The formula seldom is entirely valid, but it at least gives a rough idea of the effects of hole size and wall thickness. If two or more holes are open, there is no such simple formula; you may read Benade to learn more about the effects of an entire row of open holes.

To play an ascending scale, we can start with all tone holes covered and lift one finger at a time beginning from the bottom end of the pipe. Each time the player uncovers another hole, the natural mode frequencies all shift upward to correspond to the new effective tube length. A normal fingering pattern has all holes closed above some point and all holes below that point

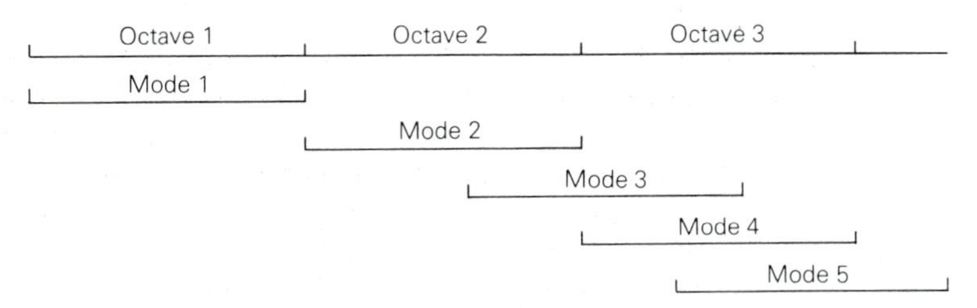

FIGURE 12.20 Fundamental frequency ranges covered by repeating fingering patterns from the first octave while overblowing an open pipe.

open. To produce all notes in the chromatic scale over a range of approximately two octaves (for the recorder) or three octaves (for the flute), it would appear that dozens of holes are needed; yet only 10 fingers are available to cover them. There are several ways around this problem and one of them is overblowing. Once we get beyond the first octave we can just repeat the same fingering patterns (or, in reality, nearly the same) over again while blowing harder; we do not actually need additional tone holes above the halfway point along the pipe. Once into the upper half of the second octave, some notes can be played by using the third mode; in the third octave the fourth mode becomes available, and so forth (Figure 12.20).

To make the overblowing more reliable and less forced in tone quality, it helps to open a tiny leak halfway along the tube. This is done on the recorder by partially uncovering the thumbhole to make a *register hole* much smaller than the tone holes. The leak kills off the first mode by allowing air to escape whenever the pressure rises. But the second mode has a pressure node near that point anyhow, so is practically unaffected and still can resonate strongly. Similar consideration shows that all the even-numbered modes will participate in producing the tone an octave higher.

But even the first octave alone still seems to call for 12 tone holes. The recorder has only seven (not counting the thumbhole), but it still manages a chromatic scale by using split holes (two small holes close together that can be covered by the same finger) and more complicated fingerings. These finger patterns (called cross-fingerings or forked fingerings) leave one hole open but then close others below it (Figure 12.21).

12.6 THE TRANSVERSE FLUTE

The flute takes a more straightforward approach to getting 12 notes in the first octave. It simply has 12 tone holes (plus one extra that serves as a duplicate for easier fingering). Its key mechanism allows one finger to control two or more holes, which in some cases are far beyond direct reach. Although most of the holes are "normally open," the key mechanism allows a few to be "normally closed" instead, so that pushing a finger down lifts the pad off the hole instead of covering it.

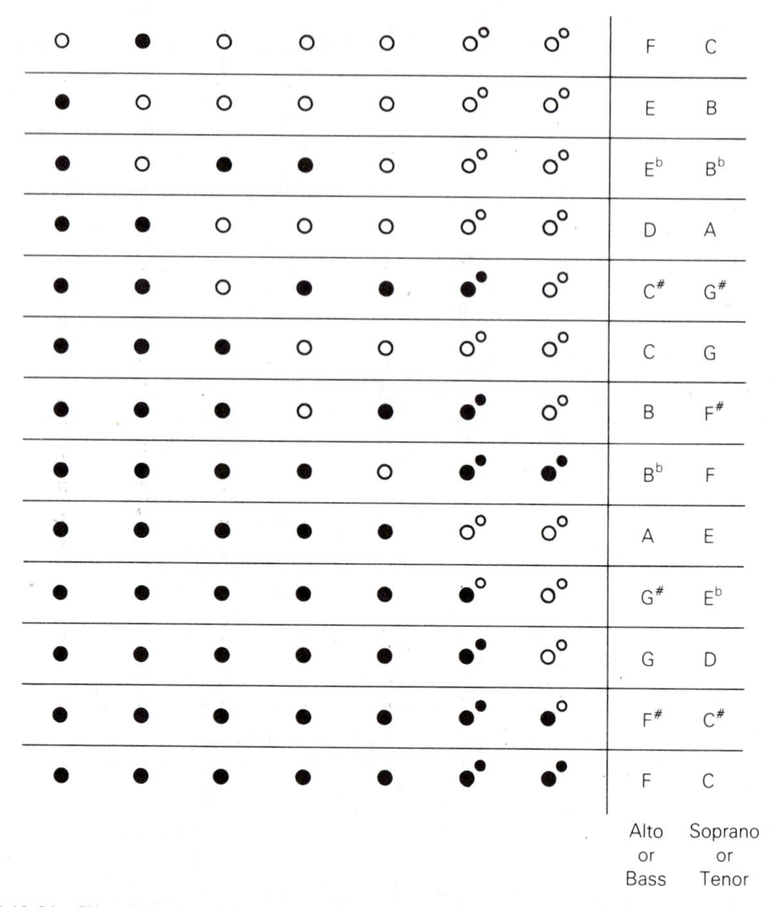

FIGURE 12.21 Fingering patterns for the first octave of a recorder. Lowest note is at the bottom, ascending scale is toward the top; mouthpiece is to the left (as in Figure 3.7, page 46); covered holes are shown by ●, open holes by ○. Note that it is mainly the first (leftmost) open hole that determines the pitch.

There is an intimate connection here between hole size and placement, one that may not be apparent at first sight. Placement of the flute tone holes where they are, relatively evenly spaced over a distance of approximately 30 cm, is tightly bound up with the large hole size (hole diameter approximately 70–80% of pipe diameter, compared to 30–50% for recorder). This, in turn, means that the tube is effectively terminated at the first open hole, and it hardly matters whether any lower holes are covered or not. Thus cross-fingerings will not work on the lowest octave of the flute, and it must have a separate tone hole for each chromatic note. But placement of recorder holes within reach of the fingers is bound up with their relatively small size. That means the first open hole lacks enough of being a true pressure node that there still is appreciable air vibration in the lower part of the pipe, which, in turn, is influenced by whether or not various lower holes are covered. Here

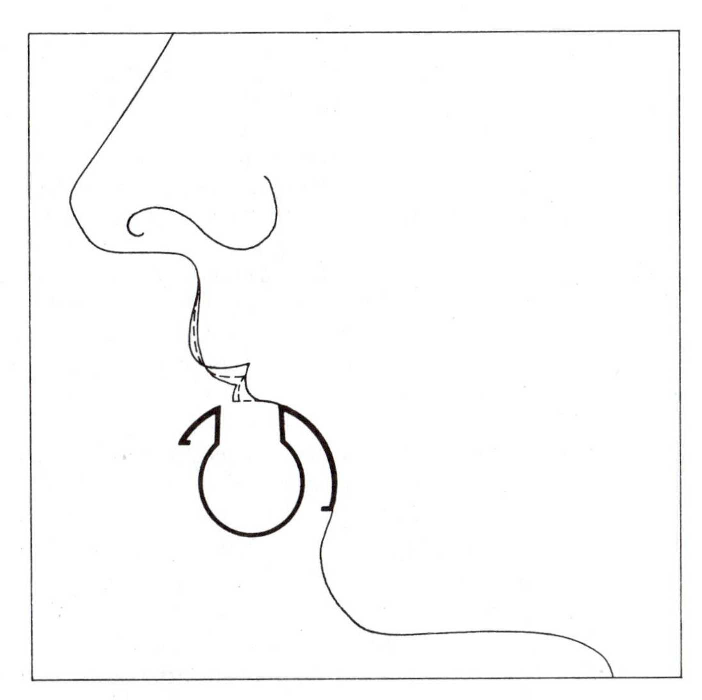

FIGURE 12.22 Flutist's lip positions over the mouth hole, retracted for lower notes (or harder blowing), advanced for higher notes (or softer blowing or brighter tone). (From Coltman, 1968)

cross-fingerings are effective both in pulling some notes in tune that otherwise would not be and in changing resonant frequencies enough to provide all the sharps and flats on the recorder.

The recorder's fipple mouthpiece (Figure 3.7, page 46) has a fixed gap, and the player can control only the speed of the jet. The condition $f \cong 0.1v_j/b$ for optimum pipe control of the edgetone must be met by blowing harder on higher notes. But the flute player's lips are right over the mouth hole as they direct the airstream across it (Figure 12.22), and the *embouchure*, or lip configuration, is flexible. This gives the flutist an advantage over the recorder player (but one that makes the instrument more difficult to learn) in being able to change the gap distance by moving the lips or by rolling the instrument slightly toward or away from the lips. Having independent control over gap distance and jet speed, the flutist can obtain a variety of relations among pitch, loudness, and timbre, whereas all three must change together on the recorder.

The adjustable gap also makes it easier to control overblowing. Only skilled players can complete even the second octave on recorder, and an instrument that speaks well in the low register tends to be difficult in the high and vice versa. The third octave is much more accessible on the flute, where the gap can be narrowed to encourage speaking in the higher modes. In fact, on the flute (unlike other woodwinds) embouchure control suffices *instead* of a vent hole for most second- and third-octave notes.

*For most purposes, our tacit assumption that the flute is cylindrical serves quite well. But in fact the change in embouchure tends to flatten the second mode, leaving it less than an octave above the first mode. To get them back in tune with each other, the second mode frequency is raised slightly by tapering the head joint. A similar purpose is served on piccolos and old conical flutes by having a cylindrical head on a body that otherwise tapers. Pre-Boehm flutes with smaller tone holes have an additional need for the taper, because small tone holes also tend to leave the second mode less than an octave above the first. Recorders also are tapered rather than cylindrical. Our comments in Section 12.3 about enhancement of the second mode in the spitz flute thus illuminate the fact that Baroque recorders, with a relatively large taper, play better in the high register, while Renaissance recorders, which are more nearly cylindrical, are better on low notes.

The flute, when viewed as an energy-conversion device, is extremely inefficient. Coltman estimated in one case that only 2.4% of the airstream energy was converted to energy of the standing-wave vibrations inside the pipe. And only 3.5% of that, in turn, was radiated away as sound; most of it was lost in the form of heat, through the viscous drag exerted on the air in a layer approximately 0.1 mm thick immediately adjacent to the pipe walls. So out of perhaps 0.1 W of power expended by the player's lungs, only some 10^{-4} W ends up in the form of sound; because our ears are so sensitive, that is quite enough.

We will encounter further illustrations of several of the points studied here as we continue with reed woodwinds in the following chapter.

SUMMARY

The natural modes of vibration for an air column bounded by a cylindrical pipe are sinusoidal standing waves. These waves must have pressure antinodes and displacement nodes at a closed pipe end but pressure nodes and displacement antinodes at an open end. To a good approximation, the natural mode frequencies form a complete harmonic series (with $f_1 = v/2L$) for an open pipe and odd harmonics only (with $f_1 = v/4L$) for a stopped pipe.

Energy to maintain such standing waves at large amplitude can be supplied by an air jet blown against a sharp edge at one end of the pipe. The general tendency toward instability in the jet flow is exploited by positive feedback signals to make the fluid jet drive the pipe at its natural resonant frequency. While the pitch of an organ pipe is determined by its length, the timbre is affected by the diameter, as well as being highly sensitive to the exact mouth configuration. Other things being equal, a thinner pipe will have stronger high harmonics that make for stringier tone. The cooperation of thousands of individual pipes is required to achieve the full versatility of a large pipe organ.

Fingerholes supplement overblowing to allow many notes to be played on a single pipe by modifying the standing waves. It is a combination of hole size and placement that determines how much the pitch is altered by a fingerhole. Pitch is controlled primarily by the first open hole (the one closest to the blowing end), especially if that hole is large. The recorder offers several interesting contrasts to the modern transverse flute: fixed-geometry fipple mouthpiece versus variable embouchure; tapered pipe versus cylinder; plain fingerholes versus keypad mechanism; small tone holes versus large; tone holes placed for finger accessibility versus evenly spaced; fewer tone holes supplemented by cross-fingerings versus many with mechanical assistance.

REFERENCES

The physics of organ pipes is presented by Neville Fletcher and Suzanne Thwaites in *Scientific American, 248,* 94 (January 1983). Further explanation of edgetone action, with elegant experimental support, is given by John Coltman in *JASA, 60,* 725 (1976); for a technical article, this is unusually readable and interesting. Other important research that supports our current understanding of pipe tones can be found in the following articles: S. A. Elder, *JASA, 54,* 1554 (1973); N. H. Fletcher, *JASA, 56,* 645 (1974)*; N. H. Fletcher, *JASA, 60,* 926 (1976); R. T. Schumacher, *Acustica, 39,* 225 (1978); N. H. Fletcher and L. M. Douglas, *JASA, 68,* 767 (1980); S. Thwaites and N. H. Fletcher, *Acustica, 51,* 44 (1982). A good treatment of organ-pipe scaling is provided by Fletcher in *Acustica, 37,* 131 (1977). Interesting work by A. W. Nolle on pipe voicing adjustments appears in *JASA, 66,* 1612 (1979). For a good but brief introduction to different types of organ pipes and their tone colors, see Strong and Plitnik, Section 35. This and general organ design are treated in greater depth by H. Klotz, *The Organ Handbook* (Concordia, 1969), and P.-G. Anderson, *Organ Building and Design* (New York: Oxford University Press, 1969).

Resonance frequencies of the recorder have been studied by D. H. Lyons, *JASA, 70,* 1239 (1981). Benade presents his theory of tone hole placement in Chapter 21, with the full mathematical treatment in his article in *JASA, 32,* 1591 (1960).* This problem also is treated in depth from a different viewpoint by Cornelis Nederveen in *Acoustical Aspects of Woodwind Instruments* (Frits Knuf, Amsterdam, 1969). Nederveen's approach is applied to the Boehm flute by Coltman in *JASA, 65,* 499 (1979).

Benade treats the flute in Chapter 22; in particular, he presents further information on the role of the tapered head joint and on the effect of cork adjustment. Coltman wrote on the flute for *Physics Today, 21* (November 1968), p. 25, with a more detailed account of the same work in *JASA, 44,* 983 (1968).* Some further relations of physical parameters to aspects of flute performance technique are discussed by Fletcher in *JASA, 57,* 233 (1975).* Additional recent work on flutes includes another article by Coltman in *JASA, 69,* 1164 (1981); one on the role of the head joint by N. H. Fletcher, W. J. Strong and R. K. Silk, *JASA, 71,* 1255 (1982); and impedance calculations by Strong, Fletcher and Silk, *JASA, 77,* 2166 (1985).

*Can be found, along with several others, in Earle L. Kent, ed., *Musical Acoustics: Piano and Wind Instruments* (Halsted Press, 1977).

SYMBOLS, TERMS, AND RELATIONS

L_o open-tube length
L_c closed-tube length
f_n pipe mode frequency
f_I Stage-I edgetone frequency
P period
p pressure
v sound speed in air
v_j jet velocity
b jet gap width
D pipe diameter
$f_n = n(v/2L_o)$

$f_n = (2n - 1)(v/4L_c)$
$f_I \cong 0.2\, v_j/b$ (pure edgetone)
$f_I \cong 0.1 v_j/b$ (normal pipe operation)
$N \cong L/D$ critical mode number
closed and open ends
nodes and antinodes of pressure and displacement
laminar, oscillating, and turbulent flow

dynamic instability
flue and jet
edgetone
organ pipe ranks and scaling
tone holes
vent hole and overblowing
embouchure
recorder versus transverse flute

EXERCISES

1. Consider standing waves in air confined in a long pipe. Explain why displacement antinodes do not occur at the same locations as pressure antinodes.
2. Find in the graphs of Figure 12.1 points representing the greatest displacement, greatest velocity, and greatest pressure in that standing wave. Which two of these three occur at the same time (although at different places)? Which two occur at the same place (although at different times)?
3. For an open tube of length $L = 57$ cm, what are the wavelength and frequency of the fundamental mode?
4. For a closed tube of length $L = 1.72$ m, what are the wavelength and frequency of the fundamental mode?
5. For an open tube of length $L = 17$ cm, what are the wavelength and frequency for the first mode? For the fifth mode? Sketch the standing-wave patterns (both displacement and pressure) for the fifth mode, as in Figure 12.2.
6. For a closed tube of length $L = 43$ cm, what are the wavelength and frequency for the first mode? For the fifth mode? Sketch the standing-wave patterns (both displacement and pressure) for the fifth mode, as in Figure 12.3.
7. For fundamental vibration frequency 860 Hz, what length open tube would be required? What length closed tube?

*8. Two open tubes both have actual length 1 m, but their diameters are $D_1 = 1$ cm and $D_2 = 10$ cm. Allowing for end corrections, what are their fundamental frequencies?

9. If a simple edgetone setup has gap $b = 0.5$ cm and jet velocity $v_j = 40$ m/s, what should be the approximate frequency of the edgetone?

*10. Suppose you have an organ pipe whose length is appropriate for frequency $f = 250$ Hz. What is the vibration period P? What is the required transit time for jet disturbances to cross the gap? If $v_j = 50$ m/s, what is the optimum gap width b?

11. Which of the graphs in Figure 12.13 would you expect to be most closely related to the waveform picked up by a pressure-sensitive microphone some distance away from the pipe? (Suppose the experiment is done outdoors to avoid confusion from sound reflections off the walls of a room.)

12. Two open pipes both have length $L =$ 40 cm, but one has diameter $D_1 = 10$ cm and the other $D_2 = 2$ cm. What is the *maximum* number of harmonics that could possibly be strong in the spectrum of each?

*13. Suppose an 8′ rank of organ pipes has diameters ranging from 10 cm at C_2 down to 8 mm at C_7. Use the critical N to estimate the cutoff frequency beyond which you would expect to find no strong harmonics in the spectrum for the C_2 and C_7 pipes.

*14. Which harmonic is reinforced by a $1\frac{1}{3}'$ organ stop? Figure out what role might be played by a $5\frac{1}{3}'$ stop. (Hint: It is more likely to be found in a pedal division.)

15. Suppose you drill a large hole 44 cm down from the top of a 66-cm open pipe for C_4. What note should you get with the hole open?

16. Suppose you have a 66-cm open pipe for C_4. Where would you expect to drill a large hole in order to get $C_4^\sharp$? Where for A_4? Compare if possible with a real transverse flute.

*17. In Exercise 15, suppose you still want to get the same higher note with the same pipe but with a small hole. If the pipe diameter is 2 cm, the hole diameter 1 cm, and the wall thickness 4 mm, where would you drill the hole?

18. Suppose your flute has a leaky keypad about halfway down to the first open hole. What difficulty may you have in playing?

19. Where would you locate a small vent hole to encourage the third mode of an open pipe? The fourth mode?

20. If you go from any node in a standing wave to the next adjoining antinode, how many wavelengths have you gone?

21. Explain how the air motions graphed in the left side of Figure 12.2 are analogous to the pendulum motions on the right side of Figure 9.9. Explain why the air motion corresponding to pendulum mode 1 was not counted as one of the pipe modes.

22. Other things being equal, what difference in timbre do you expect between a fat organ pipe and a thin one? Why? What other things may not necessarily be "equal"?

23. A spiderweb located midway along an open organ pipe has collected enough dirt and dead insects to slightly resist any flow of air past that point. How does this affect the first several natural modes? Explain why they would be affected in the opposite way by air leakage through a small hole in the pipe wall midway down its length. Which of these two situations is more closely analogous to resting a finger lightly on the midpoint of a guitar string?

PROJECTS

1. Use fingering charts to study the patterns of open and closed holes for the entire range of the flute. Verify as far as possible that these patterns accord with the general principles presented in this chapter. (You should find cross-fingerings in the third octave that make more sense after you study the next chapter.)

*2. Obtain access to a pipe organ, giving assurance that you will take great care not to actually disturb any of the pipes. Identify which rank of pipes corresponds to each stop. Make note of their shape (open versus stopped versus chimneys, and so forth), length, width, and mouth size; correlate these as far as you can with the speaking pitch and timbre of each stop.

3. Obtain cardboard mailing tubes or PVC water pipe. Prepare two or more pipes of equal length (50 or 60 cm may be convenient) but different diameters. Use a sine-wave oscillator to drive a small loudspeaker placed close to one end of a pipe, while a microphone near the other end sends its signal to an oscilloscope. Watch the amplitude on the oscilloscope while tuning the oscillator to find the resonant frequencies of the first several natural modes of each pipe. Measure these carefully with a frequency counter and check whether they are indeed harmonic series. If you measure carefully enough, you should be able to show that the fundamental frequency is affected by the pipe diameter. Are your results compatible with the prediction that the end correction should be 0.6 times the pipe diameter? You should measure the ambient temperature and use the corresponding speed of sound rather than merely assuming 344 m/s. Optional extension: Make similar measurements on two pipes of the same diameter but with a large "tone hole" drilled in the side of one and a small hole in the other. What is the "effective length" of each pipe now? Think about whether you want to leave the microphone at the pipe end or put it near the tone hole.

CHAPTER

13 Blown Reed Instruments

What does the blare of a trumpet in a marching band have in common with the reedy squawk of a clowning bassoon or the hollow-sounding low notes of a clarinet? Our first task in this chapter is to recognize the same basic sound-generating mechanism in all these instruments and to understand how it works. Then we must explain how further details of instrument construction, especially tube shape, modify the vibrations to produce a variety of recognizably different tone colors.

Although we shall broaden the definition later, let us begin by thinking of a *reed* as a long, thin strip of flexible material clamped at one end. If you pluck the free end, the reed vibrates with a natural frequency determined by its mass and stiffness; removing mass near the tip raises its frequency, but thinning near the base mainly affects the stiffness and so lowers the frequency. But this vibration dies away as energy is gradually lost to friction and sound. To maintain steady vibration, some source must continually replenish the energy, and an airstream can do the job.

We will study first the air-driven vibrating reed in its simplest form, as found in organ reed pipes. This is followed by consideration of the reed woodwinds and then the brass instruments, for which the player's lips serve the function of a double reed. Along the way we must delve into topics such as natural mode shapes and frequencies; hard and soft reeds; fingerholes, keys, and valves; and the acoustical functions of mouthpieces and bells. In Section 13.4 we present the concept of *regimes of oscillation*, or cooperative agreements negotiated among the many natural modes within an instrument to determine steady sound recipes. Finally, we will consider mutes, radiation efficiency, and the difference between sound inside and outside the instrument.

13.1 ORGAN REED PIPES

Many—perhaps a fourth or a third—of the pipes in a good organ work in an entirely different way from the flue pipes we studied in the last chapter. Hidden inside the pipe foot is a metal reed, usually made of brass (Figure 13.1). The tuning wire is springy and holds part of the reed firmly against the open side of the air channel called the *shallot*, leaving just enough free to have

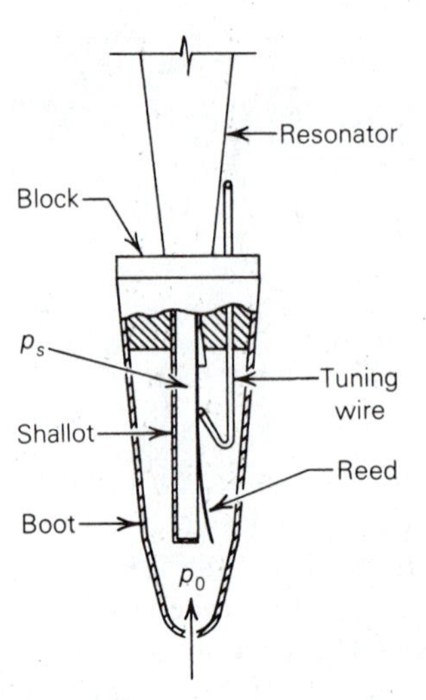

FIGURE 13.1 Mechanism of an organ reed pipe. The reed normally is manufactured to have a slight curvature when relaxed; then it will close smoothly against the shallot with a rolling motion. The shallot is tubular, open along one side next to the reed and at the top end into the resonator. If the tuning wire is moved downward, the active portion of the reed is shortened and its pitch raised.

the desired vibration frequency. When air is admitted to the pipe foot from the windchest under excess pressure p_0, its only way of escape back to the atmosphere is through the shallot. To force a continuing stream of air through the narrow opening between reed edges and shallot, the pressure p_0 out in the boot must be greater than the pressure p_s inside the shallot. If the size of the opening changes, the flow rate and the pressure difference also change, as shown in Figure 13.2.

But this difference in pressures means there is a nonzero force pushing the reed inward and becoming stronger as the opening narrows. That pressure-difference force is opposed by the spring-type restoring force from the

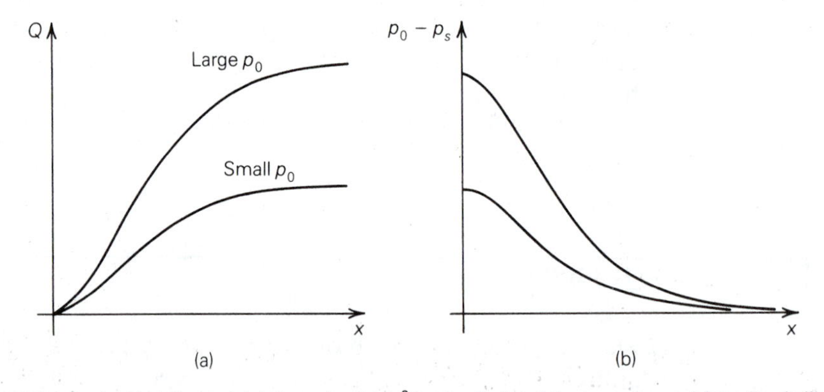

FIGURE 13.2 (a) Variation in total flow rate Q (cm^3/sec) with the distance x of a reed from its shallot, for two different values of applied pressure p_0. (b) Pressure difference across the reed; for large x this approaches zero, as most of the pressure drop moves to whatever other part of the system has more resistance and limits the flow.

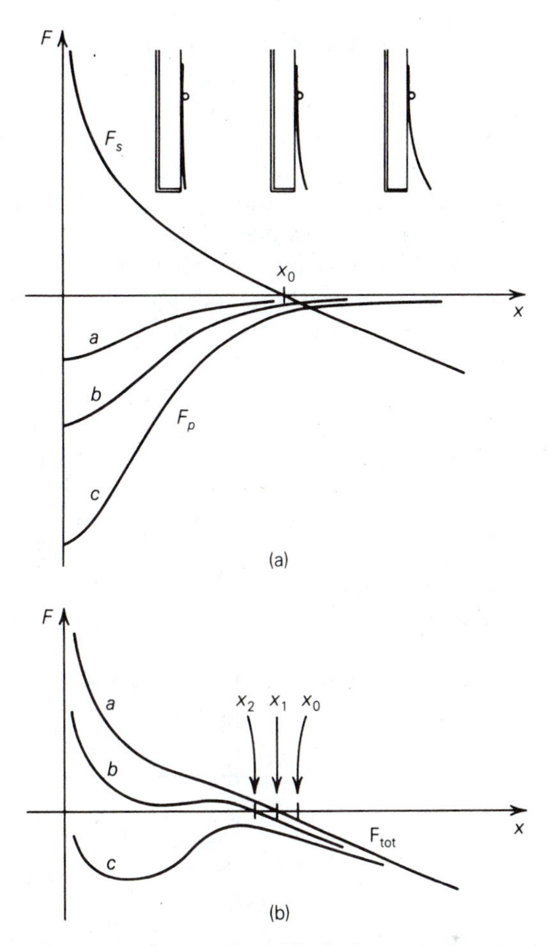

FIGURE 13.3 Forces acting on a reed as functions of its displacement x from the shallot; its relaxed position with no airflow is x_0. F_s represents stiffness, F_p air pressure difference, and F_{tot} the sum of both; a positive force is one tending to pull the reed away from the shallot. (*a, b*) Low boot pressure and small average flow; the reed can remain at a stable equilibrium position such as x_1 or x_2. (*c*) Higher boot pressure overcomes the spring force and moves the reed against the shallot.

reed stiffness, which pushes it back out. The dependence of both forces on reed position is shown in Figure 13.3a.

Here we encounter echoes of the last chapter—another fluid-flow instability! But this time there is an important difference: It is not just instability intrinsic to the fluid flow itself within rigid boundaries, but an instability depending on interaction between the fluid and a flexible boundary. (Examples 4 and 5 in Box 12.2 are the most relevant now.) As with other fluid-flow instabilities, there is a critical amount of flow (and corresponding boot pressure) beyond which the reed cannot just move over slightly to a new stable equilibrium position (Figure 13.3b). Instead it closes more or less completely against the shallot, cutting off the airflow. The stiffness of the reed may need the aid of a positive pressure pulse in the shallot before the reed will finally spring back. When the reed opens and the flow resumes, so do the forces that bring the reed over to the shallot again.

The *beating reed* is slightly larger than the opening in the side of the shallot and is the kind normally encountered in organ pipes, as well as in the reed woodwinds. It also is possible to use *free reeds* cut slightly smaller than the opening so that they practically block it without actually hitting the sides; these are found occasionally in organ reed pipes, as well as in the harmonium (a pipeless reed organ with foot-operated bellows), harmonica, and accordion. Because beating reeds shut the opening more suddenly and completely, they create more jagged waveforms that are richer in high harmonics and so more harsh or penetrating in tone quality.

The buzzing sound of the air-driven reed can be greatly strengthened by having the shallot open into a resonating pipe. The resonator usually is conical, because this will reinforce all harmonics of the reed sound. The resonator does not have an entirely passive role, for a reflected wave returning from the open end may help open the reed to admit the next puff of air. This positive feedback is most effective if the natural frequencies of reed alone and pipe alone are matched. Then the flow is qualitatively as shown in Figure 13.4.

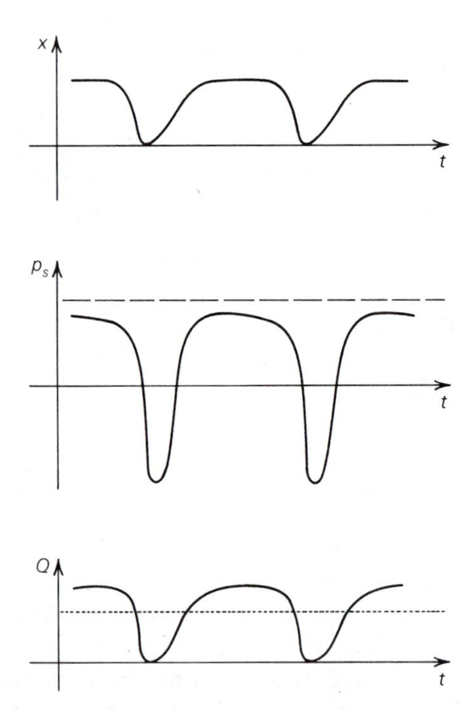

FIGURE 13.4 Phase relations among dynamical variables in typical reed instruments. The curves are generally nonsinusoidal, but their exact shape depends greatly both on instrument design and on how loudly it is played. x = reed opening; t = time; p_s = pressure inside the shallot, which always remains less than pressure p_0 in the boot (dashed line); Q = volume flow rate (cm^3/s) through the opening, which is proportional to air velocity a short distance inside the tube. The flow should be regarded as consisting of a steady average part (dotted line) and an acoustic part that goes both positive and negative.

But what happens if the reed and pipe frequencies do not match? Will the reed stick to its own frequency, or will it go along with the pipe? The answer depends on whether we have a hard or soft reed. We use the word *hard* not only because such reeds are made of metal, but also more specifically to indicate that the inertial and elastic properties of the material dominate its behavior. Pluck a metal reed and it vibrates for several seconds because it loses minimal energy to friction on each cycle. A hard reed has a very high and narrow peak on its resonant response curve (Figure 11.16), and so a strong preference to vibrate at one well-defined frequency.

A *soft* reed, on the other hand, is strongly influenced by dissipative processes—pluck it and the vibration lasts only a fraction of a second. Cane reeds used on woodwinds fall into this category, because woody materials are less ideally elastic than metal, and even more so because the player's soft lip clamped against the lower part of the reed absorbs much of its energy. The resonance curve for a soft reed is low and broad, so it can respond almost equally well over a wide range of frequencies.

If a single reed must be capable of producing many different notes, as it does when connected to a pipe with fingerholes, it is essential that it be a soft reed. Then the pipe becomes the main determiner of vibration frequency through its feedback, and the reed merely follows along. But a reed that is called upon only to play a single note, as in an organ pipe, may be a hard reed. Then the reed insists stubbornly on its own frequency, whether or not it gets any help from the pipe. Tuning a reed pipe means getting the right pitch by adjusting the reed length and then getting the fullest tone quality by adjusting the pipe length for best cooperation. If the resonator is out of tune with the reed, the tone becomes weak or may even be stifled altogether.

A variety of sounds, including some that are imitative of brass and woodwind instruments, can be obtained by varying the details of reed and shallot design (for instance, the degree or the suddenness of closure), as well as by differences in resonators. By the end of the chapter, you will see why broad cones favor loud sound poor in high harmonics, narrow cones favor softer sound rich in harmonics, and cylinders favor clarinetlike sound.

One additional interesting trick is the use of fractional-length resonators for reed stops such as the *regal* or *vox humana.* Take a set of reeds tuned to normal pitch, which would ordinarily call for resonators up to 8 feet long, and suppose you install instead quarter- or eighth-size resonators (longest pipe 2′ or 1′, respectively). Then the first few harmonics of the reed sound receive no encouragement at all from the resonator, and you hear a nasal or buzzing tone quality whose fundamental is very weak; practically all the sound energy is in those higher harmonics that can resonate in the short pipes.

We can make an instructive contrast between flue and reed instruments by thinking of a steady incoming stream of air being modulated in both cases by a valve that opens and closes repeatedly to release puffs of air into the resonating tube. The flute (or anything else based on an edgetone) has a *flow-controlled valve*: It is the acoustic flow in and out through the mouth hole that carries the jet with it and thereby determines when the airstream goes inside or outside. Reed action, on the other hand, constitutes a *pressure-controlled valve*: It is the acoustic pressure at the entrance of the pipe that directly determines when the reed will open to admit another puff of air.

The fact that flue and reed pipes are opposite in that sense suggests that the condition required of all standing waves in the pipe at the mouth end also should be opposite; where the flute had a pressure node or displacement antinode at its mouth hole, *a reed instrument must have a pressure antinode or displacement node at the reed.* Thus, to predict what standing waves will be allowed in a pipe, we treat the reed end as a closed end.

We may argue that this is reasonable by noting that the closure of the reed during even part of the cycle forms a wall against which a large pressure difference can build. It is not precisely true that no flow is allowed; but even when air is flowing past the reed, that opening is much smaller than the rest of the pipe. Comparable pressures cannot produce nearly as much flow through the reed opening as they do elsewhere in the pipe, so comparatively speaking it is almost a displacement node.

The same idea may be cast in the language of *input impedance*, which is defined as the ratio of acoustic pressure to acoustic volume flow at the entrance to the pipe. The input impedance is a function of frequency determined by the length and shape of the tube (as we will exhibit later in Figure 13.15 and others). To feed maximum energy into the resonator, a reed instrument requires a high input impedance (large pressure at the entrance), while a flute requires a low input impedance (large volume flow). In each case, the requirement is met only for a few special frequencies—those of the instrument's natural modes.

13.2 THE REED WOODWINDS

Let us first describe the main features of the several types of reed woodwind instruments and then make a series of comparisons to see which properties are responsible for which differences in sound.

1. The *clarinet* has a single reed and a cylindrical bore with some flare at the lower end (Figure 13.5a). Except in soft playing, the reed beats against the mouthpiece edge (Figure 3.8a) just as organ reeds do against the shallot. The same name is used for the entire family, with the B♭ soprano being most common. All later specific statements refer to the B♭ clarinet, because it has been studied extensively.

2. The *saxophone* has a single reed on a beak-shaped mouthpiece just like the clarinet but a conical bore with only a small flange at the lower end (Figure 13.5b). Here, too, the same name is used for all members of the family, and several different sizes are common, especially in stage bands.

3. The *oboe, English horn,* and *bassoon* all have double reeds (Figure 3.8b) and conical bores (Figure 13.5c) and so work on the same acoustical princi-

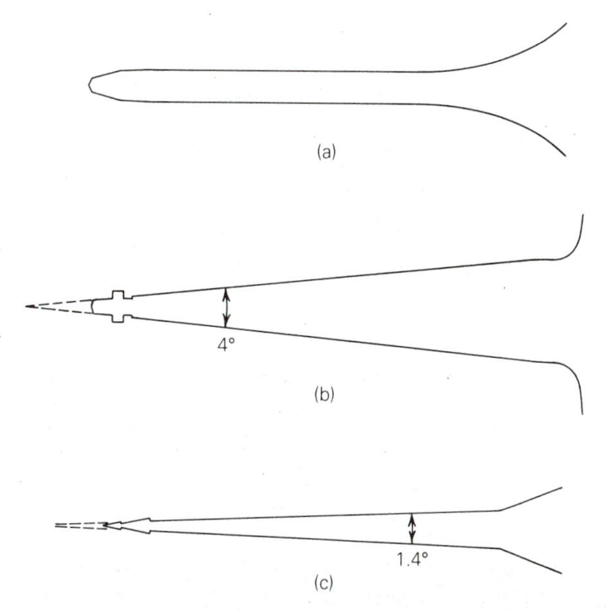

FIGURE 13.5 Profiles of three woodwinds with width exaggerated three times compared to length. (a) Clarinet: total length, 66.5 cm; inner diameter of cylindrical portion, 1.5 cm; diameter at end of bell, 6.0 cm; lowest note, D_3. (b) Soprano saxophone: length, 69 cm; maximum flange diameter, 9 or 10 cm; lowest note, $A^{\flat}_3$. (c) Oboe: length, 64 cm; maximum diameter, 3.8 cm; lowest note, $B^{\flat}_3$. Missing cone tips (dashed lines) are compensated for by extra volume in reed cavity.

ples. They are the soprano, alto, and bass members of what we shall call the oboe family. The *shawm* belongs to the same family and sometimes still is heard in concerts of medieval and Renaissance music. Its tone is less refined, with the reed often inserted fully into the mouth, whereas the oboe player uses his lips against the lower part of the reed to maintain more control over its operation. The *bagpipe* chanter also belongs to this family, as do numerous instruments from Asia and the Middle East.

4. The fourth logical combination, a double reed and a cylindrical tube, has no common modern representative, but the *krummhorn* is found in Renaissance consorts. Because it has a capped reed (Figure 3.8c), the player has even less control than with a shawm; in particular a krummhorn cannot be overblown reliably, so its playing range is limited to only a little more than one octave.

The two halves of a double reed beat against each other instead of a mouthpiece, but that is of little acoustical importance. The choice is largely a matter of the relative convenience of mounting on large or small tube ends; the oboe family can be played perfectly well with miniature single-reed mouthpieces, although they have never become popular.

The most obvious acoustical differences have to do with bore shape. Clarinets have a cylindrical tube open at one end and effectively closed at the other. A glance at Figures 12.2 and 12.3 (pages 228 and 229) reminds us that

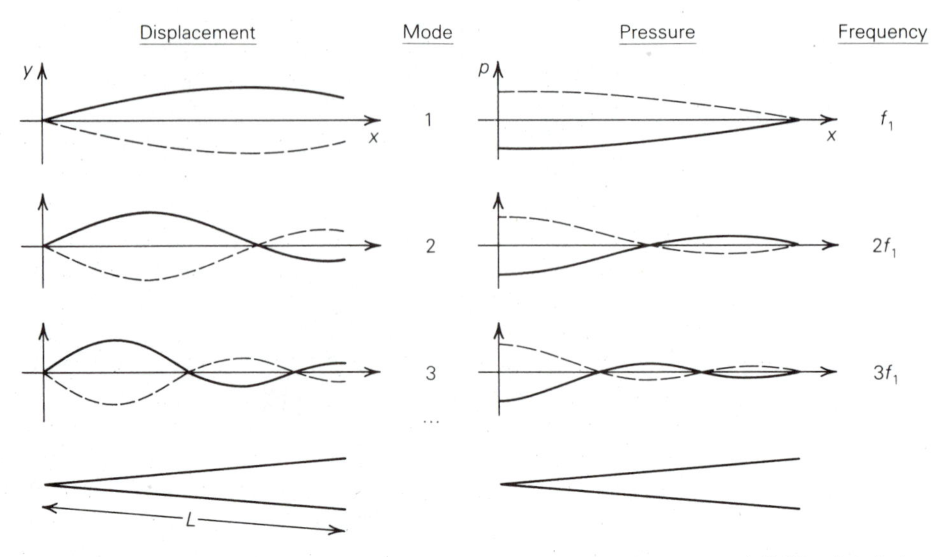

FIGURE 13.6 First three standing-wave patterns in a complete cone, open at one end. Solid and dashed lines represent times a half-cycle apart. The frequencies are $f_n = n(v/2L)$, the same as for an open cylindrical tube of the same length, in spite of the difference in mode shapes. Compare with Figures 12.2 and 12.3.

the fundamental wavelength for a closed tube is four times the tube length, compared with only twice the length for an open tube. This explains why the clarinet sounds about an octave lower than the flute for similar pipe lengths. We also see from Figure 12.3 that (to a first approximation) the clarinet's natural modes supply only the odd-numbered members of a harmonic series. But conical tubes open at one end and closed at the other have natural modes, shown in Figure 13.6, whose frequencies happen to form a complete harmonic series with fundamental $f_1 = v/2L$.

---

*It is natural to ask whether there are shapes other than these two that would have natural mode frequencies forming part or all of a harmonic series and thus be musically useful. There is one, called a Bessel horn of order one, about which you may read in Benade's Chapter 20. It has a gradually increasing flare; it is not used as the basis of any complete instrument, but is related to the bell shapes used on brass instruments.

The first important consequence of this difference is in tone color. It is the absence (or relative weakness) of the first few even-numbered harmonics that makes the characteristic hollow timbre of the clarinet's low notes, which set it apart from most other instruments. The oboe and saxophone families have all harmonics at least potentially present, and other reasons must account for their differences in timbre from each other or from violins, for instance.

A second important consequence is in overblowing. For an oboe or saxophone, we may expect overblown notes to sound an octave higher (and an octave plus a fifth for third mode, two octaves for fourth mode, and so on). Thus, if fingerholes or keyholes can be worked out to provide the lowest octave, all higher notes come with overblowing; so we can expect fingering schemes along the same lines as for flutes or recorders. But when a clarinet is overblown, the first available higher mode is at the fifth beyond an octave, 19 semitones rather than 12 above the fundamental. This means that the clarinet must have enough tone holes, interlocking keys, and fingering patterns to produce a lower register of 19 notes before overblowing begins. (It actually has 24 tone holes, some of which allow getting the same note with alternate fingerings and one of which doubles as a vent hole or register hole.) It is theoretically possible that flute, oboe, and saxophone families could all have a common fingering scheme (although they do not at present), but clarinets can never fit in such a scheme.

Now let us consider the major difference between oboes and saxophones. Both are conical, but typical cone angles are several times larger for the saxophones (soprano sax 4°, oboe 1.4°; tenor sax 3°, bassoon 0.8°, according to Nederveen). The same remarks made earlier about organ flue pipes apply here: narrow tubes have sharp resonances for many modes, whereas higher frequencies readily escape from wide tube mouths without forming strong standing waves. The especially reedy timbre of the oboe family comes precisely because its narrowness gives a spectrum rich in high harmonics, while only the first few harmonics can resonate well in the saxophone. Just as with organ pipes, the lower-pitched members of each family need to be a little narrower to have enough high harmonics for a satisfactory timbre that sounds as if it belongs to the family.

The larger bore of a saxophone allows correspondingly large tone holes. These make it a more efficient radiator of sound, and it can be played quite loudly. The smaller holes severely limit the sound output of the oboe, even when the player blows hard.

We may also connect tone hole size with the presence or absence of a bell at the lower end. The large tone holes of a flute or saxophone make such an effective termination of the tube that most of the sound energy radiates from the first open hole. The standing-wave amplitude at the second open hole is less than half as much as for the first hole and smaller yet at the third open hole and beyond. All notes are radiated much the same way as the lowest note, which comes out the end of the instrument, and all will sound similar if the end resembles a tone hole. Hence no need for a bell.

But clarinets and oboes have relatively small holes, so an appreciable portion of the oscillatory airflow continues on down the bore instead of out through the first open hole. The amplitude at the second hole may be as much as 70%, and at the third hole 50%, of that at the first hole, so it is the first *several* holes that make important contributions to the radiated sound of the fundamental. This cooperation enables many low and medium harmonics to radiate with comparable efficiency, whereas radiation through a single hole favors only very high harmonics. Thus, the lowest two or three notes will tend to have a different timbre, as if they did not belong to the same instrument—unless, that is, some other device can mimic the effect of a long row of open holes. And that is what the bell is for—it has no effect on the higher notes at all. From this viewpoint, the recorder ought to have a bell but does not, and indeed its lowest notes are anemic in tone quality and hard to play convincingly.

13.3 THE BRASS FAMILY

We would like to apply much of what we learned in the last two sections to brass instruments as well, and the key is to recognize them as reed instruments. The fact that the shape of the "reed" is quite different from a strip of metal or cane is nonessential; what is important is that the *lip reed* is also a very small opening with flexible boundaries through which air is admitted to the instrument so that a related fluid-flow instability operates.

Excess pressure inside the mouth pushes outward on the lips, tending to open them with a rolling motion. But once they are open the pressure drop decreases, and tension in the lip muscles tends to close them again (Figure 13.7). The maintenance of lip oscillations is aided further by aerodynamic effects (discussed in Box 14.1).

Like other reeds, the lips can vibrate alone this way and produce a buzzing sound; but positive feedback from a resonating tube can strengthen this sound greatly and give it a more stable and musical quality. Look back at Figure 13.4, and regard the graphs as showing lip separation and mouthpiece pressure and flow.

As with other reed instruments, the blown end of the tube is effectively closed. We might expect to have two families, one with cylindrical and one with conical tubes. We can identify the cornet, alto and baritone horns, and tuba as being derived from cones, and the trumpet, trombone, and French horn as cylindrical for at least a major portion of their length. But neither group actually is built closely to those theoretical idealizations.

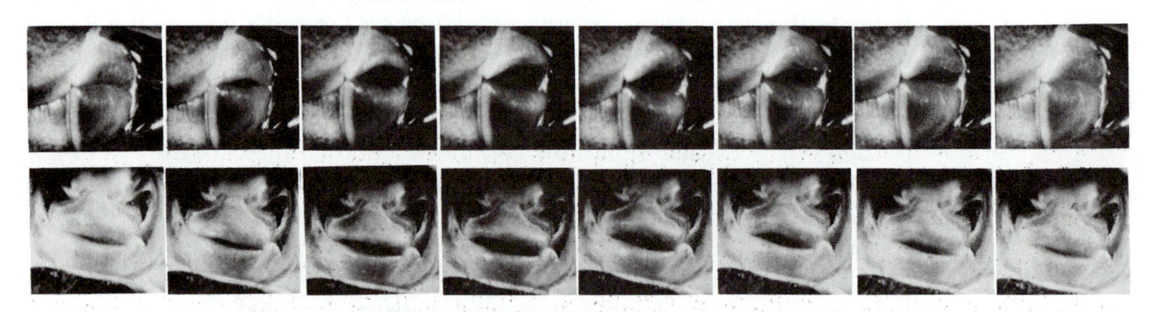

FIGURE 13.7 Side and front views of one cycle of lip motion for a trombone player. The lizard-eye appearance in the front view is an artifact of optical distortion in the transparent mouthpiece. Taken from high-speed movies by Dr. H. Lloyd Leno of Walla Walla College.

Because in the case of a cone the entire instrument already is flared, the open end is quite large and needs little more in the way of a bell to radiate efficiently. So the cornet family, even though not completely conical, has relatively inconspicuous bells. But a purely cylindrical trumpet would have an extremely small open end and would put out a thin, stringy tone that could just compete with a flute in loudness. So the bell is a major feature of the trumpet and trombone, completely altering the character they would otherwise have—not only in loudness but also, as it turns out, in other remarkable ways.

The cornet family, like the oboes and saxophones, should have natural modes whose frequencies closely approximate a harmonic series. So it is no surprise that (for a given valve setting) overblowing produces bugle-call patterns (harmonics 3, 4, 5, 6 of some fundamental), as well as the second harmonic and several above the sixth. It is the great difficulty of producing the sound of that fundamental that is puzzling.

The puzzle is much more striking for the trumpet family, for we would expect by analogy to clarinets to get only odd-numbered members of a harmonic series by overblowing. Yet here, too, we get a complete harmonic series except for an absent fundamental. The mystery is real—by blowing directly on a truly cylindrical tube, such as a length of garden hose, you can verify that notes closely approximating harmonics 3, 5, 7, 9, . . . are produced. There are two crucial differences between this and a real trumpet: the flaring bell and the constriction in the mouthpiece. Through trial and error, artisans long ago came up with bell and mouthpiece combinations that modify the standing-wave patterns (Figure 13.8, p. 266) and frequencies in just the right ways to get a complete harmonic series of overblown notes, including both even and odd members.

Let us emphasize that these alterations do not just leave original odd-numbered modes there and somehow insert other modes in between. Rather, a minor shift of all modes together can turn their frequencies into multiples 2, 3, 4, 5, . . . of a *different* fundamental, as shown in Figure 13.9 (p. 267). The lowest modes feel the flare of the bell sooner and so are shifted toward the higher frequencies appropriate to a shorter effective length. Resonance in the mouthpiece cup (typically around 800 Hz) makes this end seem several centimeters longer for the higher modes and so shifts their frequencies downward. Notice the price paid for getting the higher modes approximately aligned this way: The first-mode frequency is entirely different from the fundamental frequency of the series; this mode thus is not musically useful.

Should we consider the lip reed to belong to the hard- or soft-reed category? It is actually in between. The tension in the lip muscles can cause a strong enough preference in frequency to ensure that the instrument will be excited in the sixth mode, for instance, rather than the fifth or seventh, which are only 17% away on either side. Yet for each mode the lip reed is soft enough to be controlled by the pipe resonance. If you try to change pitch by adjusting the lips, particularly on midrange modes such as 3 or 4, you generally get less than a semitone away from the right pitch for that mode

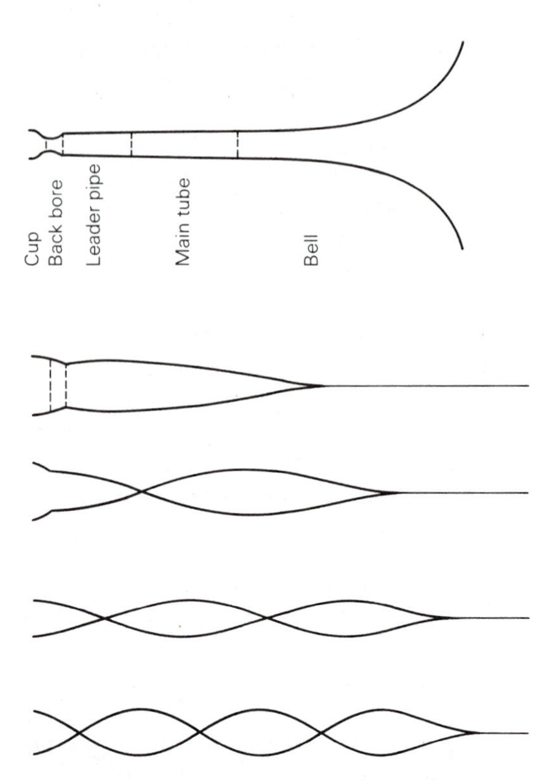

FIGURE 13.8 Pressure standing waves in a trumpetlike instrument. Notice how the first mode is largely confined to the cylindrical portion, while the higher modes penetrate farther out into the bell. (From "The Physics of Brasses" by Arthur H. Benade. Copyright © 1973 by Scientific American, Inc. All rights reserved.)

before you break up or down to the adjoining mode. The high modes (say, 8, 9, 10) are not so well separated, and present the danger of jumping back and forth or even sliding continuously from one mode to another. They require much better lip control and player sensitivity to what the tube wants to do. This is one of the reasons why French horns are particularly difficult to play; they normally are used much of the time in such high modes, up to the twelfth or even beyond.

Lack of comparable control over the reed prevents playing oboes and clarinets in such high modes; it is their use of first mode that requires more notes for a complete scale than a three-valve system possibly could provide. But the trumpet begins its normal range with the second mode and needs only seven second-mode semitones to get up to where the third mode can take over. The straightforward way to achieve this is with a slide trumpet—just like the trombone, its tube is easily adjustable to any length within the necessary range (Figure 13.10). But this allows less quick and accurate jumps from any note to any other than the fixed lengths of extra tubing on a regular trumpet that are inserted in the flow path by depressing each of the three

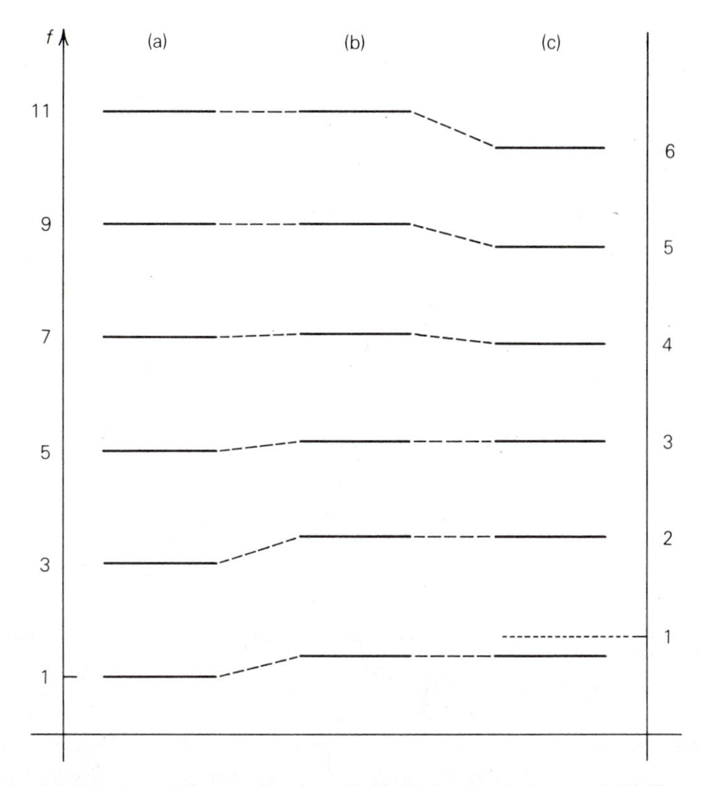

FIGURE 13.9 (a) Natural mode frequencies of a cylindrical tube closed at one end. (b) The upward shift of frequencies for the lower modes caused by replacing an appropriate portion of the tube by a bell. (c) The downward shift for the higher modes caused by replacing another portion on the opposite end by a mouthpiece. The dotted line shows the missing fundamental of the newly formed harmonic series.

piston valves (Figure 13.11 and Box 13.1). Modern French horns use rotary valves, but the acoustical function is identical.

Unfortunately, lengthening the cylindrical section with a valve loop calls for slight changes in the bell to keep all modes in tune. Because that is impractical, the bell design must be some compromise that will work moderately well for several different overall lengths. We will explore certain consequences of the unavoidable imperfections in mode-tuning in the next section.

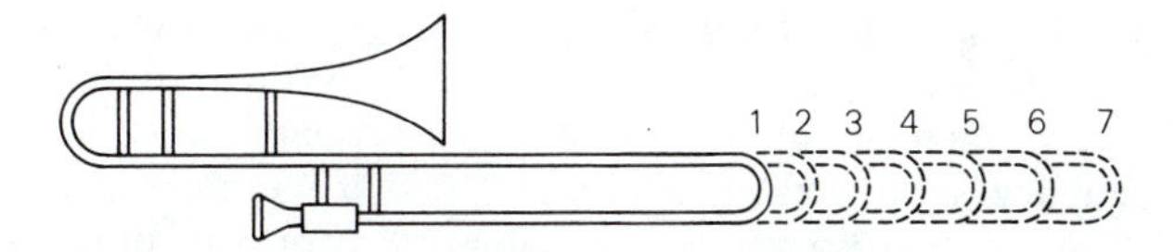

FIGURE 13.10 Trombone slide positions. Note that they are not equally spaced, because each must represent a 6% increase over the effective length for the preceding position to lower the pitch another semitone. (From Backus)

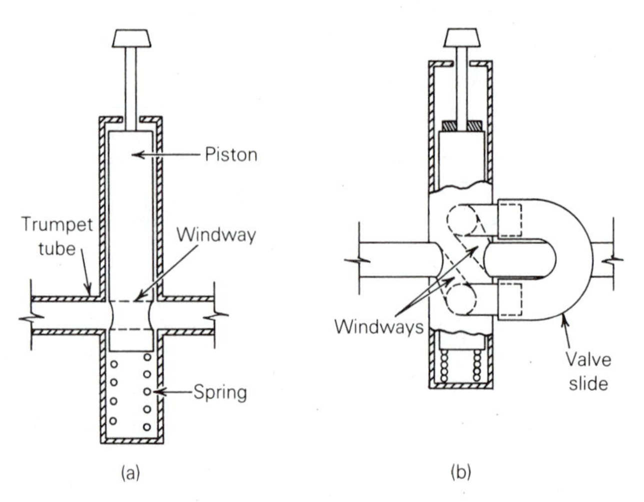

FIGURE 13.11 A trumpet valve. (a) When raised, the air goes straight through; (b) when lowered, it diverts the air through the extra loop of tubing. (From Backus)

*BOX 13.1 A CHROMATIC SCALE FOR THE TRUMPET

For woodwinds, we found it convenient to begin with the maximum tube length and build an ascending chromatic scale by opening tone holes one by one. But the normal starting point for the trumpet is its shortest possible length, with all valves up. This gives its *highest* note for each mode, and we will obtain a *descending* chromatic scale by adding successively longer pieces of tubing.

How long should the loops of valve tubing be? Take first the middle valve (number 2); it adds the shortest piece, which is supposed to lower the pitch one semitone. That means a 5.95% decrease in frequency (assuming equal temperament for this purpose; see Box 7.1); so the extra length should be just 5.95% of the effective length (call it L_0) of the basic trumpet. Next, valve 1 alone (closest to the player's face) is supposed to lower the pitch two semitones. Because $(1.0595)^2 = 1.1225$, there should be a 12.25% decrease in frequency; so we make this loop 12.25% of L_0. So far, so good—no valves, B$^\flat$; valve 2 down, A; valve 1 down, A$^\flat$ (concert pitch).

Now all our hopes for getting by with three valves instead of seven lie in the expectation that valves 1 and 2 together will lower the pitch three semitones to G. The 5.95% and 12.25% together give a total added length 18.20% of L_0—but that is *not* the correct amount, which would be 18.92%; it is about a tenth of a semitone too high. To see why, imagine that valve 1 already is down for a total length of 1.1225 L_0. It is 5.95% of that quantity, *not* just 5.95% of L_0, that needs to be added to go down one more semitone. So it is theoretically impossible to have all three notes properly in tune. The actual lengths used are compromises, close enough that the well-trained player can coax each pitch up or down into its proper place with lip control.

The same problem becomes more severe on combinations involving valve 3. Valves 2 and 3 together are supposed to provide the next lower note. This F$^\#$ is four semitones or 25.99% below B$^\flat$, indicating a length 20.04% of L_0 (25.99 − 5.95) for the third loop. (Notice that valve 3 alone is a poorly tuned alternate fingering for G.) But only if we make it closer to 21% will we get usable approximations for F (valves 1 and 3, ideally 33.48% total) and E (all three valves down, ideally 41.42%).

The severity of the compromises can be re-

(continued)

BOX 13.1 *(continued)*

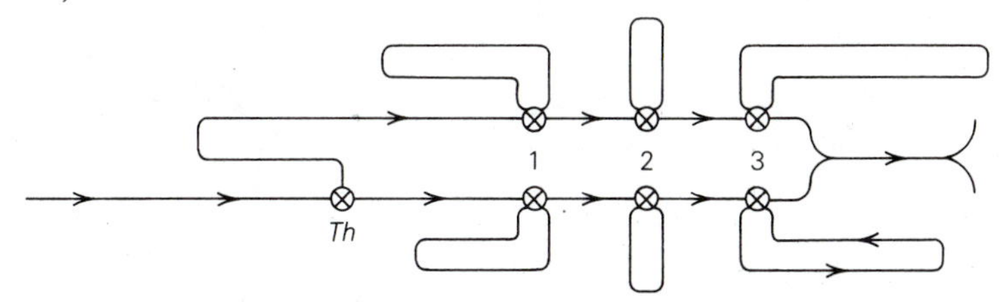

FIGURE 13.12 Schematic airflow through a French horn; F horn on top, B♭ horn on the bottom. Each numbered valve pair is operated by a single lever.

duced by having a small section of the valve 3 tubing that can slide trombone-style. It is moved in or out as needed by the little finger of the left hand, to let this loop be a little shorter or longer depending on what other valves are down. Yet another solution you sometimes see on high-quality instruments, particularly the low brass, is a fourth valve. This provides a separate loop for lowering the pitch a perfect fourth (five semitones) and substitutes for the combination of valves 1 and 3.

It may occur to you that the extra valve extends the instrument's range downward. But if you attempt to go down $2+1+3+5=11$ semitones by using all four valves at once, you will end up closer to 10 semitones down for the same reasons as above. This situation can be remedied by duplicating each of the first three valves, so that a longer or shorter valve loop is added depending on whether valve 4 is down or not. By this time, you practically have two separate horns that just share the same mouthpiece and bell and whose valves operate in tandem from the same fingers. This (Figure 13.12) is how French horn players view their instrument—as an F horn that can be traded off for a higher B♭ horn by squeezing the thumb-operated fourth valve.

13.4 PLAYABLE NOTES AND HARMONIC SPECTRA

We come now to a crucial issue, one that we have sidestepped thus far. What does a reed do when it gets contradictory feedback messages from different standing waves as to how soon it should repeat its vibration to get good cooperation? And what determines how much of each harmonic is present in the recipe for steady oscillation?

Recall our motivations for first introducing the concept of natural modes: (1) Whatever happens in a complex system can be described as a combined motion of all these modes, and (2) each individual mode performs simple harmonic motion. There was also a third idea, which is now called into question: (3) Each mode was supposed to oscillate independently in response to driving forces, completely oblivious to whether or how much any other mode may be moving at the same time. That was true to the extent that we dealt only with *linear* systems—in fact, that is one of the best ways to define what linear means. But now we must face up to nonlinearity: Each mode can influence the others after all.

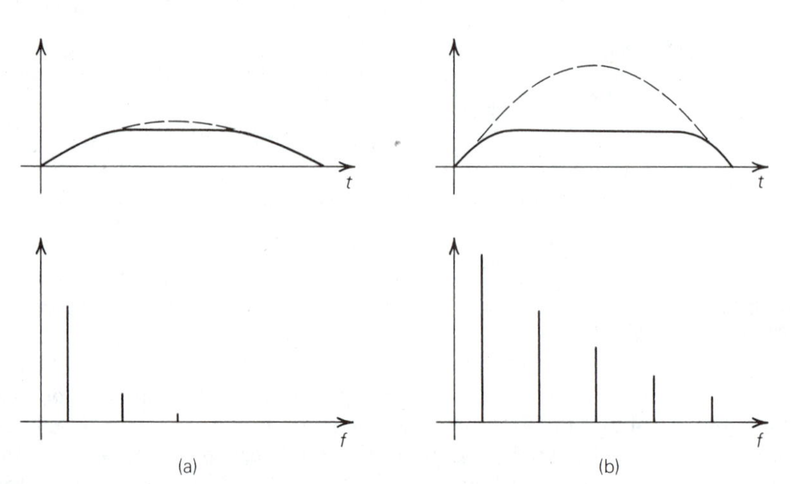

FIGURE 13.13 A half-cycle of edgetone jet motion for (a) small and (b) large acoustic amplitude. Dashed lines show jet position; solid lines, resulting jet flow past edge; vertical lines, Fourier components of jet flow waveforms. Even if jet motion is sinusoidal, jet flow is not, because once fully switched to one side of the edge further displacement does not change the amount of flow.

If we consider an air column alone, it is quite linear and all the standing waves are indeed independent. Make a sudden disturbance inside a pipe, say, by snapping a key over a tone hole, and all the natural-mode concepts apply to the transient "pop" just as well as they did to percussion instruments.

The steady excitation mechanism is what provides the nonlinearity that makes one mode aware of what another is doing. Try to imagine a single natural mode oscillating alone in a blown flute or oboe, with all the other modes quiescent. In the flute, this means the acoustic flow in and out of the mouth hole would be sinusoidal; yet the jet carried back and forth by this flow would deliver a nonsinusoidal disturbance (Figure 13.13) that would immediately excite other modes as well. In the oboe, it means the pressure at the reed would vary sinusoidally; yet the resulting variation of reed opening (Figure 13.2) would permit a highly nonsinusoidal airflow, which brings additional modes into play.

Because each mode has some effect upon all the others by way of the jet or the reed and is, in turn, affected by them, the production of a steady tone requires a compromise as to just how much of the available energy is apportioned to each mode. We call this delicate balance a **regime of oscillation**; the name was given by Benade, who has done most to develop this important concept. A regime of oscillation has relative mode strengths such that each mode cooperates with all the others in maintaining the overall pattern. Specifically, the total effect of the pressure oscillations of all modes upon the reed (or flow oscillations upon the jet) must produce a reed (or jet) motion that maintains all modes at those levels, so that the entire arrangement is self-consistent.

The mathematical formulation of these nonlinear effects is quite complex,

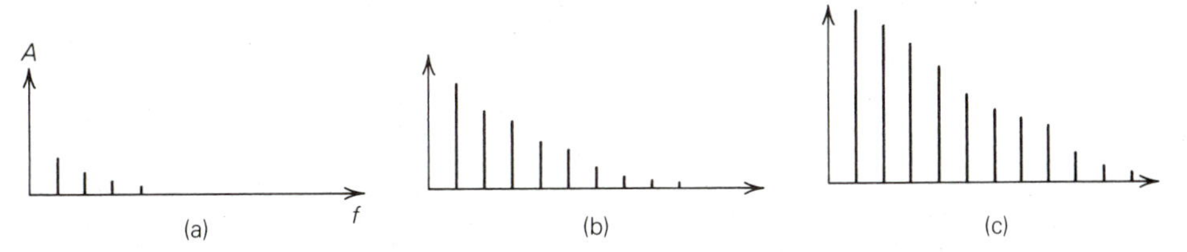

FIGURE 13.14 Spectra of pressure amplitudes inside the mouthpiece of a trumpet playing $B^{\flat}_3$; (a) *pp*, (b) *mp*, (c) *fff*. (Adapted from Benade, p. 419)

but we can describe several useful qualitative generalizations. First:

Nonlinear coupling generally increases with amplitude.

This is perhaps most easily seen in the case of the flute (Figure 13.13). As a consequence, the best chance of playing mainly on a single mode with minimal participation by the others comes at pianissimo levels; fortissimo levels bring on strong coupling that tends to share the energy around so that several modes all have comparable amplitudes. This explains the general tendency in most wind instruments to have a richer spectrum and brighter tone color when played loudly, as illustrated for a trumpet in Figure 13.14. The oboe shows less variation in timbre than most, because its narrow reed opening tends to snap completely shut and thus introduce strong nonlinearity even at lower playing levels, whereas clarinet reed motion may not close off its larger opening completely until forte levels are reached.

Our second and third generalizations have to do with the availability of various modes to participate in regimes of oscillation, with regard both to the strength of their resonance and to their being in tune with one another. To explain these points, we will use *input impedance curves* such as Figure 13.15. These show how strong a pressure fluctuation would accompany a given amount of alternating airflow through the mouth of an instrument at each different frequency. Recall from the end of Section 13.1 that reed instruments have the proper positive feedback only for high input impedance. The input impedance curves are roughly analogous to the multimode resonant-response curves introduced in Figure 11.17 (page 212). We can see at a glance the strength and choosiness of each mode resonance, because they correspond to the height and sharpness of a peak on the input-impedance graph. And the location of these peaks along the horizontal axis tells the natural mode frequencies.

Consider now the availability of resonances in trumpetlike instruments. There are two striking differences between the input impedance curves for a simple cylindrical tube (Figure 13.15) and for a cornet (Figure 13.16). First, the presence of the bell on the cornet makes it so easy for all frequencies above about 1500 Hz to escape that the high-frequency response of the instrument shows no appreciable preference for any particular frequency. So any attempt to play a note above mode 10 on this instrument depends entirely on frequency control by the player's lips; as far as the cornet is concerned, the pitch

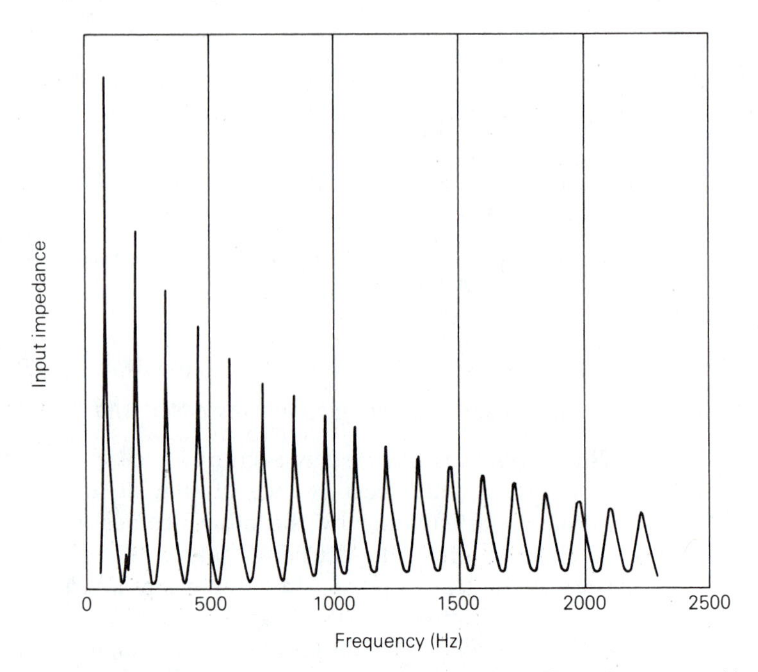

FIGURE 13.15 Input impedance of a cylindrical pipe of length 140 cm. The peaks occur at odd multiples of 63 Hz. (From "The Physics of Brasses" by Arthur H. Benade. Copyright © 1973 by Scientific American, Inc. All rights reserved.)

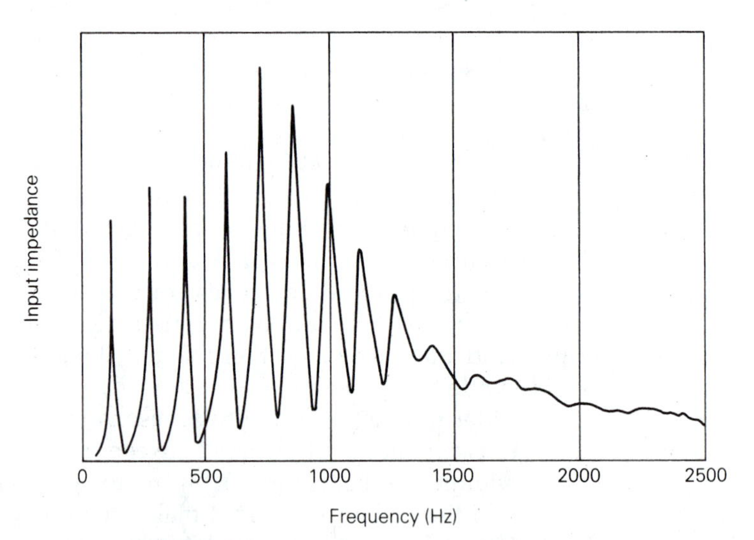

FIGURE 13.16 Input impedance of a cornet (Henry Distin, 1865), measured by Benade (all valves up). (From "The Physics of Brasses" by Arthur H. Benade. Copyright © 1973 by Scientific American, Inc. All rights reserved.)

can slide anywhere you please. Second, mouthpiece resonance strengthens everything between approximately 500 and 1000 Hz in comparison to the lower modes. This means that the mouthpiece shape is responsible for increasing the amount of modes 4–7 in the vibration recipe, making brass timbre more rich and solid than it otherwise would be.

The last statement was an application of the second general principle:

> ***The strength of participation of any mode tends to be proportional to the height of its resonant peak.***

Another example of this is shown in Figure 13.17: In agreement with our earlier, less-sophisticated argument, the input impedance curves indicate that the low notes of an oboe will have many harmonics of comparable strength, while those of a saxophone will be dominated by the first few harmonics.

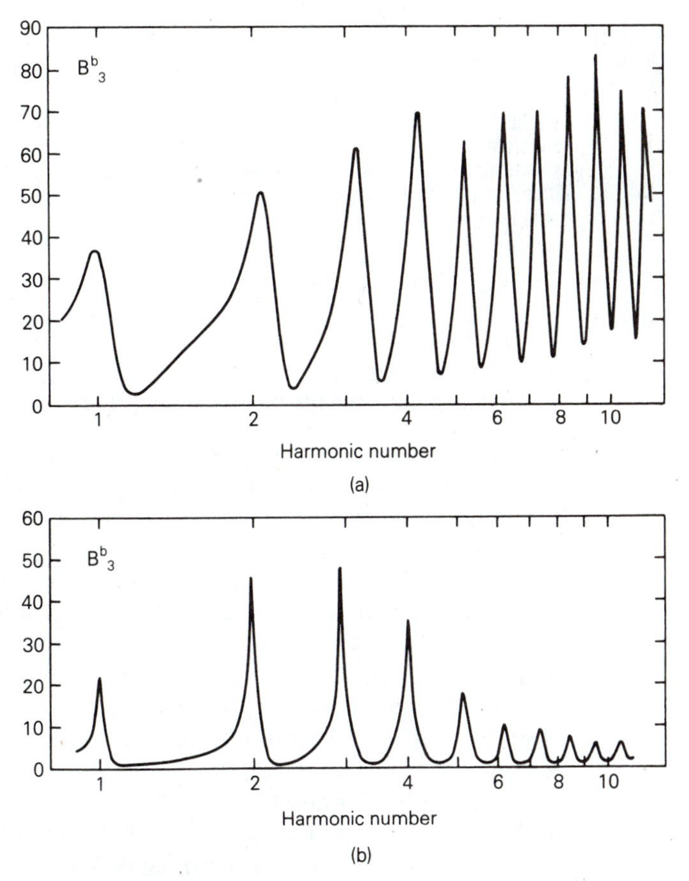

FIGURE 13.17 Input impedances of (a) an oboe and (b) an alto saxophone, both with all holes covered to produce the lowest note. Actual sounded pitch on the sax is $D^{\flat}_3$. Peaks appear unevenly spaced only because the horizontal axis has been distorted to represent pitch in octaves instead of frequency in Hz. (From Backus, 1974, with permission of *JASA*)

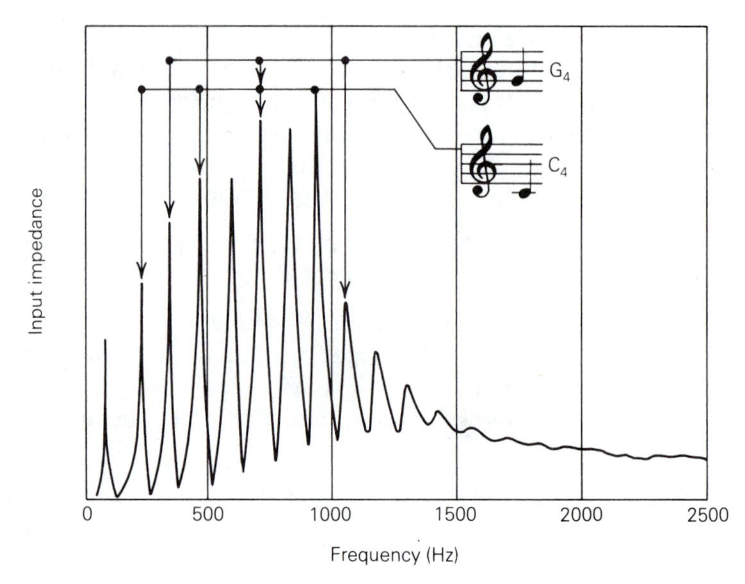

FIGURE 13.18 Regimes of oscillation used by a trumpeter blowing second- and third-mode notes (all valves up). They are strong and stable, involving the cooperation of several high peaks. Notes shown are as written; because the trumpet is a transposing instrument, the actual sounded pitches are $B^\flat_3$ and F_4. (From "The Physics of Brasses" by Arthur H. Benade. Copyright © 1973 by Scientific American, Inc. All rights reserved.)

Our third general principle comes in two parts:

(1) The strength and stability of a regime of oscillation are increased if more and stronger modes can participate in it, and (2) the strength of participation of any mode depends on how nearly it is in tune with the others.

The first part is illustrated by Figures 13.18 and 13.19. The trumpet's $B^\flat_3$, based on mode 2, has strong participation by modes 4, 6, and 8 as well (forming the members 1, 2, 3, 4 of the harmonic series of this note); it is quite easy to play. But F_5 is harder because mode 6 gets only the slightest help from mode 12, and all higher notes must be accomplished by one mode single-handedly. $B^\flat_3$ also is rather mushy at pianissimo level where higher mode participation is weak (Figure 13.14) but becomes much steadier when played loudly.

To illustrate the effect of mistuning of modes, consider Figure 13.20, where we imagine a trumpet player says, "I can see a peak indicating the existence of mode 1, so I will try to play the corresponding note." First let us state clearly why modes 1, 2, 3, 4, . . . cannot all participate in full proportion to the height of their peaks. If they did, we would have component oscillations with frequencies of approximately 85, 233, 349, 466, . . . Hz; but that is *not* a harmonic series. So there is no period of time after which all components would be ready to repeat an identical pattern of vibration; this would *not* be a periodic vibration. Its unsteady sound would represent a failure to arrive at a cooperative regime of oscillation.

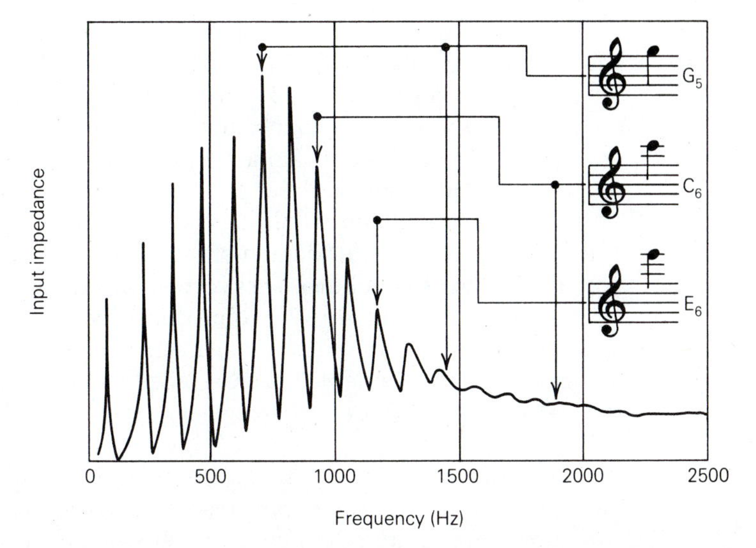

FIGURE 13.19 Regimes of oscillation for sixth-, eighth- and tenth-mode notes on a trumpet. They are progressively harder to play reliably, with the latter two depending on a single impedance peak. Sounded pitches are F_5, $B_5^{\flat}$ and D_6. (From "The Physics of Brasses" by Arthur H. Benade. Copyright © 1973 by Scientific American, Inc. All rights reserved.)

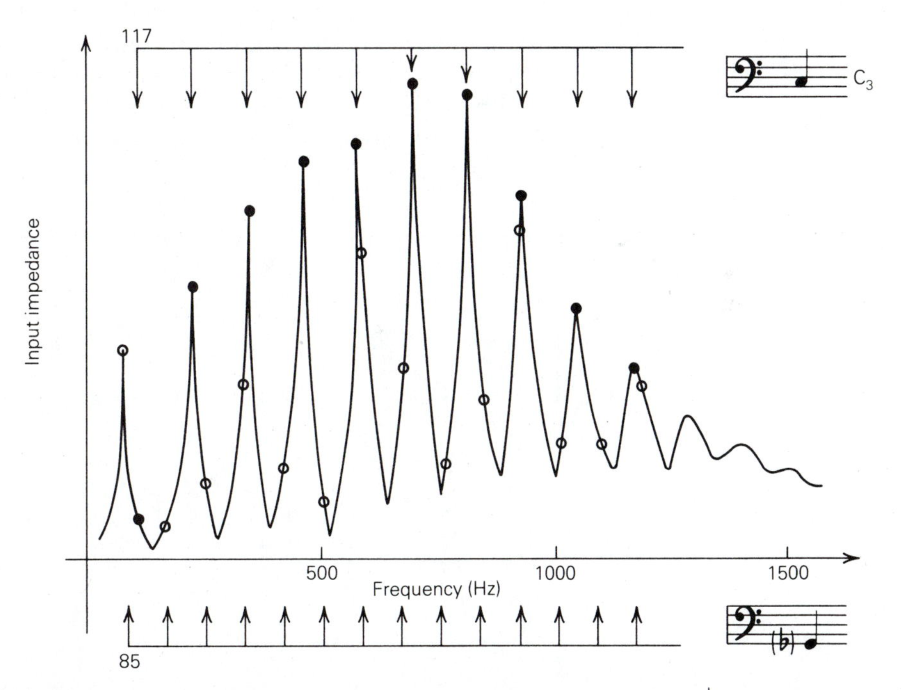

FIGURE 13.20 Solid dots: production of the trumpet's pedal tone (sounded pitch $B_2^{\flat}$) involves the cooperation of second and higher modes, but the fundamental of the harmonic series is nearly absent. Open circles: An attempt to obtain the note corresponding to mode 1 fails for lack of cooperation from the higher modes.

A steady tone *must* have components that form a harmonic series. What about the possibility of the series 85, 170, 255, 340, . . . Hz? The lone asset of this series is the full participation of mode 1; all the rest fall in the valleys where the tube's poor response tends to stifle rather than feed them. Mode 1 cannot carry the burden alone against all that opposition; such a note is practically unplayable.

Another possibility would be the harmonic series 117, 233, 349, 466, . . . Hz. Even though its fundamental falls in a valley and represents an energy drain, the next several members correspond to high peaks. The note actually can be played, although with some difficulty and inferior tone quality; trumpet players call it the *pedal tone.* Even though the sound output contains practically no fundamental, the presence of the other members of the harmonic series means your ears perceive it as having the pitch $B\flat_2$ corresponding to 117 Hz.

Now we can illustrate briefly how these principles apply to woodwinds, whose input impedance curves (especially when many tone holes are open) are more complicated than the trumpet's. Figure 13.21 shows the case of the clarinet's lowest note, D_3. Oscillation with a fundamental frequency exactly matching the first-mode resonance (150 Hz) does not get good cooperation from the higher modes. It is advantageous to sacrifice some of the first-mode strength by operating at a compromise frequency of 147 Hz. Note that mistuning of the modes means our previous statement about expecting only odd-numbered harmonics in the clarinet was only a rough approximation. In truth, the fifth mode is enough flat that it makes a strong eighth harmonic rather than ninth. Only in the low ("chalumeau") register does the clarinet have even approximately odd-harmonic–only spectra. In the high ("clarino")

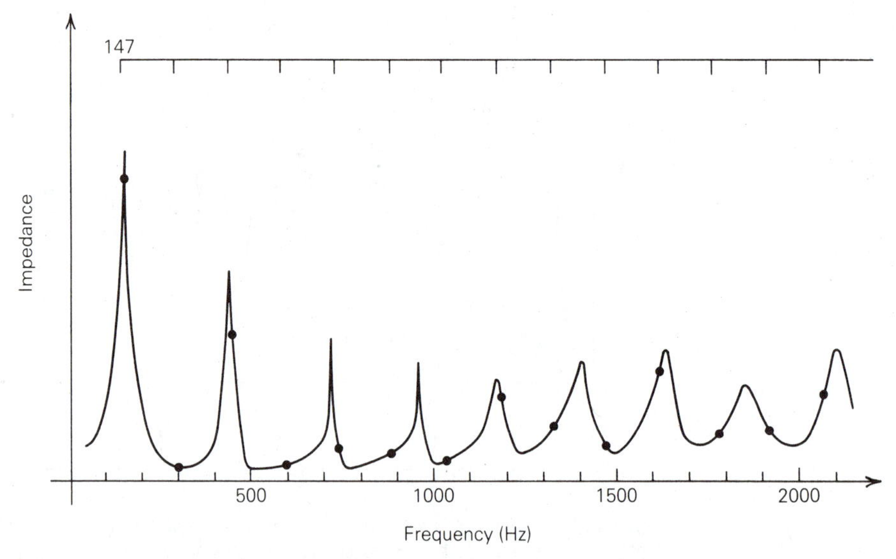

FIGURE 13.21 Input impedance curve for D_3, lowest note on the clarinet. Modes 1 and 7 would prefer a little higher frequency; modes 2 and 5 would prefer lower. (Adapted from Backus, 1974)

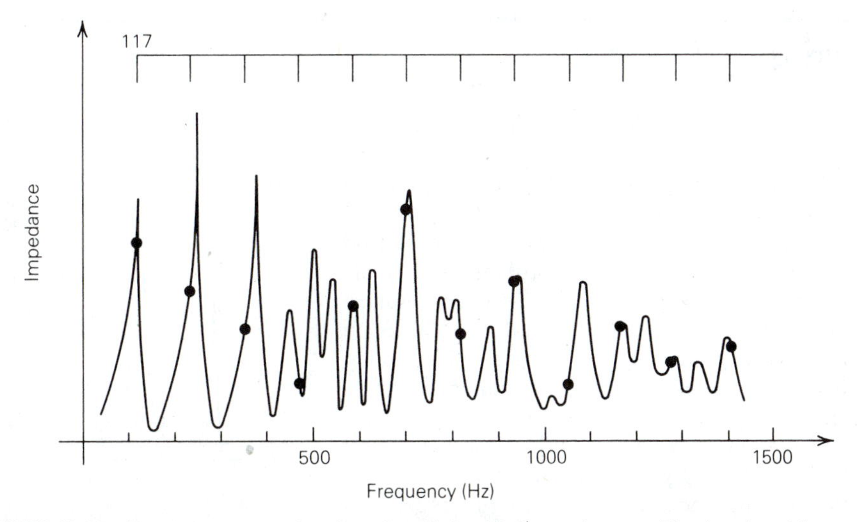

FIGURE 13.22 Impedance curve and regime of oscillation for $B^{\flat}_2$ on a bassoon. The damping of the reed tends to pull the playing frequency a little below what would otherwise appear to give the most favorable regime of oscillation. (Adapted from Backus, 1974)

register the mode mistunings are enough that even and odd harmonics participate comparably, and indeed the characteristic hollowness is missing and it is much harder to tell the clarinet apart from other instruments.

A more complex example provided by the bassoon is shown in Figure 13.22. In spite of many secondary resonances due to bumps and bends interrupting the smooth cone shape, it can be seen quickly that the instrument will play a little below the first-mode resonance to get support from peaks 7, 9, and 13 for harmonics 5, 6, and 8. All those harmonics are fairly strong in the actual sound (along with 1, 2, 3). But the ninth harmonic (at 1049 Hz) is weak, having been sacrificed to bring several others into the alliance instead.

To sum up this section so far, we can say (1) when several natural mode frequencies already fit a harmonic series, they cooperate to set up a stable regime of oscillation, and (2) when they are only slightly out of tune, they still may set up a good regime of oscillation by each playing a bit off to one side of its resonant peak. It would also be interesting to know what happens when such regimes are not available. One possibility is that little or nothing happens (Box 13.2). Another is that two peaks each are strong enough to make a sustained sound, and yet they do not manage to set up a cooperative regime. The result is not a truly steady sound; that is, it is not a periodic vibration.

That second possibility accounts for the phenomenon of *multiphonics*, which some people consider musically interesting. These can be produced on any woodwind by using unorthodox fingerings. Some sound like a combination of two or more tones with distinct pitches, others like a single tone with a rough or beating quality. Interested readers are referred to the Backus article for detailed explanation of the spectra of these multiphonics.

BOX 13.2 THE TACET HORN

An experiment by Arthur Benade in 1964 provides an amusing illustration of a good scientist's mind at work. He reasoned that if the idea of regimes of oscillation was a valid explanation for successful musical instruments, it also should correctly explain an unsuccessful instrument. He deliberately designed a tube with an unusual nonuniform taper, so that it would *avoid* as much as possible having in-tune relationships among its lower modes. Its input impedance curve is shown in Figure 13.23.

We must expect from this curve that it will be impossible to get a tone out of the horn at or near the first-mode frequency; in this respect it is much like the trumpet. Furthermore, it should be rather difficult to play in the second or third mode, because you must go considerably off the left side of those peaks to get the cooperation of mode 6. And indeed, in Benade's words, "When a clarinetist attempts to play upon this horn with the help of a standard reed, he eventually succeeds in obtaining a raucous sound whose fundamental frequency is 692 Hz. This is considerably lower than the 724 Hz frequencv of the third normal mode of the horn. . . . It is possible also to start an oscillation whose fundamental frequency lies at 472 Hz, just below the second resonance peak. . . ."

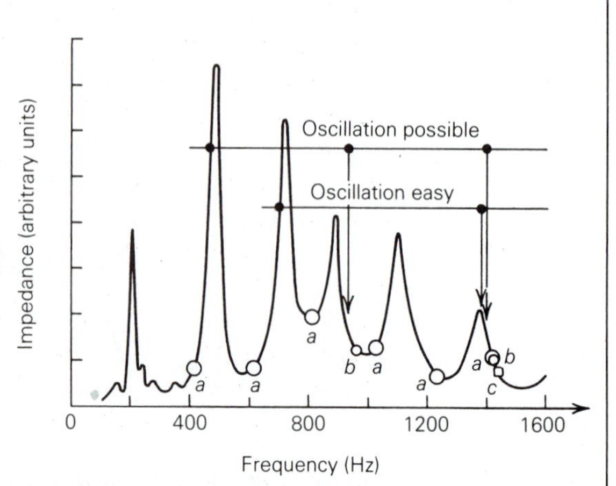

FIGURE 13.23 Input impedance of the tacet horn. (a) Harmonics of the first natural mode frequency. (b), (c) Similarly, harmonics of the second and third natural mode frequencies, respectively. Their deliberate placement in valleys instead of on peaks makes it difficult to find good regimes of oscillation. (From Benade and Gans)

Many a poorly designed musical instrument is a cousin to this tacet horn and owes its poor tone or difficulty in playing to a failure to have resonance peaks fit closely enough to a harmonic series.

13.5 RADIATION

Most of this chapter, and especially the last section, has dealt with vibrations *inside* a horn. We must be sure to take into account the ability of these vibrations to escape from the horn before trying to match up explanations of harmonic spectra with what we actually hear.

Let us define *radiation efficiency* as the ratio of acoustic pressure just outside the end of an instrument to that just inside. If a traveling wave keeps right on going past the end, it has 100% radiation efficiency; if most of it is reflected at the end and stays inside to build up a standing wave, that means small radiation efficiency.

The radiation efficiency of a small opening (either the end of a narrow pipe or a tone hole) is very small for low frequencies. It gradually increases

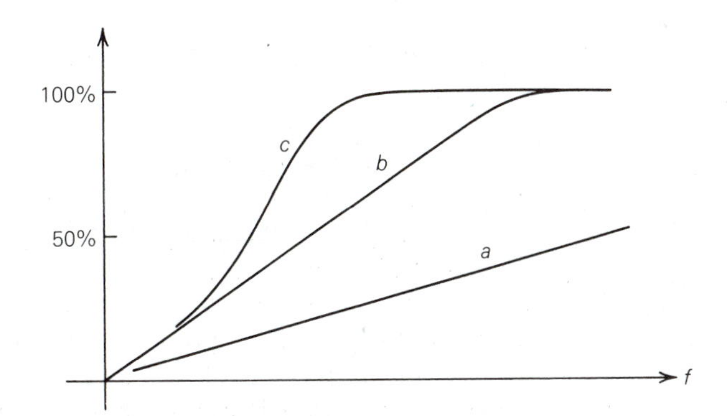

FIGURE 13.24 The dependence of radiation efficiency on frequency for (a) a small hole, (b) a large hole, and (c) a flaring bell. The efficiency reaches its maximum at approximately 1600 Hz for a typical trumpet. Woodwind radiation efficiencies also have a leveling-off point, typically around 1700 Hz for clarinet, 1400 Hz for oboe, and 400 Hz for bassoon.

with higher frequencies, finally leveling off at 100% for wavelengths shorter than about twice the hole diameter (Figure 13.24*a*). A large hole lets more sound out and reaches 100% efficiency for lower frequencies (Figure 13.24*b*).

A trumpet bell keeps the lowest modes back mainly in the cylindrical pipe; they effectively radiate from a minute opening, so the radiation efficiency is very small. But the bell helps the higher modes escape; Figure 13.8 suggests that they effectively radiate from a much larger opening and thus with correspondingly high efficiency. The overall behavior of the trumpet is then as in Figure 13.24*c*.

In direct analogy to response curves for the violin (Figures 11.18 and 11.19, pages 214 and 215), we may now use the radiation efficiency curve to convert spectra of the internal vibrations of the trumpet into spectra of the sound we should hear nearby. Figure 13.25 illustrates how the bell shape helps make the bright, incisive tone of the trumpet by boosting the strengths of the harmonics at approximately 1000–1500 Hz as compared to the lower ones.

Radiation efficiency from rows of open tone holes on woodwinds is qualitatively similar to that of the trumpet, because the lowest harmonics must come out mainly through the first open hole while the higher harmonics penetrate farther down the tube and are radiated cooperatively by several holes. So woodwinds also enjoy a treble boost in creating their external spectra. This is illustrated in a perverse way by another of Benade's experiments: He and Patterson recorded musical passages with a tiny probe microphone mounted in a bassoon reed cavity to pick up the *internal* sound. When this recording was played back, they heard "curiously muted tones that sound vaguely woodwindlike, but are not (to many listeners' ears) recognizable as originating from a bassoon."

What we hear also may depend strongly on our listening location. Sound radiated through a single small hole goes out almost equally in all directions, and this is the case for the lowest modes of both brass and woodwinds. But

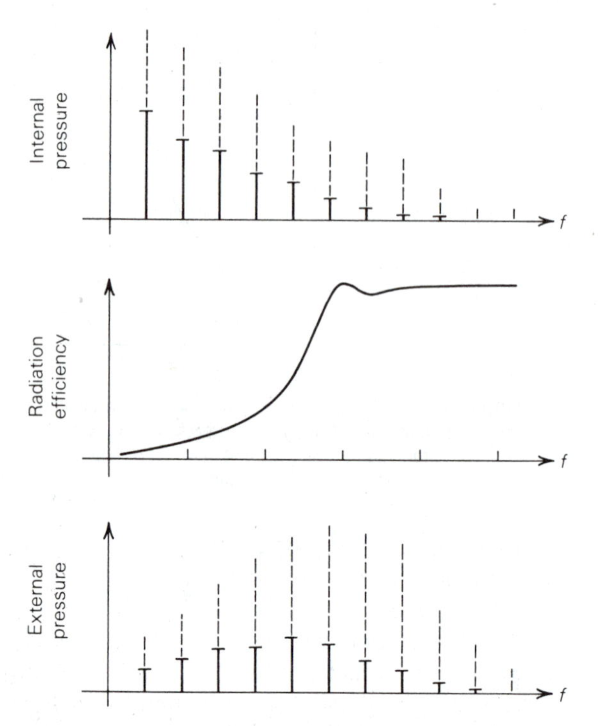

FIGURE 13.25 Transformation of internal to external spectrum by the radiation efficiency for $B^{\flat}_{3}$ on a trumpet; same data as Figure 13.14. Solid bars indicate *mp*; dashed extensions, *fff*. Compare with Figure 11.19 (page 215).

sound radiated cooperatively by a series of holes emerges in a complex pattern, and sound whose radiation is aided by a bell is concentrated along the direction in which the bell points. So the higher harmonic strength, and thus the overall timbre, depends on whether you are to the front, side, or back of a woodwind or brass player. In particular, this explains why trumpet tone is brighter when the horn is pointed at you.

Brass players occasionally insert *mutes* into the bells of their instruments, thereby cutting down the total sound output while also changing its timbre to what can best be described as a muted quality. Can we understand precisely what sort of change this is and how it comes about?

First, the sound is not cut off completely, because there is always some opening left either through the mute or around its edges. Second, after all we have said about how important bell shape is in tuning up the modes so that trumpet notes will be playable, you may fear the presence of the mute would ruin all that. But it is located far enough out in the bell that it does not make major changes in either the strength or location of the standing-wave reflections; it merely helps the bell a bit in confining the higher modes but does not really contradict what the bell is supposed to do in tuning. It is true that each mute has some favorite resonant frequencies of its own, but successful mutes

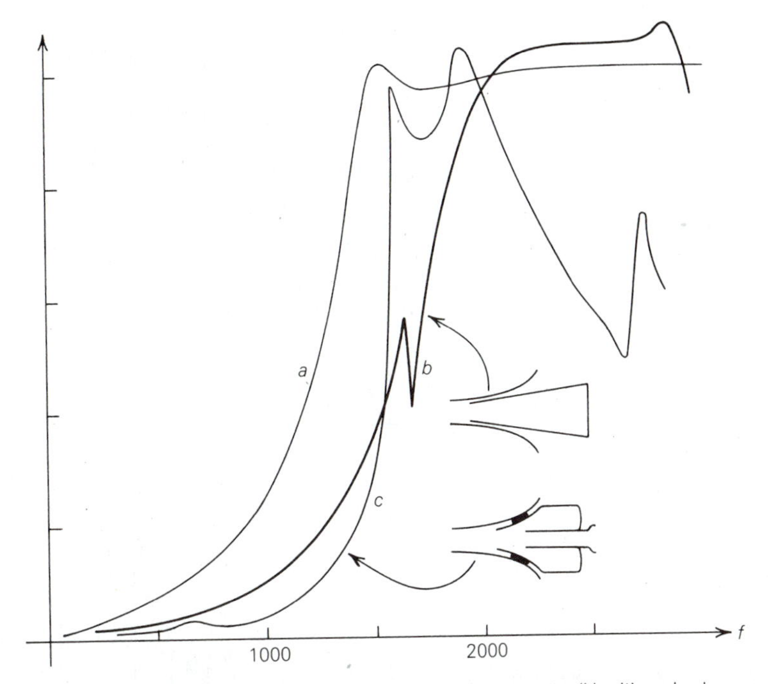

FIGURE 13.26 Radiation efficiency curves for a trumpet (a) without mute, (b) with a simple cone mute, and (c) with a Harmon (''wah-wah'') mute. The cone mute is held in place by three small pieces of cork and leaves a little space between it and the bell most of the way around. The Harmon mute seals against the bell all the way around but has an opening through the middle. (From Backus, 1976)

affect the total input impedance of the instrument only with an extra peak at approximately 100 Hz or less and with a next higher resonance not lower than approximately 1000 Hz. The trumpet's modes numbered 2 through 10 have their tuning practically undisturbed, and the usual regimes of oscillation still are playable.

The main effect of the mute is on the radiation efficiency. It is easy to explain the curves of Figure 13.26 as a combination of (1) a cutback in overall efficiency simply because the area through which the air escapes has been reduced, and (2) enhancement of some band of high frequencies by resonance in the mute. The Harmon mute in Figure 13.26c, for example, resonates at 2 KHz, because that is the frequency for which a half-wavelength just fits its short open cylindrical tube. These radiation efficiency curves modify the external spectrum as shown in Figure 13.27.

A related topic is the placement of a French horn player's hand in the bell of the instrument. This not only provides a mild muting of the tone quality but also cuts down the radiation of the higher modes enough to add more usable peaks to the horn's input impedance curve. This enhances the playability of higher notes. The interested reader is referred to Backus or Benade for further discussion.

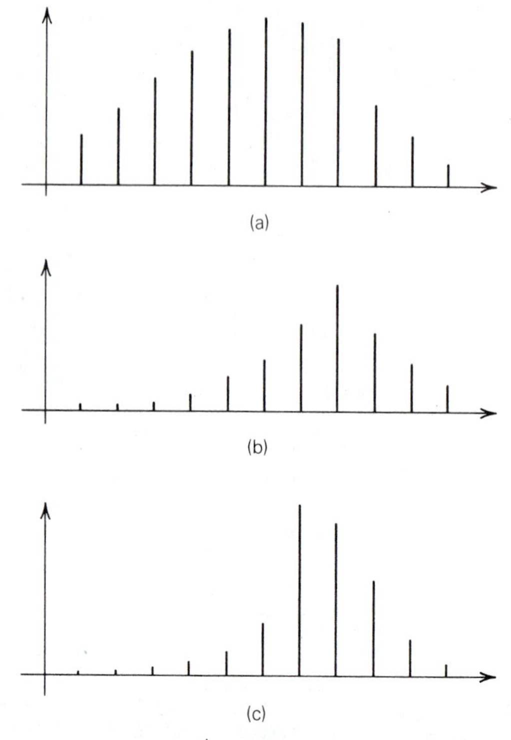

FIGURE 13.27 Trumpet external spectra for $B^{\flat}_3$ played *fff*, using the radiation efficiency curves of Figure 13.26 in the same way as shown in Figure 13.25: (a) no mute, (b) cone mute, (c) Harmon mute.

SUMMARY

Organ reed pipes, reed woodwinds, and brass instruments all depend on the operation of a pressure-controlled valve to admit air to a resonating pipe in rhythmic bursts. The vibration frequency is determined in some cases by the reed and in others by the pipe. Standing-wave patterns must correspond to a closed pipe end at the reed.

Clarinets are basically cylindrical, while the other common woodwinds are conical. The resulting natural mode frequencies correspond approximately to a harmonic series with all even-numbered harmonics missing for the clarinet, compared with a complete series for the others. This accounts both for the clarinet's characteristic timbre and for its overblowing at the twelfth instead of the octave. The essential difference between oboes and saxophones is in their bore sizes; smaller bores allow more modes to resonate, leading to stronger high harmonic content in their sounds.

The large bell and the constricted bore in the mouthpiece are both essential to the trumpet and its relatives, for they make it possible to achieve mode frequencies fitting a harmonic series along with high radiation efficiencies.

This not only provides notes that fit into a chromatic scale but also gives each of them its rich spectrum and robust timbre.

The tone color and playability of each note depend on the availability of several modes to cooperate in forming a regime of oscillation. These form best when (1) the instrument is played louder (which increases the nonlinear mode interaction), (2) the modes involved resonate strongly (have high input impedance), and (3) the natural mode frequencies are close to harmonics of some fundamental.

Wind instruments generally radiate their higher harmonics more efficiently, so that the sound we hear is brighter and richer than that inside the instrument. For brass, this may be carried a step further by adding a mute, which reduces radiation efficiency for the lower harmonics even more.

REFERENCES

On the topic of blown reed instruments especially, there is a great wealth of useful information beyond the brief introduction given here. I urge every player of brass or woodwind instruments to delve at least one layer farther into this fascinating lore to understand what he or she is really doing.

The first level you should explore is represented by Chapters 11 (woodwinds) and 12 (brass) in Backus's book, and by Benade's *Scientific American* articles on woodwinds (October 1960)* and on brass (July 1973),* both reproduced in the Hutchins collection. Benade on brass instruments is especially interesting and informative; among other things you will find there is a good discussion of the meaning of regimes of oscillation, and of the French horn player's hand in the bell. Differences in design and in playing techniques for the baroque trumpet are discussed by D. Smithers, K. Wogram, and J. Bowsher in *Scientific American, 254,* 108 (April 1986). You also may read about numerous obsolete members of these families by looking in the *Harvard Dictionary of Music* under "Clarinet," "Oboe," and "Brass."

The second level is found in Benade's book, Chapter 20 on brass and Chapters 21 and 22 on woodwinds.

The third level is in the wealth of original research articles. Among the most useful of these are:

A. H. Benade and S. N. Kouzoupis, "The Clarinet Spectrum: Theory and Experiment," *JASA, 83,* 292 (1988).

N. H. Fletcher, "Excitation Mechanisms in Woodwind and Brass Instruments," *Acustica, 43,* 63 (1979).

J. Backus, "Input Impedance Curves for the Reed Woodwind Instruments," *JASA, 56,* 1266 (1974).

J. Backus, "Input Impedance Curves for the Brass Instruments," *JASA, 60,* 470 (1976); this includes explanation of mutes.

S. Elliott, J. Bowsher and P. Watkinson, "Input and Transfer Response of Brass Wind Instruments," *JASA, 72,* 1747 (1982).

*The papers marked with an asterisk also are reprinted in *Musical Acoustics: Piano and Wind Instruments,* edited by Earle L. Kent (Halsted Press, 1977).

R. Causse, J. Kergomard and X. Lurton, "Input Impedance of Brass Musical Instruments," *JASA*, *75*, 241 (1984).

G. R. Plitnik and W. J. Strong, "Numerical Methods for Calculating Input Impedances of the Oboe," *JASA*, *65*, 816 (1979).

*J. Backus, "Small-Vibration Theory of the Clarinet," *JASA*, *35*, 305 (1963).

J. Backus, "Multiphonic Tones in the Woodwind Instruments," *JASA*, *63*, 591 (1978).

C. J. Nederveen, *Acoustical Aspects of Woodwind Instruments* (Frits Knuf, Amsterdam, 1969). An exhaustive treatment of the problem of hole size and placement, including extensive tables of measured data.

J. Backus and T. C. Hundley, "Harmonic Generation in the Trumpet," *JASA*, *49*, 509 (1971).

*A. Benade and D. Gans, "Sound Production in Wind Instruments," *Ann. N.Y. Acad. Sci.*, *155*, 247 (1968). Includes account of the tacet horn.

*A. Benade and E. Jansson, "On Plane and Spherical Waves in Horns with Nonuniform Flare," *Acustica*, *31*, 80 and 185 (1974).

*A. Benade, "On the Mathematical Theory of Woodwind Finger Holes," *JASA*, *32*, 1591 (1960).

R. T. Schumacher, "Ab Initio Calculations of the Oscillations of a Clarinet," *Acustica*, *48*, 71 (1981).

D. Luce and M. Clark, "Physical Correlates of Brass-Instrument Tones," *JASA*, *42*, 1232 (1967).

SYMBOLS, TERMS, AND RELATIONS

p pressure
F force
v sound speed
L air column length
f_n mode frequency
clarinet, saxophone, and oboe families
registers and overblowing
lip reed
trumpet versus cornet
$v = 344$ m/s
$f_n = n(v/2L)$ (cone)
$f_n = (2n - 1)(v/4L)$ (reed-closed cylinder)
shallot
free and beating reeds
hard and soft reeds
flow-controlled and pressure-controlled valves
input impedance curves
regime of oscillation
multiphonics

EXERCISES

1. Taking inside pressure, outside pressure, and stiffness separately, describe the direction of each force on a reed and how its strength changes as the reed moves toward the shallot or mouthpiece.

2. Tell whether each of the following uses hard or soft reeds: (a) oboe, (b) harmonica, (c) saxophone, (d) accordion.

3. The typical natural frequency of a clarinet reed is approximately 2 KHz. Explain why a clarinet is likely to squeal rather than play the proper note when the player clamps his teeth directly against the lower part of the reed instead of cushioning them with his lower lip.

4. Comparing a reed and shallot, both prac-

tically flat, against another pair curving strongly away from each other, what difference would you expect (a) in the manner of closing and (b) in the resulting sound?

5. Suppose (even though it is only approximately true) that a bassoon is simply a precisely scaled-up oboe. Given that the oboe's lowest note is $B_3^\flat$ and its length approximately 64 cm, what length would you expect for the bassoon, whose lowest note is $B_1^\flat$? What length for the contrabassoon, lowest note $B_0^\flat$?

6. The clarinet's lowest note is D_3, 147 Hz. What is its effective length—that is, the length of an idealized closed pipe that has this frequency for its first mode? Explain why this is less than the actual length of the instrument. (Hint: Look at Figure 13.5.)

7. The oboe's lowest note is $B_3^\flat$, 233 Hz. What is its effective length—that is, the length of an idealized complete cone that has this frequency for its first mode? Explain why this is greater than the actual length of the instrument. (Hint: Look at Figure 13.5.)

8. Consider a clarinet fingered to sound F_3 in its low register. What pitch would you expect to get by overblowing to the second mode with the same fingering? What pitch for the third mode?

9. Where would you look on a clarinet for a register hole (small vent hole to encourage speaking in the second mode)? Where on an oboe? (Hint: Look at Figures 12.3 and 13.6.)

10. Suppose the critical mode number (introduced in Chapter 12 for cylindrical pipes) is also $N \cong L/D$ for cones, where L is the length and D the diameter at the open end. What is the approximate value of N for an oboe? For a soprano saxophone? Are your answers compatible with Figure 13.17? (Hint: Use Figure 13.5, being sure to allow for the exaggeration of width.)

11. In view of the size of their tone holes, what do you expect to be the relative importance of cross-fingerings on the oboe and the saxophone? Can you verify that your prediction agrees with actual practice?

*12. The trombone's low $B_2^\flat$ is played with the slide in first position (see Figure 13.10). What is its effective length—that is, the length of an idealized closed tube with the same second-mode frequency? Estimate how far the slide should be moved to reach second position in order to play A_2, remembering that each cm of slide motion adds 2 cm to the tube length. If already in sixth position (F_2), how much farther must you move the slide to reach seventh position (E_2)?

*13. Tell the mode number and valve positions for two different ways of playing the note $C_5^\sharp$ (written $D_5^\sharp$) on the trumpet.

*14. With valve tubing lengths expressed as percentages of L_0, try various possibilities and choose one you feel is a good compromise for the tuning of all notes on an ordinary three-valve trumpet. Do the same for a four-valve instrument. (Hint: Your goal is to come close to all six correct total percentages mentioned in Box 13.1.)

15. Given Figure 13.21, sketch the sort of spectrum you expect for the *internal* vibrations of a clarinet D_3 (a) pianissimo and (b) fortissimo.

16. Using information from the caption of Figure 13.24, convert your answers from Exercise 15 to *external* spectra.

*17. Suppose you have a trumpet whose second mode is a little flat and fourth mode a little sharp, so that instead of an accurate harmonic series you have impedance peaks at 230, 350, 480, ... Hz. In terms of mode participation in regimes of oscillation, describe what happens to the pitch of $B_3^\flat$ (written C_4) as you increase the dynamic level from pianissimo to fortissimo.

18. Discuss the prospects for using a mute on a clarinet.

19. Suppose you locate the first open hole for the note C_2 (which is not the lowest note) on a bass clarinet. You would expect an adjoining hole to function as first open hole for $C_2^\#$. Which of these is closer to the mouthpiece end? Assuming both holes are the same size, about what distance do you expect to separate them?

20. A clarinet and a flute both have nearly cylindrical bores with similar length and diameter, yet they are quite different in both tone quality and range of pitch. Discuss these differences more specifically, explaining the underlying reasons.

PROJECTS

1. Make accurate dimensional measurements of a cornet and a trumpet, and of some output spectra for both. Compare and try to explain the differences.

2. Measure frequency ratios (by comparing piano notes) of a garden hose section alone, with funnel, with mouthpiece, and with both. Discuss your results.

3. Learn to play multiphonics (see Bartolozzi, *New Sounds for Woodwinds*, Oxford University Press, 1967) and to explain them (Backus, 1978 article).

CHAPTER 14 The Human Voice

In a contest for versatility among musical instruments, the human voice wins easily. (Coming from a pipe organist, that is no mean compliment!) This is all the more remarkable because the same apparatus serves other purposes as well, such as when we eat and breathe.

After a brief introduction to the relevant anatomy, we will take up the mechanism of sound production in speech and singing, two activities that are both covered by the term *phonation.* This requires consideration of both transient and steady sounds, which correspond only roughly to consonants and vowels. We will take this opportunity to explain the role of a fluid-dynamical force called the *Bernoulli effect* in the operation of vocal cords and lip reeds.

We will find ourselves particularly concerned with the concept of *formants* or frequency ranges in which harmonic components are especially strong. These are central to our recognition of vowels in speech and song. The chapter will close with brief remarks on miscellaneous aspects of singing, such as vibrato and carrying power.

14.1 THE VOCAL APPARATUS

In common with all wind instruments, the human voice has (1) an air reservoir with a means of maintaining pressure above atmospheric, (2) an outlet channel with a narrow constriction (or in this case several) where airflow can be interrupted or modulated, and (3) a resonant cavity to strengthen some aspects of the resulting sound waves.

The air reservoir is located in the *lungs*, where you normally hold approximately 3 to 4 liters of air, with half a liter moving in and out with each breath. Air is drawn in by raising the rib cage to expand the lungs, and expelled by contracting abdominal muscles, which force the abdominal contents upward against the diaphragm. The lungs of an adult taking a very deep breath may hold as much as 5 to 6 liters of air, and still will contain 1 to 2 liters after maximum exhalation. The difference of 3 to 5 liters represents the maximum available air for singing from a single breath, although most people will not use more than 1 to 2 liters without some vocal training. A tube called the

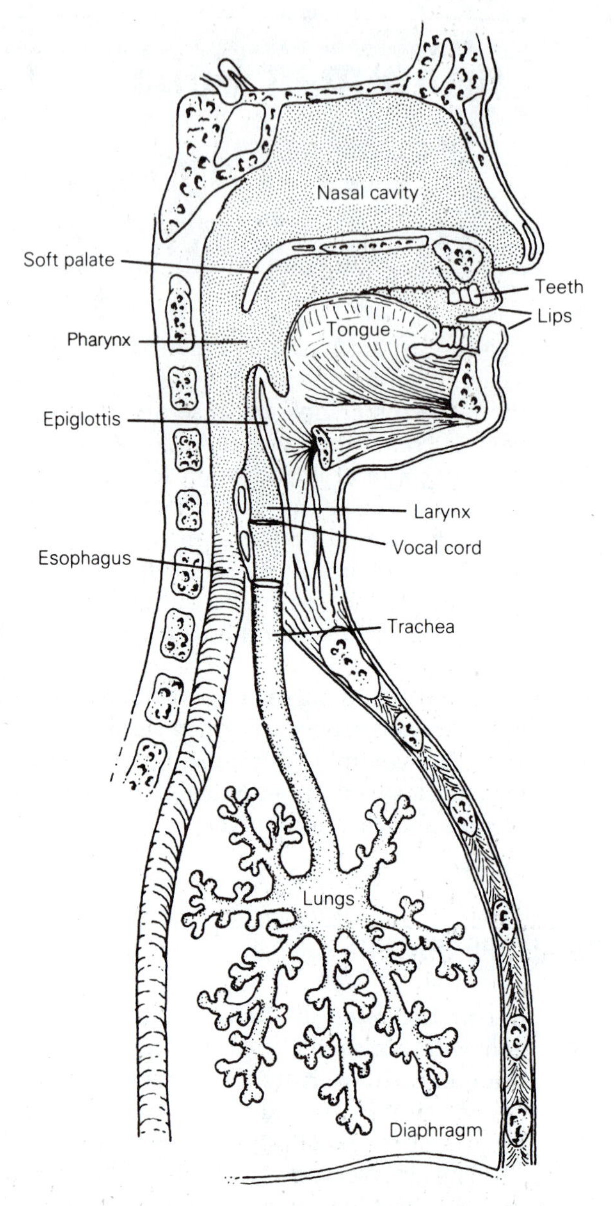

FIGURE 14.1 The human vocal apparatus. See Figure 14.2 for more detail on the larynx.

trachea leads from the lungs up to the *vocal tract*, a term that includes the throat, mouth, and nose (Figure 14.1).

At the top of the trachea, serving as a switchyard to join it with the esophagus and the vocal tract, is a hollow boxlike structure of cartilage called the *larynx* (Figure 14.2). The *epiglottis* is a flaplike valve on top of the larynx that drops down during swallowing to prevent food from entering the

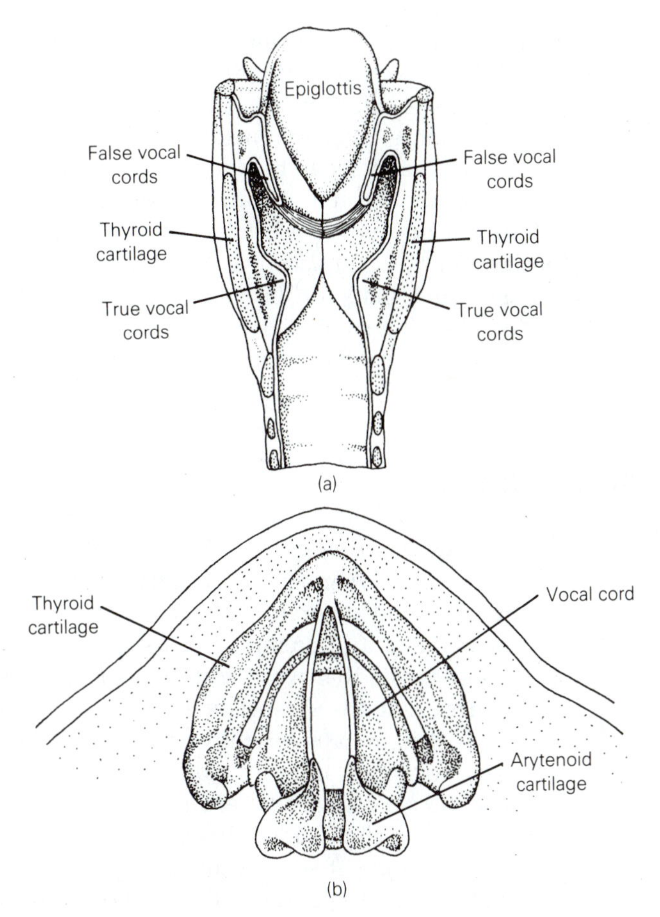

FIGURE 14.2 The larynx, which is approximately 7 cm high and 5 cm across. (a) Cutaway view, looking from back to front. (b) View from above, with front at top, showing arytenoid cartilage whose movements open and close the vocal cords.

trachea, although it is open for phonation. The airway also can be blocked by the *vocal cords* (or *vocal folds*), a pair of ridges of soft layered tissue on the inside walls of the larynx, whose shape and rigidity can be changed by several small muscles. The opening between the vocal cords is called the *glottis* and is V-shaped because the vocal cords stay together in front while moving apart in back. The glottis is approximately 2 cm long and 1 cm across when open wide.

The vocal cords close for swallowing, as a backup in case anything gets past the epiglottis, and they are open during normal breathing. For phonation the vocal cords close, or nearly so, and the lungs apply a pressure equivalent to a column of at least 5 and up to 40 or more cm of water; that is, .005 to .04 atmospheres. This excess pressure forces the cords to open and admit bursts of air into the vocal tract. The cords vibrate at a frequency controlled by the tension applied in their muscles. For normal speech these frequencies typically extend over a range from 70 to 200 Hz for a man's voice or 140 to 400 Hz for a woman's; the difference is due to the longer and more massive vocal cords of adult males. These ranges may be extended upward another octave or more when singing.

Immediately above the larynx is the *pharynx* or throat cavity. This opens to the outside through the mouth, with the tongue, teeth, and lips providing additional means of restricting or blocking the airflow. Depending on the position of the *soft palate*, the throat may or may not open also into the nose. The size and shape of the vocal tract can vary greatly and thereby produce widely differing sounds. This occurs largely because of the tongue's ability to change both its position and its shape in several ways.

14.2 SOUND PRODUCTION

Let us now describe in more detail how this mechanism works to produce audible sound. We must in fact describe several mechanisms, because speech includes several different types of sounds. Each distinct elemental speech sound is called a **phoneme**. The usual distinction between consonants and vowels suggests that we might explain them as corresponding to transient and steady sounds, respectively. But that is overly naive; for acoustical purposes it is more informative to classify the phonemes into five groups: plosives, fricatives, other consonants, pure vowels, and diphthongs.

The *plosive* consonants are those produced by completely blocking the vocal tract and then suddenly opening it to let a single burst of air through. This can be done at several different points; when the blocking and release occur at the lips we have *p*, at the front of the tongue *t*, and at the soft palate *k*. These are clearly transient sounds (Figure 14.3); there is no way you can "hold" a plosive. Acoustically, the plosive is a simple sudden pulse of higher pressure followed by a brief interval of whatever damped vibration this sets

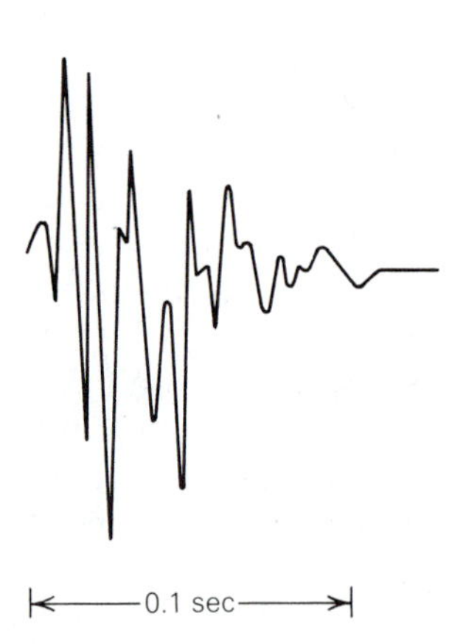

FIGURE 14.3 An oscilloscope trace for the plosive *k*.

up in the vocal tract together with the breathy sound of air continuing through the opening until the excess pressure is fully relieved. This nonsteady, nonperiodic sound has a continuous spectrum of frequencies and thus no definite pitch.

Each of the *unvoiced* plosives described above has a *voiced* version as well—*b, d,* and *g*. In these, the vocal cords are set to begin steady vibration with some vowel sound immediately (meaning within approximately 30 ms) after the air is released instead of leaving a larger gap between (compare *coal* and *goal*). They also may be voiced very briefly when occurring at the end of a word (compare *lack* and *lag*). The transient nature of plosives means they carry little total sound energy, so singers are taught to exaggerate them.

The *fricatives* also come in unvoiced/voiced pairs: *f, v; th* (as in *thin*), *th* (as in *them*); *s, z; sh, zh* (*measure*). The last four are sometimes called the *sibilants*. To this list we add one more, unvoiced only: *h* (*hat*). These are quite unlike the plosives in that they *can* be sustained steadily for any length of time (even though in normal speech they are not). Yet their unvoiced versions have no identifiable pitch; they are steady only in the sense that white noise is steady; their waveforms are nonperiodic (Figure 14.4) and their spectra include a continuous range of frequencies, not just a harmonic series. The same statements can be made about that part of the complex sound of the voiced fricatives that distinguishes them from vowels. (To hear for yourself that a voiced fricative is a mixture of two distinct sounds, only one of which has pitch, try gradually opening your mouth while saying "zzzzz"; then say it again while gradually closing your lips. In the first case you are left with a vowel, in the second with the unvoiced *s*.)

The "frying" sound of the fricatives is merely the turbulence in a fluid flowing through a small opening at greater than critical speed. As with the plosives, this opening may be formed at several different points along the vocal tract. These *places of articulation* are the lips and teeth (*f/v, th*), tongue and palate (forward for *s/z*, farther back for *sh/zh*) and glottis (*h*). The different vocal tract shapes help emphasize certain frequency ranges (around 4 to 6 KHz for *s* versus 2 to 3 KHz for *sh*, for example) but no particular individual frequency within those ranges. There is insufficient positive feedback (resonance) to control the flow and force it to be periodic instead of random.

Under the heading "other consonants" we group several kinds that accomplish transitions whose acoustical properties are not fundamentally

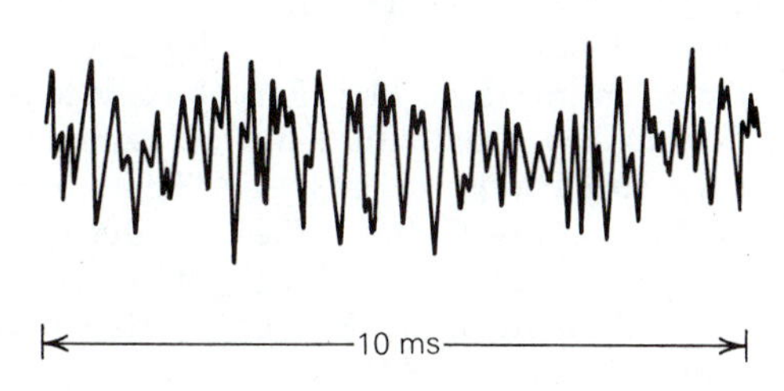

FIGURE 14.4 An oscilloscope trace for the fricative *sh*.

different from those described elsewhere. These include the *semivowels* (or *glides*) *w* and *y* and the *liquids l* and *r* (both loosely related to diphthongs) and the *nasals m*, *n*, and *ng*, which encompass a vowel-consonant transition. As a nasal begins, the vocal cords already are vibrating. Only the nose is open, but that allows a steady voiced sound—the sound of humming. That sound can be held indefinitely before articulating the consonant, and it has definite pitch and a harmonic-series spectrum. It could be called "the nasal vowel" and added to the others below as far as acoustics is concerned; it is practically the same for all three nasal phonemes. The consonant ending of the nasal results from the opening of the mouth passage to prepare the way for some following phoneme. The nasals differ from each other only in place of articulation of their final consonant—lips for *m*, tongue for *n*, and soft palate for *ng*.

The *vowels* are steady, voiced sounds with definite pitch; their waveforms are periodic (aside from small imperfections such as unsteadiness in muscle control or superimposed hissing noise from airstream turbulence) and their spectra are harmonic series. So it is appropriate to try to characterize the vowels according to the relative strengths of these harmonics, which we shall do at length in the following section. Here we will concentrate on how voiced sounds in general are produced by the vocal cords.

When a stream of air is sent between nearly closed vocal cords, there is a critical speed (or critical pressure driving it) above which the flow cannot be steady. Fluid-dynamic instability makes the cords vibrate back and forth so that the flow becomes intermittent. The resemblance to a trumpet player's lips is obvious. We can classify the voice as another reed instrument and apply to it some of the concepts developed in the last chapter. But there are certain features of the flow instability that deserve further explanation, and these are presented in Box 14.1.

*BOX 14.1 THE BERNOULLI EFFECT

Understanding the vibration of vocal cords and brass players' lips requires consideration of the Bernoulli effect. This concerns the change in pressure at different points along a stream of flowing fluid, and has many other applications in such devices as sailboats, airplane wings, aspirator pumps, and airspeed indicators.

Consider the airstream in Figure 14.5. For the same amount of air to flow steadily through the entire channel, its speed must be greater in the constriction *B* than at *A* or *C*. But if the air speeds up in going from *A* to *B*, something must be pushing it toward the right; that is, the fluid pressure at *A* must be greater than at *B*. Similarly, to slow it down again, the pressure at *C* must be greater than at *B*. In general, the pressure along any fluid streamline is reduced wherever the speed is increased; that is the Bernoulli effect.

Now consider hypothetical oscillations of a lip or vocal cord (Figure 14.6a). The springlike force of tension and elasticity in the tissues tends to return the mass to its equilibrium position (Figure 14.6b). The Bernoulli force, coming from a reduction of pressure at *B* below atmospheric, pulls the mass downward; but its strength changes during the motion. For a narrower con-

(continued)

BOX 14.1 *(continued)*

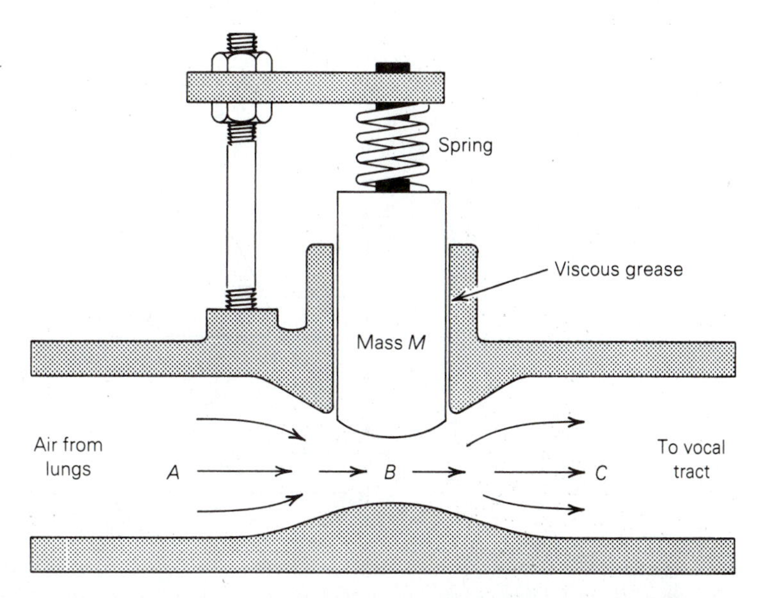

FIGURE 14.5 The mass M is a physicist's analogy to a vocal cord; it is merely a minor detail that it is not a pair. For sufficiently large airflow, M will undergo steady oscillations, thus periodically changing the flow; that is, creating sound waves. (From *Fundamentals of Musical Acoustics* by Arthur H. Benade. Copyright © 1976 by Oxford University Press, Inc. Reprinted by permission.)

striction the velocity and resulting downward force are reduced; if the channel widens, they both increase (Figure 14.6c). The average Bernoulli force moves the equilibrium position down a bit; it is the changes above and below average that pull effectively sometimes downward and other times upward.

As described thus far, the Bernoulli effect has the same result as if the spring were made stiffer. The mass would oscillate only if "kicked," and that oscillation would gradually die away as energy is lost to friction. How can the oscillation spontaneously grow to large amplitude and continue indefinitely? Only if it arranges to receive a continual supply of energy. This comes about because the inertia of the air means that it takes a little while after a change in pressure is applied before the flow attains its new speed. So the least and greatest flow speeds, and the accompanying least and greatest Bernoulli forces, occur a little later than the times of widest and narrowest openings (Figure 14.6d). This is precisely what is needed to make the oscillating part of the Bernoulli force act downward during the greater part of the downward motion of the mass and similarly upward for the upward motion, thus delivering more energy to the motion than it takes away. This extra energy helps make up for frictional losses so that the oscillation can be sustained at large amplitudes. (See also Box 11.1.)

Lips and vocal cords also are subject to forces that are parallel to the flow and in most cases stronger than the Bernoulli forces; these produce rolling motions so that the total picture is somewhat more complicated. But again the inertial time lag between applied force and resulting motion is an essential ingredient in making self-sustained oscillation possible. The Bernoulli component must be relatively minor for organ or woodwind single reeds, for which most of the reed area feels only a quasi-static pressure difference; only a narrow strip around the edge of the reed is close enough to the shallot or mouthpiece facing to experience a significant reduction in pressure. Benade reports an estimate

(continued)

BOX 14.1 *(continued)*

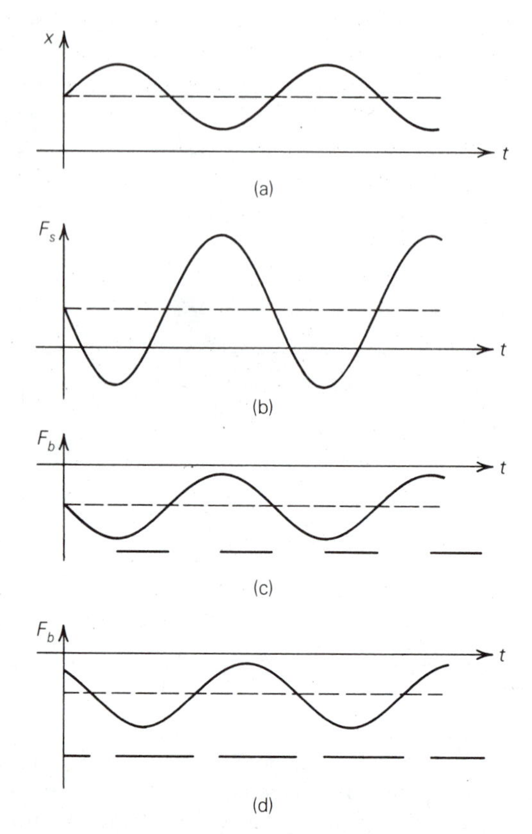

FIGURE 14.6 (a) Width of opening in a channel during vibration. All dashed lines show time-average values. (b) Restoring force exerted by the spring; note this is downward when the position of the mass is high and upward when it is low. (c) Bernoulli force, always down but stronger downward than average when mass is high and weaker when it is low. The *alternating part* of this force (difference from the average value) is negative for mass high and positive for mass low. (d) Bernoulli force with a hypothetical one-eighth-cycle lag of flow in response to pressure. Bernoulli force is in same direction as motion (delivers energy to the system) three-fourths of the time in this case (horizontal bars) instead of only half.

by Worman that the Bernoulli effect contributes only a few percent to the forces on a clarinet reed. This still can be important, however, in determining the exact way the reed finally closes against the mouthpiece. The Bernoulli effect plays a somewhat greater role for the oboe family's double reeds.

There is also a critical difference in the relation between reed and resonator. In all the brass and woodwind instruments, the feedback from the tube is strong enough to have a major influence on the reed frequency, so bore and bell shape are especially important in determining the sounded pitch. The vocal tract differs in having soft, yielding walls that absorb much of the vibration energy. Whatever resonances occur in the vocal tract are relatively weak—like the broad, low curve in Figure 11.16 (page 212) rather than the extremely high, sharp peaks of Figures 13.15 or 13.17 (pages 272, 273). So the feedback from vocal tract to vocal cords is much too weak to influence them. Even though made of soft tissue, the vocal cords acoustically must be viewed as hard reeds whose vibration frequency and waveform are determined almost entirely by their own tension, mass, and separation, and with some slight influence from the lung pressure.

So regardless of what vowel may be involved, the pitch of the voice is determined by muscular control in the larynx. Learning to sing pitches in a musical scale accurately on demand is a matter of training the mind by repetition to control these muscles precisely. And vowel identity is determined by an entirely separate set of muscles controlling vocal tract configuration, as we shall see in Section 14.3.

The *diphthongs* are quick transitions, starting as one vowel but ending as another because the tongue changes shape. There is no long *i* sound (as in *might*), for instance, that can be sustained. The essence of the diphthong is in the transition, in this case *ah-ee*. Others in English are *eh-ee* (*mate*), *aw-oo* (*moat*), *a-ou* (*mount*), and *aw-ee* (*oil*). Diphthongs are a veritable minefield for singers, because on long notes they raise the problem of how much to lean toward the transition and how soon to finally go through with it. Trying to hold the diphthong at the halfway point may change the meaning altogether (for instance, *mate* to *mitt*).

14.3 FORMANTS

Of all the phonemes, the pure vowels are of special concern to musicians, because it is these that are sustained for the assigned length of each musical note. We should like to understand how different vowels are produced and what acoustical properties make each one distinguishable from the others.

We already have stated that vocal tract resonance is too weak to control the oscillations of the vocal cords. We may carry that line of reasoning one step further: Not only is the frequency of cord vibrations determined almost exclusively by the larynx, so also is their waveform. That is, for a given lung pressure and vocal cord opening and tension, practically the same cord vibration takes place regardless of vocal tract shape (as long as it is at least open). Producing different vowels by moving the tongue around must mean producing different filtering actions on one and the same sound from the vocal cords.

What is the sound input to the vocal tract? If you could hear it unaltered, you would discover it is a buzzing sound rather like that from a trumpet player's lips separated from the trumpet. It is a series of puffs of air, whose exact nature depends on how forcefully the air is being sent through the larynx. For gentle sounds the vocal cords may never close completely and the waveform may be fairly smooth (Figure 14.7a). More commonly, the flow is shut off during some portion of the cycle, which may be as much as a third for

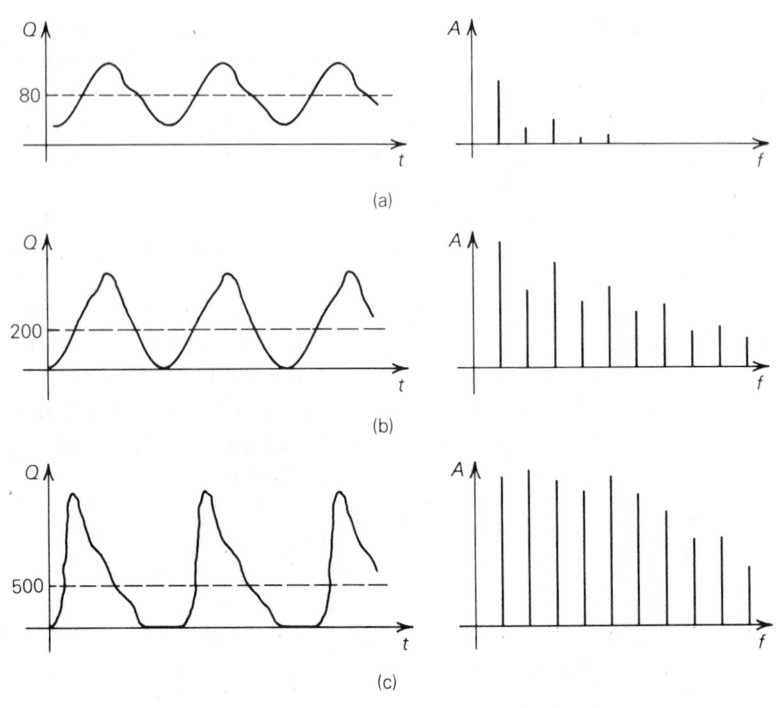

FIGURE 14.7 Approximate waveforms created in the larynx (left) and the corresponding harmonic spectra (right). Q is the total flow rate through the vocal cords in cm^3/sec; dashed lines represent average flow. Note that vertical scales differ for the three graphs. (a) A very soft sound for which the glottis never completely closes; the harmonic content is extremely poor. (b) An intermediate case, with a smoothly decreasing spectrum. (c) A very loud sound, with the glottis staying closed about one-third of each cycle; the lowest half-dozen harmonics have comparable amplitudes, and the spectrum drops off beyond them.

high breath pressure and close initial cord spacing (Figure 14.7b,c). For purposes of understanding how these sounds are modified by the vocal tract, it is most helpful to translate each one into its recipe of Fourier components. For subsequent discussion let us take Figure 14.7b as typical of moderate intensity.

How does the vocal tract modify the sound spectrum? It is roughly in the form of a tube about 17 cm long, closed at the inner end (as for all reed instruments) and open at the mouth. Suppose as a first rough approximation we pretend that it is a uniform cylinder. Then Figure 12.3 (page 229) reminds us that the natural frequencies are odd multiples of $v/4L$, or 500, 1500, 2500, 3500, . . . Hz. Suppose further (to make the numbers easy) that we have a male voice singing a pitch a little above G_2 so that the frequency of vocal cord vibrations is 100 Hz and their spectrum includes all multiples of 100 Hz.

Naive application of the vocal tract resonance idea would suggest that only 500, 1500, 2500, . . . Hz come out of the mouth with appreciable strength while all others are suppressed. But we must remember that these are only weak, broad resonances because of the softness of the tube walls (Figure 14.8a). So all spectral components remain in the radiated sound, and there is

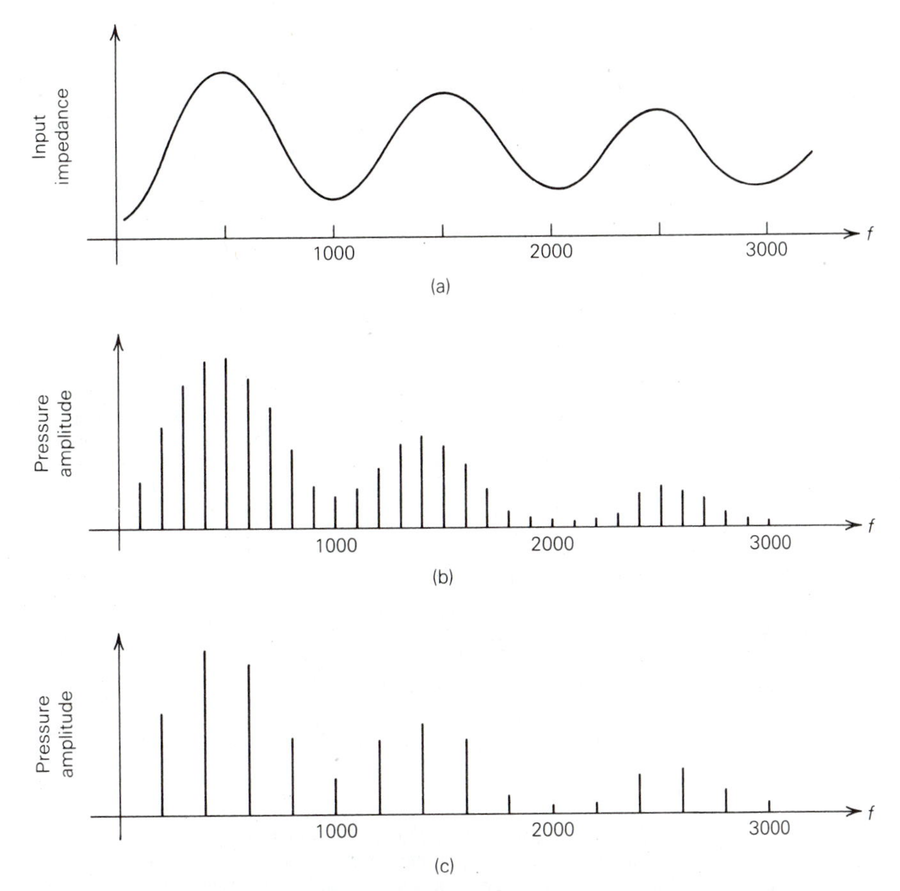

FIGURE 14.8 (a) Resonant response of the hypothetical perfectly cylindrical vocal tract. (b) Strength of harmonic components of a spectrum like that of Figure 14.7b (smooth trends only, ignoring individual even-odd differences) after transmission through the cylindrical vocal tract for fundamental frequency 100 Hz. (c) The same for 200 Hz fundamental.

merely a mild boost of *all frequencies within the general vicinity of each resonance* (Figure 14.8b). Each such frequency range, in which amplitudes of spectral components are enhanced, is called a **formant**.

Suppose our subject sings an octave higher while keeping the same uniform cylindrical vocal tract shape. Then the harmonic series contains all multiples of a 200 Hz fundamental, but the formants remain the same (Figure 14.8c). The ear somehow recognizes the locations of formant regions almost independently of whatever individual frequencies make up those formants, so that the sounds represented in Figure 14.8b,c are perceived as having approximately the same vowel quality. (Specifically, it would be a relatively neutral sound such as *ea* in *heard*.)

What about other vowels? There are a few other idealized shapes for which resonant frequencies are only mildly difficult to calculate; the one that

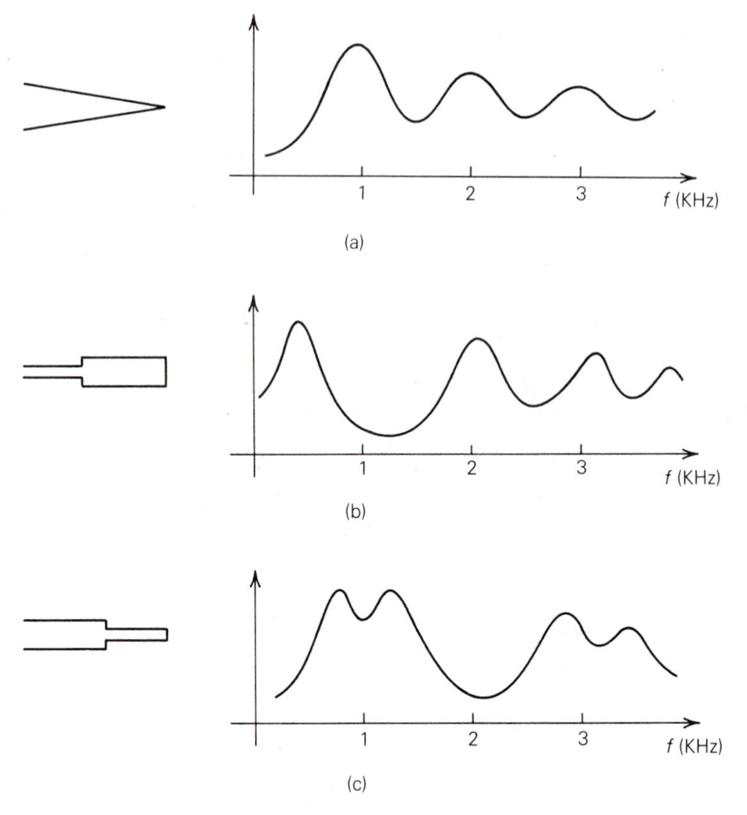

FIGURE 14.9 Formant-defining resonance curves of (a) a cone of length 17 cm, (b) a cylindrical bottle of length 9 cm and cross-sectional area 8 cm^2 with a neck of length 6 cm and area 1 cm^2, and (c) a narrow tube (length 8 cm, area 1 cm^2) opening into a wider one (length 9 cm, area 8 cm^2). (Based on data from Strong and Plitnik, Chapter 5.)

is simple enough for us to give the answer immediately is a cone (Figure 14.9a). For the same 17-cm length, this gives formants around 1, 2, 3, . . . KHz. Although these formants give a sound that would probably be identified as the short *a* in *had*, we must not jump to the conclusion that the vocal tract is really cone-shaped for that vowel; there may be other more complicated shapes whose first few formants happen to fall at similar frequencies.

Strong and Plitnik present calculated results for two cases where two cylindrical tubes of different diameter are joined (Figure 14.9b,c). These roughly approximate the vocal tract shapes, and resulting formants, for *ee* (*heed*, tongue up and forward) and *aw* (*bought*, tongue down and back), as illustrated in Figure 14.10. Additional examples of formant spectra and corresponding waveforms are shown in Figure 14.11. The first formant frequency is especially affected by the jaw opening, the second by the body of the tongue, and the third by the placement of the tip of the tongue.

We could proceed to show formant graphs for a long list of different

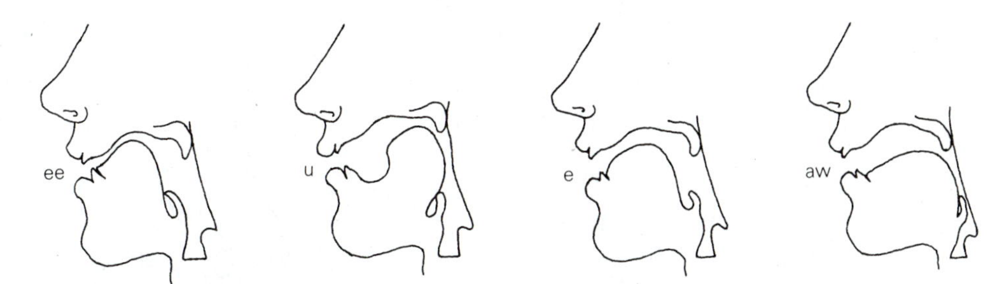

FIGURE 14.10 Vocal tract configurations for vowels in *see, put, let,* and *bought.* Tongue forward for *ee* and *e*, back for *u* and *aw*; tongue high for *ee* and *u*, low for *e* and *aw*. Compare *ee* and *aw* with Figures 14.9b,c, respectively. (From *The Speech Chain*, by Peter B. Denes and Elliot N. Pinson. Copyright © 1963 by Bell Telephone Laboratories, Inc. Reprinted by permission of Doubleday & Company, Inc.)

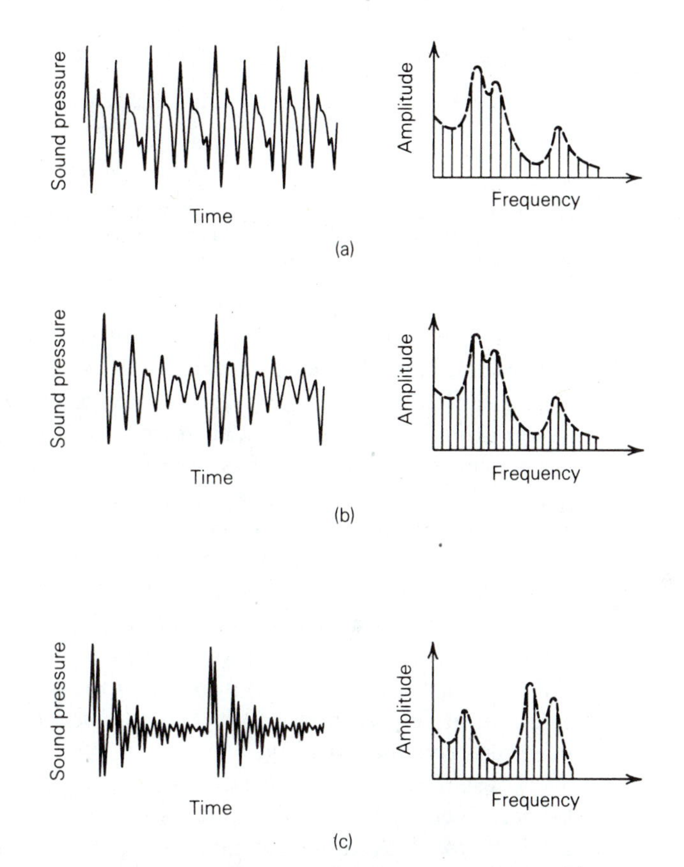

FIGURE 14.11 Waveforms and spectra for (a) *ah* at 150 Hz, (b) *ah* at 90 Hz, and (c) *uh* at 90 Hz. Compare (a) and (b) for same formants but different fundamental, (b) and (c) for same fundamental but different formants. (From *The Speech Chain*, by Peter B. Denes and Elliot N. Pinson. Copyright © 1963 by Bell Telephone Laboratories, Inc. Reprinted by permission of Doubleday & Company, Inc.)

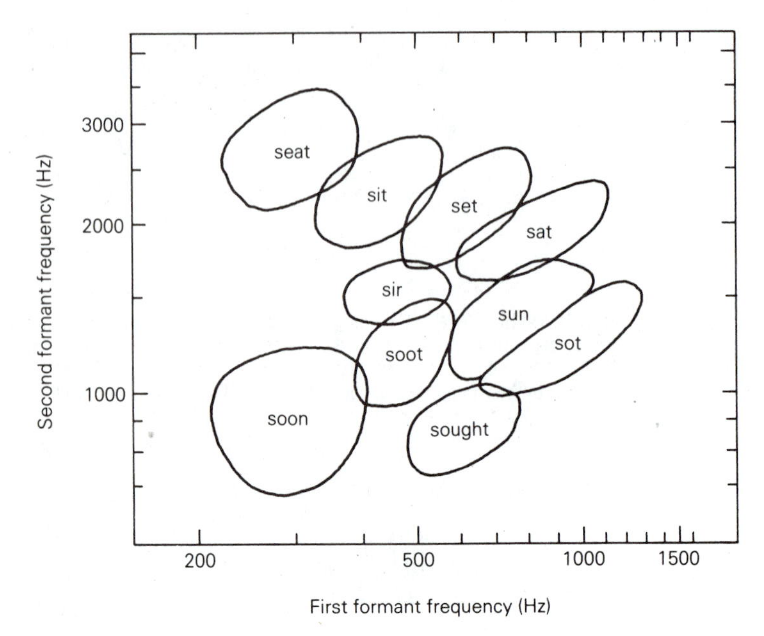

FIGURE 14.12 Regions of vowel recognition in terms of the first two formant frequencies, for standard spoken American English. Values for men tend to be toward the lower left of each region (both frequencies lower), for children toward the upper right, and for women intermediate.

vowels or a large table of their formant peak frequencies. But it is perhaps more informative to take a limited portion of this information and work it into a single picture (Figure 14.12). Although we actually are aided in vowel recognition by third and fourth formants, it is possible to distinguish them largely on the basis of the first two formants alone. If we choose to make a graph with horizontal axis representing first formant frequency and vertical axis representing second formant frequency, then every point on this graph represents a unique pair of formants. Ideally, each vowel would correspond to such a point, but of course in real life there is considerable variability not only from one speaker to another but also from time to time for each person. So there is a whole range of formant-frequency pairs (a whole region in this graph) that may be used to convey the same vowel information. A first formant in the vicinity of 800 Hz and second formant near 1500 Hz, for example, give *uh* as in *sun*.

These ranges overlap, so that in some cases precisely the same sound may be perceived in two different ways. For instance, with formants at 500 and 2100 Hz you may think you hear either *hid* or *head*. Usually only one of the two makes sense in context, and we automatically pick that one without consciously realizing the ambiguity. This flexibility in vowel perception is helpful in music: Because some vowels sound shrill when sustained, singers can be taught to "cover" their tone, shading short *e* (*head*) toward *ea* (*heard*), for example, without losing intelligibility. Thus a modified version of Figure 14.12 for sung vowels would have several of the uppermost blobs moved downward.

One of the reasons for the rather elongated area for each vowel is that men, women, and children do not all have the same size vocal tracts. They could not reasonably be asked to produce precisely the same formant frequencies. Listeners seem to readily make allowance for this: A formant pair at 700 and 1000 Hz might be perceived as *aw* (*hawed*) in a child's speech but *ah* (*hod*) in a man's, because of its position relative to other vowels heard from the same speaker.

The "long *o*" (*moat*) provides a nice illustration of the pitfalls awaiting those who want everything to fit neatly into standard pigeonholes. People who study running speech generally classify this as a diphthong, *aw-oo*, as we did at the end of Section 14.2. But singers will insist that they can sustain a long *o*, and regard it as a pure vowel. That is, they would be inclined to add another blob (*sone*) in the lower part of Figure 14.12, occupying the empty space among *soon*, *soot*, and *sought*, and partially overlapping them.

An important modern tool for analyzing the rapid succession of phonemes in running speech is the *speech spectrogram* (Figure 14.13). This uses darkness of shading to represent strength of signal so that a whole series of spectra can be displayed, with frequency of spectral components represented on the vertical axis and time elapsed on the horizontal axis. Then the acoustic features of speech production can be studied, and especially the changes in formants resulting from adjustments in vocal tract shape. Formant frequencies for pure vowels, which in singing would remain constant for a long time, seldom remain the same for even a tenth of a second. Even these vowels are modified by quick upward or downward formant shifts at the transitions to and from adjoining consonants, which generally require a different vocal tract configuration. It is the formant shifts themselves that characterize the semivowels and diphthongs.

There are two factors, dynamic level and voice range, that sometimes greatly reduce the distinctness of vowel formation. Reexamine Figure 14.7, and ask what happens to each of those spectra as they pass through the formant filtering process. The results are qualitatively as shown in Figure 14.14, which suggests that an extremely soft sound may provide so little high-harmonic input that the higher formants are hard to recognize. You can easily verify by doing a little singing that softly sung vowels are rather colorless, while loud ones are much more distinct from one another.

The other problem occurs when higher pitches are sung. Higher fundamental frequencies spread out the harmonic series of the vocal cord input, even to the point where a given formant simply may not have any candidates for resonance within its range (Figure 14.15a). This means that the best sopranos, especially on their higher notes, are prepared to shift formants, deliberately sacrificing vowel accuracy to have strong enough low harmonics to make a strong and musical tone (Figure 14.15b). Several vowels all may merge into approximately the same sound on the high notes.

To close this section, we add remarks about the comparison between voice and other instruments. The discussion of vowel recognition from formants suggests returning to the question of instrument recognition (Section 6.7) and asking whether a similar mechanism may operate there. Some

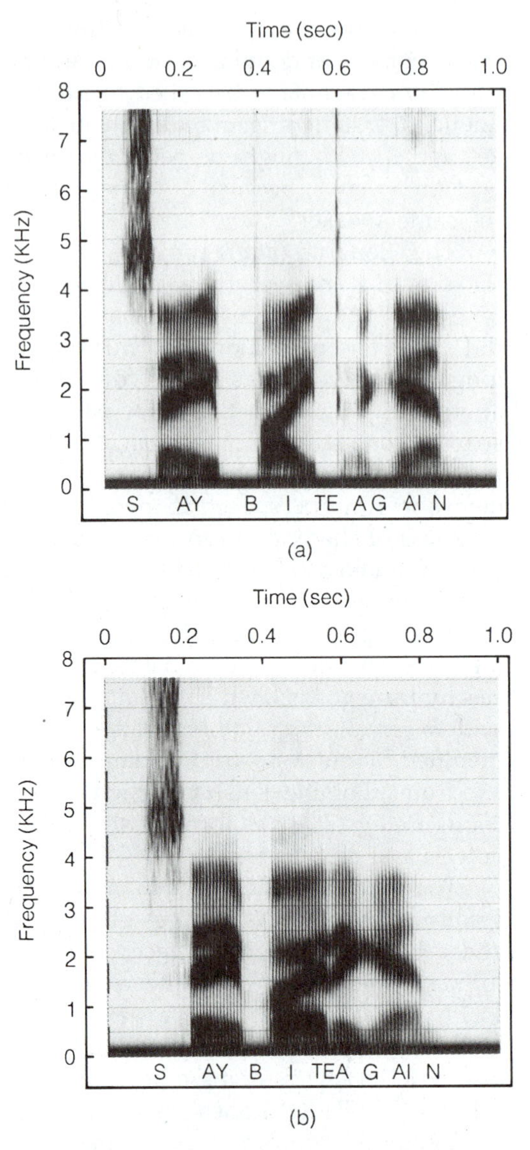

FIGURE 14.13 A speech spectrogram for a male voice saying "say bite again" (a) with short pauses separating the words and (b) in normal running speech. The broad, dark bands running horizontally represent formants. Ignore the narrow vertical stripes, which are only an artifact of the method for producing the picture. (Courtesy of Kay Elemetrics Corporation)

studies (see references on page 307) have reported identification of characteristic formant frequencies for several wind instruments. More recent work, however, led Benade to believe that they are better characterized by a single "cutoff frequency." Below this frequency the spectra are relatively flat, while above it they fall off steeply. These cutoff frequencies typically are approximately 400 Hz for bassoons, for instance, but 1400–1600 Hz for oboes and clarinets. Finally, a reminder: Transients also provide extremely important recognition clues to supplement those available in the steady sound.

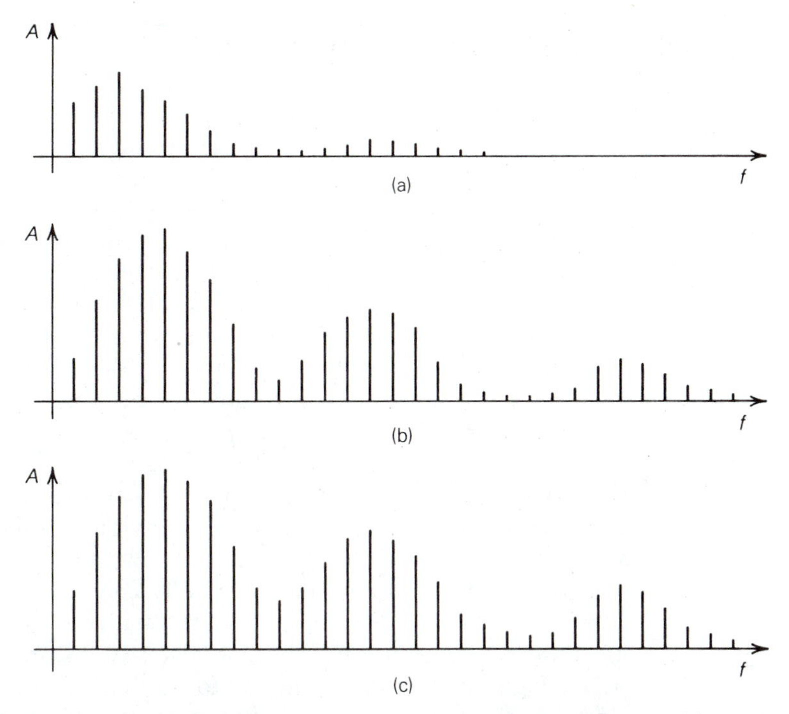

FIGURE 14.14 Application of the filtering action of a cylindrical vocal tract (Figure 14.8a) to each of the signals of Figure 14.7 for fundamental frequency 100 Hz. (a) If some harmonics are extremely weak in the original spectrum, the filter will not compensate; so higher formants are sparsely populated for a soft sound. (b) An intermediate case; this is the same as Figure 14.8b. (c) The richer input spectrum of a loud sound makes its higher formants more prominent in the output.

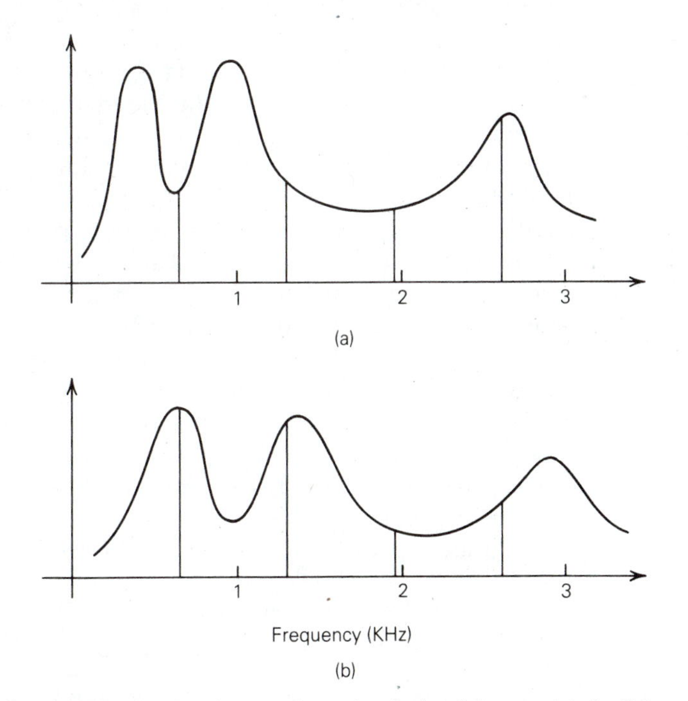

FIGURE 14.15 (a) Normal formant placement for *oo* (*cool*) gives little support to the first several harmonics of 660 Hz (E_5). (b) If the soprano is willing to shift formants toward *could* or even *cull*, the first two harmonics can be strengthened to give a more solid tone.

*14.4 SPECIAL PROBLEMS OF THE SINGING VOICE

There are significant differences in the way the vocal apparatus produces steady tones in different pitch ranges. An important part of voice training lies in learning to control and utilize transitions between "chest voice" and "head voice." This is most obvious for male singers in the change from normal voice to *falsetto* on high notes. In falsetto singing, the vocal folds are stretched longer and thinner, becoming effectively stiffer and thus making it possible to vibrate at higher frequencies than could be reached otherwise. The glottis generally does not close in falsetto singing, so the waveform is relatively smooth and the tone color more pure or bland (refer to Figure 14.7a). The wider glottal opening also accounts for the singer's air supply running out more quickly in falsetto singing.

Vocal teachers sometimes use words such as *resonance* and *projection* in ways that seem vague and hard to understand to an acoustician. Remember that *resonance* means an especially strong vibration that occurs because some system is driven at a frequency close to that of its own free vibrations. It is reasonable enough to speak of carefully controlling jaw, tongue, and soft-palate positions to adjust vocal tract resonances to produce better tone quality. But resonance in the chest is a false issue. It may, of course, be helpful psychologically for voice students to think in a way about their chests that leads to producing a steady, strong, and well-controlled pressure at the larynx. But any literal resonance in the chest is quite out of the question, even if we ignore the way the glottis practically isolates the lungs from the vocal tract. The spongy lung tissue is a prime example of a region that will greedily absorb any vibrations that enter it and never return any strong reflected waves.

Similarly, talk of projection may become associated in the singer's mind with muscular controls that produce louder, steadier sound at the vocal cords or a wider mouth opening. But intensity will always fall off with distance in quite the same way, and directional spreading will be controlled entirely by diffraction. Our statements in Chapter 4 make it clear that audible wavelengths from approximately 15 m down to perhaps 0.5 m (frequencies up to approximately 700 Hz) are radiated almost equally in all directions, being well able to get around the barrier formed by the head. For higher frequencies the sound goes largely to the front half-space rather than the back but still spreads well throughout that half. Only at extremely high frequencies (say, 6 KHz or more, corresponding to wavelengths below 6 cm) could a singer reasonably expect to literally project a beam of sound in one particular direction by using a very wide mouth opening. Such frequencies, of course, would have to be higher harmonics; singing a fundamental well beyond C_8 is entirely out of the question.

Sundberg presents an interesting explanation of how a singer—say, an operatic tenor—can make himself heard above an orchestral accompaniment. Even though the total sound output of the singer can hardly match that of the

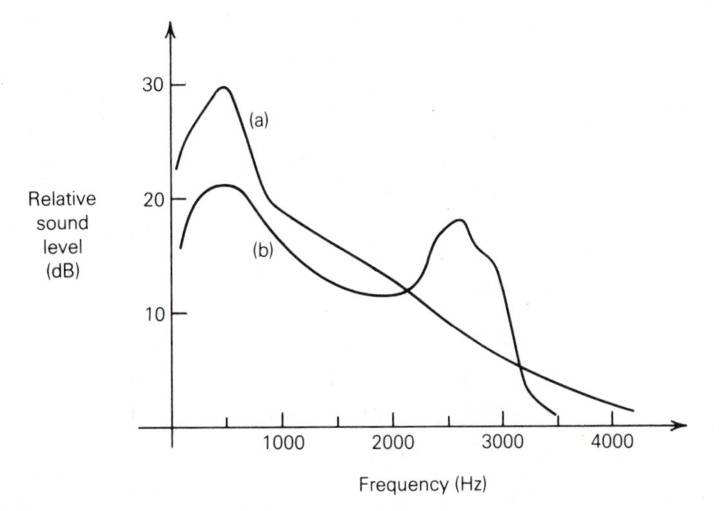

FIGURE 14.16 Long-term average sound output at different frequencies for (a) typical orchestral music and (b) an operatic tenor. The tenor is heard above the orchestra partly because of his strength in the singer's formant at 2500–3000 Hz. (After Sundberg)

orchestra, the singer still can draw attention to his part by concentrating much of his acoustic power in a part of the spectrum where the orchestra is not so strong (Figure 14.16); the result may be perceived by listeners as "good projection." This apparently is accomplished by a lowering of the larynx and accompanying expansion of the throat immediately above it, which makes a discontinuity in the cross section of the vocal tract. This enables the larynx to have some standing waves of its own, nearly independently of the rest of the vocal tract, and the first of these has a frequency of approximately 2500 to 3000 Hz. This resonance provides the *singer's formant,* enabling the singer to be heard well at the expense of some distortion of vowel production.

Vibrato (frequency modulation) is a voice characteristic that is sometimes cultivated; fashions often change. It is also possible to develop *tremolo* (amplitude modulation, sometimes confusingly called "amplitude vibrato"; refer to Section 8.3 for this distinction in terminology). The best training for a singer probably is to learn to sing both with and without vibrato, so that it can be used deliberately for special effect rather than being insistently and remorselessly present all the time.

How often should the modulation occur? Modulation frequencies of approximately 5 to 7 Hz usually are judged to be most pleasing; as low as 3 Hz or as high as 10 Hz definitely calls for retraining. How much should the audio frequency be changed by the modulation? Excursions of half a semitone on either side of the central pitch (3% in frequency) are not uncommon. Much more than that is distracting; much less than 1% will hardly be noticed; around 2% seems to be musically pleasing.

Why is vibrato considered desirable not only for the voice but for many other instruments as well? A cynical answer from evolutionary biology would be that many human voices do waver when they sing; therefore we have come to believe it is a good thing. And because the voice does it, perhaps we think other musical sources also should. But we also could say that vibrato lends warmth to the tone, as well as drawing attention to the solo lines, where it is often strongest. There also may be good acoustical reason to have more modest amounts of vibrato in ensemble music, where it blends in with the chorus effect discussed at the end of Chapter 5. If two or more voices sing together, a little vibrato can camouflage defects in the ability of the singers to stay perfectly in tune with one another. Similarly, the sound of a string section in an orchestra may blend better when there is some vibrato. The problem encountered in the opposite case of rigidly fixed frequencies becomes quite clear if you ever hear two organs (or to a lesser extent two pianos) together; modest amounts of mistuning between them become prominent and distracting.

Finally, why is 6 Hz more desirable than other values for the modulation frequency? At least two theories have proposed that modulation is most easily produced at this frequency, but there is no strong proof for either one. Certain natural brain rhythms occur at similar frequencies, so perhaps it is at these frequencies that it is easiest for the brain to send the series of electrical control signals to whatever muscles are used to produce the vibrato or tremolo. Alternatively, crude estimates of the abdominal mass and the springiness of the air in the lungs suggest that the natural vibration frequency for the body contents is approximately 5 or 6 Hz; this makes it easier to maintain a good tremolo at such frequencies, and so we have come to prefer it. Finally, there may be a perceptual reason, too: Any sounds within one- or two-tenths of a second tend to become merged during processing by the ear and brain. Thus, modulation frequencies much above 5 Hz will become harder to perceive as such; there will seem to be a homogenized rough sound instead of a rhythmically varying simple sound.

SUMMARY

The human voice produces three distinct types of sounds. The plosive consonants are intrinsically transient and are made by blocking and then suddenly opening the vocal tract. The fricatives depend on turbulence in air forced through a narrow opening to generate a continuous noise. Only the vowels have periodic waveform and definite pitch, and they depend on vocal cord vibration. Those vibrations are of the lip reed type and are aided by the pressure reduction that occurs where an airstream passes through a constriction (the Bernoulli effect).

The pitch of a sustained vowel is determined entirely by the vocal cords. But the vowel identity is determined by the shape of the vocal tract, which determines the location in frequency of the broad resonance bands called formants. A formant enhances those components of the harmonic series of the vocal cord vibration that happen to fall within the formant region. Distinct vowel formation is more difficult for soft sounds or high pitches, in both cases because of a dearth of harmonics to populate the formants and make them recognizable.

REFERENCES

An excellent article on the acoustics of the singing voice is that by Johan Sundberg in *Scientific American* (March 1977), which is also reproduced in the Hutchins collection. The singing voice is treated at greater length in Benade's Chapter 19, and in the important new book by Sundberg, *The Science of the Singing Voice* (Northern Illinois University Press, 1987). Voice use in solo and choir singing is compared by T. D. Rossing, J. Sundberg, and S. Ternstrom, *JASA, 79,* 1975 (1986). Circumstances in which vocal tract shape can influence the vocal cords are studied by Sundberg in *Acustica, 49,* 47 (1981).

Excellent further reading on speech production and recognition may be found in Denes and Pinson; Chapter 4 is especially relevant, but 1, 2, 7, and 8 also may be of interest. Strong and Plitnik treat speech more extensively than we have done here; their section on machine processing of speech deserves special mention.

The possibility of using formants to describe wind instrument spectra was studied by W. J. Strong and M. Clark, *JASA, 41,* 39 and 277 (1967), and by D. A. Luce, *J. Audio Engr. Soc., 23,* 565 (1975). Some differing results of Sirker are quoted by Backus, p. 120.

SYMBOLS, TERMS, AND RELATIONS

phonation
phoneme
vocal tract
larynx
vocal cords
plosive
fricative
vowel
diphthong
voiced/unvoiced
formant
harmonic series
vibrato
tremolo
modulation frequency

EXERCISES

1. Use information from Figure 14.7 and from the text to estimate maximum singing duration on a single breath. (1 liter = 1000 cm^3)
2. What pairs of elementary phonemes make the consonant sounds of *ch* (as in *church*), *x* (*fix*), and *j* (*jump*)?
3. What string of phonemes is used to say "musical acoustics"?
4. What happens to the distinction between voiced and unvoiced plosives in whispering?
5. Explain how vowels still can be recognized in whispering, even though there is no voicing (that is, no vocal cord vibration). (Hints: Is there some continuous sound? What kind? Would it be altered by vocal tract resonance?)

6. Why is a whisper more directional than a voiced sound?

7. Suppose that representative vocal tract lengths for man, woman, and child are 17, 14, and 11 cm, respectively. For any given vowel, what approximate percentage differences would you expect to characterize all their formants when you compare man versus woman or child?

8. Let a 300 Hz vocal cord signal be sent through (a) a cylindrical and (b) a conical vocal tract, each 17 cm long. Draw expected output spectra one above the other for comparison. (Hint: Your procedure should be analogous to Figure 14.8.) It is important to locate the harmonics accurately on your frequency axis.

9. Draw output spectra as in Exercise 8 for 200 Hz sent through two different cylindrical tracts, one 17 cm long and the other only 14 cm.

10. Identify the frequency, musical pitch, and vowel identity of the sung tone whose spectrum is shown in Figure 14.17.

FIGURE 14.17

11. For first and second formants at 450 and 2300 Hz, what vowel would be understood?

12. For first and second formants at 400 and 1000 Hz, what two vowel interpretations are possible? What clues are likely to decide the choice?

13. Estimate the frequencies of the first two formants you would use in saying *e* as in *red*.

*14. Suppose you sing with your vocal tract filled with helium, in which the speed of sound is approximately 930 m/s. Does this change the frequency of vocal cord vibration? Does it change the formant frequencies? By how much?

15. Sketch a pair of spectra to indicate the changes that take place during the diphthong in *oil*.

16. To get solid tone on a high C (C_6), a soprano might do well to position her tongue as if pronouncing what vowel?

17. If the singer's formant is due to resonance of that part of the larynx above the vocal cords, use the cylindrical approximation to estimate the length of the upper-larynx tube.

*18. The Bernoulli pressure reduction in a narrow constriction is approximately $\frac{1}{2}\rho v^2$. The density ρ for air is 1.2 kg/m^3. Suppose we estimate the glottis opening during phonation as $1\text{ cm} \times 1\text{ mm} = 10^{-5}\text{ m}^2$ and take $500\text{ cm}^3/\text{s} = 5 \times 10^{-4}\text{ m}^3/\text{s}$ for a flow rate. What is the average velocity v in the glottis under these conditions? What is the pressure reduction in N/m^2, and in atmospheres?

19. Making use of Figure 14.14, discuss what happens when you are asked to "speak up" in front of an audience. Will your intelligibility be helped as much by turning up the gain on the microphone amplifier as by literally speaking more energetically?

20. Suppose you make a tape recording of a male voice singing slowly and play it back at doubled speed. What will happen to the pitch of each note? What will happen to the formant frequencies of the

vowels? Will it merely sound like a woman singing fast?

21. Consider the hypothesis that the vowels with the greatest tendency to sound shrill or harsh would be those with the greatest gap separating their second formant from the first. Which vowels would those be? Are they indeed the ones a voice student is taught most need "covering"?

22. In Figure 14.13, what frequencies are strongest for the sibilant *s*? Why is the *t* so short? Relate the formant frequency changes for the two diphthongs in Figure 14.13a to Figure 14.12, and discuss why they are different in Figure 14.13b. Explain the changing formants for the two nominal vowels in *again*.

PROJECT

1. Use an electronic spectrum analyzer to study vowel spectra produced by volunteer singers, and compare your results with Figure 14.12. Discuss how much variation you observe from one subject to another or one trial to another.

CHAPTER

15 Room Acoustics

Think about how different a brass band sounds when crowded into a small room for practice, when performing in a large concert hall, and when marching in the half-time show at a football stadium. It is clearly important for musicians to understand not only the way sounds are created by their instruments, but also the way these sounds are modified by the environments in which they are heard.

We must first describe several criteria for pleasing sound, and point out how these may differ depending on the type of music. We will try to understand which aspects of room design control each of these important acoustical properties. We will find that overall room size and shape, location of critical reflecting surfaces, and use of sound-absorbing materials are all important.

The property most amenable to precise measurement or calculation is the reverberation time; that is, the length of time it takes for sounds to effectively die away in a room. We shall address in particular the practical question of how to plan or predict modifications of existing rooms to improve their reverberation characteristics.

Consideration of reverberant sound levels leads to the need for public address systems to provide artificial sound reinforcement. Satisfactory PA systems depend on an understanding of how our ears judge the location of a sound source, so in the final section we will review the clues we use in estimating both direction and distance of sound sources.

15.1 GENERAL CRITERIA FOR ROOM ACOUSTICS

Let us list several different standards by which we judge the acoustics of an auditorium for musical purposes:

1. *Clarity.* Each note should arrive cleanly, crisply, and unobscured. This is especially important if the room is used for speech as well as music, because the intelligibility of words depends quite directly on clarity of articulation.
2. *Uniformity.* Listeners in all parts of the hall should hear as nearly the same sound as possible; there should be no dead spots.

3. *Envelopment.* The listener should not feel separated from the source but rather bathed in sound from all sides; yet at the same time the sound must identifiably originate on the stage to match the sense of sound with that of sight.
4. *Freedom from echo.* Even though there must be repeated reflections of sound off walls, none of these should be perceived as a separate echo; all reflections must blend together smoothly.
5. *Reverberation.* This continuation, or hanging of the sound in the room, must have both an appropriate loudness relative to the original sound and a pleasing rate of decay.
6. *Performer satisfaction.* The stage also must be free from distracting echoes and at the same time provide enough enclosure that performers in a group can feel they are in good communication with one another.
7. *Freedom from noise.* Soft passages in the music should not be disturbed by traffic noise outside or by noise from the auditorium's ventilating system.

To understand how these criteria can be achieved, let us describe the total sound experience resulting from the playing of a single note. Consider first an extremely short percussive sound, such as the impact of a stick on a snare drum head. The shortest possible path from source to listener is a straight line, and the **direct sound** traveling that path arrives first (Figure 15.1a). It is followed shortly by several reflected sounds from walls or ceiling (Figure 15.1b); the distance traveled for each reflecting path determines how much later that reflection arrives. Those sounds arriving within approximately 50–100 ms after the direct sound qualify as **early reflections**. If they follow closely enough after one another, all are perceived together as a unified acoustical event. Later on we continue to receive sound from more and more different multiple-reflection paths. Each individual component is weaker and weaker as the number of reflections increases (Figure 15.2), and they all merge together in a continuously decaying **reverberant sound**, which is perceived as a stretching out and gradual decay of the original event.

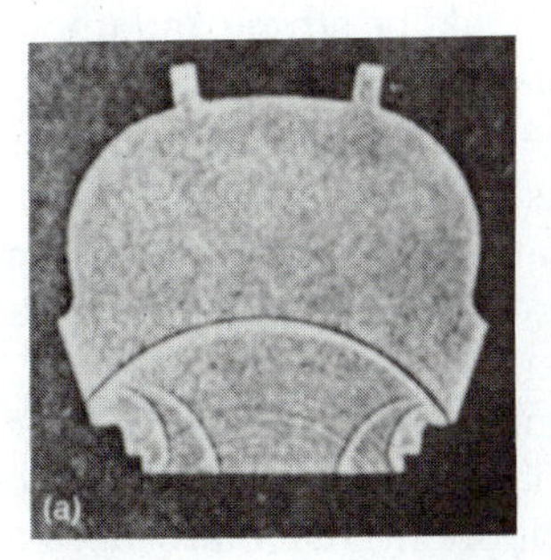

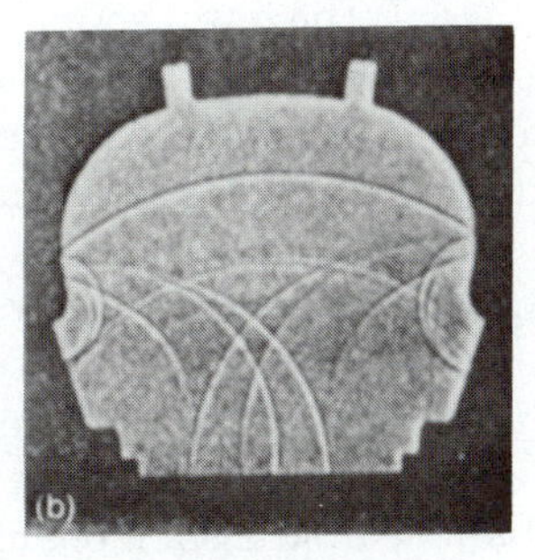

FIGURE 15.1 A sound impulse (made visible by Schlieren photography) in a scale model of a theater, in studies by Sabine (J. Franklin Inst. *179*, 1, 1915). (a) Direct wave arrives at center of seating area. (b) Early reflections spread across the audience from both sides. (c) The pattern of repeated reflections rapidly becomes increasingly complicated. This design needs changes in its rear wall, which sends a strong echo back toward the stage and focuses too much sound at two spots near the rear exits.

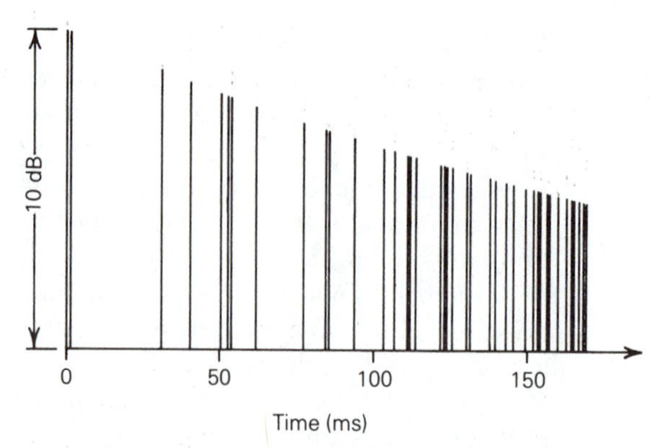

FIGURE 15.2 Schematic sequence of early reflections for a particular choice of source and detector locations in a rectangular room 40 × 25 × 8 m. (From Kuttruff, by permission of Halsted Press)

For a sustained note each contribution continues after its initial arrival, as in Figure 15.3. Only after some time sufficient for many reflected waves to arrive does the total sound approach a steady level. When the note ends, it is the direct sound that seems to the listener to stop first; for the various reflected waves it takes just as long for news of the stopping to reach the listener as it did in starting.

---

*The detailed pattern of steps in Figure 15.3 changes drastically if either source or listener location changes, but in a good room the smoothed general tendency remains exponential. Small amounts of performer movement, listener movement, and vibrato are all important in smoothing unavoidable irregularities in room response so that sound characteristics remain recognizable. (See Benade for extended discussion of these interesting problems.)

Let us now describe in these terms which features of an auditorium primarily determine how well it meets each acoustical criterion:

1. *Clarity*. The direct sound should be strong and unobstructed, so we try to get everyone as close to the stage as possible, and we raise the stage and often seat the audience on a slope or in balconies. If every listener has a good unobstructed sight line, the acoustical clarity is more likely to be good. Clarity also is enhanced if new sounds are not obscured by too much reverberation hanging over from previous words or notes, and the greatest clarity can be achieved only by sacrificing reverberation.

2. *Uniformity*. Again, to minimize differences in direct sound we use slopes and balconies so that the last row will not have to be so much farther away than the first. Even more important, the total reflected sound must have similar strength everywhere. This requires extreme caution in allowing concave walls, because they tend to focus sound rather than diffuse it. Rectangular rooms with plain flat walls also are undesirable, because they make it too easy for sound to bounce back and forth repeatedly over the same path. The

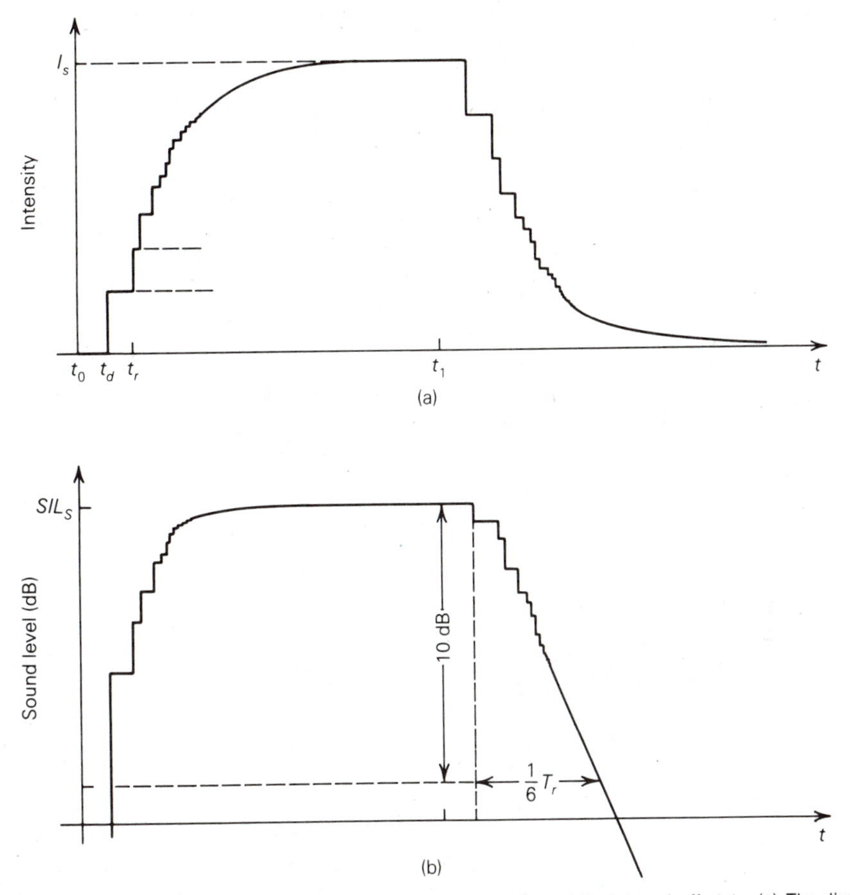

FIGURE 15.3 Total sound received from a steady source turned on at time t_0 and off at t_1. (a) The direct sound arrives at t_d, first reflection at t_r, and so on; the eventual total, as many reflections accumulate, approaches steady intensity I_s. After the source is turned off, reflected contributions drop out one by one and the intensity gradually approaches zero. If individual jumps are ignored, these are approximately exponential curves. (b) When the same information is plotted in terms of sound levels instead, it appears that the initial approach to SIL_s is very quick, but the final decay continues steadily to ever-lower levels. The arrow shows the time to decay 10 dB below SIL_s; the 60-dB standard reverberation time is six times as long if the decay is well behaved.

sound is more thoroughly mixed and distributed over the whole room if it has irregular shape, nonparallel walls, convex surfaces, and many protruding edges (see Box 15.1). Coffered ceilings, balcony fronts, open beams, chandeliers, and large three-dimensional decorations all help break up the sound and distribute it more uniformly. There should be small-, medium-, and large-scale structures, each helping especially to diffuse those sounds with wavelengths comparable to its size.

3. *Envelopment*. Early reflections should arrive not just from front or back walls, but also from the ceiling and especially the side walls. Preferably the sides and ceilings are not just flat, but include enough structure to provide

BOX 15.1 TRAVELING AND STANDING WAVES

As with violin strings and organ pipes, we can gain different insights into room acoustics by describing the air vibrations in two seemingly different but equivalent ways.

The traveling-wave picture is probably a little easier to grasp intuitively. We think of a bit of sound energy starting off from the source in a particular direction, bouncing back and forth between the walls and gradually getting weaker with each reflection. Meanwhile, other bits that started off in different directions follow different paths. This picture suggests that irregular room shapes are better because they ensure that the sound reflection paths are also quite irregular, and taken all together are almost equally likely to visit any part of the room and so achieve the goal of uniformity.

The standing-wave picture describes all possible sounds in the room as combinations of natural-mode vibrations. Just as one-dimensional vibrating string modes have nodal points and drumhead modes have nodal lines, so room modes have nodal surfaces subdividing the entire room into smaller regions (somewhat like cardboard dividers in a packing carton for delicate glassware), with air motion in adjoining regions being in opposite directions. We think of a musician on stage as providing the energy to

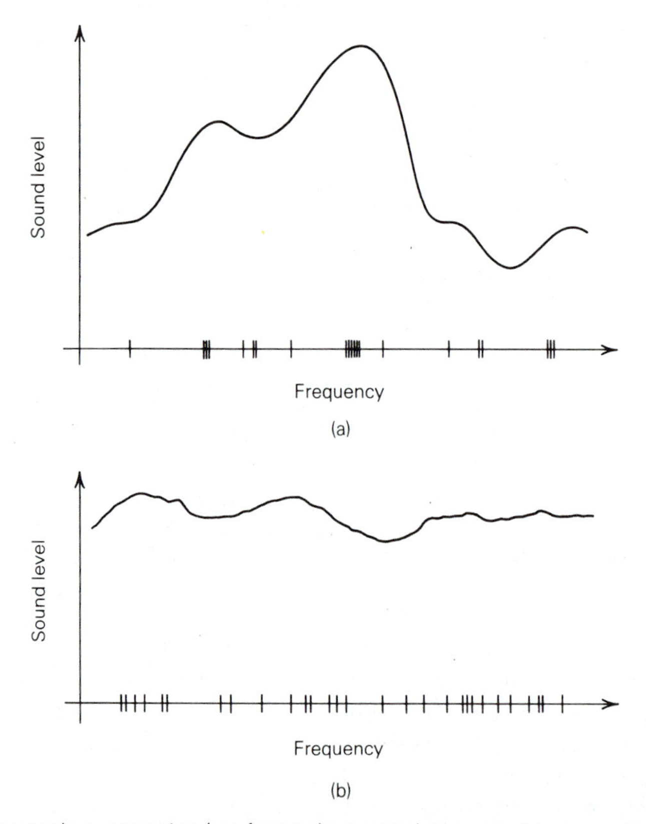

FIGURE 15.4 Total response to a source at various frequencies averaged over many listening positions throughout a room. This response is due to excitation of natural modes whose frequencies are indicated by tick marks. Total response includes contributions from many overlapping mode resonances, as in Figure 11.17b. (a) Mode clumping in a symmetrical room. (b) Randomization in an irregular-shaped room.

(continued)

BOX 15.1 *(continued)*

excite vibration of these natural modes, especially of those whose frequencies are close to that of the source.

This picture helps us understand another advantage of irregular room shapes: A symmetrical small room may have many natural modes all clustered about one frequency, but few cooperating modes available at certain other frequencies. In such a room one bass note may resonate impressively and another seem extremely hard to project (Figure 15.4a). Introduction of randomness in the room design tends to spread the natural mode frequencies more uniformly (Figure 15.4b) and so allows the room to have more nearly even response at all frequencies.

The number of modes whose frequencies are within a 1-Hz range around frequency f in a well-randomized room of volume V is approximately $3 \times 10^{-7} f^2 V$. All modes within a range of about $(4/T_r)$ Hz around f make important contributions to the total room response. The product of these two numbers tells about how many room modes respond strongly to a given source. (See Exercise 6 for an example.) Except for bass notes in small rooms, this is generally dozens or hundreds of modes.

several early reflections from each that will truly surround the listener with sound (Figure 15.1c).

4. *Smoothness.* A poorly placed concave surface (such as the back wall in Figure 15.1) or even a large flat, hard surface may provide a particularly strong reflection more than 100 ms after the direct sound. This will be perceived as a distinct echo, which of course should be avoided. But even a delay of more than 30 or 40 ms can result in an unpleasant roughness. To ensure that they blend together smoothly so that the reflected sound seems only to strengthen and lengthen the direct sound, we should keep that time gap less than 30 ms. Because sound travels 0.34 m per ms, this means the path length for the first reflection should not be more than approximately 10 m longer than the direct path, and this condition should be obeyed for every seat in the room. This may require careful placing of reflecting panels toward the front of an auditorium (Figure 15.5). Shorter time lags (say, 10–20 ms) encourage a sensation of intimacy. There also should be no gaps of more than 30 ms between first and second reflection, second and third, and so on.

5. *Reverberation.* A happy medium must be struck somewhere between dead and muddy sound. Reverberation is controlled by the size of the auditorium and by the relative amount of absorption or reflection of sound by the materials placed on walls, ceiling, and floor. Strong reverberant sound of low frequencies is responsible for warmth, and high frequencies for brilliance. We shall consider in detail how reverberation times can be measured and calculated in the following sections.

6. *Performer satisfaction.* The rear wall should not return a single strong echo to the stage. But a blend of many reflections should return to the stage strongly enough to give the performer some sense of what the audience is hearing. A good stage usually has more or less of a shell-type structure to bind together the members of a group and enable them to hear one another as well as help project the sound out to the audience. The stage should not have

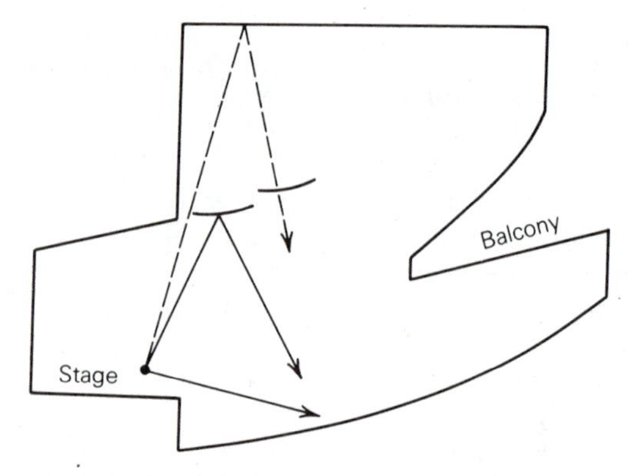

FIGURE 15.5 Use of suspended reflectors (sometimes called *clouds*) for early reflections when a ceiling is too high to provide a first reflection soon enough after the direct sound.

hard parallel side walls, which cause a very annoying problem called *flutter echo* in which you can hear any sharp percussive sound bounce back and forth many times. Flutter echo is doubly bad because it means part of the sound is trapped on stage and is not reaching its intended audience.

7. *Freedom from noise.* Substantial construction, double doors, and felt stripping for airtight closure of all doors are important in keeping extraneous sounds outside. Noisy ventilating systems may require redesigned outlet grills, quieter machinery, or acoustic filters in the duct work. Total background noise much above 40 dBA (with no audience present) makes a room unsatisfactory for sensitive musical performance, but a 30 dBA maximum usually is acceptable. Any standard less than approximately 20 dBA is difficult to achieve and not of further noticeable benefit anyhow.

While many of these criteria remain generally the same, there is one important respect in which the goal may be quite different for different kinds of music. That is the relative amount of direct sound, early reflections, and reverberation in the total mixture. As shown in Figure 15.6, speech is at one

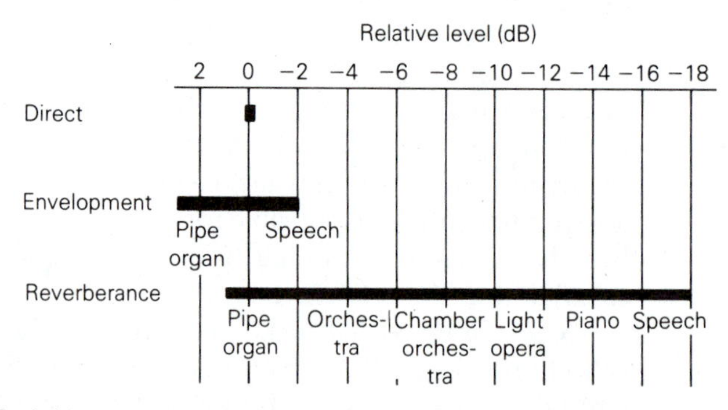

FIGURE 15.6 Desirable strength of early reflection and long-term reverberation relative to direct sound, for different types of music. Relative level is stronger toward the left, weaker toward the right. (From an article by Veneklasen in *Auditorium Acoustics*, ed. R. Mackenzie, Wiley-Halsted, 1975)

extreme; clarity is paramount and very little reverberation is desired. At the other extreme, some pipe organ music demands a full sound that is so live that the reverberant sound outweighs the direct and one chord flows gradually into the next.

All of these criteria are most important when we rely on the natural acoustics of an auditorium. Many popular-music concerts depend on powerful electronic amplification to create their characteristic sound in a wide variety of rooms that are often entirely unsatisfactory for classical music.

15.2 REVERBERATION TIME

We should like to use specific numbers to describe how long sound reverberates in a room. Your first thought naturally may be to define this as the length of time after a source stops putting out a sound until the sound is all gone. Unfortunately, this does not work; as we noted in our discussion of percussion instruments (Section 8.7), the vibration gets weaker and weaker without ever reaching zero energy. So instead we define **reverberation time** T_r to be *the time in which the sound level drops 60 dB below its original level.*

That is a precise definition useful to the physicist or acoustical engineer, but how is it related to human perception? First, if each note (or word) is followed immediately by another, the reverberation from the preceding note will no longer be noticed once it has dropped more than 10 or 15 dB below the present note; so even when we speak of a 2-second reverberation time, it may mean that only notes within about a half or a third of a second of each other effectively overlap. Only a final note followed by silence gives you a chance to hear the complete reverberation. Even then, if you make a casual estimate you are likely to judge only on the basis of the first 30 or 40 dB of decay. Only under ideal conditions—negligible background noise and extremely attentive and purposeful listening—do you agree that you hear the reverberation continuing as long as the official definition. With training a person can learn to judge 60-dB reverberation times well enough that an average of several trials will be within a couple tenths of a second of the correct value. A pistol firing blanks (or even a handclap) serves as a convenient sound source for rough estimates.

The standard method of reverberation-time measurement is sketched in Figure 15.7. Because different frequencies are not absorbed equally well by the walls, they do not decay equally fast. Accurate characterization of a room requires determining its reverberation time for several different frequencies. But you should not use a pure sine wave signal, because you might happen to pick a frequency corresponding to one particularly live (or particularly dead) combination of natural modes of the room. What is musically important is the average characteristics of the room over a whole band of frequencies (over a large group of neighboring natural modes), so narrow-band noise should

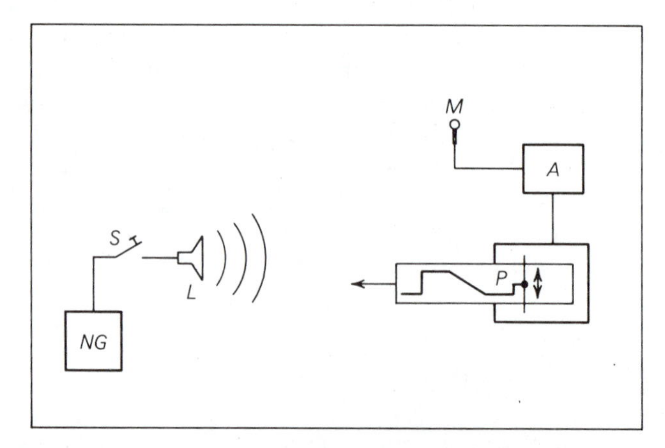

FIGURE 15.7 Reverberation-time measurement: The narrow-band noise generator *NG* drives the loudspeaker *L* long enough for sound to reach a steady level, and then is interrupted by opening a switch *S*. The microphone *M* sends its signal through an amplifier *A* (preferably including a narrow-band filter) to the strip-chart recorder where the pen *P* moves up and down while the chart paper moves to the left, to draw a curve such as in Figure 15.8a.

be fed to the loudspeaker. To determine reverberation "at 1000 Hz," for instance, you might use a mixture of all frequencies between 890 and 1120 Hz. (Noise generators and analyzers commonly offer such one-third-octave–wide standard bands.) The microphone signal is preferably sent through a filter that rejects all frequencies except that same narrow band to keep both room background noise and electronic noise from the amplifier to a minimum. The filtered and amplified signal then drives a strip-chart recorder to produce a graph like Figure 15.8a.

Today one can buy devices that use digital electronics to automate this process. A timer is started and stopped when the momentary decay level passes two prescribed values (for example, 5 dB and 25 dB down, followed by a tripling of the electronically measured time interval). To not be fooled by random peaks or dips on the decay curve, you should always take the average of several readings with this type of instrument.

Common faults in room reverberation are revealed by decay curves that are not smooth or that have a double decay rate like piano strings (Figure 15.8b,c). Besides representing less pleasing types of reverberation, these are signs that the sound energy is not uniformly distributed either. Such problems can arise if one part of the room is too isolated from the rest (Figure 15.9a) or if absorbing material is not well distributed around the room (Figure 15.9b). In terms of traveling waves, those starting out in certain special directions may reflect back and forth many times in one part of the room without visiting the other parts; if the parts they do visit have hard reflecting surfaces, these waves may last a long time and gradually leak out into the rest of the room after the main reverberation is over to produce the tail of the double-decay curve. In terms of standing waves, the same situation is described by saying that some natural modes push mainly on hard walls and

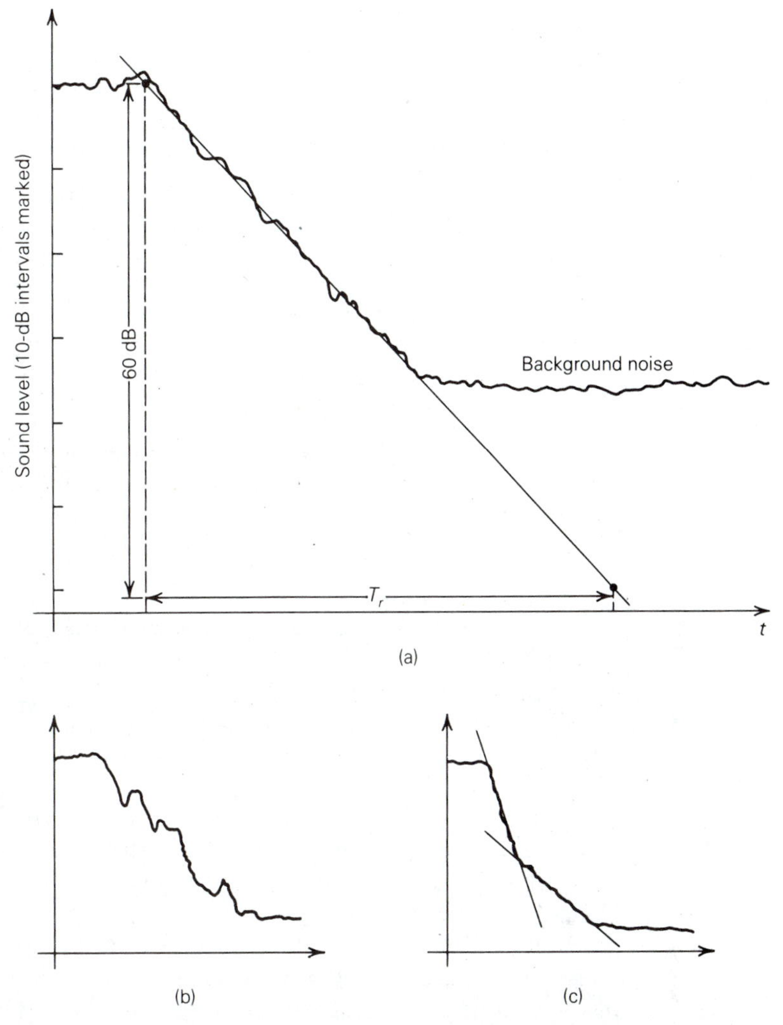

FIGURE 15.8 (a) Chart record of reverberation-time measurement (CSUS Recital Hall, at 1000 Hz). Even though background noise often prevents recording a full 60-dB decay, the slope observed for the first 30 or 40 dB can be extrapolated to make a good estimate of the 60-dB decay time. (b) Ragged decay resulting from nonuniform sound distribution. (c) Double-slope decay resulting from some modes dying out more rapidly than others.

have a long decay time, while others push on soft surfaces and die out quickly.

To have smooth decay of reverberation (and uniform sound distribution) it helps to have nonparallel walls, many obstacles and irregularities to scatter the sound in all directions, and absorbing materials distributed over many surfaces so that all natural modes have similar decay times. To prevent some modes from sneaking in between, absorbing materials (such as acoustical tile) should be placed more or less randomly, and *not* in highly regular patterns.

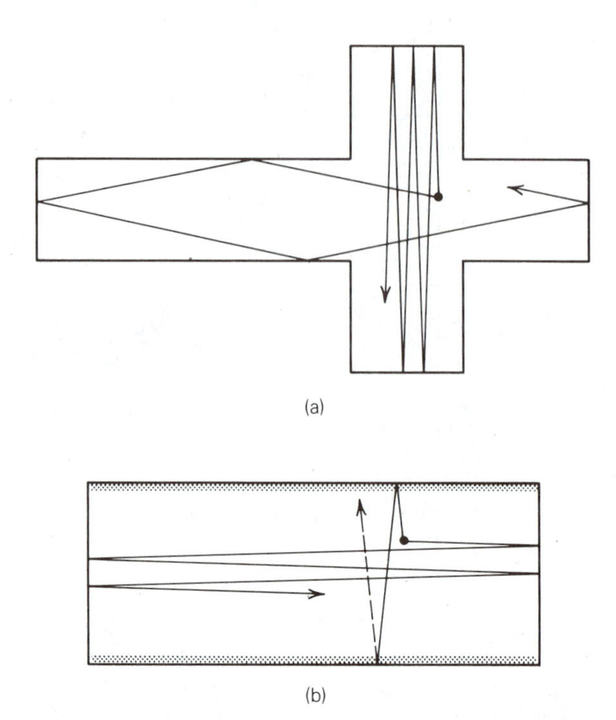

FIGURE 15.9 (a) A problem often arising in churches: Some reflected sound components tend not to visit all parts of the room equally. (b) Another common problem: Nonuniform distribution of absorbing material (shaded area) makes some modes die out quickly while others last much longer.

Just as the desired relative strength of reverberant sound depends on the type of music, so also does the reverberation time. As indicated in Figure 15.10, the larger forms of music tend to call for longer reverberation. Notice that speech and organ are again at opposite extremes; this makes it particularly difficult to find acceptable compromises for church buildings. These figures are only typical; what is desirable depends a lot on individual pieces of music. An organist might be quite happy with 1.5 s for a baroque fugue in order to keep the counterpoint crisp; yet for a Romantic work he might prefer 2.5 s.

Our opinions about reverberation time are strongly molded by cultural experience. Catholic churches, for instance, often have longer reverberation than do Protestant churches, a factor stemming partly from the differences in nature and purpose of their traditional liturgies. Readers who have traveled in Europe and visited the great cathedrals where T_r is 6 or 8 s will realize that the typical American church with only 1 to 2 s is acoustically quite dry. From a musician's point of view, 2 to 3 s often is more attractive, and would not be at all unusual for a smaller church in Europe. With a properly designed sound-reinforcement system and careful planning to make the most effective use of early reflections, this much reverberation is not incompatible with good speech intelligibility.

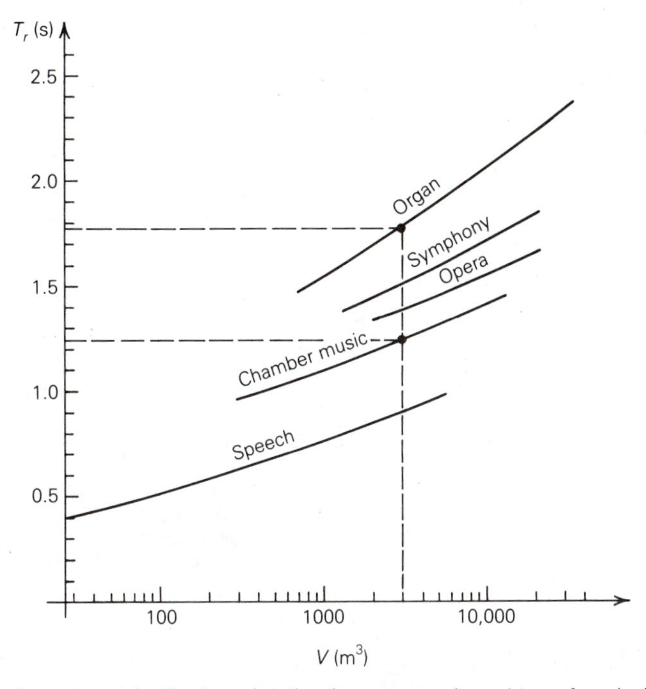

FIGURE 15.10 Dependence of optimal reverberation time on room size and type of music. Numbers indicated here are only typical, and individual preferences may well be 10–20% higher or lower. Even in the same room, seventeenth- and eighteenth-century music usually calls for shorter reverberation and greater clarity, while late-nineteenth-century music with its greater concern for sheer sonority calls for longer reverberation times. Dashed lines represent example of a 3000 m^3 recital hall in which $T_r = 1.2$ s would be good for a string quartet, but an organist would prefer much more than 1.5 s.

Notice that even for the same piece of music the desired reverberation time depends somewhat upon the size of room. This is at least partly to avoid the "cognitive dissonance" of conflicting visual and aural cues; the listener is not comfortable unless the reverberation heard through the ears indicates a room size reasonably close to what is seen by the eyes. In addition, in a larger room the only way to maintain the desired *level* of reverberant sound (Figure 15.6) is to allow a longer T_r.

If a single number is quoted for reverberation time, it ordinarily refers to mid- and high-frequency ranges (say, 500 Hz and above). Although opinion is not unanimous (see Kuttruff, p. 195) it often is said to be best if measured 60-dB reverberation times gradually increase toward lower frequencies, so that they are approximately 50% longer in the extreme bass than in the treble, as shown in Figure 15.11. The reason for this is quite simple—recall the Fletcher–Munson diagram (Figure 6.12, page 105). A 60-dB drop in the bass represents a greater decrease in loudness level (phons) than in the treble, so equal reverberation times at all frequencies would make it seem to your ears as if the bass had dropped out too quickly.

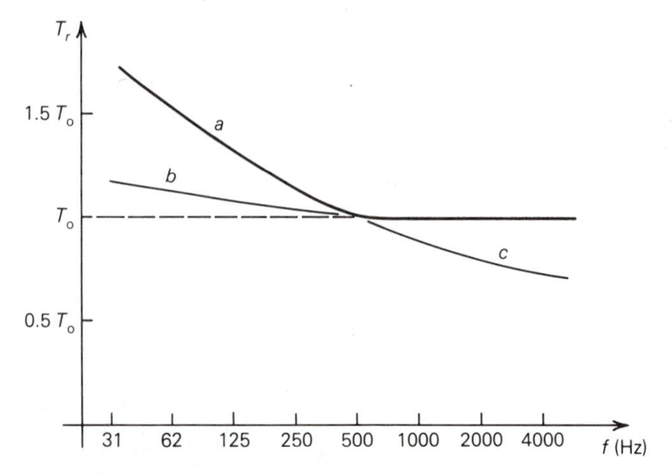

FIGURE 15.11 Dependence of reverberation time on frequency. Curve *a*: Most desirable is an approximately constant T_0 (determined from Figure 15.10) for all frequencies above 500 Hz, with substantially longer reverberation for low frequencies. Curve *b*: Too short reverberation at low frequencies makes the room lack warmth. Curve *c*: If high-frequency reverberation is too weak, brilliance is lacking.

15.3 REVERBERATION CALCULATION

Let us consider now the precise relation between reverberation time and room size and materials. The reverberation time is the result of competition between two factors: The more the sound can travel in the open air where practically no energy is lost, the longer it will last; but the more often it hits any solid surface and the softer that surface, the sooner it will reach 60 dB below its initial level. Larger room volume means more air to travel in and longer reverberation time; more surface area and softer material covering that area mean shorter reverberation time.

This is summarized in a simple formula first used by Wallace Sabine, who laid the foundations of scientific room acoustics in a series of studies at Harvard around 1900. Sabine's formula says

$$T_r = (0.16\ \mathrm{s/m})\, V/S_e,$$

where T_r is the reverberation time in seconds, V the room volume in m^3, and S_e the *effective absorption area* in m^2. This, in turn, is calculated from

$$S_e = \alpha_1 S_1 + \alpha_2 S_2 + \alpha_3 S_3 + \ldots,$$

which says that each different surface area in the room (for example, S_1 for ceiling, S_2 for carpet on the floor, etc.) contributes to the total absorption of sound in proportion to the *absorptivity* of its material. That absorptivity (or *absorption coefficient*) is represented by α (Greek alpha).

---

*Technically, these α's represent a combination of true absorption (sound energy converted to heat) and transmission through the wall to adjoining rooms; but we ignore this distinction, because we are presently concerned only with the fact that both represent loss of sound energy from the room interior. In very large auditoriums and for frequencies above 1 KHz, S_e may have a significant volume-proportional contribution from absorption in the air. The formula above is only an approximation, tending to overestimate T_r, and is less accurate when the absorptivities are large. A better approximation is $T_r = 0.16V/S_e\,(1 + \frac{1}{2}\bar{\alpha})$, where the average absorptivity is $\bar{\alpha} = S_e/S_t$ and S_t is the total surface area.

In older books you often will find the numerical coefficient 0.049 instead of 0.16; that is because they were measuring in feet instead of meters. For those who wonder where the number 0.16 s/m comes from, it is $(4/v)\ln(10^6)$, where v is the speed of sound and ln the natural logarithm. The 10^6 is evidence that 60 dB is part of the definition. One m^2 of effective absorption area sometimes is called a *sabine* unit. Unfortunately, other sources use sabine for 1 ft^2, so we shall try to prevent confusion by avoiding the term altogether.

A perfect reflector has $\alpha = 0$ and contributes nothing to S_e, because it removes no energy from the reverberant sound. A perfect absorber has $\alpha = 1$ and reflects nothing back into the room. All real materials have absorptivity somewhere between 0 and 1, because it represents the *fraction of all incident sound energy that is lost* on each reflection. An area of 10 m^2 covered by material with $\alpha = 0.4$, for instance, has the same acoustical effect as if there were 4 m^2 of perfect absorber and 6 m^2 of perfect reflector; it contributes 4 m^2 to the total effective absorption area S_e. One example of a perfect absorber (as far as the room interior is concerned) is an open window; any sound falling on it goes right on through and never comes back. An "equivalent open window area" means the size window that would remove as much sound from the room as the surface in question (in the example above, 4 m^2).

Let us do an easy sample calculation. Suppose there is an empty rectangular room 6 m high, 10 m wide, and 20 m long; let its walls and ceiling all be made of the same material with $\alpha = 0.1$ and its floor with $\alpha = 0.3$. Then the volume of the room is $V = 6 \times 10 \times 20 = 1200\ m^3$, the floor area is $10 \times 20 = 200\ m^2$, the ceiling also is 200 m^2, each side wall is $6 \times 20 = 120\ m^2$, and each end wall is $6 \times 10 = 60\ m^2$. The total wall and ceiling area is 560 m^2, so the effective absorption area is $S_e = 0.1 \times 560 + 0.3 \times 200 = 116\ m^2$. Then the predicted reverberation time is $T_r = 0.16 \times 1200/116 \cong 1.7$ s.

For more realistic problems we need a list of absorption coefficients for common construction materials. Typical values are given in Table 15.1. You will notice that smooth and rigid materials reflect most of the incident sound, while porous or yielding materials absorb more energy from it, because their impedance is not as different from that of the air. We must emphasize that any calculations with these values are only rough estimates, because there obviously will be wide variations in individual samples of carpet, or in the thickness and mounting rigidity of plywood or plasterboard sheeting.

TABLE 15.1 Approximate typical absorption coefficients of various surfaces. Individual examples may vary considerably from these values.

Surface Treatment	Absorptivity at Frequency					
	125	250	500	1000	2000	4000
Acoustic tile, rigidly mounted	.2	.4	.7	.8	.6	.4
Acoustic tile, suspended in frames	.5	.7	.6	.7	.7	.5
Acoustical plaster	.1	.2	.5	.6	.7	.7
Ordinary plaster, on lath	.2	.15	.1	.05	.04	.05
Gypsum wallboard, $\frac{1}{2}''$ on studs	.3	.1	.05	.04	.07	.1
Plywood sheet, $\frac{1}{4}''$ on studs	.6	.3	.1	.1	.1	.1
Concrete block, unpainted	.4	.4	.3	.3	.4	.3
Concrete block, painted	.1	.05	.06	.07	.1	.1
Concrete, poured	.01	.01	.02	.02	.02	.03
Brick	.03	.03	.03	.04	.05	.07
Vinyl tile, on concrete	.02	.03	.03	.03	.03	.02
Heavy carpet, on concrete	.02	.06	.15	.4	.6	.6
Heavy carpet, on felt backing	.1	.3	.4	.5	.6	.7
Platform floor, wooden	.4	.3	.2	.2	.15	.1
Ordinary window glass	.3	.2	.2	.1	.07	.04
Heavy plate glass	.2	.06	.04	.03	.02	.02
Draperies, medium velour	.07	.3	.5	.7	.7	.6
Upholstered seating, unoccupied	.2	.4	.6	.7	.6	.6
Upholstered seating, occupied	.4	.6	.8	.9	.9	.9
Wood/metal seating, unoccupied	.02	.03	.03	.06	.06	.05
Wooden pews, occupied	.4	.4	.7	.7	.8	.7

SOURCES: Backus (p. 172) and L. Doelle, *Environmental Acoustics* (McGraw-Hill, 1972), p. 227.

Table 15.1 expresses absorption coefficients for seating areas also as fractions (m^2 of effective absorption area per m^2 of actual area). In some sources you will find these presented instead as m^2 (or ft^2) of effective absorption area per person or per seat. If you should need to convert from one to the other, a range of reasonable seating densities is 1.5–2.0 people per m^2. Upholstered seats that absorb approximately the same fraction as a person (including clothing) are desirable so that the room acoustics will remain nearly the same regardless of whether the seats are empty or full. Then conditions will not change drastically between rehearsal and performance.

Now let us take a more interesting example. Suppose you are on a committee responsible for planning a small church building with dimensions as shown in Figure 15.12. A preliminary plan proposes wooden pews to seat some 200 people, tile floors, plastered end walls, plywood-sheeting side walls, and an open-beam wooden ceiling. Before investing too much money, we want to predict how the acoustics will turn out.

First we compute the volume by imagining it separated into two parts (dashed line, Figure 15.12b). The lower volume is rectangular, so $V_1 = 5 \times 12 \times 25 = 1500\ m^3$. The upper part has the shape of a triangular prism, and its

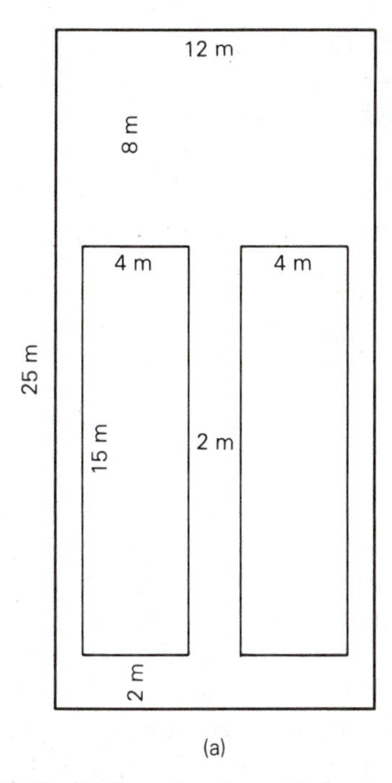

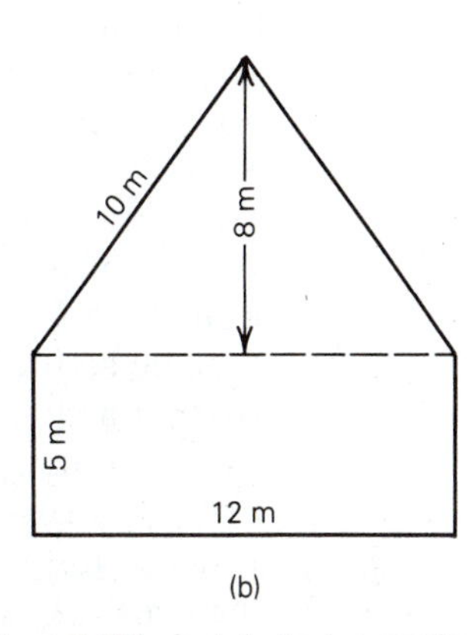

FIGURE 15.12 (a) Floor plan of a small church with pews to seat 200 people in the two smaller rectangles. For simplicity, there is no raised platform. (b) End wall.

volume is just half that of a rectangular solid of the same height and base: $V_2 = \frac{1}{2} \times 8 \times 12 \times 25 = 1200\ \text{m}^3$. So the total volume is $V = 2700\ \text{m}^3$. Each end wall similarly has area $5 \times 12 + \frac{1}{2} \times 8 \times 12 = 108\ \text{m}^2$. Each half of the ceiling has $10 \times 25 = 250\ \text{m}^2$. Each side wall has $5 \times 25\ \text{m}^2 = 125\ \text{m}^2$, out of which we will take $10\ \text{m}^2$ for windows. The total floor area is $12 \times 25 = 300\ \text{m}^2$, out of which $2 \times 4 \times 15 = 120\ \text{m}^2$ is occupied by pews.

We make a work sheet (Table 15.2) to find the total effective absorption area, using first the absorptivities for $f = 1000$ Hz. It is typical of real life that

TABLE 15.2 Calculation work sheet for reverberation time $f = 1000$ Hz

Surface	S	α	αS
Pews (occupied)	120	.7	84
Tile floor	180	.03	5
Windows	20	.1	2
Side walls	230	.1	23
End walls	216	.05	11
Ceiling	500	.2	100
		Total:	$S_e = \overline{225}\ \text{m}^2$

$T_r = 0.16 \times 2700/225 = 1.92$ s

we cannot find an entry in Table 15.1 corresponding exactly to every type of surface in the room; we must guess that this kind of ceiling structure might absorb sound in somewhat the same way as a wooden platform floor. Because of the uncertainties about correct values for the α's, we must allow that our answer may well be off by several tenths of a second. While the estimated 1.9 seconds may please the organist, it may seem too long for the spoken parts of the church service.

But before blithely throwing in more absorptive materials such as acoustical plaster or tile, we should check out what happens at other frequencies. A similar calculation for $f = 125$ Hz indicates approximately only 1.0 s reverberation time; this room already is very dead in the bass. To remedy that we would want to make sure the roof construction is extremely solid to keep its absorptivity at 125 Hz down to 0.3 or even 0.2 instead of 0.4. We also would suggest increasing the thickness of the plywood sheets for the side walls from $\frac{1}{4}''$ to $\frac{1}{2}''$ or more, or mounting them on studs at only half the ordinary spacing so they would not be so yielding at low frequencies; or, even better, we might change to brick walls if possible. Once these basic changes in construction have made the room reasonably live for low frequencies, the high-frequency reverberation can be brought down by adding a modest amount of acoustical plaster or tile, choosing the type carefully for minimal effect on the bass.

More often than not, we are concerned with already existing buildings. Then we need not rely on attempting to predict the reverberation time; we actually can measure it. Even though we may not be sure of the absorptivity for each separate material, we can use a measured reverberation time to find the total effective absorption by turning Sabine's formula around: $S_e = 0.16\ V/T_r$. For instance, suppose the church described above finally is built with modifications and actually has a reverberation time at 1000 Hz of 1.5 seconds when occupied. Then, even though we may not know precisely how much absorption is due to which surface, we at least know that the total is $S_e = 0.16 \times 2700/1.5 = 288\ \text{m}^2$.

That enables us to say what would happen if we adopt a suggestion that the aisles and nave be carpeted. A change of α from .03 to .4 for this 180 m^2 area would increase its contribution to S_e from 5 m^2 to 72 m^2. The new S_e of $288 + 67 = 355\ \text{m}^2$ would reduce the reverberation time to $0.16 \times 2700/355 = 1.2$ s, short enough that the organist is sure to complain. The debate over whether to have carpets is a perennial one afflicting many American churches.

Here is a nicely practical piece of advice. It is relatively easy to design a room (and its furnishings) with somewhat excessive reverberation and then add a bit of carpeting, draperies, or acoustical tile later to reduce the reverberation time. It is much more difficult to take an originally dead room and find inexpensive ways to liven it up! If you are ever in a position to influence the construction or furnishing of a church or auditorium, try to have final installation of absorptive materials delayed until after the acoustics are tested by full-scale musical performances. You often will find that you do not wish to deaden the room as much as would result from carrying out original plans.

Let us add this reminder: A correct reverberation time alone does not ensure satisfactory acoustics. All the other criteria of Section 15.1 also must be met. Auditorium design still is as much an art as a science, and even the best architects sometimes encounter difficulties.

*15.4 REVERBERANT SOUND LEVELS

When a steady sound source operates long enough for the reverberant sound to approach a constant level (as in Figure 15.3), how high is that level? It depends on the source strength, of course, but it also depends on the room. The longer the reverberation time, the greater the number and combined strength of the reflected waves and the higher the overall level of a sustained sound.

We should point out that a uniform sound level is not to be expected for a pure sine wave source. Those few room modes responding strongly to such a source combine in a complicated standing-wave pattern, as in Figure 15.13a. Only if a wide range of frequencies is present can the sound energy be distributed uniformly as in Figure 15.13b. Even then, all modes must have pressure antinodes at the walls, so the sound level is as much as 3 dB higher there than in the rest of the room.

Close to the source the direct sound predominates; the sound level is about what it would be for the same source outdoors, and it becomes weaker as you move away from the source. Only beyond a distance called the reverberation radius ($R_r \simeq 0.06\sqrt{V/T_r}$) does the reverberant sound predominate (Figure 15.13b).

> *Here is a case in which sound intensity level *SIL* is ill-defined because waves travel in all different directions, and we should recognize that what our meters actually measure is sound pressure level *SPL* (Section 5.2). We shall gloss this over in the following paragraphs by using I_{rev} to mean the intensity of a unidirectional traveling wave that would have the same *SPL* as the actual diffuse sound field.

Now we can rephrase our original question more meaningfully. If we are not especially close to either the source or any wall, and if the sound includes a wide range of frequencies, what is the reverberant sound level? Consider the total power output P of the source in a room where the average absorptivity is $\bar{\alpha}$. A fraction $\bar{\alpha}P$ of the direct sound is lost when it first reaches the walls, and only the remaining $(1-\bar{\alpha})P$ becomes reverberant sound. If the equivalent average intensity is I_{rev}, in the steady state we expect the reverberant power

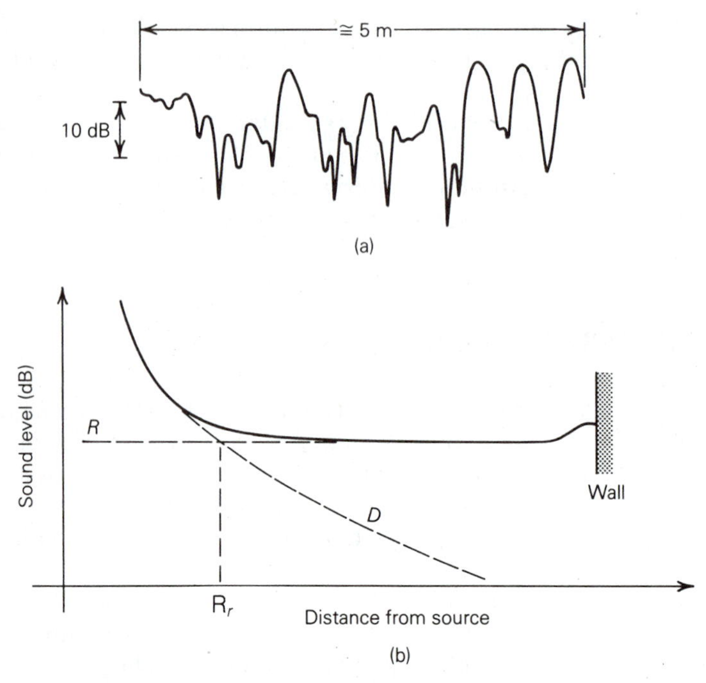

FIGURE 15.13 (a) Measured steady sound pressure level of a steady 1000 Hz sine wave in a small lecture room as the microphone moves slowly along a line. Each different frequency would produce a similarly complicated standing-wave pattern with its details completely different from this one (From Kuttruff, p. 70). (b) Steady sound pressure level averaged over a wide range of frequencies. Dashed lines indicate contribution of direct (*D*) and reverberant (*R*) components. The enhancement near the wall ranges from zero for a perfectly absorbing surface to a maximum of 3 dB for a perfectly reflecting surface.

input $(1-\bar{\alpha})P$ from the source to match a total energy loss rate $(\frac{1}{6})I_{\text{rev}}S_e$ at the walls. (The factor $\frac{1}{6}$ appears because sound travels east, west, north, south, up, and down; yet for a point close to a wall only one of these six is carrying energy into the absorber.) This suggests that we must have $I_{\text{rev}} = 6(1-\bar{\alpha})P/S_e$, or equivalently $I_{\text{rev}} \cong 6PT_r/0.16V = 36PT_r/V$. The approximation is good if $\bar{\alpha}$ is small, which is true whenever there is a smooth, extended reverberation.

For a given room geometry, the reverberant sound level depends on wall absorption as shown in Figure 15.14. For example, a source with $P = 0.1$ W in a room with $S_e = 200\ \text{m}^2$ (such as for $V = 1200\ \text{m}^3$, $T_r = 1.0$ s) generates $I_{\text{rev}} \cong 3 \times 10^{-3}\ \text{W/m}^2$; that is, a sound level of 95 dB. If the absorption were cut in half so that $S_e = 100\ \text{m}^2$ in this room, I_{rev} would rise to 98 dB (and T_r to 2.0 s). The same source in a larger room with $S_e = 2000\ \text{m}^2$ (such as for $V = 18{,}000\ \text{m}^3$, $T_r = 1.5$ s) produces approximately only $3 \times 10^{-4}\ \text{W/m}^2$, or 85 dB, of reverberant sound.

In quiet conversation a human voice may produce only 10^{-5} W; but in loud speech this can be raised to 10^{-3} W. Measured maximum total power outputs have been reported for various musical instruments; these range from around 0.05 W for woodwinds through 0.4 W for piano and 6 W for trombone, and up to 60 W or more for a full orchestra.

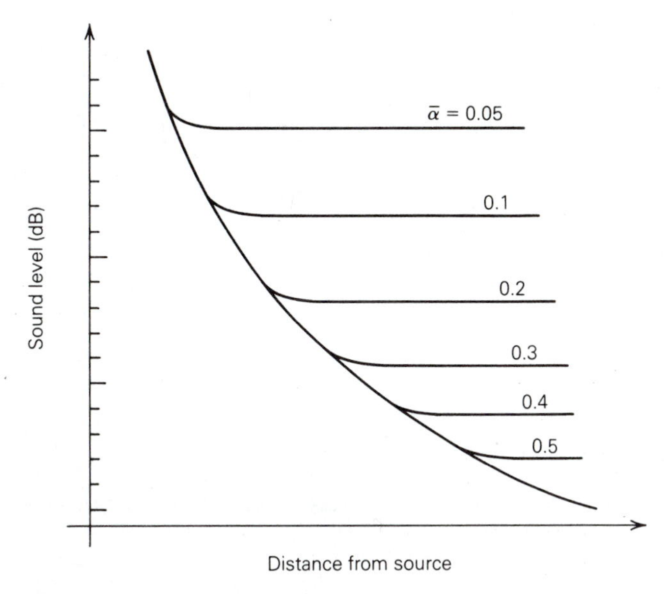

FIGURE 15.14 Increasing the average absorptivity $\bar{\alpha}$ of a room causes a decrease in reverberant sound level. It also increases the size of the region near the source where direct sound is louder than reverberant. (Compare with Figure 15.13b.)

15.5 SOUND REINFORCEMENT

Satisfactory listening levels for speech or small musical groups in large auditoriums may require artificial reinforcement. Sound picked up by a microphone close to the source is amplified electronically and broadcast through one or more loudspeakers. All components should meet high-fidelity standards in order not to change the musical qualities of the original sounds. Whether this is successful depends a great deal on judicious placement of the loudspeakers and adjustment of the amount of amplification.

One common problem we can readily understand is *ringing*. This is most likely to happen if a loudspeaker is behind the microphone, as in Figure 15.15a. It takes only a modest amount of amplification to make the direct signal from the loudspeaker even louder when it reaches the microphone than was the original voice or musical sound. The microphone picks this up, amplifies it again, and causes a still louder signal to arrive a very short time later. Here is a very undesirable form of *positive feedback*; the resulting *instability* generates a self-sustaining, ear-splitting howl that will not stop until the amplifier is turned down. Even when not quite strong enough to cause instability, positive feedback still may produce serious distortion in the amplified signal.

Location of the loudspeaker in front of the microphone (Figure 15.15b) is much safer, for now it is primarily reflected loudspeaker sounds (weaker than the direct sound) that reach the microphone. Even so, these still can lead to

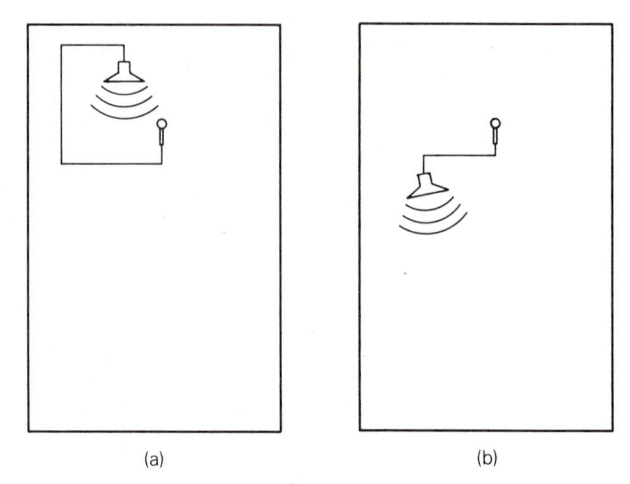

FIGURE 15.15 (a) Loudspeaker placement behind a microphone sends a strong feedback signal that is likely to cause ringing. (b) Loudspeaker placement in front of the microphone reduces the danger of ringing.

instability and ringing if the amplification is set too high. Other strategies to avoid ringing include keeping the original source as close to the microphone as possible (so that less amplification is needed) and using more highly directional microphone and speakers. Sufficient directionality even may allow loudspeakers to be placed behind the microphone, as long as they are somewhat off to one side or the other.

Now let us consider whether it seems to the listener that the sound comes from the original source or from the loudspeaker. It is certainly more pleasant if your eyes and ears both derive the same message about source location. It appears at first sight as if this would be difficult to achieve, because the loudspeaker signal typically is louder than the original signal. A conservative way to avoid any problem is to mount the loudspeaker close to the source (as is sometimes done in the front panel of a speaking desk) so that both signals come from practically the same place.

Another useful strategy takes advantage of the *precedence effect* to fool the listener's ears; that is, whichever component of an extended sound arrives first predominates in the judgment of source location. Ordinarily this simply means that we perceive a sound source as being in the direction from which we receive the direct sound rather than any of the other directions from which we receive early reflections.

But electrical signals travel much faster than sound, so a loudspeaker broadcasts its contribution *before* the corresponding direct sound traveling through the air reaches its location. When a loudspeaker signal is added, any of the following might happen: (1) If the source-to-listener and loudspeaker-to-listener distances differ by much more than 10 m, then the arrival times differ by more than 30 ms and there will seem to be an echo, which certainly must be avoided; (2) if the loudspeaker is closer (we now assume less than 10 m closer) it will seem as if that is where the entire sound originated; (3) if

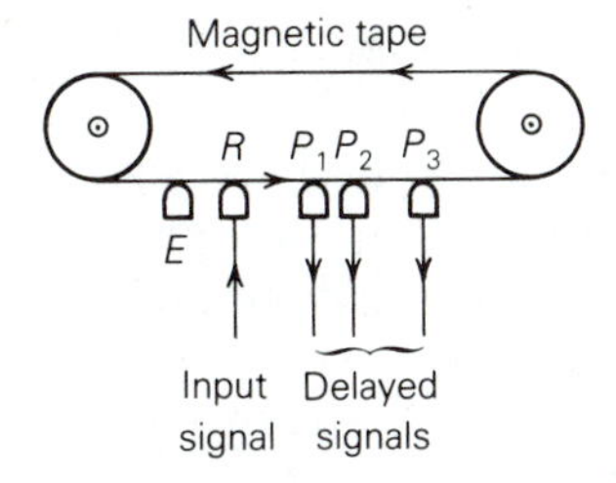

FIGURE 15.16 One of several ways of delaying acoustic signals. A tape loop runs continually past an erasing head *E*, recording head *R*, and several pickup heads *P* which supply speaker signals. (From Kuttruff)

the source is closer than the loudspeaker so that its signal arrives first, it will seem as if the entire sound comes from the original source. It is especially remarkable that this remains true even if the first direct sound from the source is as much as 10 dB weaker than the loudspeaker signal! As long as the reinforcement arrives within approximately 30 ms *after* the direct signal, it only strengthens it without changing the apparent place of origin.

Several strategies can be used to keep the loudspeaker signal later than the direct signal: (1) If the speaker and microphone are quite directional, place the speaker a bit behind the microphone; (2) put the speaker in front of the microphone, but far enough to the side or high enough that the sound path from the loudspeaker is longer than the direct path; or (3) use a delaying device (Figure 15.16) to feed the signal to the loudspeaker only after enough time has passed that this signal will arrive second even though the loudspeaker-to-listener path is shorter. In recent years, tape delays have been supplanted by digital delay systems, which are now both cheap and reliable. They offer the best solution in many large auditoriums where part of the audience is unavoidably served by loudspeakers much closer than the stage (Figure 15.17).

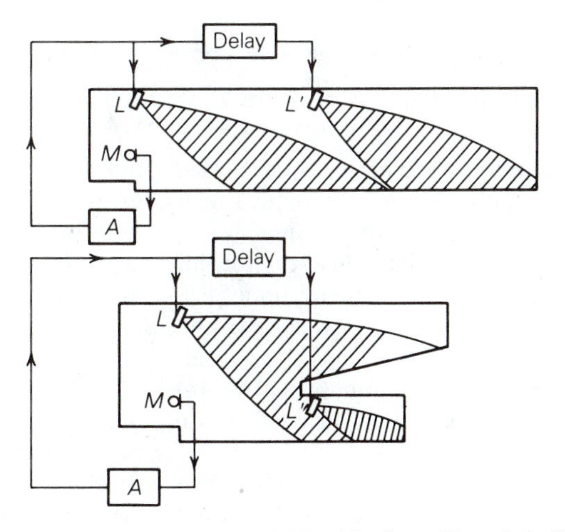

FIGURE 15.17 Two situations calling for speakers fed by delay lines. (From Kuttruff, p. 278)

Notice the opportunity for a clever trick here: Directional loudspeakers can aim their sound preferentially into a highly absorbent region; namely, the audience. The reinforced sound then does not get a fair chance to visit other, harder surfaces in the room. Reinforced speech can have a weaker effective reverberant component (and correspondingly higher intelligibility) than unamplified musical signals in the same room. Similar effects can be achieved by amplifying bass frequencies less than the rest, because they are less important for speech intelligibility.

15.6 SPATIAL PERCEPTION

For deeper understanding of certain problems in room acoustics and sound reinforcement, it is worth commenting further on the subject of *sound localization,* or how we arrive at judgments of the distance and direction from which a sound comes.

You might suppose at first that you judge source *distance* from how loud or soft the sound is when it reaches you. But overall loudness in fact is a clue of limited usefulness, because it requires some prior knowledge of source strength. Although hardly anyone is conscious of using such a trick, human ears can judge quite nicely the *relative* strength of direct sound as compared with early reflections. Making certain allowances for the size of room, relatively strong direct sound tells you the source is close and relatively strong reflections tell you it is farther away. This is beautifully illustrated by a demonstration we often conduct in our anechoic chamber (Box 15.2), where all sound reflections are reduced to a very low level. When I turn out the lights in this chamber so that there are no visual cues and then talk for a couple of minutes to people on the opposite side of the room, they often think

BOX 15.2 ANECHOIC CHAMBERS AND REVERBERATION CHAMBERS

In many engineering or scientific experiments, reverberant sound only confuses the issue. If you want to accurately measure the frequency response of a high-fidelity loudspeaker, for instance, you need to have your microphone pick up only direct sound from the speaker. If any reflected sound is included, as it will be in an ordinary room, your measurements will represent a mixture of room characteristics with those of the speaker.

So sensitive acoustical measurements often are taken in a room completely lined with material of absorptivity as high as 0.95, such as large fiberglass wedges. The chamber at the university in Sacramento (Figure 15.18) has been used not only for scientific experiments by students, but also by loudspeaker manufacturers to test new designs and by the California Highway Patrol to certify the performance of sirens.

The opposite extreme would be a room with highly reflective walls. Such reverberation chambers are useful for scientific measurements of power output from sound sources or of absorptivities of various materials, as well as for adding artificial reverberation to recorded music.

(*continued*)

BOX 15.2 *(continued)*

FIGURE 15.18 The author and a student inside the Sacramento anechoic chamber. (Photo courtesy of Sam Parsons.)

I have walked over close to them when in fact I have remained some 5 meters away. Their brains try to interpret the predominance of direct over reflected sound in the same way that would work well in an ordinary room with its stronger reflections.

Directional perception also is complicated by our use of at least four different complementary mechanisms. First is the detection of *onset times*, which is especially relevant to our previous remarks about judging direction from the first-arriving sound according to the precedence effect. This applies only to transient sounds, mainly those of duration less than approximately 100 ms, or to the transient attack portion of sustained notes. Because your two ears are separated by some 15 cm (measured straight through the head), sound from a source on your right or left may travel as much as 19 cm farther to get around the head to the far ear. So it may arrive there as much as 0.6 ms later than at the near ear. Your brain can judge this accurately enough to *lateralize* the sound (judge how much toward right or left) within a few degrees; but you get no information at all about location in the *median plane* (front versus back, above or below) or, more generally, on the *cone of confusion* (Figure 15.19).

A second method is the detection of *phase differences* between continuous signals arriving at both ears (Figure 15.20). The phase difference is another form of time lag resulting from one ear being farther from the source than the other. Phase detection works mainly below approximately 1500 Hz, because for higher frequencies the auditory nerve fibers cannot fire fast enough to retain phase information. Furthermore, this method would be ambiguous at higher frequencies, because lags exceeding one full cycle would become possible. Phase detection, too, leads only to lateralization, not to any discrimination in the median plane.

A third method is the detection of interaural *intensity differences*, which can

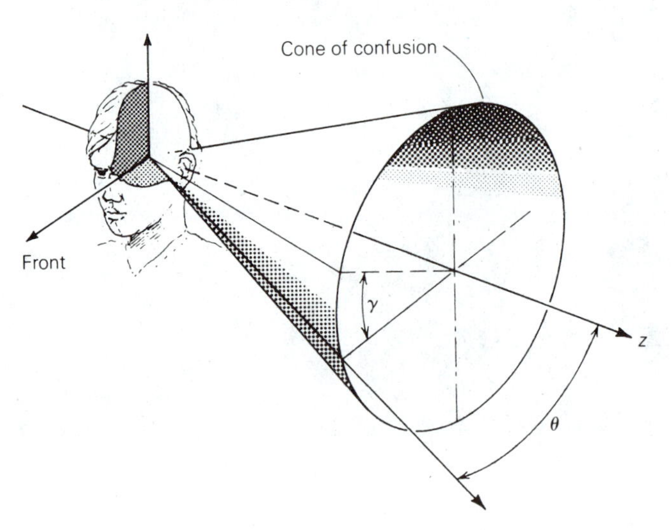

FIGURE 15.19 Pure lateralization determines only the angle θ; unambiguous localization requires determining γ as well.

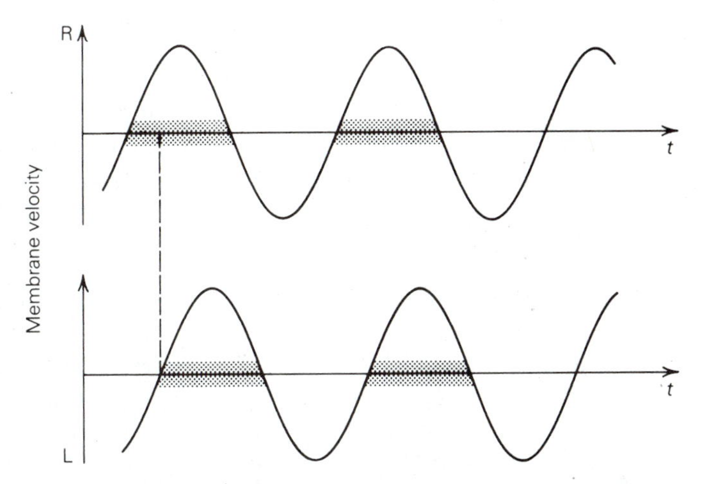

FIGURE 15.20 Sinusoidal disturbance of both ears by the same source, located closer to the right ear than to the left. For low frequencies, nerve firings occur preferentially during upward motion of the basilar membrane (shaded on time axes) and preserve phase information for further processing in the brain.

operate with either transient or continuous sounds. It is effective mainly for high frequencies—say, above 2 or 3 KHz—because sounds with wavelength much larger than your head diffract around it so well that intensities are practically the same at both ears regardless of source location. In its simplest form this also achieves only lateralization. These first three all are *binaural* detection mechanisms; that is, they depend on hearing with both ears.

A fourth mechanism is used to distinguish front from back. As we hinted in Exercise 2 of Chapter 6, the shape of the outer ear gives it a direction-dependent collecting efficiency for high frequencies, mainly those above 5 KHz. Thus the relative strengths of high-frequency components may seem to change somewhat if the source switches from front to back. This opens up possibilities (although only modest ones) for resolving the ambiguity of the cone of confusion.

Finally, you ordinarily enhance the usefulness of all four methods by moving your head around to see how the sound changes. If you hear only one brief sound, or if your head is held in a rigid frame, you are seldom sure of the direction from which it came. But if you have a chance to move your head while listening to several repetitions or to a continuous sound with complex timbre, you probably will be able to locate its source quite accurately.

Let us close with one more comment about speaker placement. The preceding paragraphs indicate that incorrect lateral placement is much more likely to cause problems than is location in the median plane. That is, if the listener normally faces the stage and does not move his or her head around too much, loudspeakers too far right or left may be distracting but it will hardly matter whether they are straight ahead on the stage or mounted on the ceiling directly overhead. In fact, you may even get by with close overhead speakers without delay lines in spite of the precedence effect, whereas speakers the same distance to right or left would be extremely annoying.

SUMMARY

A good auditorium for musical performance provides a clear direct path for the original sound to reach every listener, followed within approximately 30–50 ms by several early reflections from both sides as well as above and then by smoothly decaying reverberation lasting around 1.5–2.0 s depending on the type of music. The room should be free from noticeable individual echoes and from extraneous noise.

Although several other room characteristics are at least equally important, reverberation time is the one most amenable to precise measurement and calculation. It is the length of time for reverberant sound to decay 60 dB below its original level, and is related to room volume and effective absorption area by Sabine's approximate formula $T_r = 0.16V/S_e$. It is desirable for bass frequencies to reverberate substantially longer than treble. Useful estimates can be made with this formula of probable effects of proposed changes in room furnishings, such as addition or removal of carpeting.

Large auditoriums (as well as smaller ones with poor acoustical design) require sound reinforcement for speech or (less often) chamber music. Careful placement of loudspeakers and conservative use of amplifier gain will keep the feedback from loudspeaker to microphone weak enough to avoid ringing. Properly placed loudspeakers are unobtrusive and take advantage of precedence effect to strengthen the sound while still leaving the listener's perceived image of source location undisturbed. Interaural differences in arrival time, phase, and amplitude all provide prime clues for lateralization. Full localization of a source depends mostly on head movement, although pinna directionality also provides some help.

REFERENCES

Room acoustics is treated by Backus in Chapter 9 and by Benade in Chapters 11 and 12. There are some interesting and readable remarks on auditorium design, and particularly on noise suppression, in Strong and Plitnik, Chapter 4. A good complete review of the physics of this field is *Room Acoustics*, 2nd ed., by Heinrich Kuttruff (London: Applied Science, 1979). The relation of various physical characteristics of sound fields to our subjective preferences is discussed by Y. Ando, *Concert Hall Acoustics* (New York: Springer, 1985).

You will find an enjoyable and informative article about the checkered history of New York's Philharmonic Hall (now Avery Fisher Hall) in Lincoln Center by Bruce Bliven in *The New Yorker* magazine, November 8, 1976, p. 51; I highly recommend this. There is a general introductory article on "Architectural Acoustics" by Vern Knudsen in *Scientific American*, November 1963; it is noteworthy for interesting comments on the Mormon Tabernacle in Salt Lake City, and is reproduced in the Hutchins collection. A classic study of the acoustical characteristics of many of the best auditoriums in both Europe and America was made by Leo Beranek in *Music, Acoustics and Architecture*

(Wiley, 1962). A more recent survey of the acoustics of historic auditoriums and one that is excellent reading is *Buildings for Music* by Michael Forsyth (MIT Press, 1985).

A readable study of local reverberation onstage for good performer ensemble is given by R. S. Shankland in *JASA, 65*, 140 (1979). Measurements supporting similar conclusions are reported by A. H. Marshall et al. in *JASA, 64*, 1437 (1978).

For careful examination of the history and limits of validity of Sabine's formula for reverberation time, see W. B. Joyce, *JASA, 58*, 643 (1975) and *64*, 1429 (1978).

A review article on auditory localization by A. W. Mills appears in *Foundations of Modern Auditory Theory*, edited by J. V. Tobias (Academic Press, 1972), Vol. 2, p. 301. This subject is treated at greater length by Jens Blauert in *Spatial Hearing* (MIT Press, 1983).

SYMBOLS, TERMS, AND RELATIONS

T_r 60-dB reverberation time
V enclosed room volume
S surface area
S_e effective absorption area
direct sound
early reflections
reverberant sound
flutter echo
α absorptivity

$\bar{\alpha}$ average absorptivity of all room surfaces
P total sound power output
I_{rev} equivalent intensity of reverberant sound
$T_r \cong 0.16V/S_e$
$S_e = \alpha_1 S_1 + \alpha_2 S_2 + \alpha_3 S_3 + \cdots$
$I_{\text{rev}} \cong 36PT_r/V$
ringing and positive feedback

precedence effect
localization
lateralization
cone of confusion
binaural arrival-time differences
binaural phase differences
binaural intensity differences

EXERCISES

1. Churches often have lights mounted in hanging lanterns that are six-sided and approximately 30 cm across and 50 cm tall. How effective are these in diffusing sound for frequencies (a) much less than, (b) much greater than, and (c) in the vicinity of 1000 Hz?

2. Explain why ceilings in newer auditoriums commonly consist of several large panels slanting different ways to make a wrinkled or corrugated pattern, instead of just being entirely flat.

3. In the room shown in Figure 15.21, How long does it take for (a) the direct sound, (b) the ceiling reflection, and (c) the back wall reflection to get from source S to listener L? Discuss the acceptability of these delays.

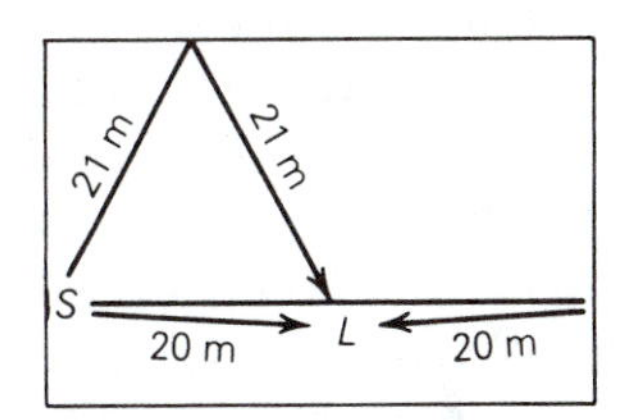

FIGURE 15.21

4. If you stand at one side of a stage with hard parallel side walls 17 m apart and

make a handclap, how often do the flutter echoes arrive?

5. Suppose city traffic may generate sound levels as high as 90 dB outside an auditorium. The building should cause an *attenuation* or reduction of outside noise to a much lower level inside. Recommend how many dB of attenuation should be assured in construction of this building. What percentage of the noise energy are you requiring to be excluded? (Hint: It is far more than 99%.)

*6. In a room with $V = 10^4$ m^3 and $T_r = 2$ s, approximately how many modes are strongly excited by a signal at (a) 100 Hz and (b) 1000 Hz? What about a room with $V = 10^2$ m^3 and $T_r = 0.5$ s? For which room and which frequency range is there some danger of uneven response? (See Benade, p. 196, for further comments along this line.)

7. Read a short passage aloud to estimate a typical speaking rate in terms of (a) syllables per second and (b) phonemes per second. Suppose intelligibility requires each syllable (or phoneme) to drop 10 dB by the time the next one comes along; that is, that these time intervals should be at least $\frac{1}{6} T_r$. Use this criterion with your estimated syllable and phoneme rates to make two estimates of the maximum acceptable T_r for speech. Discuss whether you think it more appropriate to apply such a criterion to syllables or to phonemes.

8. What T_r would you recommend for a Beethoven symphony in a hall with $V = 5000$ m^3? What if $V = 30{,}000$ m^3?

9. What T_r would be desirable for the sermon in a church with $V = 10^4$ m^3? What T_r for the music? What kind of music do you have in mind? What T_r would you recommend as a compromise?

*10. For the church example discussed in the text, what T_r's would you get (instead of 1.9 s at 1 KHz and 1.0 s at 125 Hz) if you used the alternative formula including $(1 + \frac{1}{2}\bar{\alpha})$?

11. Explain why the absorptivity for $\frac{1}{4}''$ and $\frac{1}{2}''$ plywood is practically the same at high frequencies but quite different at very low frequencies. Put this in the context of a discussion of any thickness in general, including the extreme cases of very thick and paper-thin. Think about the ability of the plywood to yield to the sound pressure.

12. Let a living room be 5×7 m, with a ceiling 3 m high; the floor is covered with padded carpet and the ceiling with acoustical plaster; the rest is gypsum wallboard except for two open doorways 1 m wide. Calculate an expected reverberation time at 1000 Hz. How satisfactory will this be for (a) conversation, (b) playing live chamber music, and (c) listening to recorded symphonic music? Part (c) is tricky, for you must remember that the recording already includes the reverberation of the auditorium where the original performance took place.

13. A small recital hall is basically rectangular (20 m long, 10 m wide, 5 m high), but with a stage 5 m deep raised 1 m above the rest of the floor. Let all vertical walls (including the stage front) be plywood on studs, the ceiling plaster on lath, the main floor vinyl tile on concrete, and the stage floor hardwood. Predict the reverberation time at 500 Hz for the empty unfurnished hall and compare with what would be desirable. What would your answer become if the ceiling were changed to acoustical plaster? Estimate a reasonable number of people to put in the 10×15 area, and let them have upholstered seats. With the audience in

place, what will T_r be for each ceiling type? Which ceiling do you recommend?

14. If every inner surface of a hollow cube 10 m on a side had the same absorptivity 0.1, what would be the reverberation time? What if every surface had absorptivity 0.5? If two opposite faces have $\alpha = 0.5$ while the other four have $\alpha = 0.1$, what T_r does the usual formula predict? But what in fact will be the nature of the reverberation?

15. If a $30 \times 20 \times 8$ m room has $T_r = 1.2$ s, what is S_e? Is this reasonable when compared with the total actual surface area?

16. You measure a reverberation time of 2.0 s in an empty church with $V = 5000\ \text{m}^3$. What is its S_e? How much will S_e increase, and what will T_r become, when 250 m^2 of wooden pews are fully occupied? (Suppose the relevant frequency is 500 Hz.)

*17. A certain band rehearsal room with $V = 2000\ \text{m}^3$ and $T_r = 2$ s at 500 Hz is annoyingly live. Estimate how much plastered wall area must be covered with draperies to bring T_r down to 1 s. How many dB will this reduce the reverberant sound level?

*18. In a room $5 \times 5 \times 3$ m with $T_r = 0.5$ s, how far must you be from the source for reverberant sound to dominate over direct? How far in a room $50 \times 40 \times 25$ m with $T_r = 2.5$ s? Is it usually fair to consider most of a room as characterized by the reverberant level?

*19. An orchestra putting out 10 W in a $10^4\ \text{m}^3$ auditorium with $T_r = 2$ s can fill it with sound at what level in dB? How about a piano generating 0.1 W in the same auditorium? How about the same piano in a small practice room with $S_e = 30\ \text{m}^2$?

*20. Estimate the largest room in which a 10^{-4} W voice can provide a minimum 65 dB listening level without the aid of a PA system.

21. What difficulty will you have with a PA system with all omnidirectional components (that is, speakers that send out equally strong waves in all directions, and a microphone that picks up equally well from all directions)?

*22. From our ability to lateralize within a few degrees, estimate roughly the minimum interaural time lag our brains can detect.

23. Would you guess that intensity-difference lateralization is more or less useful to a bird than to a human? What difference would there be in useful frequency range for this mechanism?

24. A recital hall with volume 18,000 m^3 has reversible "acoustic panels" that can expose either a reflective or an absorptive side to the room. (a) With the hard side out, T_r of the room is 1.8 s; what total effective absorption area does the room have? (b) If the total area of the reversible panels is 200 m^2, what is the shortest T_r you could get just by turning the soft sides out, even with the most extreme imaginable choices of panel materials?

25. An empty hall with volume 10,000 m^3 has $T_r = 1.6$ s. Suppose it has 500 m^2 of seating area for which the absorptivity is 0.5 when empty, 0.7 when occupied by lightly dressed people, and 0.9 when occupied by people wearing heavy winter wraps. What is the effective absorption area of the empty hall? How much is it increased when people are present? So what T_r do you predict when the seats are filled with lightly dressed, or with bundled-up, people? Is it acoustically significant whether they check their coats in the lobby?

PROJECTS

1. Listen to several concerts in a local auditorium and judge it according to the seven suggested criteria. Critically examine its design to explain its good and bad points.
2. Compare the acoustic characteristics of several local auditoriums, and explain what factors may be responsible for the differences.
3. Measure dimensions, identify materials, and calculate a worksheet for reverberation-time prediction of some local auditorium. If possible, compare with an actual measurement.
4. Make a thorough study of PA system design and performance in some church or auditorium, with special attention to whether the speaker placement is successful.
5. Clap your hands to create echoes from a building wall, as suggested in Project 2 in Chapter 1. See how close you can stand to the wall before you can no longer perceive the echo separately from the original sound. Compare with maximum allowable first-reflection lags as described in the text.
6. Try a lateralization experiment with a friend. Take turns being experimenter (stealthily moving around and occasionally making some sharp noise such as a click or a clap) and subject (blindfolded, trying to point toward the location of each sound). (Studying lateralization with continuous sounds is trickier, because you must be sure they are turned on so smoothly that the location is not given away by the transient.) Is it any harder or easier to judge with the same degree of accuracy when the source is toward front or back than when it is well to one side?

CHAPTER

16 *Sound Reproduction

The cultural role of music has been revolutionized by twentieth-century technology. Previously, the only way to hear music was to be present at the time and place of performance. Now with radio, television, satellite communication, and sound recording, we can hear live performances from anywhere in the world or listen at our convenience to sounds recorded years before. Previously, the type and amount of music a person heard might be severely limited by geographic location or social class. Now we all have easy, almost unlimited access to music of every kind—access of a type no king could have commanded before.

We can make more intelligent use of our devices for recording and reproducing sound if we understand a little of the physical principles governing their operation, so we first present general perspectives on the interconversion between sound and other forms of vibrational energy. We then will take up one by one the usual components of an electronic recording and reproduction system—microphones for converting sound into electrical signals, amplifiers for increasing the strength of these signals, recorders for storing this information for later use, playback devices for recovering stored signals, and loudspeakers for reconverting electrical vibrations into sound. In the final section, we will consider the esthetic improvement that can be obtained with multiphonic reproduction.

It should be stressed that this chapter gives only a brief introductory background and makes no pretense of competing with other books devoted entirely to this topic. You should consult them for many practical details and advice on choosing hi-fi equipment (see the references at the end of this chapter).

16.1 ELECTRIC AND MAGNETIC CONCEPTS

Let us briefly review the terminology that we must use to talk about electric and magnetic devices. *Electrons* are those tiny elementary particles that orbit around the nucleus of each atom of the various chemical elements. The electrons are called negative, purely as a matter of convention, while the protons in the nuclei are positive. In ordinary matter the total number of electrons corresponds precisely to the total number of protons to maintain overall electrical neutrality. But in conducting materials (principally metals) it is sometimes possible for large numbers of electrons to migrate far away from their original positions. Whenever we speak of a *negative charge* in an electrical circuit, we mean an accumulation of extra electrons; a *positive charge* occurs where there is a shortage of electrons.

Electric **current** is the flow of charges from one place to another; we say there is one *ampere* of current flowing through a wire when approximately 6×10^{18} electrons per second pass a given point on that wire. **Voltage** (or electric potential, or electromotive force) is a measure of the electrical "push" trying to force current to flow from one place to another. Voltage difference between two points in an electric circuit is directly analogous to pressure difference between two points in a pipe carrying an ordinary fluid flow, and electric current is analogous to the volume flow rate of that fluid.

A *magnet* may be defined as an object that exerts a force upon any compass needle in its vicinity. (Actually, the compass needle itself is a small magnet.) The invisible influence that pervades the space around the magnet is called a *magnetic field*. The direction of a magnetic field at any point is shown by the orientation of a small compass needle placed there and its strength by the rapidity with which the needle vibrates when you tap the compass. The metallic elements iron, nickel, and cobalt have the ability to form *permanent magnets*.

A large, steady electric current through a coil of wire constitutes an *electromagnet*, whose external effects are indistinguishable from those of a permanent magnet (Figure 16.1). This is a clue that, despite the apparent lack of connection between lightning bolts and compass needles, electricity and magnetism are in fact intimately related. Permanent magnets actually consist of myriads of individual spinning electrons, each acting as a tiny magnet, oriented together to achieve a cooperative effect.

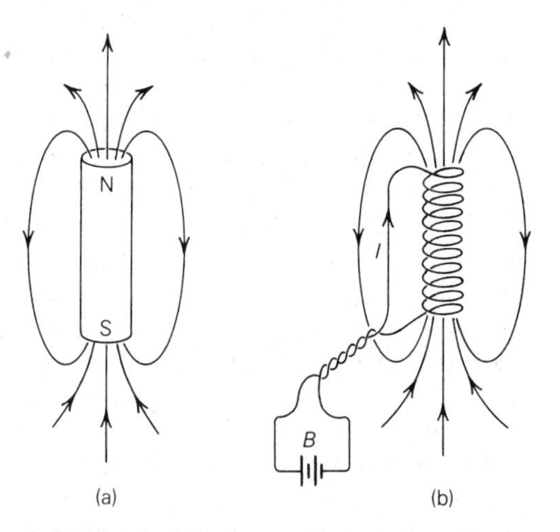

FIGURE 16.1 (a) Magnetic field lines in the space around a bar magnet. The end marked *N* will point north if this bar is mounted as a compass needle. (b) A similar magnetic field produced by a long coil of wire carrying electric current *I* furnished by battery *B*.

16.2 TRANSDUCERS

Rather than present a long list of individual devices with no clear connections among them, we will stress a few unifying principles that provide a framework for understanding each particular case.

The key to this approach is the concept of a **transducer**, meaning any device that converts a signal from one energy form to another. The same information can be contained, after all, in any similar sequence of oscillations—sound waves (air vibrations), mechanical vibrations of some solid object, alternating currents and voltages in an electrical circuit, electromagnetic wave vibrations broadcast through space as a radio signal, or permanent markings recorded on a phonograph disc or a magnetic tape.

A microphone, for instance, is a transducer that receives energy from air vibrations and converts it into energy of electric-circuit oscillations (Figure 16.2). A tape recorder transduces time-varying electrical vibrations into the "frozen" form of spatially varying magnetic patterns on the recording tape. Even an amplifier can be regarded as a transducer whose input and output both happen to carry the same kind of energy; the amplifier, however, converts a weak electrical vibration into a stronger one.

A fundamental unifying property of many transducers is *reciprocity*: For every type of energy conversion (say, from form A to form B) accomplished by a certain family of transducers, the opposite conversion (from B to A) is accomplished by operating those same transducers in reverse. Probably the most obvious case is that of radio or TV antennas: The same wire aerial may have electric currents driven through it and serve as a broadcasting antenna, or it may be attached to a detecting circuit and used as a receiving antenna. It serves equally well for either purpose: Its efficiency and its pattern of preferred directions remain the same whether it is converting circuit vibrations into radio waves or vice versa.

Now consider a less obvious case, conversion between air vibration and electric-circuit vibration. Devices that respond to sound and produce

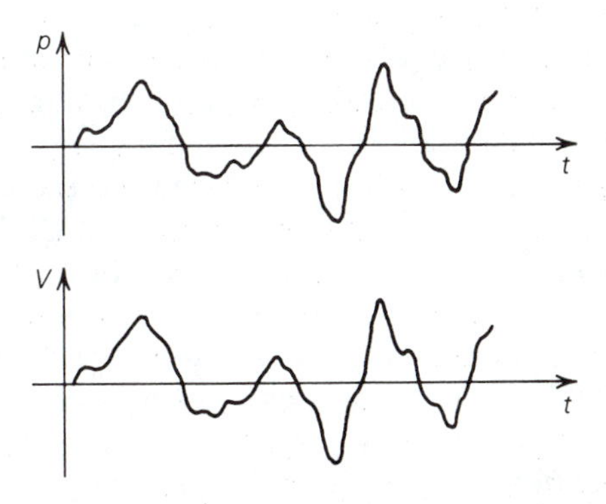

FIGURE 16.2 Air pressure p on a microphone as a function of time and the resulting output voltage V. An ideal transducer is one whose output waveform is a perfect replica of the input.

corresponding electrical signals are called **microphones**, and devices driven by electrical signals to produce air motion are called **loudspeakers**. The reciprocity principle tells us that microphones and loudspeakers constitute a single family of transducers—each is just the other operated in reverse! So if we describe one particular type of microphone, we immediately know about a corresponding type of loudspeaker that operates by the same physical principle.

---

*This reciprocity applies to the simple direct transducers discussed below, but not to parametric transducers such as the FM microphone.

This insight may be quite surprising, because microphones and loudspeakers generally do not look at all alike. But that is the consequence of a mere technical detail: A microphone intercepts only a small fraction of the sound energy in a room, producing a correspondingly small electrical signal, whereas a speaker must supply all the sound energy to fill a room and thus requires a correspondingly large electrical power input. That is, a microphone is a miniature speaker operated backward, or a speaker is a gigantic microphone operated backward. We will use the term *electroacoustic transducer* to include both. Let us describe the most common types.

1. *Electrostatic.* (Equivalent terms: *condenser* or *capacitor microphone*, including electret-type condenser mikes.) Here the diaphragm that interacts with the adjoining air is metallized and located quite close to an electrically charged metal plate (Figure 16.3a). For operation as a speaker, varying amounts of electric charge are driven onto the diaphragm, resulting in changes in the force of electrical attraction of the diaphragm toward the plate; the resulting vibratory movement enables the diaphragm to push on the air and set up a sound wave. To move an appreciable amount of air, the electrostatic speaker diaphragm must have a large surface area.

For operation as a microphone, the variation in air pressure of an incident sound wave pushes the diaphragm back and forth; the resulting change in spacing between diaphragm and plate means that additional electrons sometimes are attracted onto the diaphragm and other times are driven away by the electrical forces from the plate. This sets up the desired alternating voltages in the circuit attached to the diaphragm and plate.

The requirement of a constant (DC) voltage supply is a disadvantage in some applications. In small, cheap microphones it means room must be provided in the handle for batteries, and these often require replacement. In high-quality microphones (and speakers) it usually requires a separate external power supply with extra wires to carry this constant voltage to the transducer. One way to dispense with external power supplies is to use *electrets*, thin films of plastic that can be permanently polarized (sort of an electrical analog to a permanent magnet) and incorporated into a sandwich

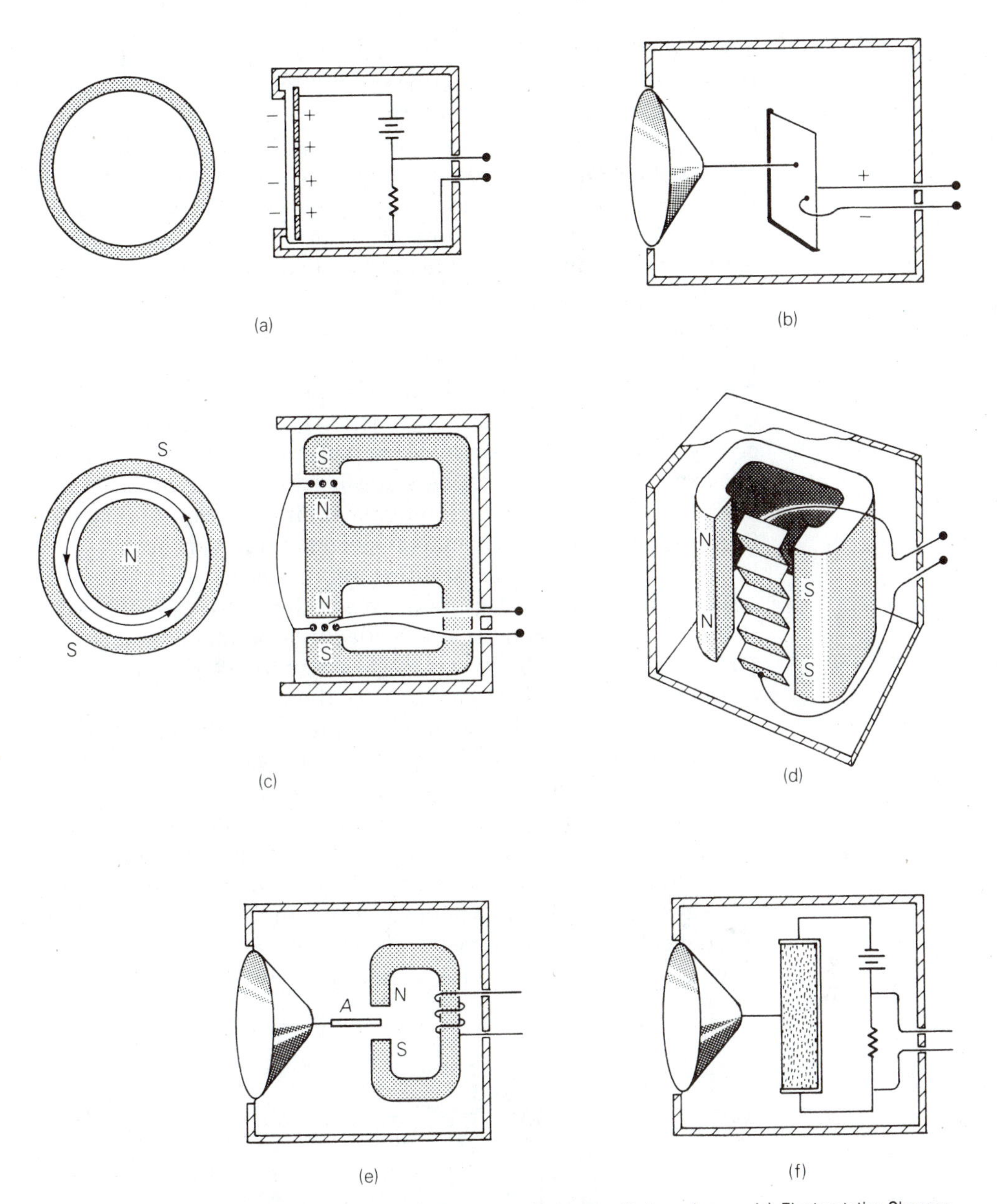

FIGURE 16.3 Principles of operation of common electroacoustic transducers. (a) Electrostatic: Charges of opposite sign flow on and off the exposed diaphragm and the back plate. (b) Piezoelectric: A thin crystal has three corners firmly mounted and the fourth attached to the diaphragm. (c) Dynamic: A current-carrying coil attached to the back of the diaphragm moves in a radial magnetic field. (d) Ribbon dynamic: A corrugated aluminum foil strip carries current between magnetic poles (not ordinarily made in this enclosed form; see Figure 16.6). (e) Magnetic: The diaphragm is attached to a small iron armature *A* moving between the poles of a magnet. (f) Carbon: Diaphragm motion is transmitted to a box of carbon granules carrying electric current.

along with the usual plate and diaphragm. In recent years there has been considerable success in building electret microphones, and this type is often sold with cassette tape recorders.

2. *Piezoelectric.* (Equivalent terms: *crystal* or *ceramic.*) These make use of a remarkable property of certain crystalline or ceramic materials, most notably Rochelle salt or barium titanate: Bending of such a crystal always is associated with an imbalance of its internal structure in which a small amount of positive electric charge appears on one face and an equal amount of negative charge on the opposite face. For operation as a microphone, air motion against a diaphragm attached to one corner bends the crystal, and the resulting electric polarization is detected in a circuit connected to two metal electrodes on opposite crystal faces (Figure 16.3b). Operation as a speaker uses the reciprocal effect: If charge of opposite signs is forced onto the electrodes, it causes the crystal to bend and thus moves the diaphragm and adjoining air. You would not expect to buy a crystal bass speaker for your living room because of the difficulty of moving a large amount of air this way, but crystal transducers find practical use both in small tweeters and in earphones.

3. *Dynamic.* (Equivalent terms: *electrodynamic, electromagnetic,* or *moving-coil.*) Here the usual diaphragm is attached to a coil of wire situated in a strong, steady magnetic field provided by a permanent magnet (Figure 16.3c) (or, in rare examples, by an electromagnet). If an amplifier output forces alternating current through this coil, there is an alternating magnetic force upon it; the resulting coil motion forces the diaphragm to operate as a speaker. It is this same principle that underlies every electric motor: A current-carrying wire in a magnetic field always experiences a force whose strength is proportional to both the current and the magnetic field and whose direction is perpendicular to both of them (Figure 16.4a).

The reciprocal effect is called Faraday's law of electromagnetic induction; this says (in part) that whenever any conducting wire is moved across a magnetic field, there appears an induced electromotive force (a voltage) between the two ends of the wire (Figure 16.4b). So in the dynamic micro-

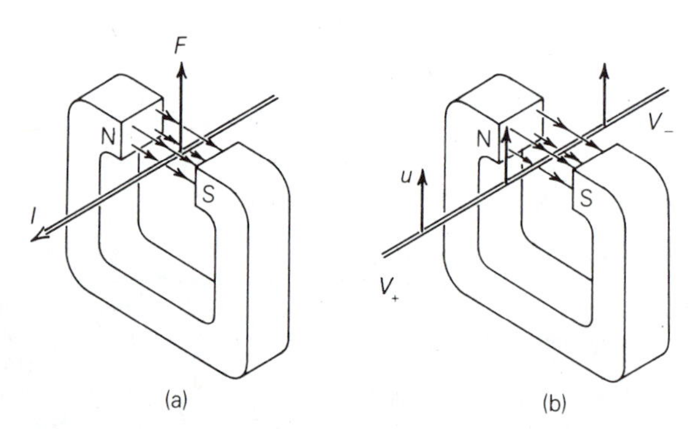

FIGURE 16.4 (a) A wire carrying current I through a magnetic field experiences a force F at right angles to both. (b) A wire moving with velocity u such as to cut across magnetic field lines develops an induced voltage that would tend to drive current from lower left to upper right in the case shown.

phone the air moves the diaphragm, which moves the coil through a magnetic field; the resulting induced voltage is available for amplification. This, incidentally, is the same principle that underlies every electric generator; you should not be surprised now at the idea that an electric generator is merely an electric motor run backward.

Most common hi-fi speakers are dynamic, and the large chunk of iron required for the magnet is a major reason why they are so heavy. For the same reason, you may recognize sometimes whether a microphone is dynamic or some other type just by picking it up. Dynamic microphones are common in routine applications such as public-address systems.

Ribbon microphones also belong to the dynamic category. Here a corrugated thin flat metal strip serves both as diaphragm and as current carrier (Figure 16.3d). Ribbon microphones are notoriously delicate, whereas their moving-coil dynamic cousins can be made quite rugged.

4. *Magnetic.* This is a rather close cousin of the dynamic type; the difference is that here it is a piece of magnetic material called the armature that moves while the coil stays still (Figure 16.3e). An alternating current forced through the coil adds an alternating component to the steady magnetic force pulling the armature down into the gap between the magnet poles; this, in turn, pulls the diaphragm along for operation as a speaker. In the microphone mode, diaphragm and armature motion causes fluctuations in the total magnetic field flux across the gap (and thus also through the coil); this induces a voltage in the coil. The main application of magnetic microphones is in hearing aids and of magnetic speakers in telephone receivers.

5. *Carbon.* Diaphragm motion presses on one side of a small container tightly packed with carbon granules (Figure 16.3f). When squeezed together, the granules increase their area of contact with their neighbors, making it easier for electric current to flow from one to another. That is, the electrical resistance of the carbon-packed box changes, and so the electric current flowing in the external circuit changes; thus we have a microphone. These do not have the good frequency response needed for high-fidelity music reproduction; we mention them only because their sensitivity and ruggedness have given them wide application in telephone mouthpieces.

An exhaustive list of electroacoustic transducers would add a few other exotic types, although none of any great importance for music reproduction. We could make a quite similar list of *electromechanical transducers,* whose job is to convert between vibrations of solid objects (rather than air) and electric-circuit vibrations. You need only bring the solid object in direct contact with the diaphragm in each picture above to see what these would be like. We will take up the example of phonograph cartridges in Section 16.5.

Electromechanical transducers also have important musical application in the form of *direct pickups,* which are an essential element of all electric guitars but are applied sometimes to other instruments also. Direct-contact transducers can be attached to a violin body or a piano soundboard, for instance. Any such setup should be treated by composer and performer as a distinct instrument with a different sound of its own. The spectrum of vibrations picked up directly from the solid body of the instrument should not be at all the same as what we would ordinarily hear through the air, because the efficiency of sound radiation depends strongly on frequency.

Electric guitars usually provide a signal directly from the strings rather than the body. Any type of transducer—crystal, for instance—would be all right on the body. But anything touching a guitar string would hamper its vibration, which rules out carbon or crystal guitar pickup directly from the strings. In principle, electrostatic or dynamic string pickups are possible, with the string itself taking the place of the metallized diaphragm or wire coil, respectively. But these both require that the string be part of the electrical circuit (which rules out nylon or gut), and that raises the issue of shock hazards. This leaves as prime candidate the magnetic pickup; here the string takes the place of the armature and need not have any electrical connections at its ends. The string, however, is required to be not merely metallic, but specifically steel, to participate in magnetic forces.

Let us now take a closer look at several kinds of transducers, in turn, commenting further about details important to their musical applications.

16.3 MICROPHONES

Regardless of the particular mechanism by which it operates, every microphone may be characterized by several aspects of its response to sound waves; most important are dynamic range, frequency response, and directionality.

The *dynamic range* is the range of sound levels over which the microphone gives a usable electrical signal, accurately proportional to sound amplitude. You may appreciate the difficulty of achieving wide dynamic range by recalling that even a 60-dB range (rather marginal for high-fidelity music reproduction) includes some signals carrying as much as a milllion times greater energy than others, and so having amplitude up to a thousand times as much. It is not a trivial matter to build any device that can survive a strong signal without damage or distortion and yet respond accurately to another signal with amplitude a thousand times smaller. So you may sometimes face a choice between one microphone that can go all the way up to 160 dB (as might be needed in some scientific experiments) but down only to 60 dB and another that can go all the way down to 0 dB but, because of this greater sensitivity, must be protected against receiving anything above 100 dB.

The *frequency response* generally is presented in the form of a graph (Figure 16.5) showing how strong an electrical signal is produced for a given sound pressure amplitude at different frequencies. The ideal is a perfectly flat response that neither exaggerates nor downplays any frequency component relative to the others. In practice, accurate frequency response is limited at the low end by approaching the resonant frequency of mechanical vibration of the diaphragm and other parts. At the high end, response must fall off when

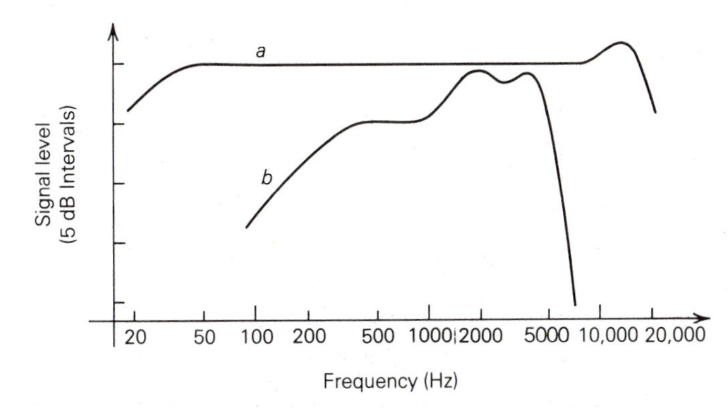

FIGURE 16.5 Frequency response of (curve *a*) a good electrostatic microphone, and (curve *b*) a typical carbon microphone.

the frequency becomes so high that the sound wavelength is smaller than the diaphragm. Notice a trade-off here: A larger diaphragm makes for greater sensitivity by intercepting more sound energy, but a smaller diaphragm is needed to maintain flat response to the highest frequencies.

Both electrostatic and dynamic microphone types are made with sufficient dynamic range and good frequency response (say, flat to within 2 dB over most of the audible range) for high-quality music reproduction. You cannot necessarily expect such quality, however, from a cheap microphone that comes along with your home tape recorder.

The forms we have shown (Figure 16.3) and discussed so far are all *pressure-driven* microphones. That is, any two waves with the same pressure amplitude, even though they may have different frequencies or come from different directions, cause the same amplitude of diaphragm motion and so the same electrical signal strength. This is a direct consequence of having the diaphragm form one side of an enclosed chamber, so that only the excess outside pressure matters. Such microphones have an *omnidirectional* response, detecting signals equally well from all directions. (This ceases to be true for wavelengths shorter than the diaphragm size; but that is where the frequency response drops off, and the microphone is not intended for use at those high frequencies.)

By removing the enclosure and exposing both sides of a diaphragm to sound waves, we can make a *pressure-gradient–driven* microphone; that is, one responding only to *differences* in acoustic pressure between front and back. In its simplest form, most easily pictured for a ribbon dynamic type, a pressure-gradient microphone has a *bidirectional* response (Figure 16.6). These also are called *velocity* microphones, because the velocity amplitude of the sound wave is directly proportional to the pressure gradients and so also to the microphone response.

By returning to a partial enclosure of one side, we also can adjust the effective path lengths to get a *unidirectional* (or *cardioid*) response pattern (Figure 16.7). Unidirectional microphones are made in electrostatic and ordinary moving-coil dynamic types, as well as ribbon dynamic.

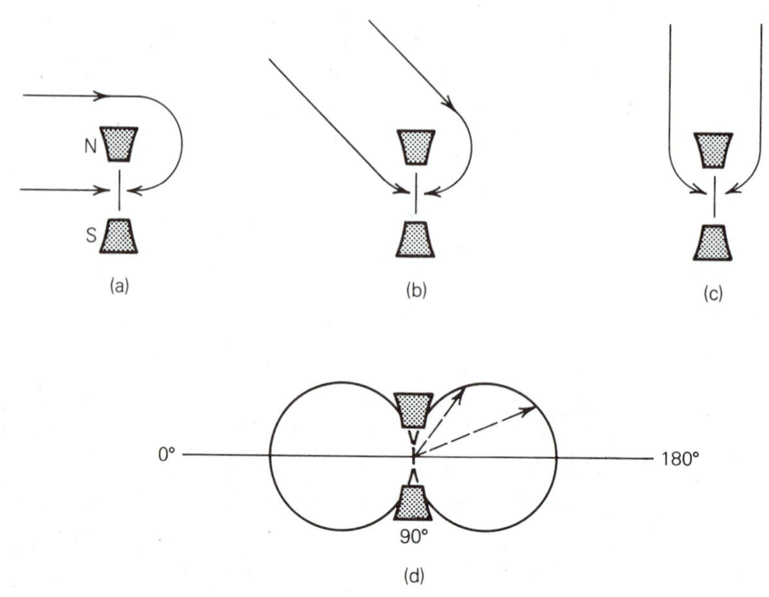

FIGURE 16.6 Bidirectional response of a ribbon microphone open on both sides. Compare with Figure 16.3d; this is a top view. (a) For waves approaching the ribbon perpendicularly, the signal diffracting around to the back must travel farther than that to the front. Even if their amplitudes are the same, they are not quite in phase, so front and back pressures are different and the ribbon moves. (b) The phase lag is less for waves approaching at 45°; front and back pressures do not differ as much and the ribbon motion is less. (c) Waves approaching from the side create equal pressure (exactly in phase) on both faces, and the microphone does not respond at all. (d) The corresponding polar sensitivity pattern; distance from center out to figure-eight in any direction tells relative amplitude sensitivity in that direction (as for dashed examples). Peak sensitivity is on-axis (0° or 180°); it falls 3 dB below maximum at 45° off-axis. These circles are cross sections through a pair of spheres representing the sensitivity pattern in three dimensions.

---

*Higher degrees of directivity can be attained only at the price of (a) lower sensitivity, as in higher-order pressure-gradient microphones, or (b) large size compared to the wavelengths in question and directionality highly dependent on frequency, as for parabolic reflectors or "shotgun" mikes.

By combining these elementary types, it is possible to make a variety of additional response patterns. One interesting example is the *stereo microphone,* which has two output signals corresponding to the sound received from two distinct directions. In its simplest form (Figure 16.8) it consists of one unidirectional and one bidirectional element mounted together. Their electrical outputs are added to obtain one of the stereo signals and subtracted for the other.

Pressure-gradient microphones respond not only to phase differences, but also to amplitude differences. When the microphone is far from the source for normal use, the front and back waves have essentially the same strength. But when close, amplitude differences become dominant; this stems from

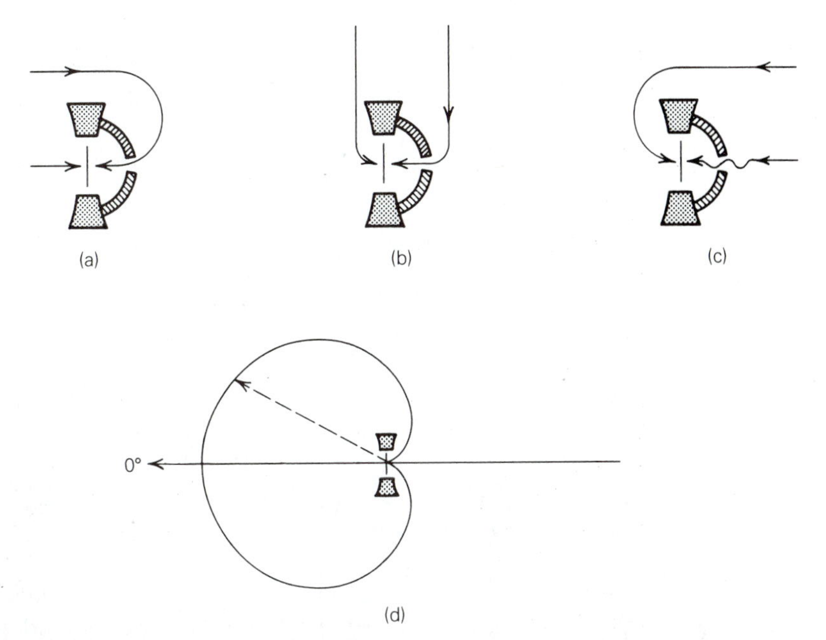

FIGURE 16.7 Effect of partial enclosure of one side of a ribbon microphone; top view as in Figure 16.6. (a) Signals coming from the front have a large path difference to reach the back side of the ribbon; the resulting phase lag means good response. (b) Signals from 90° still have some phase difference and nonzero response (unlike Figure 16.6c). (c) Although the drawing alone cannot indicate it, proper design of the enclosure can make signals arriving from the back reach both sides of the ribbon with the same phase, thus giving zero response. (d) Corresponding unidirectional polar sensitivity pattern; response is 3 dB down at 66° off-axis.

the inverse-square law (Section 5.3). Thus, a pressure-gradient microphone placed close to a source exhibits the *proximity effect*, in which bass signals are boosted relative to treble more than they would be if picked up far away. This affects components with wavelengths more than approximately six times the source-to-microphone distance and becomes increasingly pronounced as that distance decreases (Figure 16.9). This accounts for the reputed "warmth" of ribbon velocity microphones, which is cherished by many singers.

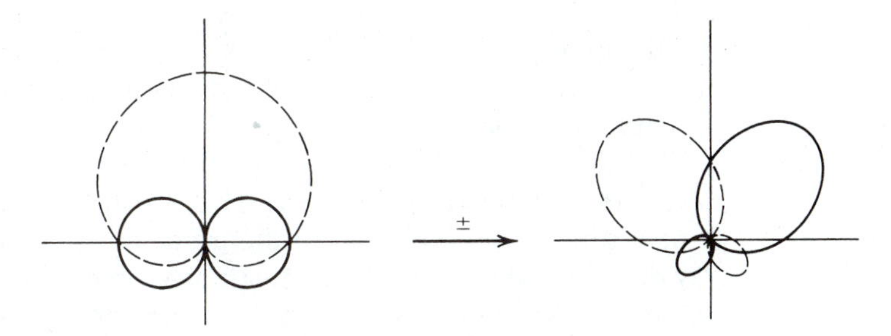

FIGURE 16.8 A stereo microphone. The cardioid unit points toward the middle of the performing group, and the bidirectional unit picks up signals mainly from both sides. Addition and subtraction of these signals gives two others that represent mainly right and left, although both include some from the middle.

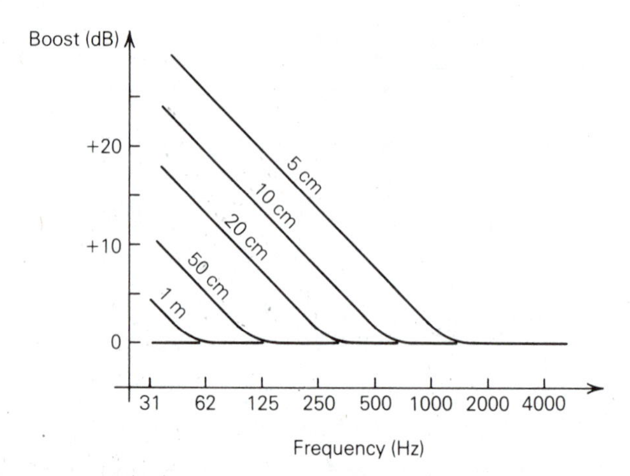

FIGURE 16.9 Flat line represents frequency response of a pressure-gradient microphone located far from the sound source. Sloping lines show enhanced response due to proximity effect when the same microphone is at various distances from the source.

Another (quite independent) effect of close placement of any type of microphone is an increased proportion of direct sound in the total signal. This gives greater clarity (and separation in the case of stereo) but at the expense of the reverberant sound that conveys an impression of the "ambience" of the performing space.

16.4 AMPLIFIERS

The basic principle of any amplifier may be understood easily with a fluid-flow analogy (Figure 16.10a). Modest flow changes in the small control jet can switch much larger flows in the power jet, so that weak alternations of the control jet produce much stronger alternations in the main jet. The classic electronic amplifier that marked the birth of modern communication technology was the triode vacuum tube (Figure 16.10b); its operation is quite similar, with small alternating voltages applied to the control grid acting to increase and decrease the large current flow between anode and cathode, causing greatly multiplied voltage alternations at the anode. Electron tubes have given way in virtually all modern audio equipment to transistors whose basic principle of amplification remains the same (Figure 16.10c).

If the amplified signal is to perfectly represent the original input, the device must be accurately *linear*. This means that the output voltage must remain directly proportional to the input throughout the oscillation; that is, any doubling of input voltage must cause precise doubling of the output, too (Figure 16.11). Nonlinearity can occur not only in amplifiers but also in all transducers involved in the reproduction process. Any nonlinearity results in a distorted output (Figure 16.12). A sine wave input, for instance, comes out as some different shape. Because it still is periodic, it can be represented by a

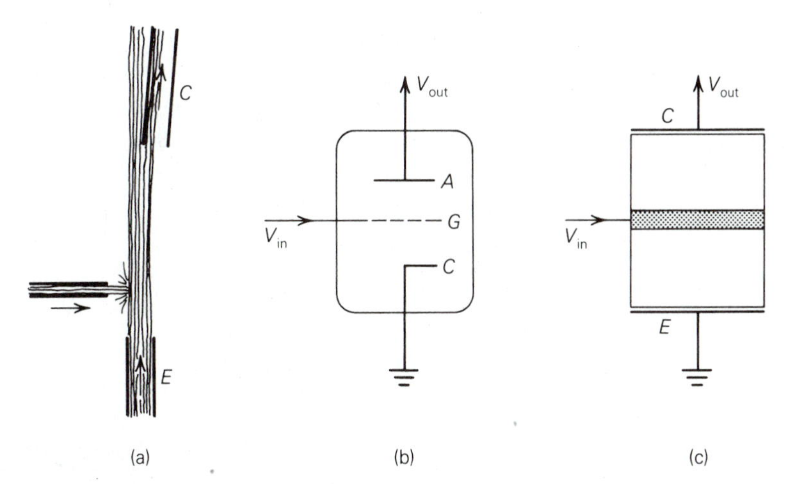

FIGURE 16.10 Examples of amplification. (a) A large jet of fluid comes out of the emitter *E*, and part or all of it may be picked up in the collector *C*. A small jet impinging from the side steers the large jet and greatly influences the amount of fluid reaching *C*. (b) A triode has external connections capable of sending a large flow of electrons from cathode *C* to anode *A*; this current, however, is easily blocked or modified by small voltages applied to the control grid *G*. (c) The most closely analogous solid-state device is a common-emitter *npn* transistor; small input voltages drastically change the number of electrons available to carry current across the shaded region and so cause large changes in the current getting from emitter *E* to collector *C*.

Fourier series (recall Chapter 8); we can think of the nonlinearity as introducing higher harmonics where they do not belong, a processs called *harmonic distortion*.

An input including two sine waves of different frequency, in the presence of nonlinearity, produces an output containing not only the original frequencies (say, f_1 and f_2, along with distortion-generated harmonics of each) but also undesired signals with frequencies $f_1 + f_2$ and $f_1 - f_2$. This is called *intermodulation distortion*, and it always goes hand in hand with harmonic distortion. Both can be kept well below 1% in amplitude (and thus 40 dB or more below program level) in a high-quality amplifier that is not overdriven.

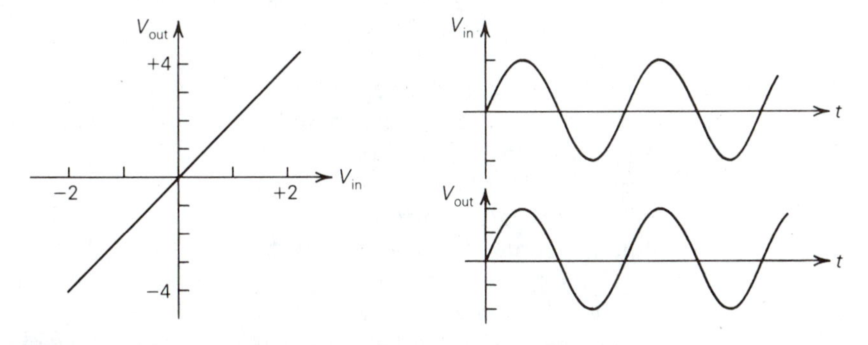

FIGURE 16.11 Input-output characteristics of an ideal linear voltage amplifier for the case of output double the input (gain 2). Although the output amplitude is larger than the input (notice the different scales), the wave shapes are precisely the same.

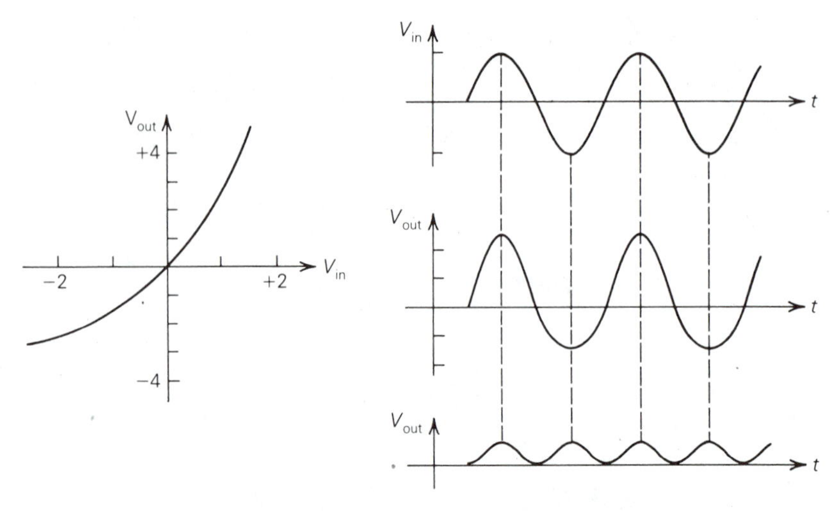

FIGURE 16.12 A gain-2 amplifier with some nonlinearity. Now a sine wave input gives a distorted output; the last graph indicates how presence of some second harmonic would account for sharpening the peaks and blunting the dips. (In general, smaller amounts of third and higher harmonics are introduced as well.)

Even when there is no harmonic distortion, it still is possible to have *phase distortion*, meaning time delays in the amplifier that are different for different frequencies. Although this is of little importance for steady signals (according to Ohm's law, as discussed in Chapter 8), it does result in audible degradation of transient sounds and should be avoided. Phase distortion usually is small over the range of frequencies for which the frequency response is flat, and for a high-quality amplifier this will include the entire audible range.

The *compressor amplifier* occasionally is useful in dealing with situations where certain sound events (say, cymbal crashes) unavoidably overload the reproduction system when amplifier gains are set at levels needed for the rest of the program. Here a monitoring circuit measures the ongoing signal strength and uses a feedback loop to automatically reduce the amplifier gain whenever the output begins to exceed a critical level (Figure 16.13). Once the loud event is past, the gain is allowed to return gradually to its original setting.

A *limiting amplifier* represents the extreme case of such strong compression that the output is effectively limited to a certain maximum level regardless of how strong the input may become. Notice that compression is not the same as merely *clipping* the peaks off the largest waves (Figure 16.13d); clipping means large harmonic distortion and generally is avoided at all costs, although sometimes used deliberately as a special effect with electric guitars.

Cassette tape recorders often have a further extension of this idea called *automatic level control*, which not only reduces the gain for loud sounds but also increases it for softer ones. The result is to make everything come out at about the same level. This may be useful for recording speech, but it is entirely inappropriate for most kinds of music.

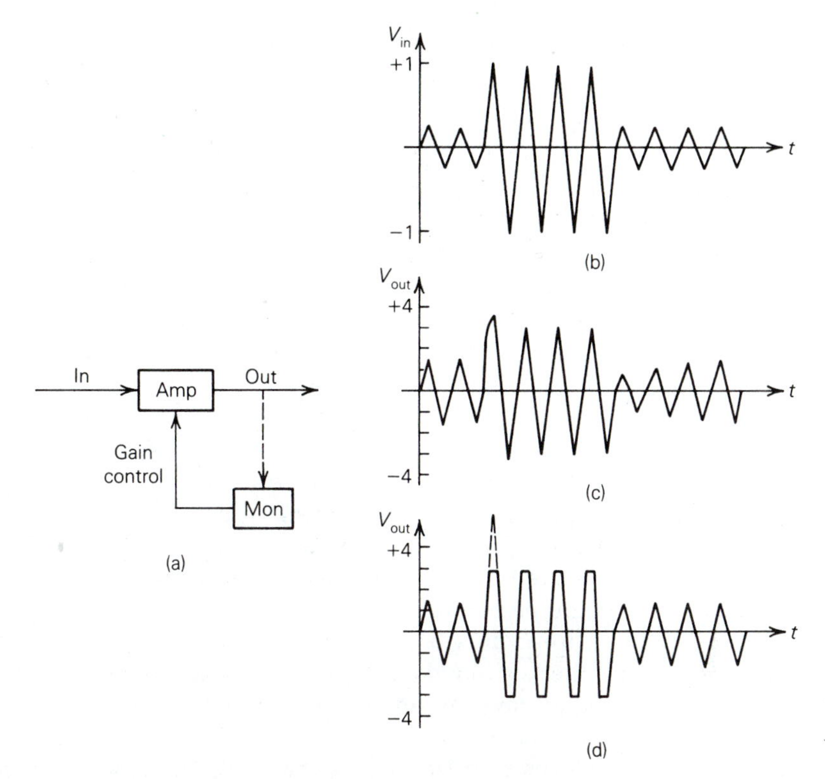

FIGURE 16.13 (a) A compressor amplifier has its gain controlled by a monitor circuit in a feedback loop. (b), (c) In this example, the gain is 6 for small signals (0.25 V in, 1.5 V out), but is reduced to 3 for the larger signal (1 V in, 3 V out). At the onset of the large signal, the gain is reduced quickly, with only very brief distortion; when the input returns to the lower level, the gain gradually and smoothly comes back up. (c) Compression only changes the relative amplitudes of steady signals, whereas (d) clipping grossly distorts the waveforms.

---

*The human ear acts somewhat like a compressor amplifier: Multiplying the incident sound intensity by 100 only increases the nerve firing rate by a factor of approximately 3 or 4. For extremely loud sounds that bring the acoustic reflex into play, the degree of compression is even greater.

16.5 RECORDING

There are several ways to store audio information for listening at a later time. The phonograph record and magnetic tape recorder are quite familiar, and the last decade has brought widespread use of digital recording processes.

Phonograph recording and playback were for many years done without electronic aid (and without the corresponding power or fidelity!) by using

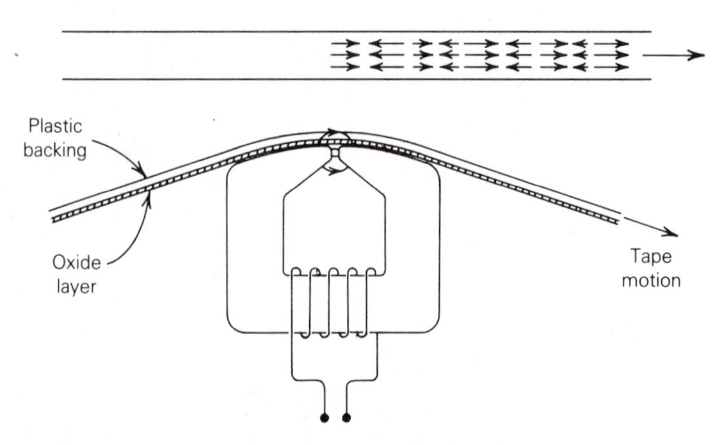

FIGURE 16.14 Tape recording or playback head and pattern of magnetization created on tape.

acoustomechanical transducers. These merely had a needle attached to a diaphragm mounted in the throat of a large flaring horn. In recording, the horn served to gather acoustical energy (like an ear trumpet) for transmission through the diaphragm to the needle. In playback, the horn served to couple the diaphragm vibrations more effectively to the air in the room, exactly as with the horn loudspeakers to be described in the following section.

Nowadays information is recorded or played back from record grooves by electromechanical transducers. In the playback process, for instance, the transducer converts the mechanical vibration of a needle tip following the undulations of the groove into electrical-circuit vibrations suitable for amplification. Crystal or ceramic pickup cartridges may be found in cheap equipment; it is the dynamic (moving coil) and magnetic (moving magnet) types that dominate the high-fidelity market.

Tape recording depends on magnetoelectrical transducers of the type shown in Figure 16.14. In the recording process, alternating electric currents through the coil cause strong magnetic fields in the gap between the poles of the electromagnet, and these create regions of alternating magnetization on the oxide-coated tape as it passes. *Magnetization* means an aligning of individual atoms so that the oxide grains in each little segment of coating become a set of tiny permanent magnets oriented in a common direction. For playback, these regions of alternating magnetization move rapidly past the transducer head gap, and their changing magnetic fields induce an alternating voltage in the coil; the underlying physical principle is the same as for the magnetic microphone or phonograph pickup. The signal is not recovered in its original form, however, until it has been equalized, as explained in Box 16.1.

Reciprocity is recognized easily in this case; cheaper tape recorders often have only a single signal head used for both recording and playback. Even on professional equipment with separate playback heads available, record heads sometimes are used in "selective synchronization" to monitor one track while recording another.

BOX 16.1 EQUALIZATION

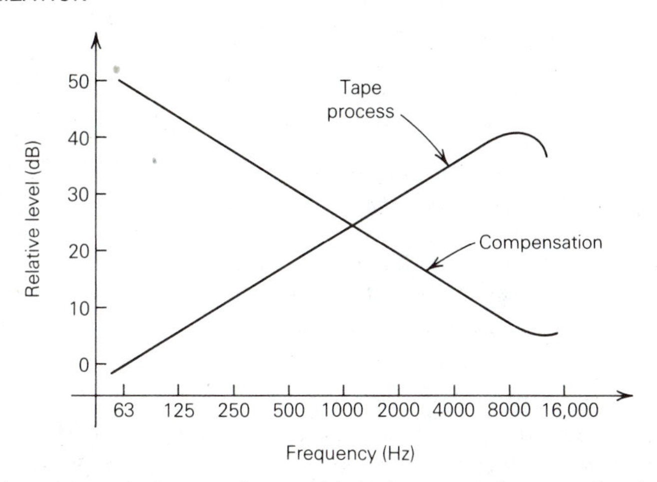

FIGURE 16.15 Treble boost inherent in the magnetic record/playback process and compensating electronic equalization that must be added to achieve accurate reproduction.

The frequency response of tape recording heads presents an important problem. If the gap width is small enough, we can achieve flat response in the recording process; that is, equal voltage amplitudes at different frequencies produce equal strengths of magnetization on the tape. But in playback, the induced voltage is proportional to how rapidly the magnetic field *changes* in the gap, which depends not only on the strength of tape magnetization but also on the frequency with which it reverses. So the playback process gives a treble boost (Figure 16.15), which must be compensated for by an electronic circuit called an equalizer before the amplified signal is used in a speaker.

Phonograph recording also requires equalization, although for different reasons. In order not to have such large excursions at low frequency that one groove breaks into the next, and still not lose the high-frequency signals in

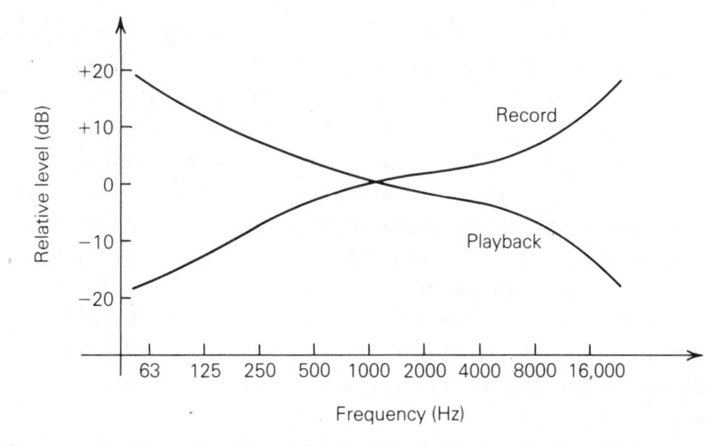

FIGURE 16.16 RIAA curves, the internationally accepted standard equalizations for disc recording and for compensation during playback.

(*continued*)

BOX 16.1 *(continued)*

the background noise, the bass is attenuated and the treble boosted during the recording process (Figure 16.16). All modern high-fidelity equipment provides the compensating deemphasis during playback to restore the original musical balance.

The term *equalization* also refers to relative level adjustments of various frequency ranges, made as a matter of technical or artistic judgment in the recording process with no expectation of compensation in playback. This can improve the esthetic quality of a recording, but it also can be overdone. I like the following comments from Benade (pp. 484–485):

> All too often the equalization is done by a man who listens to very little live music, and who may have altered his hearing by many hours of listening to monitor loudspeakers (or, worse, headphones) at sound levels that make anyone not inured to them cringe in pain. These are some of the reasons why so many records sold today have excessive treble and bass, to the annoyance of many performers and those listeners who are familiar with music played "live".

Good reproduction depends on a quiet, accurate phonograph turntable or tape transport mechanism. The transducer should pick up no detectable vibration or electrical interference from the motor, for that would give audible hum at 60 Hz or some related frequency. The rate of disc or tape movement also must be absolutely steady to avoid "wow" or "flutter."

When tapes are copied and recopied, another type of noise problem arises: Each stage of copying amplifies whatever small amount of noise may be present in the previous stage. By the time you reach fourth- or fifth-generation tape, the white-noise background called tape hiss becomes quite prominent. This problem can be greatly reduced with technical tricks such as those in the popular Dolby and DBX *noise reduction* systems. These use a sophisticated version of compression (recording softer sounds with higher amplification to keep them farther above the noise), which requires a compensating expansion during playback. A tape recorded with the Dolby process but played back on an ordinary machine (or vice versa) will not provide accurate reproduction.

In the language of Section 8.4, we could say that traditional phonograph and tape recording techniques store signals in analog form. Wide availability of computer technology suggests converting the original signals to digital form (long strings of numbers) for storage; then the playback process must reconvert to analog form. It is of secondary importance whether the numbers happen to be stored on magnetic tapes or magnetic discs (both of which are standard items of computer hardware) or in some other medium. Digital magnetic-tape storage came into widespread use in the late 1970s for original recordings in professional studios. Use of these tapes in the "mastering" process makes possible both more concise and powerful editing, and production of phonograph records with lower levels of background noise. Smaller digital audio tape (DAT) systems for the home consumer currently are being developed.

Phonograph records themselves are threatened with extinction by the new compact disc (CD) technology that invaded the consumer marketplace in

FIGURE 16.17 Photomicrograph of the surface of a compact disc recording. The laser scans along each horizontal row in turn. The bumps are approximately 0.16 micrometer high and 0.6 micrometer wide. (Photo courtesy of John Monforte)

the mid-1980s. The CD stores information about sound in digital rather than analog form, and thereby completely eliminates the problems of surface noise on phonograph records, or background hiss on analog tape. Because the rate of readout is controlled by a precise clock, this also eliminates all flutter or wow caused by imperfect tape transport or disc drive systems.

The CD master is produced on a polished glass blank covered with a thin layer of photosensitive material. The digital signal from a master tape controls a high-powered laser that vaporizes small patches of that material, leaving a sort of mask to guide a chemical etching process that produces small pits. After several stages of electroplating, the final master stamps out the mass-produced plastic CDs with a pattern of bumps corresponding to the original pits (Figure 16.17).

The final product is a 12-cm diameter disc with a protective plastic coating that makes the CD much more resistant to damage than a phonograph record. The information is read off the CD as it spins around (at a rate that varies between 200 and 500 revolutions per minute) by a small laser and optical detector. The presence or absence of a bump on the disc either diffuses the laser beam or reflects it intact so that the optical reader recovers the original stream of digital information. A digital-to-analog converter then provides an appropriate signal to send to the power amplifier and thence to the loudspeaker. By using 16 bits (each bit being either a 0 or a 1) in each "word" stored, CD systems have approximately a 90-dB dynamic range, whereas both analog tapes and phonograph discs are hard put to keep the background noise more than 60 dB below the loudest sounds reproduced. One of the most powerful features of the CD system is the inclusion of redundant information and error-correcting codes. With these codes, an occasional speck of dust or small scratch that alters an information bit in the optical readout can be tolerated because the original information still can be recovered.

16.6 LOUDSPEAKERS

The primary properties of a loudspeaker are its size, efficiency, frequency response, directional pattern, and type of mounting or enclosure; these characteristics are interrelated. Let us discuss them for the case of a dynamic speaker, because that is the basic mechanism of the great majority of high-fidelity speakers.

First of all, it is extremely difficult to make a single speaker that performs well over the entire audio range. So we generally buy *speaker systems*—two, three, or even more speakers all mounted in the same cabinet together with an electrical circuit called a *crossover network.* This divides the signal from the amplifier, sending bass frequencies to the large woofer and treble to the small tweeter; a three-way system has a separate midrange speaker as well. Producing a given sound level at low frequencies requires moving a lot of air, hence the large size of the woofer. The same sound level at high frequencies is obtained by moving a smaller amount of air more often, so a smaller diaphragm is adequate. The diaphragm may be either cone- or dome-shaped (Figure 16.18), with the difference being largely a matter of convenience in the size of magnet and coil required.

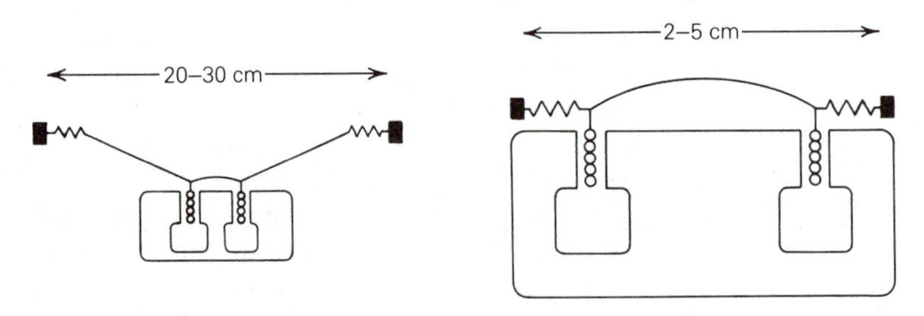

FIGURE 16.18 Cone and dome versions of the dynamic speaker, typically used as large woofers and small tweeters, respectively.

The lower frequency limit for any speaker is around the natural resonant frequency of its diaphragm. To respond well in their respective ranges, the woofer diaphragm is relatively massive and in a very flexible mounting, while the tweeter diaphragm is light and stiff.

For all wavelengths much larger than the speaker diameter, sound radiates about equally well in all directions, which usually is desirable. For shorter wavelengths, the radiation tends to come out in a beam in the forward direction. This is one reason why high frequencies usually are assigned to a separate tweeter, even in cases where the woofer could otherwise handle them as well.

We must note that a bare speaker radiates from both its front and back sides. These two signals are of comparable strength although 180° out of phase. So the two signals nearly cancel (except for wavelengths less than approximately two or three times the speaker diameter), leaving a very weak net output. (The reciprocal effect is that a pressure-gradient microphone is far less sensitive than a pressure-driven one.) Another way of arriving at essentially the same conclusion is to argue that because it is so easy for the air in front to move around and fill in behind (Figure 16.19, dashed line), it does not bother to move much outward in other directions. We recognize this as another example of "you can't fan a fire with a knitting needle" (Chapter 3).

The frontside signal will be useful only if we can somehow suppress the backside signal. One way to solve this problem is to mount the speaker in a hole in the wall (an infinite baffle, Figure 16.19b), so that the back side radiates into a different room. It even helps to mount it in a flat board (a finite baffle, Figure 16.19c), at least for those wavelengths up to two or three times the baffle size. Another approach is to enclose the speaker in a sealed cabinet (Figure 16.19d), with the idea that the backside signal cannot get out. This cabinet will function as an infinite baffle only if it is quite large and filled with sound-absorbing material so that standing waves cannot build up inside and influence the diaphragm motion.

Finally, we might say, "Why waste half the energy by absorbing the backside signal? Let's bring it out front, but in a way that gives constructive rather than destructive interference." This leads to the vented or bass reflex enclosure (Figure 16.19e), where the backside signal is supposed to be effectively delayed half a cycle in finding its way to the vent. Because that extra distance cannot be half a wavelength simultaneously for many different wavelengths, this does not solve all problems. But for a judicious choice of those wavelengths that are helped, the effect of the vent may be to extend the bass response. Sometimes the vent is covered by a passive radiator or drone cone; the idea remains the same.

Another important variation in speaker mounting is the *acoustic suspension.* Here the mounting ring is made extremely soft and flexible (high-compliance), so that the restoring force to bring the cone back to its normal position comes primarily from the springiness of the air inside the cabinet (much like a Helmholtz resonator). This, of course, requires that the cabinet

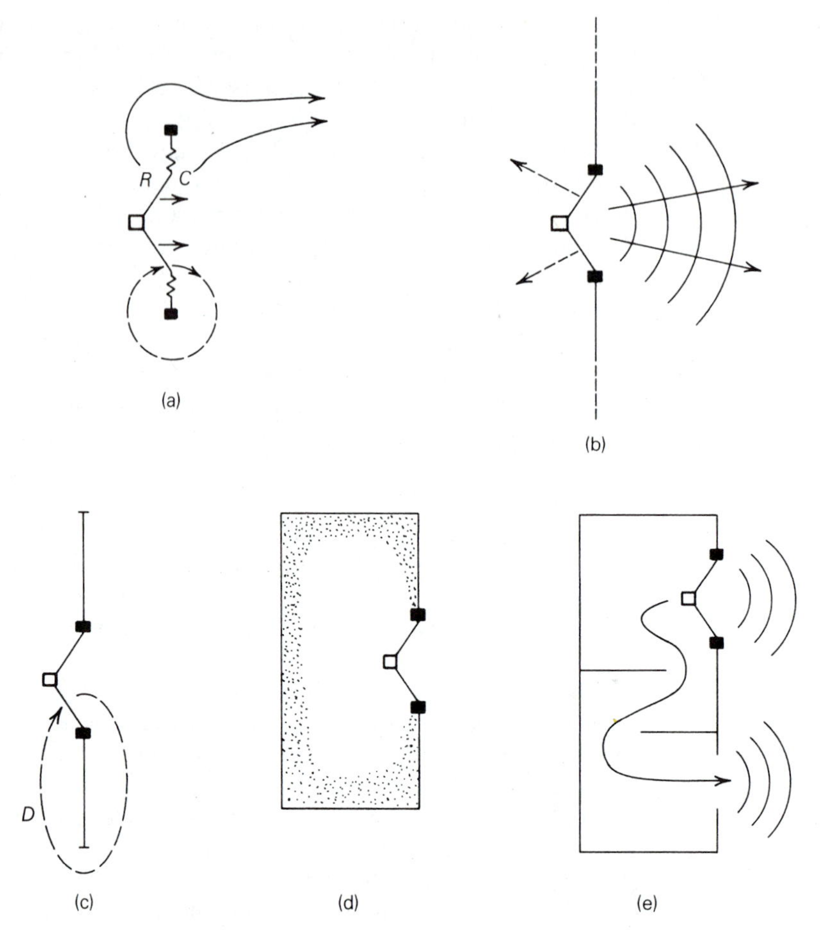

FIGURE 16.19 Speaker mountings. (a) During a half-cycle of outward motion, a bare speaker diaphragm creates a compression *C* in front and a rarefaction *R* in back. These both travel to a distant listener, and for low frequencies nearly cancel each other out. (b) An infinite baffle allows only the frontside signal to reach the listener, so there is no cancellation. (c) A finite baffle works only for sufficiently high frequencies; if *D* is much less than half a wavelength, air will again move around from front to back instead of radiating a strong wave outward. (d) An enclosed box lined with absorbing material also can prevent the backside wave from reaching the listener and so act as an infinite baffle. (e) A vented speaker cabinet reinforces certain chosen frequencies by making the effective path from backside to vent approximately a half-wavelength; then the two signals are in phase when they reach the listener and reinforce instead of cancel each other.

be sealed. Acoustic suspension is a way of getting more bass response from small "bookshelf" speaker systems, for it allows the speaker's natural resonance to occur at lower frequency. This type of speaker, however, is rather inefficient and so requires more power from the amplifier.

An entirely different approach to speaker design is represented by the *horn loudspeaker*, which is common in public-address systems. In the hi-fi market, Klipschorn speakers are well known. Cone-type speakers are quite

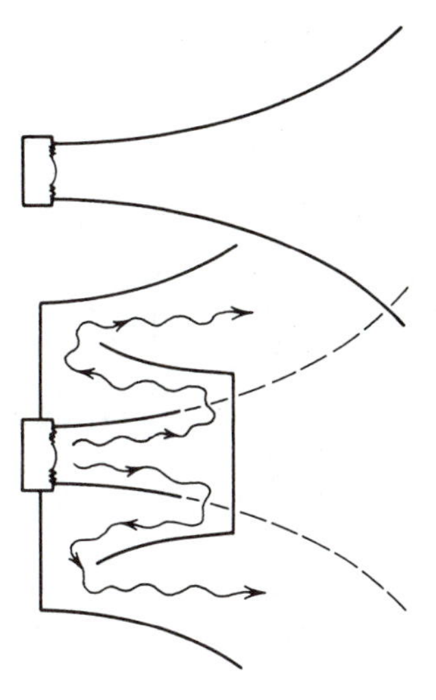

FIGURE 16.20 A straight horn loudspeaker, and a more compact folded horn of similar effective length.

inefficient in transferring energy into the surrounding air; there is no escaping this, as there is simply an impedance mismatch between the air and the much denser cone material. The gradual flare of a long horn serves to mediate between the motion of a much smaller diaphragm and the eventual air motion outside. Because the horn serves to match impedances, a horn loudspeaker is relatively efficient—up to 50% of electrical energy converted to acoustical energy, compared to 3–5% for good direct-radiator dynamic speakers.

You might think of a horn as a baffle that is folded from the sides toward the front, and this will suggest one of the horn's problems: It is effective (smooth and strong response) only when it is large compared to the wavelengths desired. Thus horns are quite practical for tweeters, but rather unwieldy for woofers unless modified to a folded-horn design (Figure 16.20).

For amplified music, as presented by touring pop-rock groups, for instance, large numbers of speakers often are used in *arrays*. The object is not only to put out more sound but also to steer it preferentially toward the crowd. This usually calls for vertical columns: If the array is only one wide, the broad dispersion of a single speaker's directional pattern still pertains in the horizontal plane. But in the vertical plane the speakers cooperate to beam the sound in a much narrower pattern (Figure 16.21). This is an interference effect: In the forward direction the signals arrive in phase for constructive interference; above or below they arrive with different phases (because they had to travel different distances) and largely cancel out. For effective operation a column should have total height greater than the longest wavelength to be beamed, but with individual speaker spacing less than the shortest wavelength going to those particular speakers.

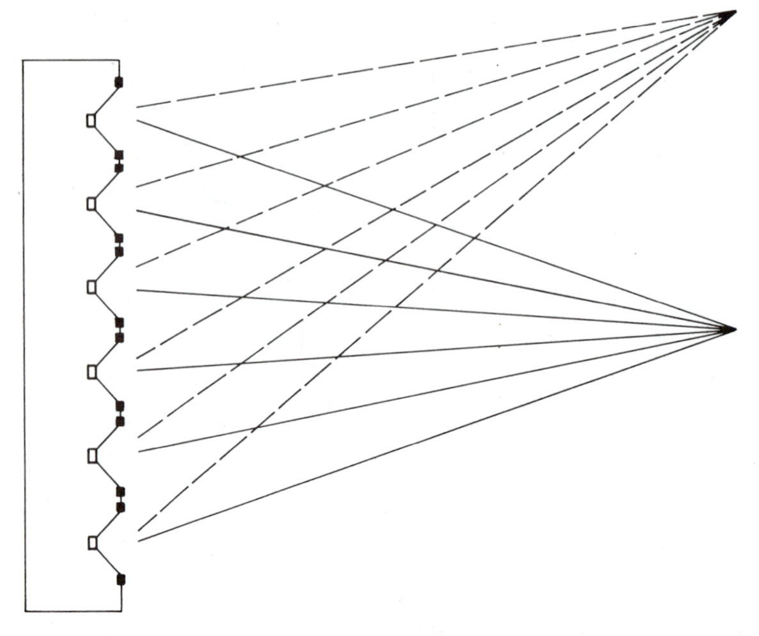

FIGURE 16.21 For a listener straight in front of a column speaker array, all signals travel nearly the same distance and arrive nearly in phase to create a strong sound. For a listener much above or below the aiming direction, different path lengths (dashed lines) mean phase differences and resulting destructive interference, and thus only a weak sound.

16.7 MULTIPHONIC SOUND REPRODUCTION

The simplest form of sound reproduction is *monophonic*—one microphone, one transmission channel, and one loudspeaker. (The term *monaural* is commonly misused here; technically it means a signal sent only to one ear but not the other.) This reproduces only a very limited portion of the total information about spatial distribution of sound in the original room. (Let us assume in this section that reproduction takes place in an acoustically dry room, such as an average living room, which adds little or no further color or reverberation to the original signal.) The reproduced sound simply is perceived as coming from the speaker; even an entire orchestra may seem confined to that little box.

To convey more information about how sound sources were originally arranged, we need to preserve some of the cues a live listener would use, such as phase and amplitude differences between two ears (Section 15.6). This suggests using two microphones, sending their signals through two independent transmission channels to two speakers. Let us distinguish between two forms of two-channel reproduction: binaural and stereophonic. *Binaural* reproduction (Figure 16.22a) obtains signals as much as possible like

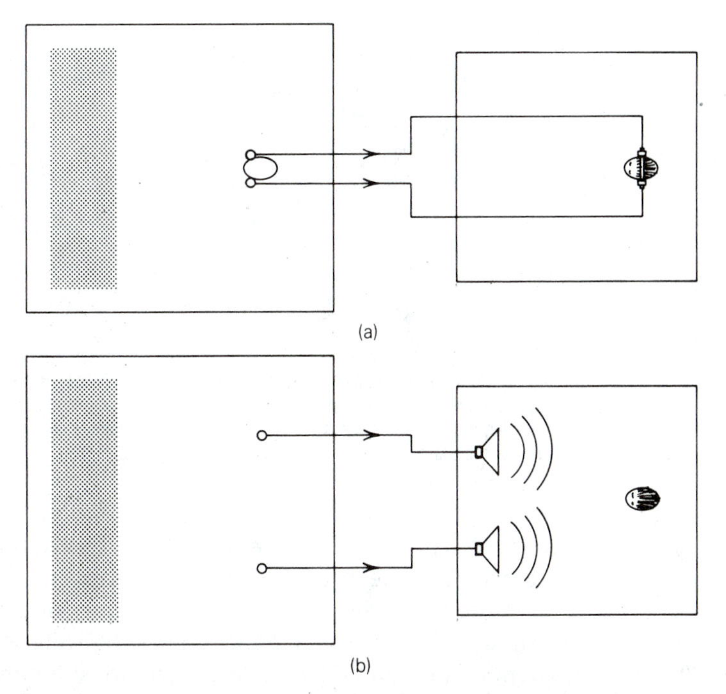

FIGURE 16.22 (a) In binaural reproduction, sound from the performing area (shaded) is picked up by microphones mounted in a dummy head so that they closely approximate what a live listener would hear at the same position; reproduction is through headphones. (b) Stereophonic reproduction replays each microphone signal through a speaker whose sound can reach both of the listener's ears.

those that would be heard by a pair of human ears and presents each of these signals to one ear only through headphones. Binaural sound is startlingly realistic except for one thing: You cannot get additional cues by moving your head, for the entire sound image simply moves along with you. The use of binaural sound is limited mainly to scientific experiments.

Stereophonic reproduction differs in two important ways: Any two microphones may supply the original signals, and their placement in the room may bear no resemblance at all to a pair of ears; and each of the two speaker signals can reach *both* of the listener's ears (Figure 16.22b). In view of these shortcomings, it is remarkable how much spatial information still is conveyed by stereo sound: Different instruments may be perceived as playing at different positions all along the line between the two speakers. Perhaps even more important than these "phantom images" is the feeling of "presence" generated by stereo; this is hard to describe but quite striking when you hear a demonstration with alternating stereo and mono modes.

Stereo must not be mistaken for simply sending the same signal to two different speakers. That sometimes makes things worse instead of better (unless the speakers are close together as part of an array, functioning effectively as a single source), because the difference in path length to the listener

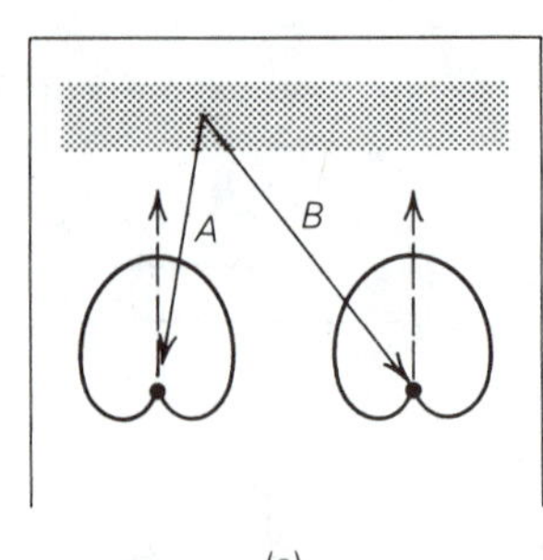

(a)

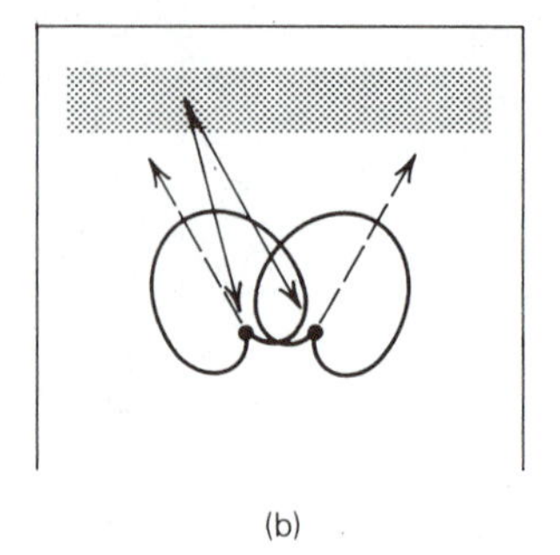

(b)

FIGURE 16.23 Directional microphone pickup patterns (as in Figure 16.7) for alternate stereo recording techniques. (a) Widely separated microphones give not only intensity-difference cues, but also vastly exaggerated phase-difference cues, because path *A* is so much shorter than *B*. (b) Closely spaced, outward-pointing microphones still give similar intensity-difference cues, but now the time delays are more realistic. A single stereo microphone (Figure 16.8) placed in the center works similarly.

introduces a phase difference that causes interference, constructive for some frequencies and destructive for others.

For precisely the same reason, mixing signals from two microphones picking up the same source introduces interference that distorts the original frequency spectrum. This altered sound sometimes is preferred over the original in popular music. But it is best when recording classical music to use a single well-placed microphone for each information channel. Even when two microphone signals are kept separate as stereo channels, time lags for large separation (Figure 16.23a) may cause trouble with both timbre and phantom image location for instruments that can be picked up by both. If the object of stereo is concert-hall realism, rather than Ping-Pong effects, it usually is better to place the microphones within a meter or two of each other (Figure 16.23b).

Movie theaters commonly broadcast sound from more than two independent tracks. Blauert reports that 20 or more channels are needed to reproduce spatial perception of sound fields "faithfully." In the early 1970s there was a brief flurry of interest in four-channel or *quadraphonic* reproduction for home hi-fi systems. It is easy enough to record four independent channels side by side on magnetic tape. But true four-channel disc recording required expensive new technology: Two channels worth of information were translated electronically up to the frequency band between 15 and 45 KHz before cutting the disc and then were decoded on playback. Both cutting stylus and pickup cartridge must then have good response up to 45 KHz, instead of only 15 KHz as for ordinary stereo (Figure 16.24). Ultimately this system did not succeed in the marketplace. Once you tire of novelty effects (like making a source pan all the way around the room) you are left with two rear speakers whose main purpose is to add only a subtle hint of reverberant sound without calling attention to themselves. The degree of improvement in realistic spatial sound perception over stereo is not nearly so clear and striking as that of stereo over mono, so most consumers could not justify the added investment.

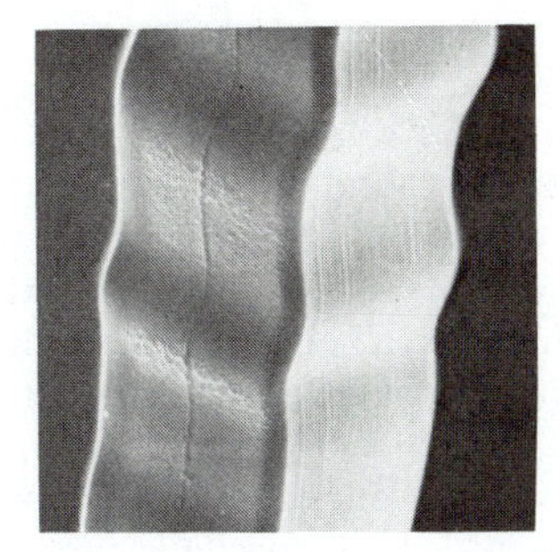

FIGURE 16.24 (a) Ordinary stereo and (b) discrete quadraphonic record grooves. Note that quadraphonic requires tracking much more rapid variations. (Courtesy of Csaba Hunyar)

SUMMARY

High-fidelity music reproduction is made possible by a variety of transducers, which convert acoustical signals to other forms for transmission or storage. Especially important are electroacoustic transducers, in the form of microphones to convert sound to electrical vibrations and loudspeakers to reconvert to sound. Most useful for musical purposes are the dynamic and electrostatic types.

Microphones and loudspeakers are characterized further by their diaphragm size, which largely determines microphone sensitivity, speaker maximum power output, and maximum effective frequency for both. The diaphragm mass and stiffness determine its natural resonance and thus its minimum effective frequency. Mounting also is important; unenclosed diaphragms mean low efficiency for speakers and low sensitivity for pressure-gradient microphones, with accompanying bidirectional pattern. Enclosure or effective baffling means higher efficiency or sensitivity and omnidirectional pattern for all wavelengths much larger than the diaphragm.

While in electrical form, signals may be easily amplified or equalized. Good amplifiers have almost no nonlinearity (as evidenced by harmonic or intermodulation distortion) or phase distortion. Compressing or limiting amplifiers automatically reduce their gain to accommodate high-level inputs.

Electromechanical transducers appear both in direct pickups (as for electric guitars) and in phonograph disc recording and playback, with magnetic and dynamic types being most important. Tape recording uses a magneto-electrical transducer to produce or detect magnetized regions on the tape's oxide coating.

Multiphonic sound reproduction can greatly improve the impression of realism, both through phantom image sources and the feeling of presence. A great deal depends, however, on judicious placement of microphones in recording and speakers in playback, as well as on the kind of mixing of different information channels that occurs in between.

REFERENCES

Of all the topics in this book, sound reproduction is undoubtedly the one for which it is most futile to attempt a meaningful survey within the limits of a single chapter. I can only hope at best to provide a few key concepts that will make your reading of other sources more meaningful. To really learn about high-fidelity sound reproduction you probably should read an entire book devoted to that subject or take a separate course.

For those with little or no technical background, *The Science of Hi-Fidelity*, 2nd ed., by K. W. Johnson and W. C. Walker (Kendall/Hunt, 1981), is exceptionally readable and practical, giving useful lists of specific standards to look for when buying hi-fi equipment. A series of articles by T. D. Rossing on the physics and psychophysics of high-fidelity sound appears in *The Physics Teacher*, *17*, 563 (1979), *18*, 278 and 426 (1980), and *19*, 291 (1981).

An excellent and up-to-date source on professional recording techniques and equipment is *Sound Recording*, 2nd ed., by John Eargle (New York: Van Nostrand Reinhold, 1980); Chapters 3 and 4 are especially informative about all the pros and cons of the various versions of stereo and quad sound, and Chapter 9 tells about compression and noise-limiting techniques. A unique source of specific information about transducers (which, unfortunately, reads somewhat like a catalog) is Harry Olson's *Modern Sound Reproduction* (New York: Van Nostrand Reinhold, 1972). An excellent article by John Monforte on digital reproduction and compact disc technology is in *Scientific American*, December 1984.

SYMBOLS, TERMS, AND RELATIONS

electron
charge
current
voltage
magnet
transducer
reciprocity
microphone
loudspeaker
amplifier
dynamic range
frequency response

electroacoustic or electromechanical transducer types:
- electrostatic (condenser, capacitor)
- piezoelectric (crystal, ceramic)
- dynamic (electromagnetic, moving-coil)
- magnetic

directionality
proximity effect

speaker systems and crossover networks
baffles and horns
acoustic suspension
harmonic distortion
intermodulation distortion
compression and limiting
noise reduction
equalization
binaural versus stereophonic
digital versus analog recording

EXERCISES

1. Suppose a more accurate criterion for flat frequency response of a microphone is that its diameter should not exceed half a wavelength. What is the maximum diaphragm size for good response up to 20 KHz?

2. If you double the diameter of a crystal microphone diaphragm, how many times larger does its area become? So, other things remaining equal, how many times greater total force acts on it? This similarly multiplies the amplitude of crystal motion, so that the voltage produced becomes four times as large as before. Using the fact that electrical oscillation energy depends on the square of voltage amplitude, show that this means an increase of approximately 12 dB in microphone sensitivity. (Hint: Remember Table 5.1.)

3. Describe specific situations that would favor the choice of an omnidirectional microphone and others that would favor a unidirectional microphone.

4. What frequency range is boosted by the proximity effect for a ribbon microphone (a) 1 m and (b) 20 cm from the sound source?

5. Explain how 1% harmonic distortion corresponds to a 40-dB margin of original signal over added components.

6. As with other waves and vibrations, it also is true in electric-circuit vibrations that the power is proportional to the square of the amplitude; that is, to voltage squared. If we consider the simple case of a preamplifier with identical input and output impedances, we may define its gain as the ratio of its output and input voltages, $G = V_o/V_i$. By what factor is the power multiplied, and so how much is the signal level boosted in dB, for amplifiers with gain (a) $G = 2$ and (b) $G = 100$?

7. For a magnetic tape traveling at 19 cm/s ($7\frac{1}{2}$ ips) past the recorder head, what is the spacing between successive regions of similar magnetization for signals of frequency (a) 100 Hz and (b) 10 KHz?

8. For a phonograph record groove of radius 10 cm, on a disc rotating at 33 rpm, what is the spacing of successive wiggles along the 62.8 cm circumference for signals of frequency (a) 100 Hz and (b) 10 KHz?

*9. Suppose a phonograph disc is punched 1 mm off center. When the needle is tracking a groove of 10 cm radius, how much pitch variation results and how often does it recur? (Recall that a 6% frequency change corresponds to a pitch difference of one semitone.)

10. Suppose a tape recording is made with the tape moving at 4.75 cm/s ($1\frac{7}{8}$ ips), and played back at 19 cm/s ($7\frac{1}{2}$ ips). What happens to (a) the pitch and (b) the tempo of the music?

11. Above what frequency would the output be confined largely to a narrow beam for a loudspeaker of diameter (a) 30 cm and (b) 3 cm?

12. Below what frequency would the output be extremely weak for an unbaffled speaker of 20 cm diameter? How large a (flat, open) baffle would it need to give good response down to 50 Hz?

13. Roughly what minimum length is required for a horn loudspeaker operating at (a) 2 KHz and (b) 50 Hz?

*14. Suppose you are putting a hi-fi system in a living room with volume $V = 100\ \text{m}^3$ and reverberation time $T_r = 0.5$ s, and you want to be able to produce reverberant sound levels up to a maximum of 100 dB. What total acoustical power capability must the system have? If the speakers are only 3% efficient in conversion, what maximum electrical power may be drawn from the amplifier? Answer the same questions for an auditorium with $V = 10^4\ \text{m}^3$, $T_r = 1.7$ s and horn loudspeakers with 30% efficiency. (Hint: See Section 15.4.)

15. How many speakers and how much spacing and total height would you suggest for a midrange column array to cover the range 500–2000 Hz?

16. Discuss the differences in what sounds your ears receive and in how those sounds are perceived when you listen to a stereophonic recording through headphones; compare both with ordinary stereo playback through speakers and with strict binaural reproduction.

17. Suppose a stereo recording is made with widely spaced microphones, so that for one solo picked up by both microphones the instrument is located 2 m farther from one than from the other. How much time delay will there be for these notes between right and left channels? In view of the time-delay cues your brain is accustomed to (Section 15.6), why may this situation produce a feeling of confusion about where the source is located?

18. To have accurate reproduction of audio signals up to 20 KHz, a digital recording would have to store at least 40,000 samples per second. (Think of needing to mark both a crest and a trough to know that an oscillation has occurred.) Compact disc systems actually read 44,100 samples per second for each of two stereo tracks, with each sample represented by a 16-bit word. How many bits per second must be read by the player? If a CD has a playing time of 67 minutes, how many bits are stored on the whole disc? If the information is stored in a total area of approximately 80 square cm, how much area is available for each bit? Does your answer agree with Figure 16.17?

PROJECTS

1. Visit a recording studio, and observe (a) room design and acoustics, (b) equipment, and (c) procedures. Can you identify an underlying esthetic philosophy? Does the approach to recording seem conditioned by the type of music involved?

2. Visit a hi-fi shop that displays speakers with the decorative grillcloths removed. Study the speakers and their manufacturers' brochures (which will involve wading through a lot of puffery to glean a few hard facts). To what extent can you account for the main features, using the concepts from this chapter? Can you explain Magneplanar speakers, for example, as belonging to one of our standard transducer categories? Explain other things you discover that were not included here.

CHAPTER

17 The Ear Revisited

We now return to the subject of human perception of musical sounds. In the remaining three chapters we will gradually move farther and farther from specific scientific measurements and toward vague esthetic judgments. Our material will change from carefully concocted and controlled sounds, sometimes very odd ones you would never encounter in normal life, to become the more complex and continually shifting sounds characteristic of real music.

Here we again take up the problem of how our ears perceive a single musical note, leaving melodies and chords until Chapter 18. We were able to study perception of steady tones only in a limited way in Chapter 6, because we had not yet fully developed such concepts as harmonic spectrum or resonance. Our further study here has a dual purpose: To find theories that are able not only to correctly predict how each sound will be perceived (loudness, pitch, and timbre) from its physical characteristics (intensity, frequency, and waveform or spectrum) but also to give us some idea of *how* our ears enable our brains to arrive at these judgments. If you look back at Figure 6.9, you will see that we are especially concerned now with the influence of sound spectrum on each percept.

The three basic percepts present interesting contrasts in our ability to make significant progress at the present level of sophistication; this refers both to the technical boundaries of this book and to the areas of current research that seem most fruitful. Loudness sensations were already relatively easy to understand in Chapter 6; we add here only a short sequel to cover the calculation of loudness for complex tones, and it will seem the natural thing to do once we have developed the concept of critical bands. Timbre is at the opposite extreme: Even with further conceptual tools, it remains a tough nut to crack, and we shall be able to say only a little more about it. Pitch is in between: Not too hard to prevent some satisfying progress but hard enough to force us to develop important new concepts and insights.

For that reason, we build the entire chapter upon our attempts to understand pitch. In Sections 17.2 and 17.3 we survey several models for pitch perception and some of the evidence for and against them. This leads to the important notions of critical bands and combination tones, which are presented in Sections 17.4 and 17.5, and to their application to the problems of loudness and timbre in Sections 17.6 and 17.7.

17.1 TYPES OF PITCH JUDGMENT

Let us distinguish three different types of pitch judgment. The most complex is *absolute pitch*, the ability to identify the pitch of a tone heard in isolation, or to sing a named pitch on demand with no external reference. This ability is limited to a small minority of people and is poorly understood. Some people simply may be left out on the basis of heredity, but it seems likely that most lose the capability through disuse. Developing this skill through practice while quite young probably is important in retaining it.

Next we have judgment of the magnitude of separation between two different pitches; that is, of *interval size*. This also remains undeveloped in many people, but well-trained musicians obviously must know whether they have played the correct notes. We will take up this subject in the next chapter.

We want to limit ourselves here to the simplest type, that of *comparative* pitch judgment. Specifically, we mean this in the limited sense of listening to two tones and then deciding whether both have the same pitch or which one is higher than the other—but with no attempt to say how much higher.

It is quite remarkable that we can make meaningful comparisons even when one steady tone is a pure sine wave while the other comes from a real musical instrument and has many strong harmonics in its spectrum. Yet steady natural tones almost invariably are closely matched to sine waves with frequencies the same as their own fundamentals. (For a qualification to this statement, see Benade, pp. 267–268.) This is something that must be explained by a good theory of pitch perception.

---

*Specifically, this is even more remarkable because the two pitches may be perceived with different mechanisms. The usual mechanism that extracts a single pitch from a complex signal is called *synthetic*. But by close and purposeful attention you may learn to exercise an *analytic* mechanism, by which you "hear out" the separate existence of several different pitches in the complex tone, identifying its first several harmonics as all present together. Your judgment of a pure sine wave probably is more closely related to the analytic mechanism, yet can be compared easily with a synthetic pitch judgment. Only in extreme or artificial cases where the spectrum is very poor in lower harmonics does this become difficult.

17.2 PITCH PERCEPTION MECHANISMS

Now let us take up, in turn, four different types of pitch perception theory: telephone theory, place theory, periodicity theory, and pattern-recognition theory. Keep in mind that each of these is really a whole family of theories, differing in many details that go beyond our present scope; we present them here only in highly simplified form.

The **telephone theory** of hearing (Rutherford, 1886) is a natural analogy, and it probably was inevitable that someone would suggest it shortly after Alexander Graham Bell's invention. In its most extreme form telephone theory proposes that the ear is merely a microphone, which converts sound waves into corresponding electrical signals. That is, the ear does nothing significant in the way of processing, filtering, or analyzing the sound, but passes on all information intact to the brain. The burden of detecting such properties as pitch is thrown entirely upon the brain.

We mention this theory only as a foil for the others, because it is quite unsatisfactory in two different ways. First, it fails to really answer the question, for it relegates all important functions to levels in the brain where there is less hope of giving any physical explanation. Second, numerous twentieth-century experiments have shown that the ear does radically transform the signals going through it, so that both the information transmitted by the cochlea to the brain and the form (nerve pulses) in which it is sent are quite different from the continuous sound signal falling on the outer ear.

The **place theory** is based upon ideas of Helmholtz (1863) as tested and developed by von Békésy. Place theory gets its name from associating each pitch with a different place on the basilar membrane. It is attractive because it accounts for pitch perception with the simplest and clearest possible physical mechanism: Different frequencies preferentially stimulate different places on the membrane and thus different nerve endings. This theory was judicious speculation on Helmholtz's part, because direct supporting evidence did not really come until von Békésy's experiments some seventy years later. Even now, the information we can get from direct examination of human cadavers or live laboratory animals is severely limited; we must depend on clever interpretation of psychophysical experiments for most of our evidence for or against any theory of what happens inside the cochlea of a live human subject.

Resonance is fundamental to Helmholtz's original place theory. Consider the analogy of Figure 17.1: If the same driving signal is applied to each of a set of oscillators, the strongest response is from those whose natural frequency nearly matches the driving frequency. Thus observation of which oscillators have large amplitude is a way of telling the driving frequency. We already have hinted in Chapter 6 that the length and stiffness of the fibers vary along the basilar membrane in a way that makes it vibrate more easily near the oval window for high frequencies and farther away for lower frequencies. So we need only suppose that the brain derives its pitch judgment from whichever nerve ending is being stimulated most vigorously. More recent versions of place theory improve the analogy of Figure 17.1 by adding coupling between adjacent masses, recognizing that vibration energy can be passed along the basilar membrane from one place to another in the form of transverse waves.

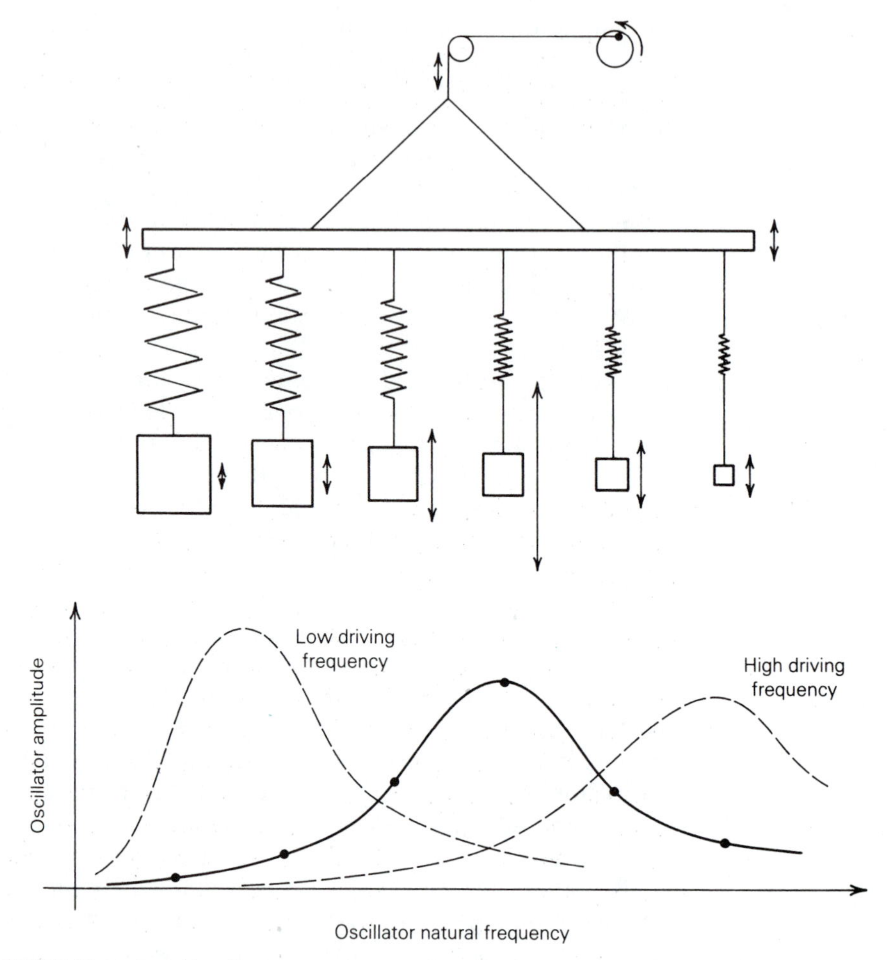

FIGURE 17.1 A set of oscillators, with large masses on weak springs giving low natural frequencies on the left, and small masses on stiff springs for high frequencies on the right. Driving all with an alternating force of medium frequency results in steady response indicated by arrows and by the solid curve below. Lower or higher driving frequency produces different response amplitudes (dashed curves). Which oscillator has maximum amplitude provides evidence of the driving frequency. The oscillators are analogous to groups of basilar membrane fibers, and the camshaft drive to the middle ear.

*Technically, this is somewhat different from resonance as explained in Section 11.3, and the analogy of Figure 17.1 is not very accurate (Green, Chapter 3). A better explanation invokes traveling-wave energy being preferentially dumped wherever the wave frequency matches the cutoff frequency corresponding to the local membrane properties. In spite of its inaccuracy, I believe Helmholtz's naive resonance model makes the best initial introduction to place theory.

In favor of place theory, we now can point to measurements mapping out membrane response in human cochleas studied within an hour or so after

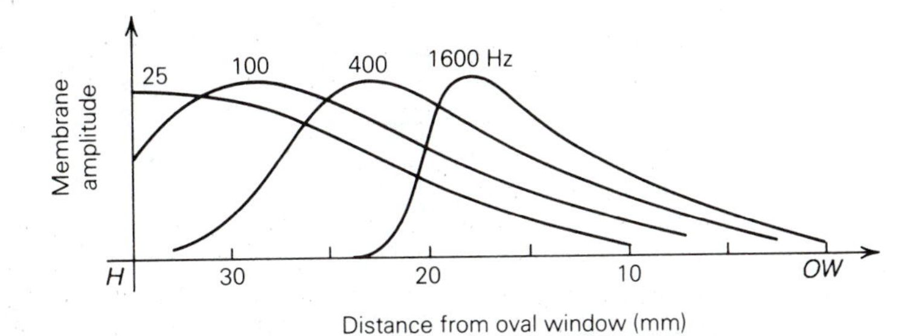

FIGURE 17.2 Human basilar membrane response for various frequencies, as determined by von Békésy (1943); *H* denotes the helicotrema. These measurements on cadaver specimens were fraught with great difficulties. Our belief that these curves are realistic is based largely on experiments with large mechanical models of the cochlea, and on measurements in the ears of live guinea pigs and squirrel monkeys.

death. Figure 17.2 shows how the amplitude of motion varies with position for several frequencies. The place of maximum amplitude serves as an index of the driving frequency and thus of the pitch; Figure 17.3 shows how it shifts smoothly from helicotrema to oval window as the frequency increases.

It also seems favorable for place theory that it naturally explains Ohm's law: Different harmonic components of a sound preferentially excite different parts of the basilar membrane. Only their relative strengths should matter and not their phases. We must admit here an assumption that each component affects the membrane the same way regardless of the presence of the other components. That is, we are supposing that the ear is a linear processor; we shall discuss certain consequences of nonlinearity in the following sections.

The widths of the response curves in Figure 17.2 are an embarrassment to place theory, because they suggest that the frequency of a sine wave must be changed by perhaps 50 or 100% to stimulate a distinct group of receptors; yet changes of only 1% are easily noticed. That is, there apparently is a *sharpening* of response by the time pitch is judged in the brain, so that a small frequency change is enough to ensure messages on a distinct set of nerves.

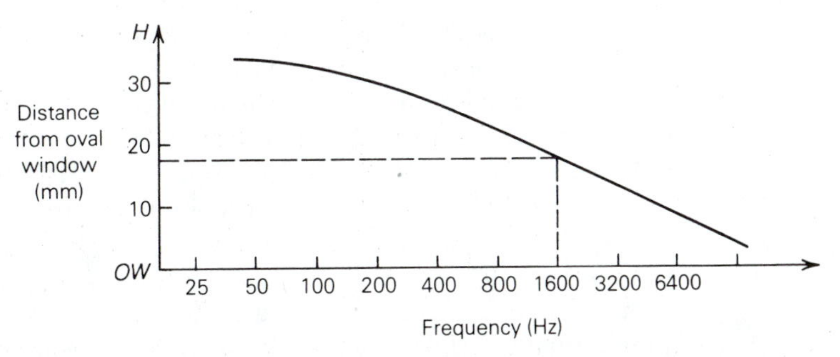

FIGURE 17.3 Typical data for the shift of the place of maximum response from the apical end of the basilar membrane (at the helicotrema) for low frequencies to the basal end (oval window) for high frequencies. The straight portion of the curve corresponds to a shift of approximately 4 mm per octave. Dashed lines indicate the 1600-Hz peak in Figure 17.2.

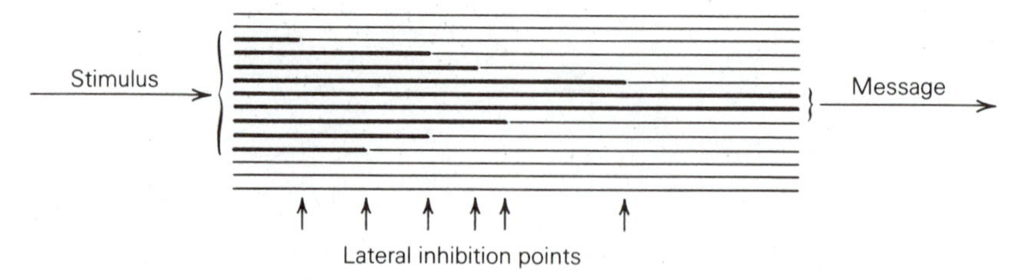

FIGURE 17.4 Lateral interactions among neighboring nerve fibers can inhibit their firing selectively, so that a few function as representatives of an originally larger group to carry the message to the brain. These few may be keyed quite accurately to the location of maximum response in the broad stimulated region. This picture applies to some phenomena of touch and vision but is *not* adequate to explain auditory sharpening.

It is tempting to save the simple place theory by supposing this sharpening occurs in the nerves after they leave the cochlea (Figure 17.4). This hypothesis is attractive, because indirect evidence from studies of touch and visual perception shows that groups of nerves can do this in humans. Lateral inhibition also has been observed directly with implanted electrodes in laboratory animals. Recent experiments show, however, that the sharpening already is present in first-order neurons before they leave the cochlea. So the actual mechanism for sharpening remains a mystery.

Other objections against the simple place theory also are hard to dispose of. Place theory gives no hint about how we assign a unique pitch to a complex tone; it suggests instead that we should perceive many pitches together, one for each region on the basilar membrane that is at the peak of the excitation pattern for one of the strong Fourier components. It does no good to claim that we merely choose the lowest of these pitches (generally meaning the fundamental of a harmonic series) and ignore all the others. We can present to the ear a harmonic recipe in which the fundamental is completely missing and the perceived pitch still will be that of the fundamental, even though the corresponding region of the basilar membrane hardly vibrates at all. The same thing happens even when the first several harmonics are missing or when they are masked by noise. This *missing fundamental* phenomenon is what enables you to "hear" bass notes from a pocket transistor radio whose small speaker reproduces practically nothing below 200 Hz.

The competing **periodicity theory** was stimulated largely by a long series of psychophysical experiments by several Dutch scientists, beginning around 1940 with Schouten. This theory supposes that the messages from cochlea to brain contain more than just information on strengths of harmonic components. Rather, some information on original waveform is retained (although not used in such a way as to cause prominent violations of Ohm's law). The simplest way to retain this information is to have each nerve ending fire preferentially during one part of the cycle of basilar membrane oscillation, as indeed we indicated in Figure 15.20 (page 335). This works well only for lower frequencies, however, up to approximately 1 KHz.

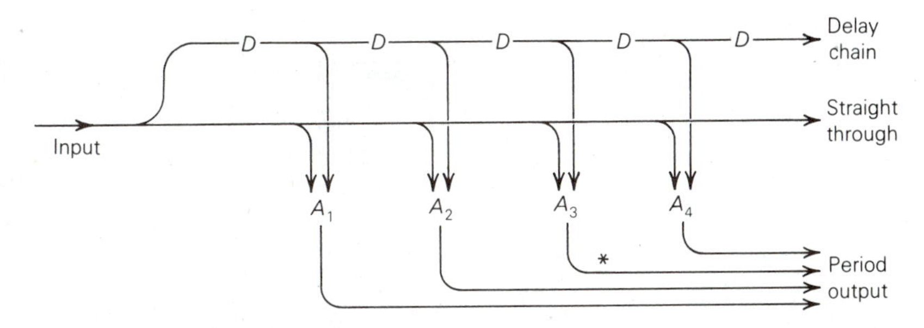

FIGURE 17.5 A hypothetical period-measuring autocorrelator (after Licklider, 1959). Certain types of nerve connections, represented by D, delay a signal in time. Others are logical "AND" units, A, which produce output only when *both* inputs are active. Signals with different period activate different output fibers. For instance, if each delay were 1 ms, then a wave with a 3-ms period would produce output on the fiber marked *, because the triply-delayed version of each input pulse would arrive at A_3 together with the undelayed version of the following pulse.

What would be the advantage for nerve impulses to arrive in bunches instead of in a steady stream? It must be supposed that the brain has some way of measuring the time separation between bunches (that is, the period of the stimulating sound wave) quite accurately. An electrical engineer can easily design circuits to perform such a task (Figure 17.5), and it is not unreasonable that analogous circuits may exist in the brain. But it is thus far impractical to obtain direct evidence as to whether the brain utilizes any one particular type of period-measuring circuit or precisely where in the brain it may be located.

An attractive feature of periodicity theory is its explanation of pitch extraction from a *residue* of high harmonics when the lower ones are missing. It makes a virtue rather than an embarrassment of the way the excitation curves overlap so that many high harmonics all strongly excite the same part of the basilar membrane. The combination of several high harmonics can make a response waveform with prominent peaks whose period is the same as that of the series fundamental (Figure 17.6), so that a period-measuring mechanism in the brain could get the same message as it would from the fundamental and thus report the same pitch.

One of periodicity theory's greatest strengths is its ability to account for pitch judgments on some strange tones you ordinarily would never hear in music. (One possible exception is ring modulation with an electronic synthesizer.) These tones consist of sets of Fourier components electronically shifted over in frequency so that they still are evenly spaced but no longer form integral multiples of a fundamental. Think, for instance, of taking ninth, tenth, and eleventh harmonics of 200 Hz and shifting them upward slightly to 1830, 2030, and 2230 Hz, or even further to 1860, 2060, and 2260 Hz. The perceived pitch also shifts upward, as shown in Figure 17.7, in a way for which place theory has no workable explanation. There are also ambiguities to be explained, with the test signal being matched sometimes to one and sometimes to the other of two distinct pitches.

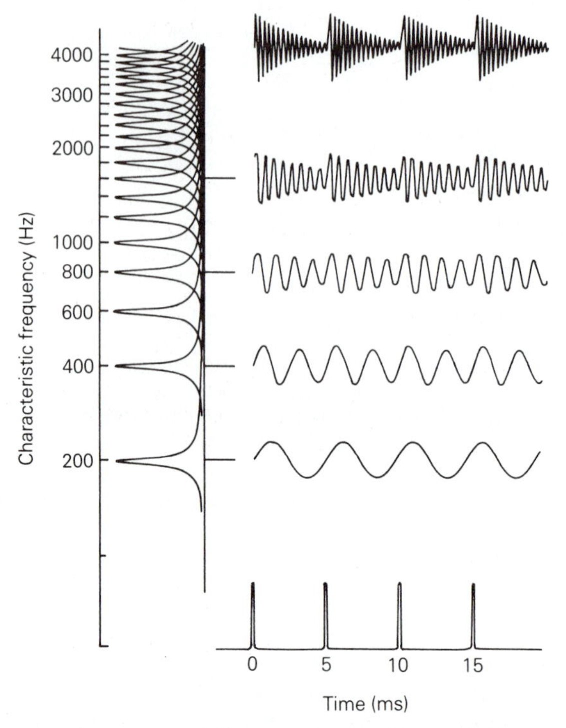

FIGURE 17.6 Response of a set of tuned oscillators to short pulses at 200 Hz (after Schouten, 1940). Low-frequency oscillators respond strongly to only a single harmonic, and this response is nearly a steady sinusoid. High-frequency oscillators feel overlapping effects of several harmonics, and their total response shows the periodicity of the pulses. (From Tobias, Vol. 1, p. 45, by permission of Academic Press)

The detailed waveform of this nonharmonic series does not repeat exactly, although it has an "envelope" that does (Figure 17.8). But the repetition frequency of this envelope remains unchanged at 200 Hz, so it apparently does not provide the basis for pitch judgment. The experimental data are accounted for by supposing that we measure the time from one high peak in the detailed waveform to another under the following bulge in the envelope. Thus for 1860/2060/2260 Hz, a high peak is often (though not always) followed by another about 4.85 ms later, suggesting a frequency of about 206 Hz. But other pairs of high peaks separated by about 5.35 ms suggest a frequency of about 187 Hz. Either frequency may come to the fore in pitch matching, thus explaining both the pitch shift and the ambiguities.

---

*It is extremely hard to compare a component cluster such as 1860/2060/2260 directly with a pure sine near 200 Hz, because their timbres are so different. The test stimuli thus are actually matched instead to a pulse wave with a frequency of approximatey 200 Hz; this, in turn, is shown by other experiments to have practically the same pitch as its fundamental.

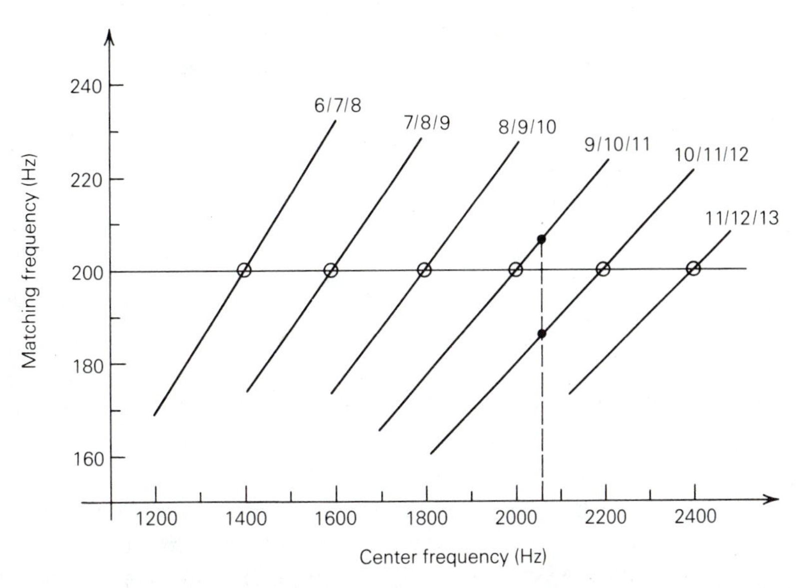

FIGURE 17.7 Pitch matching of three-component signals (after Schouten, 1962). Horizontal axis shows frequency of the center component of the test signal, with the other two being 200 Hz above and below. Vertical axis represents frequency of a pure sine wave whose pitch is judged the same as the test signal. Open circles (○): when center frequency is a multiple of 200 Hz, the test signal is part of a harmonic series; otherwise not. Solid dots (●): the example of an 1860/2060/2260 Hz test signal, illustrating ambiguous pitch perception. 6/7/8, etc.: pseudoharmonic numbers involved in the brain's attempt to find a harmonic series pattern. For more extensive data showing other alternative pitch matches, see A. Gerson and J. L. Goldstein, *JASA*, *63*, 498 (1978).

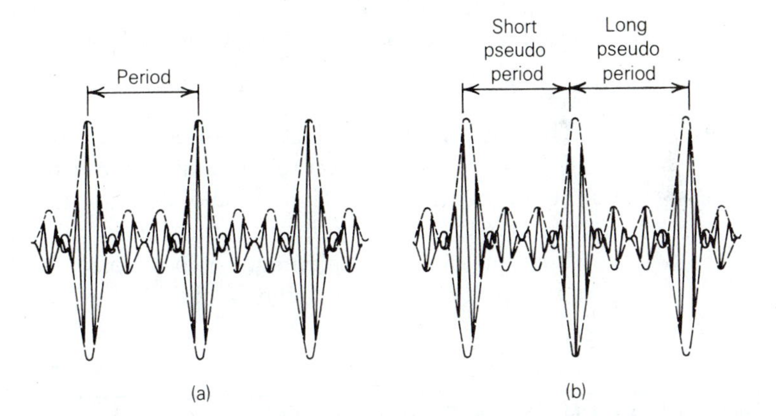

FIGURE 17.8 Waveforms of three-component signals tested by Schouten. (a) When formed by three neighboring members of a harmonic series, the total disturbance is exactly periodic. (b) When components are spaced evenly but not part of a harmonic series (as in the 1860/2060/2260 example), the actual waveform (solid curve) is not periodic, although it defines an envelope (dashed curve) that does repeat periodically. The most prominent peaks offer two different time separations that the ear might try to interpret as a repetition period, thus explaining the ambiguous perceptions in Figure 17.7.

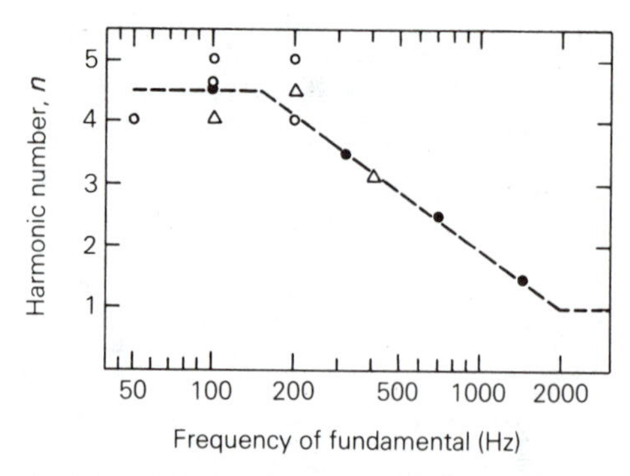

FIGURE 17.9 Harmonics dominant in determining pitch sensation for different fundamental frequencies. (With permission from *Aspects of Tone Sensation* by R. Plomp, 1976. Copyright by Academic Press Inc. (London) Ltd.)

But simple periodicity theory has recently fallen in disfavor, because it, too, is unable to account for all experimental data. It predicts, for instance, that the collective effect of high harmonics (say from seventh or eighth on up) should be more important than any of the lower harmonics in determining pitch. Yet experiments show that the lower harmonics actually are dominant (Figure 17.9). This is shown most simply by electronically generating a signal containing contradictory pitch cues. For instance, combine the third harmonic of 200 Hz with the eighth, ninth, and tenth harmonics of 205 Hz; this is perceived as having pitch corresponding more closely to 200 than to 205 Hz.

Periodicity theory also would lead us to expect pitch and timbre of complex tones to be quite sensitive to the phases of their harmonic components. So this theory is hard put to explain why Ohm's law is such a good approximation.

The place and periodicity theories both have successes that cannot be denied. We must move on to something better not because they are wrong, but because each alone is inadequate. And so currently favored theories represent not some entirely different possibility but rather a more sophisticated combination of certain elements of both place and periodicity theories together with further ideas that extend them beyond their previous limitations.

17.3 MODERN PITCH PERCEPTION THEORY

Recent work on pitch freely uses both place and periodicity concepts. These eclectic theories may be characterized as emphasizing **pattern recognition**. Several important versions were published independently around 1973, by Wightman, by Goldstein, and by Terhardt. All emphasize central or higher-level processing in the brain; like it or not, we have shades here of the telephone theory. The ears serve primarily to send spectrum information as in the place theory, and direct detection of periodicity is relegated to a minor role, although not entirely discarded.

Even though we must admit almost total lack of evidence for the actual mechanism, scientists now are persuaded that the brain somehow searches for signs of order among the frequencies of all Fourier components reported by the ears. If something resembling a harmonic series is found, the brain then assigns a pitch corresponding to its fundamental. For example, 1800 Hz alone might be a fundamental, or the second harmonic of 900, the third of 600, the fourth of 450, and so forth; 2000 Hz alone suggests possible fundamentals 2000, 1000, 667, 500, and so on. But 1800 and 2000 together rule out most of those possibilities; the first good fit is obtained by supposing they are the ninth and tenth harmonics of 200 Hz.

One of the strongest pieces of evidence for central pattern recognition is that information from both ears can be combined to arrive at the final judgment. Several experiments with *dichotic* signals (different information to each ear through earphones) show this, Houtsma and Goldstein's (1972) being especially beautiful and musically relevant. Suppose that components at 1200 and 1600 Hz are fed to the right ear and 1400 Hz to the left ear. If pitch judgments were made separately for each ear we would expect to hear two pitches, corresponding to 400 Hz on the right and 1400 Hz on the left. But Houtsma and Goldstein's subjects were able to identify a single pitch for the combined sound, although this *dichotic pitch* is rather weak and difficult to hear. For this example the dichotic pitch corresponds to 200 Hz, showing that the brain can form a combined list (1200, 1400, 1600) before deciding what harmonic series the components best fit.

The brain also tries hard to recognize orderly patterns even where they do not actually exist in the original stimulus. This has long been known for visual stimuli, through the interpretation of a wide variety of optical illusions. We have only recently begun to recognize the corresponding aural phenomena. Take again the case of a stimulus sound containing components at 1860, 2060, and 2260 Hz. For lack of any better fit, the brain usually will accept these as ninth, tenth, and eleventh harmonics of a fundamental of approximately 206 Hz, even though 9×206 and 11×206 are really 1854 and 2266 Hz. The fit, however, is not all that much worse to tenth, eleventh, and twelfth harmonics of approximately 187 Hz (1870, 2057, and 2244), and indeed sometimes the brain may decide on that lower pitch instead. These attempts to find the nearest fit between the real components and the desired harmonic-series pattern account for most of the experiments on residue pitch (such as Figure 17.7) that previously were interpreted as evidence for periodicity detection.

Pattern search and recognition account nicely for our perception of pitch in notes from piano, chimes or bells, and tympani. The spectrum of a piano note is slightly inharmonic, and the brain blithely ignores the small deviations. Bells or tympani have quite irregular spectra that are vastly different from harmonic series. Yet if the brain can find a few of these components that will even approximately make a harmonic pattern (say multiples 3, 4, and 5 of some frequency, even though that fundamental itself may be completely absent from the sound), it will latch onto them and ignore the rest in order to report a definite pitch.

As an example, suppose you hear prominent modes from a kettledrum with frequencies 202, 305, 402, 493, and 584 Hz. These resemble second through sixth members of a harmonic series. But you probably would judge the pitch to be a little below $\frac{1}{2} \times 202 = 101$ Hz, because three out of four of the remaining components are "voting" lower than the 303, 404, 505, and 606 implied by that fundamental. Similarly, the estimate $\frac{1}{6} \times 584 = 97$ Hz suggested by the highest component is "outvoted" by higher estimates from all the others. Your brain's compromise choice for the best pitch estimate probably will be around 99 or 100 Hz (or, in practice, perhaps 200 Hz). You still may make the same judgment in the presence of an extra component such as 320 Hz, which does not fit the pattern.

Goldstein's "optimum-processor" theory even has made headway toward accounting quantitatively for the salience or clarity of pitch perception; that is, the strength of the feeling that there is a definite pitch rather than just noise. It does this by identifying the pitch clarity with the closeness of fit between the actual spectrum and the harmonic series for the reported pitch, according to a specific mathematical criterion.

You may wonder *why* the brain tries to recognize harmonic series. Pattern recognition with visual stimuli clearly has survival value; although it is much less obvious, this also may be true for sound. Terhardt proposes that the central processor learns by doing: Every time we hear a steady tone with a harmonic spectrum, that experience makes it a little easier for the brain to recognize the same pattern when it occurs again. We learn early that speech sounds are especially important, and the voiced phonemes have approximately harmonic spectra, so we pay special attention and diligently practice recognizing that type of pattern. A mother's singing to her infant may be even more effective. If true, this model suggests that hearing a lot of music in early childhood would develop a stronger harmonic-pattern–recognition capability, and thus keener pitch judgment and greater sensitivity to musical intervals. Even a richer speech environment would be of some help.

Let us now move on to a couple of topics that are important for reaching a deeper understanding of competing theories of pitch perception and of the evidence for and against them.

17.4 CRITICAL BANDS

Many psychoacoustic experiments bring out a difference in response depending on whether the frequencies of two stimuli are nearly the same or quite different. If close together in frequency, they both disturb more or less the same region of the basilar membrane; but if far apart, they stimulate two regions independently—that is, there is little overlap of the two response curves (Figure 17.2).

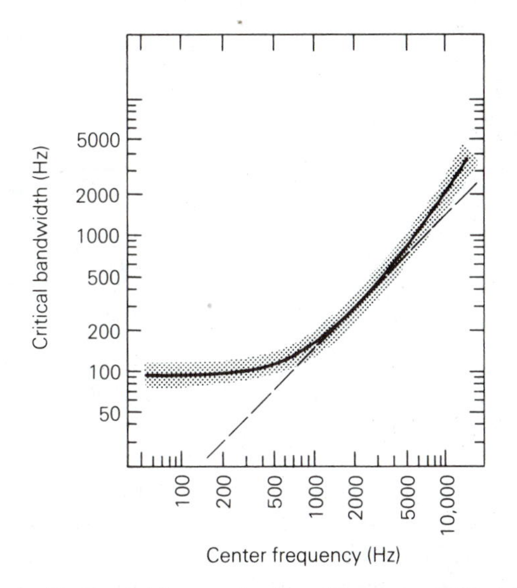

FIGURE 17.10 Measured critical bandwidths as a function of frequency. Width of line indicates uncertainties in measured data. Dashed line shows a constant 15% bandwidth for comparison.

The range of frequencies whose response curves effectively overlap is called a **critical band**. Because the overlap only gradually decreases with separation, this criterion appears somewhat arbitrary. But different experiments produce estimates of critical bandwidth similar enough for Scharf to use the definition, "The critical band is that bandwidth at which subjective responses rather abruptly change."

Typical results are shown in Figure 17.10. They may be described approximately by saying that critical bandwidths for frequencies above 500 Hz are about 15–20% of the center frequency, or about $2\frac{1}{2}-3$ semitones, or about 1–1.5 mm along the basilar membrane, enough to include some 1000–1500 receptors out of a total of 20,000 or 30,000. Lest there be any misunderstanding, we do *not* mean that there are any fixed boundaries; rather, you can pick *any* point on the basilar membrane (and corresponding frequency) and consider a critical band to be centered there.

Let us describe now some of the experiments that produce this data. Suppose first you listen to a narrow band of noise. Take, for instance, a mixture of all frequencies from 980 to 1020 Hz, which we would describe as having center frequency 1000 Hz and bandwidth 40 Hz. Now let the bandwidth gradually increase while the center frequency and total intensity remain constant; that is, spread the same energy over an increasingly larger frequency range. For a while the loudness stays the same, but once the bandwidth exceeds roughly 160 Hz, the loudness increases (Figure 17.11).

Similar critical bandwidths come from experiments on how much the presence of one tone blocks the perception of another at a different frequency. This masking phenomenon also will be discussed further in Section 17.6.

Yet another source is an experiment on pitch fusion. Suppose now you

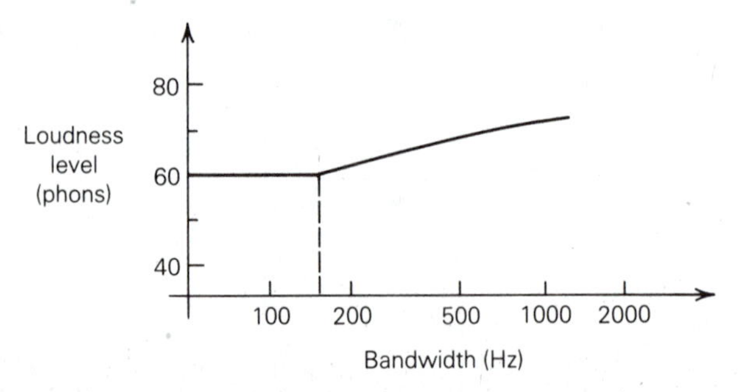

FIGURE 17.11 Change in loudness of narrow-band noise as bandwidth increases and intensity is kept constant. The slope changes at the critical bandwidth. For this example (after Zwicker and Feldtkeller, 1956) the center frequency is 1000 Hz.

listen to two sine waves of equal amplitude simultaneously. If both have precisely the same frequency and are in phase, the total disturbance is a sine wave of twice the amplitude, so you merely hear the same pitch but louder. If the two frequencies are quite different, you are conscious of two separate pitches. But if the frequencies are only slightly different, you cannot hear two pitches; you hear only one, although it has beats (recall Section 4.5, especially Figure 4.16, page 68). As the frequency separation gradually is increased, the beats become more and more rapid; even when they are too fast to be counted individually they still manifest themselves in an unpleasant *roughness*. As the two frequencies approach a critical band separation, we begin to perceive two separate tones, and when they are finally far enough apart to stimulate distinct sets of nerves, the roughness disappears (Figure 17.12).

---

*We have assumed that both sine waves go to the same cochlea, so that we get strong "first-order beats." If one sine wave goes to the right ear and the other to the left, you will hear only the much more subtle and elusive second-order beats, if any. (See the Oster article, noted in the references.)

Nature has solved a problem of conflicting requirements here in a remarkably sophisticated way. The critical bands alone provide only poor pitch discrimination, but this is improved greatly by the sharpening process discussed in Section 17.2. If we tried to design a cochlea that would achieve good pitch discrimination in a single-stage process, we would need much narrower critical bands. That is, the resonant response curves of the cochlear fibers would have to be much sharper. But that is linked inextricably with less frictional damping (refer to Figure 11.16, page 212), so that your ears would continue to ring like a bell after each disturbance. Instead of being able to follow changing stimuli within a few milliseconds, your basilar membranes would tend to continue vibrating for a large fraction of a second at high frequencies and much longer than a second at low frequencies. You may regard this as a manifestation of the uncertainty principle (see Box 17.1).

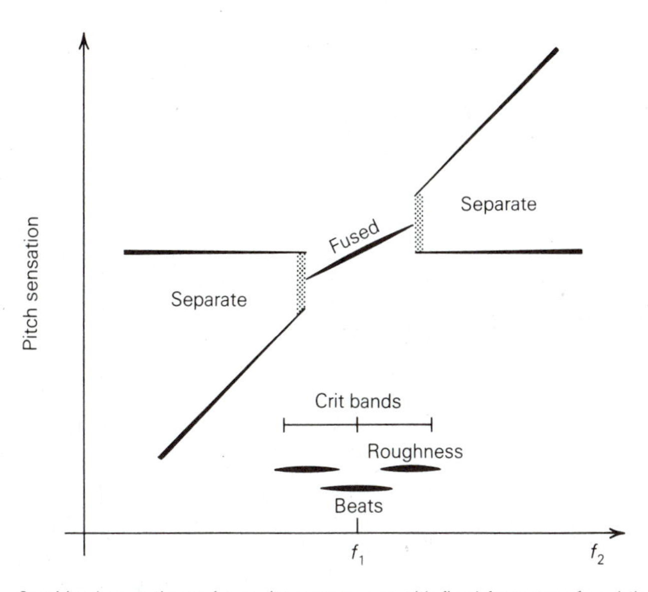

FIGURE 17.12 Combined sensations of two sine waves, one with fixed frequency f_1 and the other with variable frequency f_2. When the two frequencies are nearly the same there are audible beats, and over the remainder of a critical band on either side there is a sensation of roughness corresponding to beats that are too rapid to be perceived individually. Over most of this same range there seems to be only a single pitch, about the average of what the two pitches would be when heard individually; outside this range two distinct pitches can be heard. (After Roederer, p. 29)

*BOX 17.1 THE UNCERTAINTY PRINCIPLE

Suppose you have a tone-burst generator that can provide 10 cycles of a sine wave (frequency f_0), preceded and followed by silence. The total disturbance, including the silences, is nonperiodic (that is, it cannot be divided into identical segments that repeat unendingly), so its spectrum is not a harmonic series. Fourier analysis shows that it is instead smeared over a continuous range in the vicinity of f_0 (Figure 17.13). The width of this range can be estimated from the inequality $f't' \gtrsim 1$, where f' means the *range* of frequencies with relatively large Fourier amplitudes, and t' means the *interval* of time during which the signal remained strong. Notice this says that only a signal lasting a very long time (large t') can have a crisply defined frequency (small f').

What limitations does this place on sound production? Signals lasting a very short time necessarily contain a substantial range of frequencies, so not even an ideal detecting apparatus could assign a precise frequency or pitch to them. Fuzziness of pitch is an inherent property of short transient or percussive sounds, and the shorter the more indefinite.

The limitation applies equally well at the receiving end: A finely tuned resonant detector that responds only to an extremely narrow range of frequencies f' can do so only over a long time. Even if the input signal is short, the detector itself will ring for at least $t' \gtrsim 1/f'$.

The uncertainty principle is extremely powerful, for it applies equally well to waves of every kind. Take for instance a 600-KHz AM radio carrier wave, whose amplitude changes over time scales as short as 0.2 ms as it is modulated by audio information with frequencies up to 5 KHz. It must be regarded as occupying a frequency band at least 5 KHz wide, and this

(continued)

BOX 17.1 *(continued)*

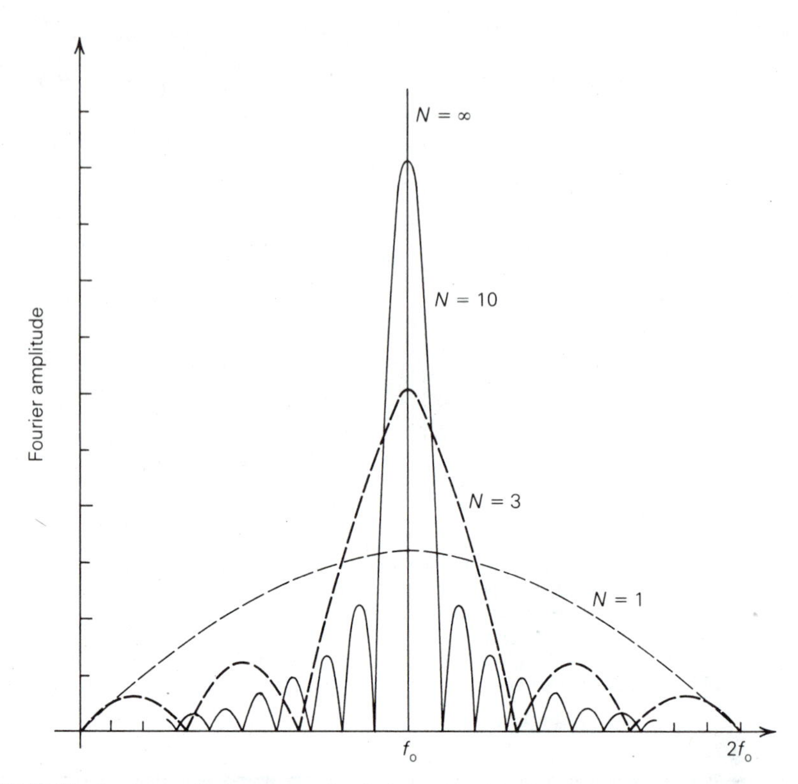

FIGURE 17.13 Fourier amplitude spectrum for N sine waves of frequency f_0, preceded and followed by silence. Only as N becomes very large does the spectrum become a perfectly sharp spike at f_0.

entire band must be selected by the tuning circuit in the receiver.

In atomic and nuclear physics, quantum theory associates frequency with energy. Then the uncertainty principle tells us that unstable states (such as radioactive nuclei) that have only limited lifetimes will exhibit inherent energy spreads in the decay processes. There is also a spatial version of the uncertainty principle, involving wavelength instead of frequency; it can be used to gain deeper understanding of diffraction, and even to explain why typical atoms are approximately 10^{-8} cm across rather than larger or smaller.

17.5 COMBINATION TONES

The violinist Tartini noted in 1714 that when playing two notes loudly he sometimes heard a third note as well. This *difference tone* appears at frequency $f_{DT} = f_H - f_L$, where f_H and f_L are the higher and lower of the two stimulus frequencies. Figure 17.14 shows a few examples in musical notation; these are easily worked out with the help of Figure E. The presence of a difference tone

FIGURE 17.14 Some examples of combination tones. ● = the two primary tones; x = the sum tone; ○ = the ordinary difference tone; ◇ = the cubic difference tone. Rules for construction of additional examples are given with Figure E.

is not too surprising in view of our remarks on intermodulation distortion in Section 16.4: if there is nonlinearity anywhere in the sound transmission path, it should introduce this additional component. (See Figure 2.11, page 211, and Figure 16.12, page 354, for a reminder of what nonlinearity means.) The air itself contributes no significant distortion unless sound levels are far above 100 dB, so we must look to the receiver.

The eardrum, middle ear, and inner ear are all candidates to contribute nonlinearity; indeed it would be surprising if any of them had an entirely linear response to loud sounds. It still is hard to pin down exactly which components in the chain contribute most, but the existence of some nonlinearity in the ear has long been accepted.

Nonlinear distortion should generate both the difference tone and a *sum tone* ($f_{ST} = f_L + f_H$) with comparable amplitude. The sum tone is much more difficult to hear, because it is always within an octave above f_H and is strongly masked by the presence of the two original tones. The difference tone also is difficult to detect when it lies between f_L and f_H, but comes through well when f_{DT} is much below f_L. As the original tones are increased in strength, the difference-tone level rises even faster (Figure 17.15.)

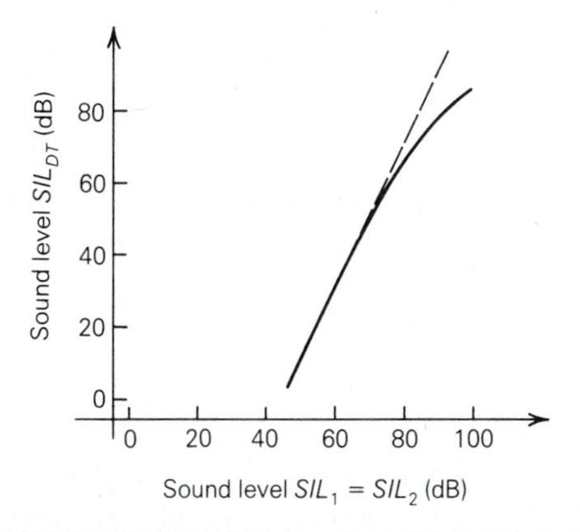

FIGURE 17.15 Simple nonlinearity should produce an ordinary difference tone whose strength rises 20 dB for every 10-dB simultaneous increase in level of both generating tones (dashed line). Data fit this prediction quite well for lower levels, but saturation prevents the difference tone from becoming stronger than the primaries at high levels.

TABLE 17-1 The hierarchy of combination tones. Those in the last column represent ordinary harmonic distortion and are present for a single primary. For simple nonlinearity and moderate amplitude, the levels in dB decrease comparable amounts in going from each row to the one below. These formulas may occasionally give a negative frequency for the difference tones, but the minus sign should be ignored.

Primaries	f_L	f_H	
Secondaries	$\mathbf{f_H - f_L}$	$\mathbf{f_H + f_L}$	$2f_L, 2f_H$
Third-order	$\mathbf{2f_L - f_H}$, $2f_H - f_L$	$2f_L + f_H, 2f_H + f_L$	$3f_L, 3f_H$
Fourth-order	$3f_L - f_H, 3f_H - f_L$, $2f_H - 2f_L$	$3f_L + f_H, 3f_H + f_L$, $2f_H + 2f_L$	$4f_L, 4f_H$
...	...	...	...

This was sometimes thought in the past to account for our ability to identify missing fundamental pitches. For example, a stimulus containing 1800, 2000, and 2200 Hz components would generate difference tones at 200 Hz (both 2000–1800 and 2200–2000), and at 400 Hz (2200–1800). This apparently is *not* the main reason for our identifying the pitch with 200 Hz, however, because (1) the impression of pitch corresponding to 200 Hz persists even at such low intensity levels that the difference tones could not be important; (2) even if that part of the basilar membrane mainly responsible for detecting 200 Hz is distracted by a band-limited noise signal strong enough to completely mask the difference tone, the perceived pitch is undisturbed; and (3) shifting components upward to 1860/2060/2260 Hz (as discussed in Section 17.2) still leaves the difference tone at 200 Hz, yet the perceived pitch shifts upward.

In general, the simplest type of distortion is a slight and gradual departure from linearity with increasing sound level. This would be expected to generate a hierarchy of higher-order combination tones (Table 17.1). Most of these are like the sum tone in being extremely difficult to detect. The main one of some importance is the *cubic difference tone*, with $f_{CDT} = 2f_L - f_H$; under some circumstances it is as easily audible as the ordinary difference tone.

Here the story becomes complicated, because the cubic difference tone does not behave as expected when sound levels vary (Figure 17.16). It may persist even at low levels where the ordinary difference tone cannot be heard. Its audibility also depends much more strongly on frequency than that of the ordinary difference tone, decreasing rapidly when f_H exceeds f_L by more than 20 or 30%. A qualitative difference is indicated: the cubic difference tone, at least for low sound levels, represents an *essential nonlinearity*. That is, the transducer behavior cannot be made to approach linearity just by limiting the signal to sufficiently small amplitudes, although this usually escapes notice unless f_L and f_H share a critical band. Recent research indicates that a great deal still remains to be understood about these combination tones. Progress is

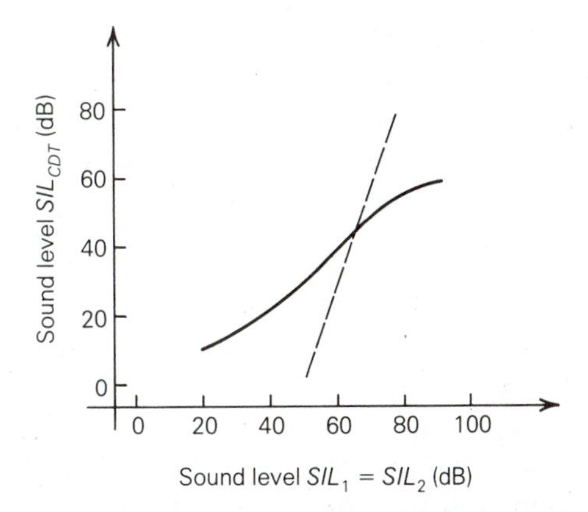

FIGURE 17.16 Simple nonlinearity predicts a cubic difference tone whose strength rises 30 dB for every 10-dB increase of both generating tones (dashed line). Actual data show a completely different trend and suggest the presence of an essential nonlinearity. (For further details, see de Boer, pp. 537–544.)

hampered even by doubts about whether the nonlinear mechanism is the same in anesthetized cats as in alert humans (see, for example, J. L. Goldstein, *JASA*, *63*, 474, 1978).

17.6 LOUDNESS AND MASKING

We considered in Section 6.4 the use of the sone scale to describe the psychophysical perception of loudness for lone sine waves. We are now in a position to extend those rules to cover complex wave combinations. Many, but not all, experimental observations are well summarized by the following rule:

> ***The impression of loudness derived from any one critical band depends only on the total energy received in that band, and any two well-separated critical bands each make an independent contribution to total loudness.***

That is, within a critical band intensity is additive, but much beyond a critical bandwidth it is loudness that is additive.

Suppose, for example, that we have one sine wave with frequency $f_1 = 1000$ Hz and loudness level $LL_1 = 70$ phons, and another with $f_2 = 1050$ Hz and $LL_2 = 65$ phons. Consulting Figure 6.13, we find that when heard separately they have loudnesses of approximately $L_1 = 14$ sones and $L_2 = 9$ sones. What happens when both are heard together? They lie within a critical bandwidth, so we must add intensities. Because we are in the vicinity of 1000 Hz, this is easy: the intensity levels are also 70 and 65 dB (see Figure 6.12, page 105), so $I_1 \cong 3I_2$ and $I_{comb} \cong 4I_2$; then the combined intensity level is 6 dB above that of I_2 alone or 71 dB. The combined loudness level of 71 phons means a loudness of approximately only 15 sones. But if f_2 were changed to 1500 Hz (that is, into a separate critical band), the combined loudness would be $14 + 9 = 23$ sones.

We may further illustrate how spreading the same energy over a wider frequency range increases the perceived loudness by considering a sawtooth wave. Let its frequency be 200 Hz and its intensity level 100 dB; this keeps the calculation relatively easy by putting us on a part of the Fletcher–Munson diagram where the loudness level in phons is about the same number as the intensity level in dB. It can be shown with Fourier analysis that the harmonic components of this sawtooth wave have approximate intensity levels 98, 92, $88\frac{1}{2}$, 86, 84, $82\frac{1}{2}$, 81, 80, . . . dB (see Figure 8.6f, page 136). The first three or four of these are in well-separated critical bands, so contribute approximately $70 + 40 + 35 + 35 \ldots$ sones to the loudness (Figure 6.13, page 106). Contributions from higher harmonics fall off more rapidly as they begin to share critical bands. But the total loudness for the sawtooth wave is more than 200 sones, compared to a mere 80 sones for a pure sine wave with the same 100 dB intensity level!

---

*This contrast can become even greater for lower levels where the ear's insensitivity to lower frequencies comes into play. A 100 Hz sawtooth wave of overall intensity level 50 dB and component intensity levels 48, 42, $38\frac{1}{2}$, 36, 34, . . . dB has component loudness levels 35, 40, 40, 38, 36, . . . phons and loudness $0.5 + 1.0 + 1.0 + 0.8 + 0.6 + \ldots$ sones. The total of more than 4 sones is to be compared with the 39-phon loudness level and 0.9-sone loudness of a 50-dB pure sine wave.

It is somewhat fortuitous that this simple addition for total loudness is modestly successful for sinusoidal components. Such calculations do not work at all for noise bands without making complicated allowances for the upward masking to be discussed next. (See Benade, Section 13.5.)

Closely related to these loudness judgments is the phenomenon of **masking**. This addresses the question, "If you already are listening to a loud sound at one frequency, what is your threshold for detecting an additional faint sound at another frequency?" As shown in Figure 17.17, the main feature is that the faint tone still is relatively easy to hear if its frequency is much different but is strongly obscured if its frequency is close to that of the masking tone.

This is not surprising if you know about critical bands, for you would expect that when the two signals are within the same critical band you would not notice the addition of the second signal until its strength is an appreciable fraction of the first. This is practically like asking for the just noticeable difference in loudness of a single tone, which typically requires an intensity increase of 15–30% (as we saw in Chapter 6), suggesting that the masking tone will effectively obscure anything more than 6 or 8 dB below it.

But when the second tone stimulates a different critical band than the masker, the report can go to the brain through a separate nerve channel that is not already busy. It is easily noticed, drawing attention to itself with its

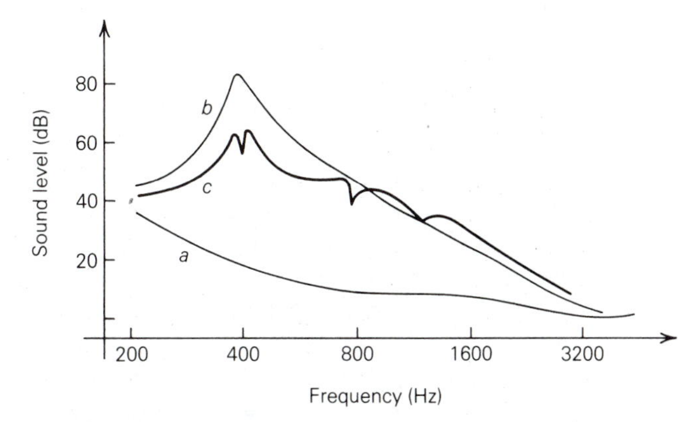

FIGURE 17.17 Curve (a) shows typical threshold for detection of a faint pure tone of variable frequency, as discussed in Chapter 6. Curve (b) shows raised threshold in the presence of a masking noise band extending from 365 to 455 Hz with sound level 80 dB. Tones with frequency above the masking frequency are more strongly masked than those below. Curve (c) shows raised threshold in the presence of a masking sinusoidal tone of frequency 400 Hz and sound level 80 dB. Notches in this curve indicate reduction of masking because audible beats give away the presence of the faint test tone at frequencies near 400, 800, 1200 Hz. (After J. Egan and H. Hake, *JASA*, 22, 622, 1950)

distinct pitch (except when its frequency could be mistaken for a harmonic of the masker), even when 30 or 40 dB below the masker.

Notice the asymmetry in Figure 17.17: The separation between curves *a* and *b* is greater toward the right than the left. This indicates that a test tone is masked more effectively when it is higher in frequency than the masker than when lower. We can understand this "upward masking" in terms of basilar membrane vibration patterns: A high-frequency tone produces motion mainly near the oval window, while the disturbance for a low-frequency tone spreads more nearly over the whole membrane. Thus, more output channels are "already busy" when the masker has low frequency than when it has high.

Masking is a familiar phenomenon in everyday life. A whisper you could easily hear in a quiet room may be totally inaudible in a noisy crowd; a car radio is played louder to overcome road noise at high speed than when at rest. You often may notice masking in musical settings, too. In a climactic *fff* orchestral passage with the entire brass and string sections giving maximum effort, the composer sometimes lets a few woodwinds play along on the same notes. It is nice to have everyone look busy and to give everyone a feeling of full participation, but as far as the sound is concerned the woodwinds might as well rest. At a more mundane level, it is all too easy for a weak soloist to be masked by a strong accompaniment. But knowledgeable composers are aware that one of the most effective ways to make a soloist stand out is to ensure that the solo part differs sufficiently either in pitch or in timbre to catch the attention of critical bands that are not tied up already by the accompaniment.

17.7 TIMBRE

When we first discussed the perception of timbre for continuous tones in Chapter 6, we could only say that it was probably determined largely by waveform. In Chapter 8 we said it might be more useful to deal with waveform information by using the spectrum description. Can we now be more specific about what spectrum aspects are most important and just what effect each has on timbre?

The concept of critical bands provides useful guidance. Without it we would have to entertain the possibility that as many as 50 or 100 harmonic components each could have a different and important effect on timbre. But now it seems more likely that if several spectral components all fall within a single critical band, it must be only their combined strength that principally determines any contribution to timbre sensation from that part of the basilar membrane. This will happen for components whose frequencies are within approximately 15% of one another. That is certainly not true for first and second harmonics; in the neighborhood of the 30th harmonic, on the other hand, the 28th through 32nd are all within 15%.

It should be at about the seventh harmonic that significant overlap begins. And indeed several experiments indicate that components up through the sixth or seventh each make an independent contribution to tone perception, while higher components merge together.

More specifically, this suggests that the number of pieces of independent information determining timbre is six or seven plus a few more for the additional critical bands occupied by overlapping components—perhaps 10 or 15 altogether. That is, a liberal estimate of the complexity of the phenomenon is that timbre perception could have a dimensionality, or number of degrees of freedom, as high as 10 or 15. Remember, for comparison, that pitch and loudness each are only one-dimensional.

But does the brain actually use even that much detail? Perhaps it pays attention only to certain combinations of these critical-band excitations, not to each one individually. There are, in fact, experimental indications that both vowel identification and musical timbre judgment depend on as few as three or four independent intensity parameters (Plomp, Chapter 6). This means it is at least theoretically possible that we may someday succeed in describing four specific aspects of timbre (or three or five—it is not all that certain), each representable by a number telling position along a line connecting two extremes. For example, a sound might be graded 7 on a scale going from 0 for "hollow" to 10 for "full," 4 on the "dull" to "sharp" scale, and so on, with four such judgments together sufficing to accurately identify almost any timbre. This would be similar to the way taste sensations can be catalogued by the degrees of their sweet, sour, bitter, and salty components.

Which is more significant, to say that a particular waveform (with fundamental frequency 300 Hz, for example) has strong first, second, and fifth harmonics, or that it has strong components at 300, 600, and 1500 Hz? It seems to be the actual frequencies, rather than the harmonic numbers, that are most important. We hinted at this in Chapter 12 in connection with organ-pipe scaling, and it follows quite explicitly from our discussion of formants in Section 14.3. Probably the most vivid demonstration of this point is to record musical material on a tape recorder and play it back at double- or half-speed. Besides the octave pitch change, the characteristic instrumental timbre also is appreciably changed, even though the *relative* distribution of energy among its harmonics remains exactly the same. Perhaps most telling of all, although more subtle, is the fact that a pure sine wave (that is, all energy in the first harmonic) has a timbre that changes from very dull at low frequencies to bright at high frequencies. It may even remind you of one or another vowel sound with corresponding formant frequency, progressing through *oo* and *ah* to *ee* as the sine-wave frequency increases.

In the light of these comments, it may be helpful to say that continuous-tone timbre in general does not depend only on spectrum but also somewhat on frequency. By this we mean (1) that any source (especially an electronic oscillator) that retains the same distribution of energy among harmonic numbers regardless of frequency must have timbre that varies appreciably with pitch, and (2) that the nearest approach to uniform timbre throughout playing range should be for instruments characterized by strong formant regions, although even then the shift in frequency of lower harmonics probably will still mean significant timbre change when the pitch moves an octave or more.

Finally, we should say that timbre will change somewhat with intensity. Take a complex waveform with fundamental frequency at approximately 100 or 200 Hz and many strong harmonics. At high levels such as 80 to 100 dB, where the ear is comparably sensitive to all the harmonics, the timbre is quite rich and solid. If the intensity level is reduced to 40 or 50 dB while still retaining the same relative distribution of energy among harmonics, the ear's sensitivity decreases far more rapidly for the lower components than for the others (Figure 6.12 again), so the sound becomes not only much softer in volume but also thinner in tone quality.

SUMMARY

Human perception of musical tones is quite a sophisticated process, one that we do not yet completely understand. Strong elements of the place and periodicity theories, and even some from telephone theory, are integrated into the pattern-recognition theories now regarded as most promising.

The important steps seem to be (1) basilar membrane response with places of maximum activity providing a code for the stimulating frequencies, (2) refinement or sharpening of this information by some process not yet understood, (3) transmission of information about the frequency spectrum by the auditory nerves, and (4) search for a harmonic series pattern, conducted high enough up in the brain to utilize combined information from both ears.

These are supplemented by limited information about individual sound wave peaks, carried by the nerve-firing patterns.

Critical bands represent regions within which neighboring basilar membrane fibers have sufficiently similar motions that they do not send independent information streams to the brain. They severely limit our ability to discriminate between signals whose frequencies are within approximately 15% of one another, including neighboring harmonics above roughly the seventh.

Important differences between the ordinary difference tone and the cubic difference tone probably give clues about the amount and type of nonlinearity present in the hearing process. Difference tones currently are regarded as far less important in pitch perception and other musical roles than they were formerly.

Loudness and timbre of complex tones both are determined by the loudness levels channeled into different critical bands. Pitch is assigned corresponding to the fundamental of the harmonic series most nearly fitting the frequency components present, within the limits set by critical bandwidth. For fundamental frequencies up to 500 Hz, the third, fourth, and fifth harmonics when present are especially important. They are often even more important than the fundamental itself, which can be completely absent without changing the perceived pitch.

REFERENCES

Roederer goes into considerably more detail on this material, and you will find many interesting tidbits in Chapters 2, 3, 4, and Appendix II. For a more complete presentation of auditory psychophysics, see David M. Green, *An Introduction to Hearing* (Halsted, 1976).

Two interesting *Scientific American* articles discuss related topics not covered in this book: "Auditory Beats in the Brain" by Gerald Oster (October 1973) and "Musical Illusions" by Diana Deutsch (October 1975). A special issue of *J. Audio Engineering Society*, September 1983, is devoted to articles on auditory illusions.

An outstanding introduction to modern research is Reinier Plomp's *Aspects of Tone Sensation* (Academic Press, 1976). More detailed articles on all aspects of sound perception are contained in *Foundations of Modern Auditory Theory*, Vols. I and II, edited by J. V. Tobias (Academic Press, 1970 and 1972). See especially the article on critical bands by B. Scharf, Vol. I, pp. 157–202. An excellent review of pitch perception, bringing in the developments of the early 1970s, is presented by E. de Boer in *Handbook of Sensory Physiology*, Vol. V/3, edited by W. D. Keidel and W. D. Neff (Springer, Berlin, 1976), pp. 479–583.

Original references for the recent pattern-recognition theories of pitch perception are: F. L. Wightman, *JASA, 54*, 407 (1973); J. L. Goldstein, *JASA, 54*, 1496 (1973) and *JASA, 63*, 486 (1978); and E. Terhardt, *JASA, 55*, 1061 (1974). A less-technical report is given by F. L. Wightman and D. M. Green in *American Scientist, 62*, 208 (1974). Key experiments are reported by A. Houtsma and J. L. Goldstein in *JASA, 51*, 520 (1972) and *66*, 87 (1979).

An important paper on timbre perception, which gives many references to earlier papers, is by John Grey, *JASA*, *64*, 467 (1978). See also H. F. Pollard and E. V. Jansson, *Acustica*, *51*, 162 (1982).

SYMBOLS, TERMS, AND RELATIONS

intensity/frequency/ spectrum
loudness/pitch/timbre
absolute pitch
cochlea
theories of hearing:
- telephone
- place
- periodicity
- pattern recognition

SIL intensity level (dB)
LL loudness level (phons)
L loudness (sones)
$f_{ST} = f_H + f_L$
$f_{DT} = f_H - f_L$
$f_{CDT} = 2f_L - f_H$
basilar membrane
sharpening
harmonic series
missing fundamental

Ohm's law
dichotic stimulus
critical bands
beats and roughness
combination tones
simple nonlinearity
essential nonlinearity
masking

EXERCISES

1. Make a summary list or table of important pros and cons for each of the main theories of pitch perception.
2. Show and explain the connection of Figure 17.2 to Figure 17.3.
3. Describe where on the basilar membrane the maximum vibration amplitude occurs when the ear receives a sine wave with $f = 6400$ Hz.
4. According to Figure 17.7, what two different pitches may be matched to the three-component stimulus 1600/1800/2000 Hz? What three pitches to 1680/1880/2080? (Give answers in terms both of frequency of the matching sine wave and the nearest pitch letter name.)
5. As a purely mathematical exercise, the three-component stimulus 1830/2030/2230 Hz could have been analyzed as the 183rd, 203rd, and 223rd harmonics of 10 Hz. Why is that of no practical relevance to our discussion of pitch perception?
6. If frequencies 275, 512, 620, 796, and 1026 Hz are all prominent in the spectrum of a certain church bell, which ones might be picked as approximating part of a harmonic series? Estimate the fundamental frequency (and corresponding pitch letter name) of the best-fitting harmonic series.
7. What pitch or pitches are likely to be perceived when you hear a sound containing strong components with frequencies 200, 230, 402, 455, 608, 685, 799, 920, 1005, and 1160 Hz?
8. If the sine-wave combination 1800/2000 Hz were mistaken by your pattern recognizer for eighth and ninth harmonics, what pitch would you report?

What if you mistook them for tenth and eleventh harmonics? (Give answers in terms of both frequency of matching sine wave and corresponding letter name.)

9. Which harmonics are especially important in determining pitch judgments for harmonic series with fundamentals in the vicinity of 100 Hz? 400 Hz? 2000 Hz?

10. What are the approximate critical bandwidths in Hz for center frequencies (a) 3 KHz, (b) 10 KHz, and (c) 200 Hz?

11. Do 100 and 150 Hz lie within a single critical band? What about 1000 and 1500 Hz? 6000 and 6500 Hz?

12. Which members in a harmonic series with fundamental 500 Hz have a critical band all to themselves, and which must share? Does the answer to this question depend strongly on whether the fundamental frequency is 500 Hz or something quite different?

13. Using information from Section 17.4 and Figure 17.12, describe the pitch and associated perceptions for an 880 Hz sine wave mixed with another of equal strength at (a) 882 Hz, (b) 890 Hz, (c) 920 Hz, and (d) 1100 Hz. (Ignore the possibility of combination tones.)

*14. What frequency range f' must be covered with Fourier components to make a signal lasting only 10 ms?

*15. To tune a cello C_2 within $\frac{1}{2}\%$, what minimum length of time would you have to listen assuming you are dealing with sine waves? Show that roughly the same conclusion follows either from the uncertainty principle *or* from listening for beats against a standard source. Show that the use of played harmonics, or even the presence of higher harmonics in a nonsinusoidal ordinary note, relaxes this limitation.

16. When a trumpet section plays loudly in its middle range, which of the following is most (and which least) effectively masked: (a) cellos playing two octaves lower, (b) clarinets playing in the same range as the trumpets, or (c) flutes playing one to two octaves higher?

17. One sine wave of fixed frequency 1 KHz is mixed with another of equal amplitude whose frequency sweeps slowly from 1.1 up to 2 KHz. Describe how both the ordinary and cubic difference tones vary, and explain why this is a good way to bring them to your conscious attention.

18. What sum, difference, and cubic difference tones would be associated with: (a) 1000 and 1150 Hz (answer in Hz), (b) B_4 and D_5 (answer with letter name), (c) E_4 and C_5, (d) C_4 and F_5, and (e) C_5 and D_5? Indicate which ones have much chance of actually being heard.

19. Suppose you have primary tones with frequencies 600 and 760 Hz, each with strong first, second, and third harmonics. List the frequencies of the nine ordinary difference tones that could be formed by these components. Do any of them form a pattern that might enhance the perception of a particular pitch?

20. Four primary tones form the major chord $C_4E_4G_4C_5$. What additional notes may arise as difference tones?

21. Three sine waves have frequencies 600, 1020, and 1100 Hz, and all have level 80 dB. What is the loudness in sones when all three are heard together?

*22. Three sine waves have frequencies 110, 150, and 370 Hz, and all have level 50 dB. What is the loudness in sones when all three are heard together?

23. One sine wave remains fixed at 400 Hz and 80 dB, while another at 50 dB level

has its frequency gradually swept from 100 Hz upward. Describe what will be heard.

*24. Suggest four pairs of opposite qualities that might serve to describe timbres. Using a scale of 0 to 10 for each, estimate the four numbers that might describe each of several familiar instrument timbres.

PROJECTS

1. Gather experimental evidence for or against the claim that all waveforms with the same period (or fundamental frequency) have the same pitch. Include sines and other electronically generated waves, and musical instrument sounds (with the aid of an oscilloscope to determine their periods).

2. Take the tune of "Happy Birthday" as a solo line. Write two different simple accompaniments, one to illustrate how a poor accompaniment may mask the solo and the other to let the solo come through clearly even when weak.

CHAPTER

18 Harmonic Intervals and Tuning

Much as there is to the perception of a single note, that is only the beginning of music. The artistic message is contained largely in the melodic succession of notes and in the harmonic structure of several parts played together. We now must bring all our resources to bear on such questions as why some combinations of notes convey a different feeling than others, which combinations should be chosen as standard musical resources, and how they function in musical contexts.

We begin this chapter by describing how our ears perceive musical intervals, and then consider the role of the harmonic series in providing special interval sizes whose sound and musical function differ from all others. In the third section we attack the problem of how chosen musical intervals can be assembled to build up various musical scales. We will especially try to understand the structure of the diatonic major scale and its chromatic extension, both of which are basic to most of our culture's music.

This leads to the problem of *tuning*, or preparing an instrument in advance to perform music using a particular scale. We will show in Section 18.4 that this is a problem for which there is no unique or perfect solution, and conclude the chapter by discussing how well various compromise schemes approximate the ideal of perfect tuning. In the following chapter we will consider how these harmonic elements are further assembled into the larger structures of music.

18.1 INTERVAL PERCEPTION

How do we quantify our perceptions of the distance of one tone from another along the pitch dimension or, equivalently, of the interval contained between them? As soon as we leave behind some of the more sterile scientific tests and bring a little musical experience into the picture, we find a complication.

As the frequency of one tone moves farther and farther away from another, the pitch does not simply continue sounding more and more different indefinitely. As the separation reaches an octave, one aspect of our perception says the moving tone has returned to its starting place. As we move on through the next octave, the tones sound so much like those of the first octave that we use the same names over again. For many musical purposes

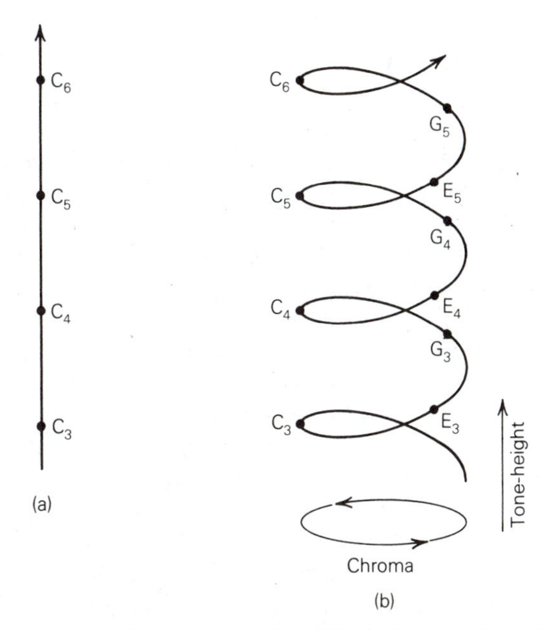

FIGURE 18.1 Contrast between (a) one-dimensional and (b) two-dimensional models of pitch perception. Notes of a scale played on an ordinary instrument spiral upward around the surface of a cylinder, but computer-generated notes can form a Shepard scale that goes around in a circle.

we have *octave equivalence,* meaning that it hardly matters if we replace any note by its namesake from some other octave.

We say the pitch has two separate aspects, a *chroma,* or color, telling where it stands in relation to others within an octave, and a *tone-height* telling which octave it belongs to. The term *chroma* allows for continuous variation of pitch, while *pitch-class* refers to the same property when we use only a finite number of discrete pitches within an octave. In our chromatic scale, for instance, all notes named C form one pitch-class, all F$^{\#}$'s another, and so on. As Figure 18.1 suggests, we should revise one of our previous statements to say that pitch is not a simple one-dimensional variable after all, but is two-dimensional in some respects.

That raises the question of whether there is any way of varying chroma while keeping tone-height constant. A fascinating demonstration of this was given by R. N. Shepard in 1964 (*JASA, 36,* 2346). He used a computer to generate sounds containing many members of a pitch class together or, equivalently, harmonics 1, 2, 4, 8, 16, . . . of low fundamental frequencies. Whichever harmonics fell in the middle of the audible range were made loudest. A sequence of these "Shepard tones" can go round and round in chroma, giving the illusion of going up the scale forever, yet without passing the limits of audibility. (This paragraph will become much clearer if you actually hear a Shepard scale—for example, on the record cited in the references.)

We also encounter the interesting (although incompletely understood) phenomenon of *categorical perception*. Suppose we prepare an experiment in which we can sound not only several neighboring standard musical intervals (say, minor and major thirds and perfect fourths) but also many others of in-between size. When these are presented in random order, there are many conditions under which listeners will assign each stimulus to one of the standard pigeonholes. That is, each is perceived as a legitimate third or fourth, with the varying degrees of mistuning being practically ignored. This helps make musical performance requirements much less stringent than they might otherwise be! A similar phenomenon occurs in speech perception. A series of sounds can be prepared that constitute a continuous transition between *b* and *p*, for instance; but listeners almost invariably will "hear" one or the other without noticing the ambiguity.

We must be careful not to assume that this means our familiar standard musical intervals are somehow certified as natural categories. Burns and Ward (*JASA, 63*, 456, 1978) conclude from careful psychophysical studies that these categories are learned in musical training. This supports the observation that people of other cultures may classify intervals in an entirely different set of categories that seems natural to them. These comments apply primarily to *melodic intervals* (two tones heard in succession), as well as to pure sine waves presented in any manner. But for *harmonic intervals* (two tones heard simultaneously) formed with nonsinusoidal signals, there are additional effects (described in Section 18.2) that give certain intervals a special claim to being natural after all.

Recall that the musical character of an interval corresponds to its *frequency ratio*. Take for instance the ratio 3 to 2, written 3:2 or, equivalently, 1.5:1. Identification of this ratio with the perfect fifth means that 600 and 400 Hz together sound like the same interval as 1500 and 1000 Hz. Now the size of a combined interval is calculated by multiplying the constituent frequency ratios. For example, let note *Z* be a minor third above note *Y*, which in turn is a major third above *X*. When discussing just tuning below, we will take this to mean that *Z* and *Y* have a 6:5 frequency ratio, and *Y* and *X* a 5:4 ratio. That is, $f_Z = \frac{6}{5}f_Y$ and $f_Y = \frac{5}{4}f_X$, so that $f_Z = \frac{6}{5}(\frac{5}{4}f_X) = \frac{6}{4}f_X = \frac{3}{2}f_X$, or, equivalently, $f_Z = 1.20(1.25f_X) = 1.5f_X$. So the interval formed by *Z* and *X* is a perfect fifth. If *Z* had been a minor third below *Y* instead of above, we would have multiplied by $\frac{5}{6}$ instead of $\frac{6}{5}$, to get a 25:24 ratio between *Z* and *X*.

We need some convenient unit (smaller than the octave) to describe precisely the sizes of our intervals. Instead of the actual frequency ratios, which are sometimes unwieldy, we would like to substitute an easy code. In Chapter 7 we used twelfths of an octave, calling them semitones; we shall see in Section 18.5 that this is too restrictive, because we want to talk about some semitones being slightly larger than others. So now let us introduce a unit of measurement called the **cent**.

The cents scale has two defining properties:

(1) The 2:1 ratio is required to be exactly 1200 cents, and (2) multiplication of frequency ratios always must translate into addition of the corresponding numbers of cents.

TABLE 18.1 Consonant interval sizes according to the Pythagorean hypothesis. For most purposes, it is quite adequate to round these off to the nearest cent. Note the four inversion pairs: For each the product of ratios is 2:1, or the sum in cents is 1200.

Musical Interval	Frequency Ratio	Size in Cents
Unison	1:1	0
Octave	2:1	1200
Fifth	3:2	701.95
Fourth	4:3	498.05
Major third	5:4	386.31
Minor sixth	8:5	813.69
Minor third	6:5	315.64
Major sixth	5:3	884.36
*	7:4	968.83
*	11:8	551.32

*The ratios involving 7 and 11 have no place (or name) in the diatonic scale, and generally are not accepted as consonant.

The first property says that a perfect octave is 1200¢, so that a cent is just one-hundredth of an equal-tempered semitone. We must *not* suppose this means there is anything acoustically special about a 100¢ interval, any more than our arbitrary definition of a kilogram means that this particular amount of mass plays any special role in nature. The second property says that the relation between frequency ratios and cents is directly analogous to that between intensity ratios and decibels. Just as we showed corresponding ratios and level differences in Table 5.1, so now we give equivalent descriptions for several especially important interval sizes in Table 18.1. (Remember that Table 7.2 and Figure E provide the necessary interval terminology and notation.)

---

*Although nearly all our discussion and problems can be formulated in a way that does not require its use, the mathematical statement of this definition is $I(¢) = 1200 \log_2 (f_Y/f_X) = 3986 \log_{10} (f_Y/f_X)$. The frequency ratio corresponding to 1 cent is 1.0005778.

We always have a choice of two ways of calculating the sizes of combined intervals: (1) multiplying frequency ratios or (2) adding cents. For the example above of a minor third added onto a major third, the same conclusion follows simply from adding the intervals expressed as cents: 386¢ + 316¢ = 702¢. In this particular example it would be easier yet to count semitones (4 + 3 = 7), but with cents we can handle any interval size whatsoever, with no limitation to those found in the chromatic scale.

If an interval goes downward instead of upward, we need only subtract instead of add. For instance, a major third upward followed by a minor third downward would give 386¢ − 316¢ = 70¢, showing that Z is only above X by a small version of a semitone, corresponding to the 25:24 ratio found above.

How accurately do our ears judge these interval sizes? The categorical-perception experiments suggest that with sine waves we may easily tolerate deviations of 50¢ either way from a standard, especially if we are not consciously trying to detect mistuning. The obviously sour notes in a beginners' band rehearsal may be 30¢ to 50¢ off, but the presence of higher harmonics in these nonsinusoidal waves makes such errors much less tolerable. And measurements in actual performance by professional musicians indicate that deviations of 10¢ to 20¢ are common. Some of these deviations represent deliberate artistic inflections, but even random errors this big do not necessarily detract from the impression of high performance standards. (Several independent studies are summarized in Ward's article.) It is a good thing our ears are so tolerant, because the very nature of wind or string instruments makes it difficult to hold pitch more accurately than this. Put the other way around, vibrato and short note durations and breathing and scraping sounds all help hide the errors in intonation.

On the other hand, there are some conditions under which the ear may be much more discriminating. Let a well-trained musician with a "good ear" listen deliberately for mistuning, with plenty of time to consider his judgments, and he (or she) may consistently reproduce an interval well within 5¢. It is reasonable to ask that the strings of a violin or piano or the pipes of an organ be tuned within 2¢ or 3¢ of the target pitches in the mid and high ranges.

Musical interval mistuning presents another relation of the type we have seen between intensity and loudness or between frequency and pitch. Here, too, a stimulus measured physically (deviation of actual from desired frequency ratio, conveniently expressed in cents) produces a response that must be measured psychologically (judgment of how unpleasant or disturbing the mistuning is). My students and I explored these responses with several musicians in our laboratory in Sacramento. In a typical experiment, subjects listened to a series of computer-generated intervals with varying degrees of mistuning, and responded by rating the tuning on a scale from 1 (very well tuned, highly acceptable) to 7 (horribly mistuned, completely unacceptable). Although specific features differ from one interval species to another, the general nature of the responses is as sketched in Figure 18.2. The presence of 5¢ mistuning is easily discriminated in many cases, and 10¢ to 15¢ is enough to make octaves, fifths, and thirds quite unacceptable for complex waveforms. One of the interesting results of this work was to find signs that some musicians depend mainly on hearing beats (as will be discussed in Section 18.2) to judge mistuning, while others use a more abstract ability to judge interval size apart from the beats. The latter subjects can judge mistuning even when one tone is presented to the left ear and the other to the right ear.

One of the more remarkable things about interval perception is that we ever perceive two separate pitches instead of just one. In view of our comments about pattern recognition in the preceding chapter, why do we not simply take the one best harmonic-series fit to the combined spectrum of the two notes? Why are the two notes not fused together and perceived as one?

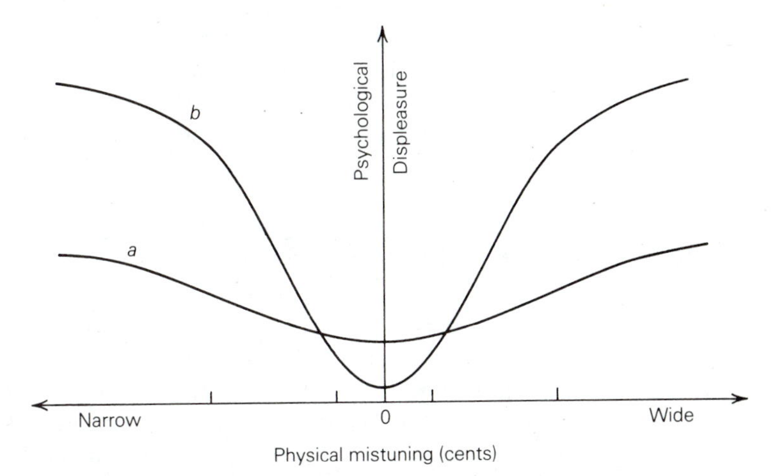

FIGURE 18.2 Qualitative features of the judgment of interval mistuning. Musically untrained observers (curve *a*) make little distinction, while experienced musicians (curve *b*) show three stages of perception. Within a rather narrow range, all sizes seem quite acceptable; beyond that, displeasure grows rapidly with increasing mistuning; finally, the mistuning is so bad that it hardly matters if it gets any worse, and the curve levels off. (Actual numbers of cents at transitions depend on species of interval, absolute pitch level, and so on.)

Our ears pick up clues from formants, from transients, and from note-to-note progressions and perform the spectacular feat of sorting out Fourier components into two or more sets and assigning a separate pitch to each. The fact that a certain group of spectral components has all shifted the same way from the previous note, or has begun an attack together in a similar way, is enough to enable the brain to identify them as belonging to a separate voice.

Here then is good acoustical reason for some traditional rules of orchestration and counterpoint. To keep two musical lines distinct, assign them to instruments of contrasting timbres or attack characteristics, and make the parts move in contrary motion. Smooth and similar timbres (especially on electronic organs) or parallel part motion (especially parallel octaves and fifths) make it too easy for voices to fuse together and lose their individuality.

Occasionally, a composer may want two parts to fuse into a single musical line and so will deliberately write for similar instruments in parallel. For a marvelous example, listen to the second movement ("Game of Pairs") of Bartok's *Concerto for Orchestra* (1944). I can almost hear each duet as a single melodic line played by some new instrument with a strange, exotic timbre.

At one extreme, two steady tones may have spectra that fit so well together that they are mistaken for a single tone. At the other extreme the spectra may produce such an irregular combined pattern that the two tones not only retain separate identity but also actively clash. Delicately balanced in between is the case in which they form a harmonious musical interval, still separately identifiable yet fitting closely in a special relationship that forms an element of our musical vocabulary.

Those intervals that sound smooth, restful, or harmonious are called **consonant** (which literally means "together-sounding") and those where the two notes clash, **dissonant**. But there are varying degrees of consonance and dissonance, with no sharp boundary separating them and with judgment strongly colored by cultural background. To understand these differences we must consider further the special role played by the harmonic series.

18.2 INTERVALS AND THE HARMONIC SERIES

One of the oldest traditions of music theory is identified with the Greek philosopher and mathematician Pythagoras (c. 500 B.C.). Although we have none of his own writings, he reportedly observed that notes produced by the two segments of a divided string on a monochord (Figure 18.3) form a unison if the two lengths are the same, an octave if the length ratio is 2:1, and a perfect fifth for 3:2. Partly on the basis of this observation, but also because Pythagoras led a mystic cult that practically worshipped numbers, his followers taught what I call the *Pythagorean hypothesis.* They claimed that there is a unique correspondence between small-integer length ratios (we would now say the corresponding frequency ratios are more fundamental) and special intervals that should serve as the foundation of all music. The smaller the integers, the more consonant and more important the interval. This is reflected in Table 18.1, where the order of listing closely resembles both the order of historical acceptance as consonant intervals and the order of importance in classical music theory. Our cultural forebears at first considered only fifths and fourths to be consonant, and only accepted thirds and sixths as consonances in the Renaissance. Ratios involving 7 or 11 still have been espoused only by a few nonconformists such as the late Harry Partch. The music of India shows much more awareness of the seventh harmonic than does ours.

There is considerable danger today that we too readily accept the Pythagorean hypothesis not on evidence but on simplicity and familiarity—or even on the appeal of mystical numerology. Is it truly an immutable natural law, or have we only been brainwashed by the overwhelming accumulation of millennia of accepted music theory and its strong selective effect on what we are taught to regard as good music? I propose that there is a considerable element of truth in the Pythagorean hypothesis, but that its application also has definite limitations. It may be used with steady complex tones, but there is little justification for applying it either to sine waves or to percussive sounds. And there is in any case little reason to carry out cal-

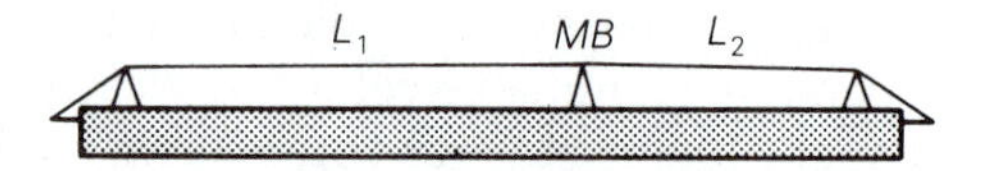

FIGURE 18.3 Division of a monochord string by a movable bridge *MB*. Because tension and density are the same in both pieces, it is the length ratio $L_1 : L_2$ that determines the frequency ratio $f_2 : f_1$.

culations to five decimal places unless you simply enjoy working with the numbers.

---

*Preferences for "octaves" of 1210¢ or more often are reported for melodic intervals and even for simultaneously sounding pure sine waves. But when complex waves are presented as harmonic intervals lasting a half second or longer, there is no doubt that 1200¢, an exact 2:1 ratio, is the preferred standard.

Singling out small-integer frequency ratios means, of course, picking intervals of the same sizes that occur in the harmonic series. Why should these intervals play such an important role in music, serving as the basis for our familiar musical scales? Physically, the spectra of complex steady tones are harmonic series. The interleaving of two such series is particularly orderly when their fundamentals have a small-integer ratio, suggesting—although only suggesting—that they somehow may sound as if they fit or belong together (Figure 18.4a). Physiologically, the imprinting of harmonic series in your spectral-pattern recognizer may prejudice your brain to eagerly grasp similar relations even between two tones whose transients or formants enable them to keep separate identities. Psychologically, this would not be surprising either because it is only one of many manifestations of human preference for the familiar, simple, orderly, and smooth.

The word "smooth" is especially important, because auditory beats give us a convenient criterion to tell when we have a *just* (properly tuned) Pythagorean interval. Strong beats are heard with two sine waves only when their frequencies are nearly equal. It is true that sine waves can form "second-order beats" (as discussed by Roederer, pp. 36–39) for frequency ratios near 2:1 or 3:2, but these are much more difficult to hear and have no direct importance in musical perception. A great many psychophysical experiments have severely restricted musical significance because they used only sine

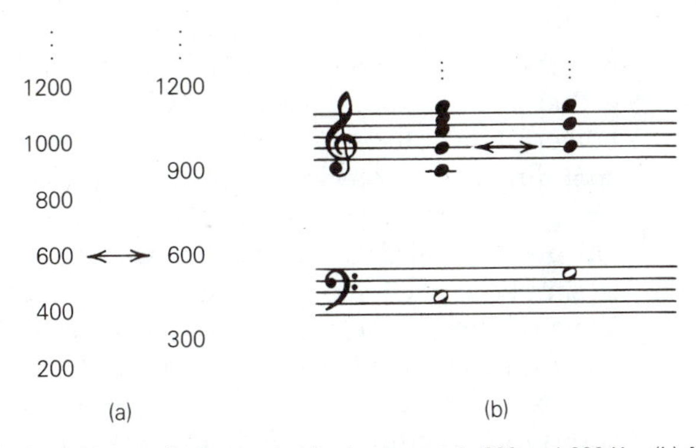

FIGURE 18.4 (a) Lists of harmonic frequencies for fundamentals 200 and 300 Hz. (b) Musical notation for a similar example of the interval of a fifth; mistuning is evidenced by beats at the harmonic indicated by an arrow.

waves, which form intervals of practically uniform acceptability with very little sensitivity to frequency ratio.

Only with complex waveforms can two notes form strong primary beats whenever their fundamental frequencies f_X and f_Y are near *any* small-integer ratio. This is because some of the higher harmonics of one note have nearly the same frequencies as some of those from the other note. A slightly mistuned perfect fifth, for example, has f_Y nearly equal to $\frac{3}{2}f_X$. Then the second harmonic of f_Y nearly coincides with the third harmonic of f_X (Figure 18.4), and beats are heard at the rate $f_b = 2f_Y - 3f_X$. For instance, complex waves with $f_X = 200$ Hz and $f_Y = 302$ Hz beat at $604 - 600 = 4$ Hz. The beating rate is reinforced further by contributions from higher harmonics—in this example, $6f_X$ and $4f_Y$, $9f_X$ and $6f_Y$, and so on. If you deliberately focus your attention at $3f_X$, you will be convinced that the beating actually is associated with this higher pitch rather than with f_X or f_Y.

Contrary to what you may see in some texts, dissonant intervals have no corresponding "correct" frequency ratio; anything in the general vicinity of 200¢, for instance, serves equally well for a major second. The dissonance of seconds, sevenths, and tritones may be attributed to the roughness of the rapid beating associated with their nearness to unisons, octaves, and fifths, respectively.

One instrument (or string) can be tuned to form a desired consonant interval with another by adjusting until the beats disappear. This is exceptionally easy for unisons, moderately easy for octaves, somewhat harder for fifths, and definitely requires careful introspective listening for the other intervals. This procedure forms the practical basis for tuning the scales whose architecture we must now consider.

18.3 MUSICAL SCALES

If our only music were for singing and violin playing and if it were all learned strictly "by ear," we could choose our notes from an unlimited continuum. But the construction of our common wind and keyboard instruments, together with the limitations of our familiar written music notation, provide powerful incentives to restrict ourselves to only a few pitches and to a corresponding list of intervals. Others are refused official recognition.

These sets of allowed pitches are called **scales**. They should be distinguished from *modes* or *keys*, which we will consider in the next chapter. A scale is simply a pitch-set, a list of possible resources, while a mode implies a use-structure as well; the mode specifies different musical roles for the different pitches, while the scale does not.

From a modern viewpoint, there is no limit to the number of different scales that might be formed. You may pick any number of pitches you like, either in some pattern or entirely at random, and call it a scale. Dozens of scales have become established as standard resources of various cultures.

TABLE 18.2 Partial scales reported for one particular Javanese gamelan (*Harvard Dictionary of Music*, p. 436). Numbers give the interval in cents from the first (lowest) note to each of the others, as in Table 18.3. Note the tendency toward stretched octaves and inexact repetition from one octave to the next. Slendro resembles $N=5$ equal temperament, but Pelog has a distinct $SSTSSS^{+}T$ pattern of wide and narrow tones T and semitones S.

Slendro	Pelog
2441*	2447*
2174	2225
1929	2021
1695	1905
1458	1778
1213*	1503
954	1360
721	1220*
473	965
218	800
0*	676
	563
	266
	125
	0*

Many of these sound quite exotic to our ears and indeed often have drawbacks if we try to hybridize them with our own instruments or musical styles.

This is partly because we make so much use of sustained tones. A culture whose music is primarily percussive has much greater freedom in choosing musical scales, because the lack of prominent beats removes the bias toward those few intervals with small-integer frequency ratios. For instance, the gamelan orchestras of Indonesia not only use several different scale types but also may have a differently tuned version of those scales in each village. They do not necessarily have equal step sizes (Table 18.2), and there is no strong tendency for available intervals to conform to those in Table 18.1.

Let us now examine what type of scales we obtain if we impose certain simple limitations. Suppose first that we require (1) that the interval of an octave be available and tuned precisely at 2:1 and (2) that the notes in the scale all be equally spaced. All that remains is to say how many scale steps it will take to make one octave; for N steps we call it an *N-tone equal-tempered scale*. Computations are easy, for all we have to do is divide 1200¢ into N equal parts. For $N=12$ each step is simply 100¢, and this is, of course, familiar as the equal-tempered version of the chromatic scale (Figure A.5). For $N=6$ we get the *whole-tone scale*, whose resources were explored so well in certain pieces by Debussy. Another case of more than ordinary interest is $N=19$,

FIGURE 18.5 The diatonic major scale, which should be thought of as extending without limit into higher and lower octaves by repeating the same pattern. Compare with the chromatic scale in Figure A.5.

with step size 1200/19 = 63.16¢. This is because five and six steps provide intervals of 316¢ and 379¢ that are much better approximations to minor and major thirds than 300¢ or 400¢, and 11 steps make a tolerable 695¢ fifth. For similar reasons, some musicians have found 31- and 53-tone equal-tempered scales attractive despite their practical performance difficulties (see Exercise 13). Most of us are so unaccustomed to microtonal resources that we can hardly judge the potential riches of such music.

Most important to us are **diatonic scales**, which have a sequence of tones (*T*) and semitones (*S*) in each octave. The term *diatonic* is often taken to mean specifically the *TTSTTTS* pattern that also is called a *major scale* (Figure 18.5). Other common arrangements of small and large steps (three standard forms of minor scale, gypsy scale, and so forth) can be regarded as being derived from the major scale. But we will not go into that here; the following sections will be couched entirely in terms of the major scale. Notice that any diatonic scale, unlike the chromatic, already has some intrinsic structure in its pitch-set, even before any use-structure is specified.

The diatonic major scale can be arrived at in several ways. One (perhaps easiest, but historically backward and theoretically inappropriate) is to regard it as a subset of the chromatic scale—the piano with the black keys left off. Another is to emphasize the pattern of tones and semitones in an octave, taking this as one particular manifestation of the tendency for vocalists of all cultures to contrast small and large melodic steps. A third is to derive the diatonic scale by attempting to build up a structure with just or beatless fifths and thirds. The last way is perhaps more abstract and complicated than the other two. But it does suggest specific tuning criteria that are appropriate to sustained harmonies, as opposed to a merely qualitative pattern in which one "fifth" would not necessarily come out the same size as another.

Figure 18.6a shows a preliminary version of the simple diatonic scale, and Figure 18.6b its chromatic extension. We must emphasize that it is only one of several possibilities; the D, for instance, could be specified as a fourth below G instead of a fifth below A. These schemes do *not* all lead to the same result, and we will deal in Section 18.5 with the problem of reconciling them.

We would like to list the exact size of all intervals in the diatonic scale. That requires asking not only which intervals we wish to include but also how precisely it will be possible to meet all the tuning requirements simultaneously. We must embark now upon what may seem like a diversion in order to come back in Section 18.5 to the problem of tuning the diatonic scale with all the conceptual tools needed for its solution.

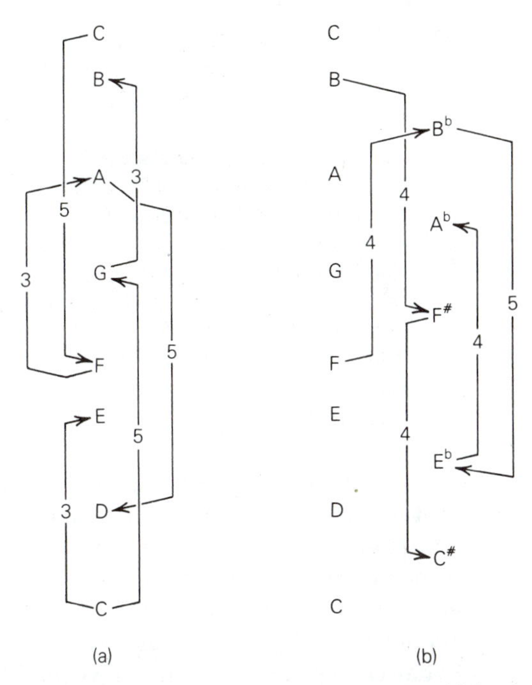

FIGURE 18.6 One particular way of building an extended diatonic scale. (a) Starting from C, tune beatless fifths to F and G, tune one or more of the beatless major thirds F–A, C–E, or G–B, then complete the list of white notes by making D–A–E–B all pure fifths. This tuning dates back at least to Ptolemy (A.D. 140). (b) Add black notes by tuning pure fifths or fourths from B (for sharps) and from F (for flats).

18.4 THE IMPOSSIBILITY OF PERFECT TUNING

There is a fundamental mathematical problem in tuning the notes of a diatonic scale, which is most easily illustrated by the following example. Suppose we start from middle C and try to add other notes. First, because we may sometimes play C major triads and want them to sound perfectly in tune (no beats), let us require that E_4 be precisely 386¢ above C_4 (that is, a 5:4 ratio). Similarly, for the sake of E major triads, tune $G_4^{\#}$ exactly 386¢ above E_4. So far, so good. But now suppose we play a piece in the key of A^b major, and so require A_3^b to be 386¢ below C_4. That gives us an A_3^b that is $3 \times 386 = 1159$¢ below $G_4^{\#}$, or an A_4^b an octave higher that is 41¢ above $G_4^{\#}$.

So what? A singer, violinist, or trombonist easily could produce either note. Even other wind instrument players can lip their notes up or down enough to stay in tune with whichever chord their comrades are playing. But keyboard instruments have pitches fixed once and for all during tuning, with no way of altering them in performance. The organ especially, with its sustained vibratoless tones, leaves all tuning discrepancies mercilessly exposed. Tuning an organ forces us as nothing else can to find out what subtleties lurk in the structure of the diatonic scale.

---

*The evidence cited by Ward seems to indicate a preference by performers for major thirds above 400¢, not below. I believe this stems from (1) too much contact with pianos, training us to substitute 400¢ in place of 386¢ as a standard; (2) a general tendency to feel that if you must err, it is less embarrassing to err on the sharp rather than the flat side; and (3) training (especially of string players) to widen thirds for melodic rather than harmonic purposes, such as leading tones. But I testify as a harpsichord and organ player that 386¢ is really a legitimate standard for the major third. The moral is that tuning theory is firmly rooted in the unique problems of keyboard music, and its application to choral or orchestral music must be viewed with great caution.

Now keyboard players are taught to push the same black key regardless of whether G♯ or A♭ appears in the score; yet we have just calculated that G♯ and A♭ should *not* have the same frequency. This is because three pure major thirds do not make an octave but fall 41¢ short. The same information in terms of ratios looks like this: $\frac{5}{4} \times \frac{5}{4} \times \frac{5}{4} = \frac{125}{64}$, whereas an octave would be $\frac{128}{64}$; the discrepancy of 41¢ corresponds to the ratio 128:125.

What can we do about this problem? (1) We can build keyboards with more than 12 keys per octave, so that we can use more than 12 pitch-classes. In fact, this has been done; many old organs and harpsichords had split black keys so that you could play *either* G♯ or A♭ (and similarly for others) with a separate set of pipes or strings for each. (2) We can tune for one and avoid the other—for instance, using G♯'s and refusing to play any pieces involving A♭'s without retuning. This was the more common solution in the Renaissance and early Baroque eras, and should be understood even now by anyone who owns a harpsichord. (3) We could tune G♯'s and play pieces in A♭ major, knowing full well that the gross mistuning (427¢ instead of 386¢) will be painfully obvious to our listeners. This is, of course, usually quite unsatisfactory. (4) We could sacrifice some of the beauty of C–E and E–G♯ by tuning them larger than 386¢, in order to make G♯–C enough less than 427¢ to be bearable. Among many possible compromises, the one we are accustomed to on the piano is to make all three intervals 400¢.

In summary, when we try to force G♯ and A♭ to be the same, we create a *circle of major thirds*: . . . G♯/A♭–C–E–G♯/A♭ . . . . If we insist on the tuning criteria of 2:1 for octaves and 5:4 for major thirds there is an inescapable 41¢ error *somewhere* in this circle. (In traditional music theory it is called the *lesser diesis*.) We have a choice whether to spread it around evenly (for example, 14¢ apiece for equal temperament) or unevenly or lump it all in one place, but we *cannot* make it go away. Each of the remaining nine pitch-classes also belongs to a major-third circle (. . . C♯/D♭–F–A–C♯/D♭ . . . , . . . D♯/E♭–G–B–D♯/E♭ . . . , and . . . A♯/B♭–D–F♯–A♯/B♭ . . .), to which the same conclusion applies.

We can make similar arguments about three *circles of minor thirds*: ... F♯–A–C–E♭–G♭ ..., ... C♯–E–G–B♭–D♭ ..., and ... G♯–B–D–F–A♭.... Because a 6:5 minor third is 316¢ wide, four of them span 1263¢, 63¢ more than an octave. (Equivalently, $(\frac{6}{5})^4 = \frac{1296}{625}$, leaving an excess ratio $\frac{648}{625}$.) Any tuning scheme limited to 12 pitch-classes *must* have a total of 63¢ error (the *greater diesis*) in each minor-third circle, regardless of whether it is spread around evenly (for example, 16¢ apiece for equal temperament) or unevenly or lumped in one place.

Perhaps most important of all is the *circle of fifths*: ... A♭–E♭–B♭–F–C–G–D–A–E–B–F♯–C♯–G♯.... Because a perfect 3:2 fifth contains 702¢, 12 of them make a total of 8424¢ (actually 8423.5¢, because there was some round-off error). Seven 1200¢ octaves account for 8400¢ of this, leaving a-discrepancy of 23.5¢ (the *ditonic comma*) between the G♯ and the A♭. (Equivalently, $(\frac{3}{2})^{12} = \frac{531441}{4096}$ but $(\frac{2}{1})^7 = \frac{128}{1} = \frac{524288}{4096}$.) Again, any tuning scheme whatsoever *must* have a total of 23.5¢ error somewhere among the fifths, whether it is spread around evenly (for example, 2¢ apiece for equal temperament) or unevenly or lumped in one place.

There is nothing further to be learned by considering fourths and sixths, because they are just the inversions of fifths and thirds. But there is still one important feature. We cannot make our choice of how to spread the ditonic comma around independently of how we are spreading the lesser or greater diesis. The circles of fifths and thirds are linked together by the way each note is supposed to participate in several different intervals. Any adjustment of tuning for the sake of thirds must disturb the tuning of the fifths and vice versa.

To see the most characteristic discrepancy in how fifths and thirds fit together, consider the interval C_4–E_4. If we tune it as a major third, it is 5:4 or 386¢. But if we tune a sequence of four fifths, C_4–G_4–D_5–A_5–E_6, we have $(\frac{3}{2})^4 = \frac{81}{16}$ or $4 \times 702¢ = 2808¢$. Subtracting two octaves brings us back down to an E_4 whose relation to C_4 is expressed by $\frac{81}{16} \times \frac{1}{4} = \frac{81}{64}$ (whereas $\frac{5}{4} = \frac{80}{64}$) or $2808 - 2400 = 408¢$. We have a discrepancy of $408 - 386 = 22¢$ (actually 21.5¢, equivalent to the ratio 81:80), called the *syntonic comma*. Like all the others above, this total error is *always* present somewhere in any sequence of four fifths and the corresponding major third.

You would easily get the impression from the circles of thirds and fifths that the difficulties lie mainly with the chromatic extension of the diatonic scale. You might think all would be fine if we would only transpose everything into C major and thus minimize the use of black keys. But that is not true, for the syntonic comma already is firmly embedded in the simple diatonic scale—that is, in the white keys.

We will find it helpful in the following section to use a diagram (Figure 18.7) that shows each pitch-class surrounded by the others with which it forms important consonant intervals. All the third circles appear as diagonal sequences, and the circle of fifths marches along the horizontal rows. Any tuning scheme can be neatly summarized by showing between each pair of

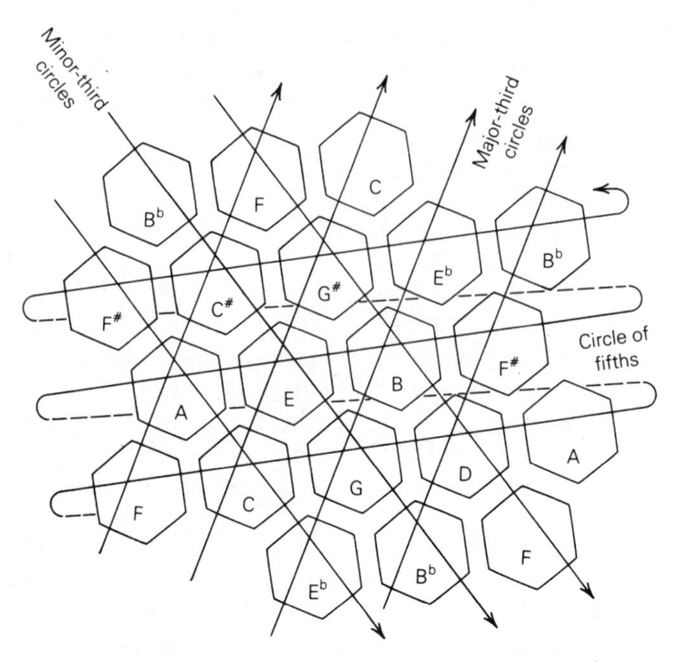

FIGURE 18.7 A tile mosaic showing important interval relationships. All major-third circles appear as diagonals from lower left to upper right and minor-third circles from upper left to lower right. The circle of fifths is the spiral that would result if the diagram were cut out and rolled up into a cylinder. To facilitate intercomparison, we will use these note names in all the following figures, even though sometimes (as in Figure 18.10) we are really tuning A♭ rather than G♯, and so on. (Mark Lindley taught me the full power of these diagrams, which are explained further in my 1974 article.)

notes the mistuning of that interval in cents, with plus or minus signs to indicate wider or narrower than ideal. In the figures that follow it will be easy to verify all the rules stated above by adding the numbers along each diagonal, and so on. The quality of any triad can be seen at a glance, for each one appears as a triangle—major triads with an apex upward and minor downward.

In summary, it is literally as impossible to arrange 12 pitch-classes so that all consonant intervals are perfectly in tune as it is to make $2+2=5$. It is a waste of time to search for a perfect tuning scheme; the appropriate attitude is to study which partially out-of-tune compromise will best serve our purposes.

18.5 TUNING AND TEMPERAMENT

Theorists over the years have studied many dozens of tuning schemes for keyboard instruments (see Barbour). Most of them have roots in the four complementary basic approaches that we will outline here. Keep in mind that we are describing theoretical ideals; there always will be leeways of at least 1¢ or 2¢ in putting them into practice.

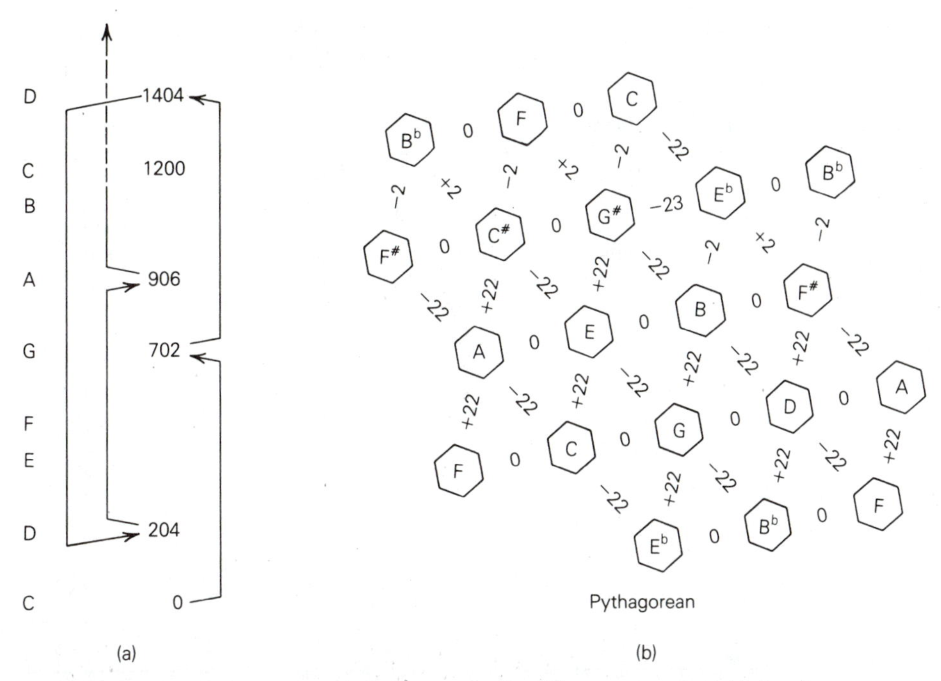

FIGURE 18.8 (a) The first few steps in Pythagorean tuning. Whenever you get outside the primary range, you tune a perfect octave to get back inside. The final result, expressed in terms of the number of cents each note is away from C, is listed in Table 18.3. The seven notes of the diatonic scale were tuned this way by Eratosthenes (230 B.C.); the chromatic extension came gradually many centuries later. (b) Mosaic showing how errors are distributed in Pythagorean tuning if the wolf fifth is placed at G♯–E♭. Positive or negative numbers indicate a fifth or third that is too wide or narrow, respectively. The corresponding inverted intervals (fourths and sixths) have the same errors but with opposite signs.

The first approach is *Pythagorean tuning*, based on the circle of fifths. Many or most of the perfect fifths are tuned to a beatless 702¢, as shown in Figure 18.8a, and the thirds are left to come out however they may. In its purest form Pythagorean tuning sets the maximum of 11 just fifths, leaving the entire 23.5¢ ditonic comma in one place. This 679¢ leftover fifth might be put, for instance, between G♯ and E♭ (as in Figure 18.8b), or between D♯ and B♭; but it can be any one of the 12 you choose. Eight out of 12 major thirds are a very wide 408¢; it is no accident that medieval theorists using this kind of tuning assigned the third to the role of a dissonance.

The second approach is epitomized by *quarter-comma meantone tuning*, based on the major-third circles. Here many major thirds (a maximum of eight) are tuned to a beatless 386¢, and the fifths are left to come out however they may (Figure 18.9). In the simplest regular form most of the fifths are a tolerable 697¢, but there is one hideous-sounding 738¢ "wolf fifth" that must be shunned in performance. Similarly, the performer must be careful not to choose music that would call for any of the four 427¢ thirds; in the case shown, any B major or F minor triad, for instance, is a disaster. In spite of these disadvantages, the presence of many pure thirds in this scheme helped

TABLE 18.3 Intervals in cents from C to each other pitch-class in the extended diatonic scale for each of the tuning schemes of Figures 18.8 through 18.14. At the bottom is a rough classification of the quality of the 24 triads in each scheme; it is possible to do as well as 12, 12, and 0, but impossible to make all 24 good.

Name	Pythagorean	1/4 Comma	Ramos	de Caus	Equal	1/6 Comma	Vallotti
C	1200	1200	1200	1200	1200	1200	1200
B	1110	1083	1088	1088	1100	1092	1090
B♭	996	1007	996	977	1000	1003	1000
A	906	890	884	884	900	895	894
G♯	816	773	792	773	800	787	796
G	702	697	702	702	700	698	698
F♯	612	579	590	590	600	590	592
F	498	503	498	498	500	502	502
E	408	386	386	386	400	393	392
E♭	294	310	294	275	300	305	298
D	204	193	182	204	200	197	196
C♯	114	76	92	71	100	89	94
C	0	0	0	0	0	0	0
Triad Quality							
Good	6	16	6	12	0	16	10
Dubious	16	0	16	0	24	6	14
Awful	2	8	2	12	0	2	0

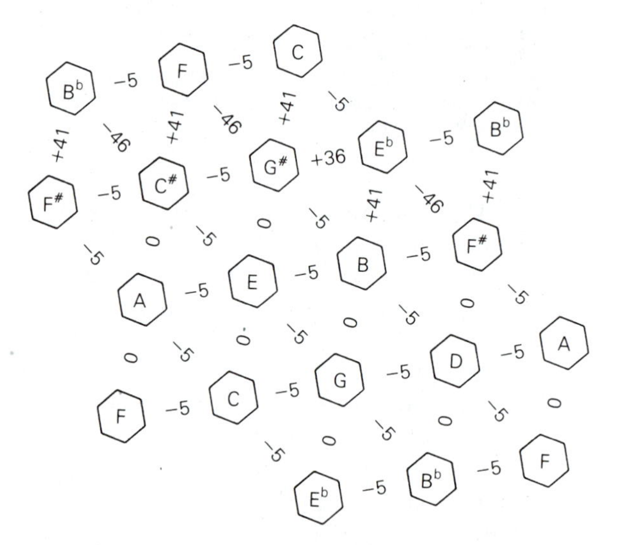

FIGURE 18.9 Distribution of errors in quarter-comma meantone tuning. Note positions relative to C listed in Table 18.3. Tuning is begun by distributing error evenly among C–G–D–A–E by counting beat rates or by successive approximations until C–E is a pure major third, and completed by tuning seven more pure thirds. Total error around the circle of fifths appears to be −19¢ instead of −23¢ because each −5¢ is really −5.4¢ rounded off.

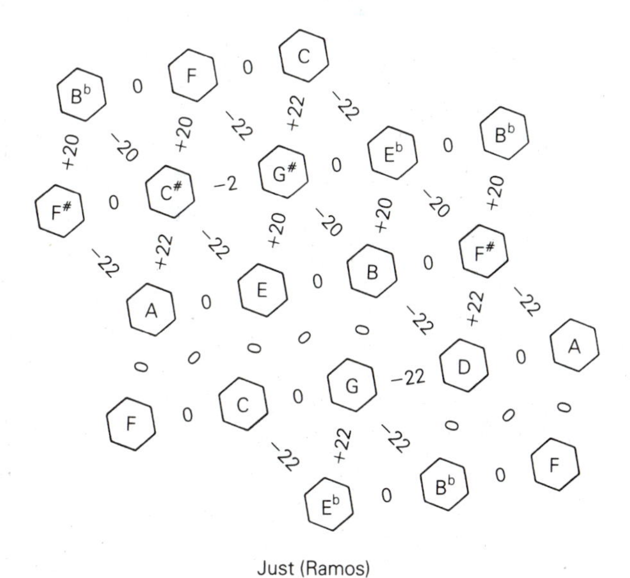

FIGURE 18.10 Distribution of errors in a just tuning scheme traditionally (but inaccurately) associated with the name of Ramos (1482). The procedure is the one shown in Figure 18.6 and the results are listed in Table 18.3.

gain acceptance in the Renaissance for the triadic music we now take for granted. You must actually hear this tuning (for example, on some Elizabethan harpsichord music) to appreciate the restful sweetness of its pure thirds; going back to equal temperament afterward can be quite distressing.

The third approach is called *just tuning*, and tries to make both some fifths and some thirds pure. As shown by the example in Figure 18.10, the price of obtaining six entire triads perfectly in tune is that all the rest have thirds about as bad as in Pythagorean and, especially, that one extremely bad fifth sits stubbornly in the midst of the most important triads. Figure 18.11 shows a more extreme example, with 12 perfect triads purchased at the expense of making all the other 12 awful.

Textbooks often mistakenly present one particular version of just tuning and call it *the* diatonic scale. These schemes are primarily mathematical exercises whose main value is to demonstrate the impossibility of any completely just keyboard tuning; they are entirely impractical for actual performance. The distinct term *just intonation* refers to the performance practice of shifting intonation with each new chord so that each can be justly tuned. This is theoretically possible for a choir or string or wind ensemble, but according to most evidence just intonation is *not* what actually is used by even the most skilled performers.

Technically, the word *tuning* might be reserved for schemes such as those above based on pure intervals. The fourth basic approach tries to compromise the conflicting demands of the fifths and thirds by deliberately mistuning

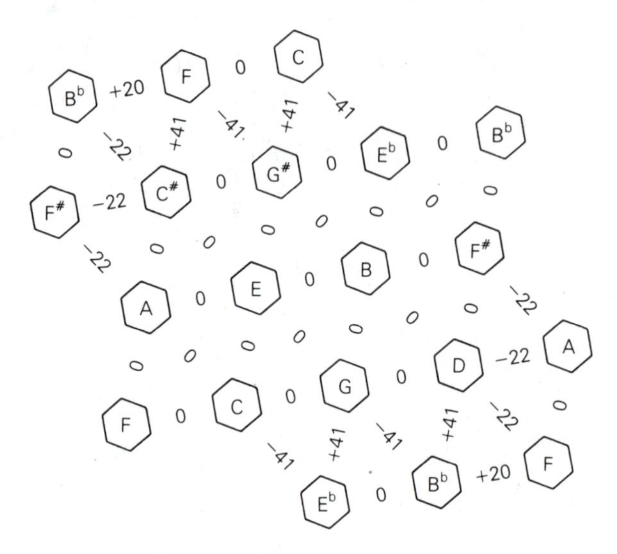

FIGURE 18.11 Another just tuning scheme, essentially that of de Caus (1615). Most easily tuned by setting beatless major thirds F–A and A–C♯, then strings of three perfect fifths upward from each of those notes.

(tempering) some or all intervals in judicious amounts; these schemes are called *temperaments.* It is, of course, possible to hybridize approaches; for example, by tuning in Pythagorean first and then tempering three notes while leaving the other nine the same.

---

*As far as theory goes, this distinction is somewhat artificial, for Pythagorean tuning is a tuning of the fifths but a temperament of the thirds. Similarly, quarter-comma meantone has tuned thirds but tempered fifths. But for practical application the distinction is quite important: To whatever extent you can tune strings (adjust for no beats) the job is easy, but insofar as you must temper (achieve a particular nonzero beat rate) it is considerably harder.

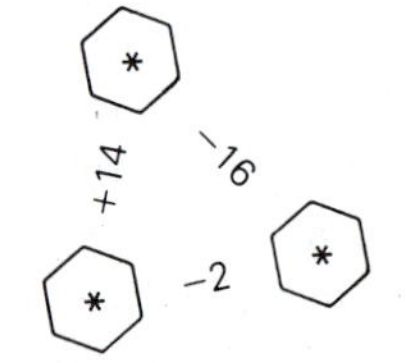

FIGURE 18.12 We need not show the entire mosaic for equal temperament, for every triad has exactly the same mistunings. This temperament must be set by using tables of calculated beat rates.

There is no limit to the different ways a temperament can distribute the various errors around, generally unevenly. There is a particularly important class of *regular temperaments* (most of them associated with a more general use of the term *meantone*) that sets as many intervals as possible to the same size (11 of 12 fifths, 8 of 12 major thirds, 9 of 12 minor thirds), whatever that size may be. These customarily are labeled according to how much the fifths are tempered to the flat side, expressed as a fraction of the syntonic comma. For example, Pythagorean tuning could be called zero-comma meantone (although it is not ordinarily), because the fifths are not tempered at all; equal temperament (Figure 18.12) could be called 1/11-comma meantone because 21.5¢/11 = 1.95¢ is the amount of tempering that brings a fifth down to a

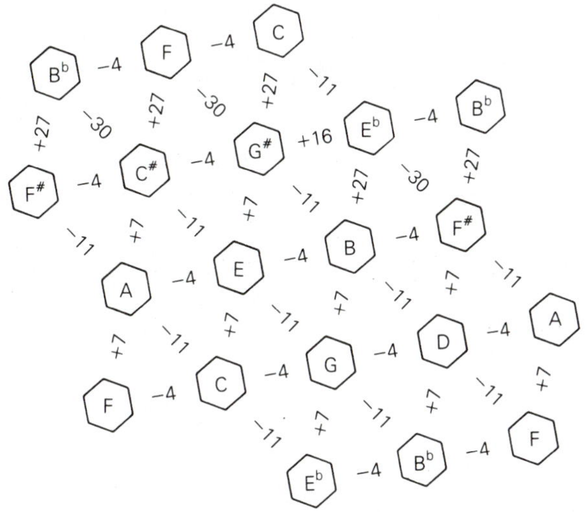

Sixth-comma meantone

FIGURE 18.13 Distribution of errors for the simplest version of sixth-comma meantone temperament. This also must be set with tables of calculated beat rates; results are listed in Table 18.3. In case you try to check the ditonic comma, each −4 is actually −3.6¢.

round 700¢. Quarter-comma meantone happens to be just enough tempering of the fifths to make the major thirds pure and so turns out to be a tuning; 1/3-comma, with 695¢ fifths, happens to make the minor thirds pure and also could be called a tuning, although one of little practical importance.

An especially important case is *1/6-comma meantone temperament*, which with numerous variants was widely used in the Baroque period. Because 21.5¢/6 = 3.6¢, the fifths are tempered to 698.4¢; this allows most major thirds to be 393¢. As you can see in Figure 18.13, most important triads are significantly better than in equal temperament, while the "bad" ones are enough better than in quarter-comma that we might get by with using them occasionally. That is, 1/6-comma approaches acceptability as a *circulating temperament*, one in which you can play music that modulates into any key (uses any number of sharps or flats) without running into total disaster.

From a practical standpoint, regular 1/6-comma meantone is hard to tune because it has no pure intervals; but there are other hybrid schemes, easier to tune, whose final effect is similar or even better. Figure 18.14 shows a circulating *irregular temperament* used by the Italian organist Vallotti (mid-eighteenth century). It is related to 2/11-comma meantone, but all traces of a wolf fifth have been removed and there is a nicer gradation from sweet to pungent triads than in any of the regular temperaments (see Exercise 17). Incidentally, this same scheme (except transposed up a fifth) was rediscovered and advocated by the physicist Thomas Young around 1800. Young's fame, ironically, rests on his experiments showing interference of light waves.

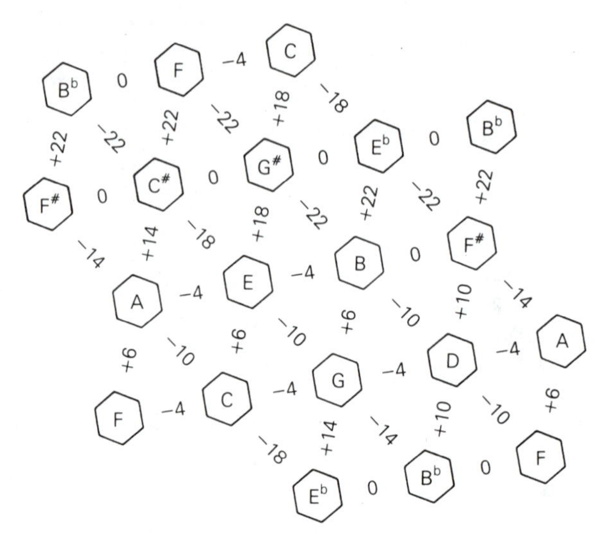

FIGURE 18.14 Error distribution for Vallotti's temperament. It is set by (a) tuning B–F♯–C♯–G♯–E♭–B♭–F all as beatless fifths, (b) setting D so that the thirds B♭–D and D–F♯ are equally wide, and (c) tempering the remaining fifths by setting C, G, A, and E so that F–C–G–D–A–E–B are all the same size. Results are listed in Table 18.3. Each −4 here is really −3.9¢, +22 is +21.5¢, and so on.

---

*In reading other sources on tuning theory you will find they are often inconsistent in labeling, sometimes using a fraction of the syntonic comma (*SC*) as I have done, but other times a fraction of the ditonic comma (*DC*). Thus, for instance, the Vallotti–Young scheme is loosely referred to as 1/6-comma rather than 2/11, and Werckmeister's (Exercise 20) as 1/4-comma instead of 3/11. Because *DC* is almost precisely 12/11 of *SC*, less than 2¢ different, this discrepancy may seem insignificant in practice; but for theoretical clarity one should always state precisely which is meant.

It was probably for the better-circulating variations on the 1/6-comma scheme that J. S. Bach wrote his famous sets of 24 preludes and fugues in all keys called *The Well-Tempered Clavier.* Most musicologists do not see that work as the harbinger of equal temperament. Bach was undoubtedly aware of equal temperament as a theoretical abstraction. But what he tried to demonstrate with these pieces was "good" temperament, which did not at all necessarily mean equal temperament.

The widespread adoption of equal temperament in the nineteenth century meant the deliberate discarding of a useful musical resource (see Box 18.1). It also greatly disturbed William Pole for another reason. In *The Philosophy of Music* (1879) he wrote:

> The modern practice of tuning all organs to equal temperament has been a fearful detriment to their quality of tone. Under the old tuning, an organ made harmonious and attractive music. . . . Now, the harsh thirds, applied to

BOX 18.1 MOODS, KEYS, AND TRANSPOSITION

Many people attach special significance to the different keys in which music is written: C is white, F♯ is brown, F is martial, A is sad, and so forth. Is this justified? If we assume that everything is played in equal temperament, it would appear not. With equal temperament a piece can be transposed into any key (the diatonic scale can begin on any note) and its acoustical structure remains exactly the same.

But two or three centuries ago, with irregular temperaments more common, there was a pattern of nuances from having some intervals farther out of tune than others, and this pattern would change with transposition. A piece usually was written to be played in a particular key, and its character would be altered by playing it in a different one. Many Baroque musicians espoused an esthetic philosophy called the *doctrine of affections*, with which they tried to explain how different keys preferentially brought forth different emotional reactions from the listener. There are even a few examples of deliberate use of a wolf interval to evoke great distress, such as "The Death of Goliath" in Johann Kuhnau's *The Combat of David and Goliath.*

The tradition of writing certain types of pieces in certain keys was strong enough to persist well after the original acoustical reasons were vitiated by equal temperament. The modern musician still can be so affected by playing many of these old pieces that he comes to strongly associate particular keys with particular colors or emotions even though using equal temperament (see Exercise 21).

the whole instrument indiscriminately, give it a cacophonous and repulsive effect.

It is unfortunate that most of us, growing up with the wildly beating thirds of the piano, are unaware of what a pure third really sounds like.

---

*Another interesting byproduct of equal temperament is the tendency to have fewer mutation or mixture stops (or even omit them altogether) on late-nineteenth- and early-twentieth-century pipe organs. The more chordal style of Romantic composition and the equal temperament both led to marked clashes between the true harmonics of the mixtures and the false ones from the tempered intervals, so the emphasis in registration shifted to large assortments of imitative 8′ stops. Unfortunately, Baroque counterpoint becomes bland and unexciting on such instruments, and this realization has prompted a return in recent years to original principles of organ design and tuning.

In view of my negative remarks about equal temperament, you may expect me now to tell you which other scheme is best. But that involves the false assumption that it is possible for a single scheme to be best under all circumstances. It is more appropriate to ask which tuning or temperament would be best for a particular piece, or group of pieces, or for the bulk of music of a particular historical period. After all, if one piece never uses the interval G♯–E♭ it is perfectly all right to hide a wolf fifth there, while for some other piece written in a different key that would be a disaster.

As a first approximation we can say that Pythagorean tuning usually is appropriate for Medieval music, 1/4-comma meantone for Renaissance and early Baroque, 1/6-comma meantone or a circulating variant for late Baroque and Classical, and equal temperament for the Romantic and Modern periods. The nearly universal adoption of equal temperament, the replacement of the harpsichord by the modern heavy-frame piano (with its less clear tone and greater difficulty in retuning), and the composition of highly chromatic music in the late nineteenth century are all interwoven; each helped make the others possible.

For a given piece, is there any way of finding an optimum temperament, one whose total tuning error for the intervals used often in that piece is as small as possible? This can be done with a fairly simple computer program. I can imagine now, with small personal computers so common, a harpsichordist punching in information about the pieces she plans to play and letting the computer suggest a temperament to use for the concert. But I am also confident that a good musician will take this only as advice, and further temper it with her own artistic judgment! Mathematical theories of tuning may abound, but only the human ear can be the final arbiter of what is musically correct.

SUMMARY

Each musical interval size corresponds to a frequency ratio, which can be described by an equivalent number of cents. The cents scale is particularly convenient because it is additive. Unless specifically asked to judge tuning carefully, we tend to assign any interval we hear to a category. These categories are instilled by musical training, and in our culture correspond to those intervals available in the 12-tone chromatic scale.

Music with long-sustained chords formed with complex waves creates conditions where intervals corresponding to small-integer frequency ratios sound noticeably smoother than others and so play a special role in scale tuning. The most prominent clue for the ears comes from beating or roughness among those upper harmonics that have nearly the same frequency for both notes in the interval.

Perfect fifths, and to a lesser extent major and minor thirds, play a special role (about which more is coming in the next chapter) in the structure of the seven-tone diatonic scale and its chromatic extension to 12 tones. It is mathematically impossible to get all intervals in such a scale perfectly in tune at the same time. The unavoidable errors are compromised in various ways by different schemes of tuning and temperament. Equal temperament is only one of many possibilities and for many types of music is not the best choice.

REFERENCES

Corresponding material is presented in Backus, Chapter 8, and Benade, Chapters 14–16. Benade's Section 15.5 includes interesting comments on the difference in our concept of "scale" and that of musicians in India. There is much interesting information on pitch and interval perception in the chapter by W. D. Ward in *Foundations of Modern Auditory Theory*, Vol. I (see Chapter 17 references), pp. 405–447.

You can hear a Shepard scale and a Shepard glissando used in a composition by Risset, on the record "Voice of the Computer," Decca DL 710180. It is extremely informative to listen to music played on differently tuned instruments. Check your music library for J. M. Barbour and F. A. Kuttner's *Theory Series A* (Musurgia Records, 1958; now out of print). Record A-1, on Greek scales, includes demonstrations of Pythagorean tuning, A-2 deals with just tuning, and A-3 with meantone. The American Guild of Organists also has produced a cassette tape demonstrating historical tunings (Tape O-50, 1988).

Intervals, Scales and Temperaments by L. S. Lloyd and H. Boyle, Rev. ed. (Macdonald & Jane, London, 1979) remains one of the most intelligent books on scales and tuning. You also will find stimulating thoughts about exotic scales and about the derivation of the diatonic scale (although I do not endorse all their viewpoints) in Harry Partch, *Genesis of a Music*, 2nd ed. (Da Capo Press, 1974) and Joseph Yasser, *A Theory of Evolving Tonality* (American Library of Musicology, 1932).

Our work on perception of musical interval mistuning is presented in D. E. Hall and J. T. Hess, *Music Perception, 2*, 166 (1984). More extensive related work by Joos Vos appears in *Perception and Psychophysics, 32*, 297 (1982), *35*, 173 (1984), and *37*, 507 (1985); also *JASA, 77*, 176 (1985), and *Music Perception, 3*, 221 (1986). Intonation sensitivity for three- and four-note chords was investigated by Linda Roberts and Max Mathews, *JASA, 75*, 952 (1984).

All modesty aside, I highly recommend two articles by Don Hall, in *J. Music Theory, 17*, 274 (1973) and *Amer. J. Physics, 42*, 543 (1974). These tell more about the way fifths and thirds are intertwined in the fundamental impossibility of perfect tuning and about computer evaluation of both historical and optimum temperaments. There is an interesting short article on practical tuning by Mark Lindley in *Early Music, 5*, 18 (1977). Lindley is also the author of the excellent article on "Temperaments" in the new 6th edition of Grove's *Dictionary of Music and Musicians* (Macmillan, London, 1979). Further information on the history of tuning appears in J. Murray Barbour's *Tuning and Temperament* (Michigan State University Press, 1951), although I must caution you that his "deviation" figures are meaningless because he used equal temperament as an ideal criterion, which it certainly is not. There is also a wealth of detail in Owen Jorgensen's *Tuning the Historical Temperaments by Ear* (Northern Michigan University Press, 1977).

For opposing views about what Bach thought of equal temperament, see John Barnes, *Early Music, 7*, 236 (1979), and the chapter by Rudolf Rasch on pp. 293–310 in *Bach, Handel, Scarlatti Tercentenary Essays* (Cambridge University Press, 1985).

SYMBOLS, TERMS, AND RELATIONS

chroma
tone-height
pitch-class
melodic interval
harmonic interval
frequency ratio
cents
octave = 1200¢
Pythagorean hypothesis
consonant and dissonant intervals
just or pure intervals

just intonation
scales:
- diatonic major
- chromatic
- whole-tone

tunings:
- Pythagorean
- quarter-comma meantone
- just

temperaments:
- regular
- irregular
- circulating
- equal

circle of fifths:
- ditonic comma = 23.5¢ or 531441:524288

major-third circles:
- lesser diesis = 41.1¢ or 128:125

minor-third circles:
- greater diesis = 62.6¢ or 648:625

syntonic comma = 21.5¢ or 81:80

EXERCISES

1. If you go upward one octave from some starting point, and then upward another perfect fourth, you will cover a total interval called an eleventh. Use this information to tell (a) how many semitones in an eleventh, (b) how many cents in a justly tuned eleventh, and (c) the corresponding frequency ratio.

2. In Table 18.1, why is there no separate entry for the ratio 4:2? For 3:1?

3. Why is 37:23 not as important as any of the ratios listed in Table 18.1?

4. What would be the ratio and the corresponding number of cents for the inversion of the 7:4 interval (one covering the remainder of an octave)?

5. If you go upward a just major sixth and then upward another perfect fifth, what total interval do you cover? (Hint: It is not an eleventh.) What is the frequency ratio for this interval? How many cents? To which interval listed in Table 18.1 is it very closely related? (Hint: Sometimes multiplying or dividing by 2 will turn an unfamiliar ratio into an old friend.)

6. Suppose you cover a net distance of a major second (one whole tone T) in two different ways: (a) up a perfect fifth and back down a perfect fourth and (b) up a perfect fourth and back down a minor third. How large is the resulting major second in both cases, expressed both as a frequency ratio and in cents? (Note: You will get two different sizes, and neither one has a unique claim to being *the* right size.)

7. Intervals are especially hard to tune with sine waves because of the absence of primary beats. Sawtooth waves are easy because all harmonics are present (as in Figure 18.4). Referring to Chapter 8, which harmonics are present in a square wave? Does this suggest that square waves would be easy or hard to tune? In the same terms, would it be easier to begin tuning a pipe organ with open or closed pipes?

8. Show the harmonics of F_3 and A_3 (as in Figure 18.4b), and point out which is common to both. Show that frequencies 176 and 220 Hz would make a justly tuned interval. If these notes are slightly mistuned, show that the beat rate is given by $f_b = 4f_Y - 5f_X$. If the frequencies are 174 and 220 Hz, how many beats per second will there be? At what pitch should you listen to hear the beats more clearly?

*9. Think about both beat rates and critical

bands, and explain why thirds become dissonant in the low register and seconds can pass as nearly consonant at high frequencies. Discuss the implications for part writing, particularly choice of chord inversion and spacing in harmonization.

10. Compare the number of harmonics in common for parallel octaves, fifths, and thirds, and use this to explain which are more prone to lose individual identity and fuse together.

11. Find the size in cents of each interval present in Table 18.2. List all in order of increasing size, grouping together those of nearly the same size. Does the list bear any resemblance to our diatonic intervals?

*12. The Touch-Tone telephone code uses one of the four frequencies 697, 770, 852, 941 Hz together with one of the three 1209, 1337, 1477 Hz to make 12 combinations. Use a calculator with a logarithm key to find the size in cents of all intervals involved. Compare the spacing with that in a seven-tone equal-tempered scale.

13. What is the step size in cents for 53-tone equal temperament? How many of these steps does it take to get a close approximation to a minor third? A major third? A fifth? How close do they come?

14. Check each circle (M3, m3, and P5, eight of them altogether) in Figure 18.14 for the correct total error. Also check a couple of cases for the correct syntonic comma total.

15. Compare the mistuning of the triad C–E–G in Figures 18.8 through 18.14.

16. List all 12 major thirds and their size in cents for Vallotti's temperament. Verify that you get the same list from either Table 18.3 or Figure 18.14. How many different sizes are there?

17. List the 12 major triads in order from best to worst for Vallotti's temperament, and compare regular 1/6-comma meantone. (My criteria are that 6¢ mistuning of a fifth or 11¢ of a major third will make a triad "dubious" and 12¢ P5 or 30¢ M3, "awful.")

18. Suppose you wanted to modify 1/6-comma meantone as simply as possible to make a circulating temperament. How much total error lies in the three fifths C♯–G♯–E♭–B♭ in Figure 18.13? (Hint: It is not 24¢, because positive and negative can offset.) How many cents should you raise G♯ and lower E♭ (leaving the other 10 notes unaltered) to distribute that error evenly over those three fifths? This changes 11 numbers altogether (not counting duplications) in the error mosaic; show the revised version.

*19. Work out an additional column for Table 18.3 representing 1/3-comma meantone. Discuss the similarity to 19-tone equal temperament.

*20. Some organists use an irregular temperament (related to 3/11-comma meantone) by Werckmeister (c. 1700) that divides the ditonic comma into four equal parts. These are used to make the fifths G–D–A–E and B–F all 5.9¢ narrow, and the remaining eight fifths are made just. Construct the error mosaic for this temperament, and its column to add to Table 18.3. (Hint: Consider the syntonic comma to show that C–E will be 3.9¢ wide; then the rest is easy.) Discuss the major thirds and major triad gradation, as in Exercises 16 and 17.

*21. For the convenience of the wind players, what key signatures are used most commonly in marches for military band? What keys are most common in string quartet music? Does this help explain why these keys are associated with certain emotions?

22. What is the frequency ratio for a just major sixth? Do C_4 and A_4 in equal temperament actually have this ratio? When these two notes are played together, what beat rate will be heard, and at what pitch?

23. One way to make a minor seventh is to go up two perfect fourths in succession; another is to put a minor third on top of a perfect fifth. How big is the resulting minor seventh in each case, expressed both as a ratio and as a number of cents? Does either one agree with the "harmonic seventh"—that is, the 7:4 interval provided by the harmonic series?

*24. An interval with frequency ratio 7:5 is how many cents wide? (Hint: You may find the answer either directly with the logarithm formula or by simply taking the difference between two of the intervals that are in Table 18.1.) What is the frequency ratio and number of cents for the inversion of 7:5 (that is, the interval that, when added onto 7:5, completes an octave)? These two different versions of a tritone might tend to resolve differently, because one leans slightly toward P4 and the other toward P5. Which one might better play the role of diminished fifth, and which would then be an augmented fourth?

PROJECTS

1. Set up two electronic oscillators and a frequency counter, and ask several subjects to make repeated settings of some target interval. Think carefully about your choice of waveform (see Exercise 7). How close is the average setting to the theoretically ideal ratio? How much spread do you find in repeated trials?

2. Do violinists actually tune their strings a 3:2 perfect fifth apart? How accurately? Devise a way to measure the frequency (frequency counter alone probably will need to be aided by something such as low-pass filtering or Lissajous figure tricks on a scope), ask violinists to tune and retune for you, and interpret the results.

3. Make a complete and orderly exploration, perhaps with the aid of a computer program, of N-tone equal temperament for all $N \leq 60$. Look, among other things, for availability of good approximations to Pythagorean small-integer–ratio intervals. Decide to what extent 12, 19, 31, and 53 really stand out among the others, and then look up my answer to this problem in *Amer. J. Physics, 56,* 329 (1988).

4. Pick some unusual subset of the 12 notes in the chromatic scale, and write some music limited to this new scale. Alternative: To escape the piano's limitation to multiples of 100¢, use an electronic music synthesizer; with some it is quite easy to explore things such as N-tone equal temperament for N other than 12. Another alternative to escape 12-tone limitations: Rummage around a percussion storage room for various wood blocks, cowbells, etc.; let these define a scale and write some music for it.

CHAPTER

19 Structure in Music

It is the business of science to analyze—to break things down into their component parts, to understand them one piece at a time. And so we have spent most of this book studying one elementary aspect of sound after another—the working of one instrument, the early echo from one wall, the pitch and timbre of a single note.

But the scientist's task is not complete until the pieces are reassembled and verified to be in working order. We must consider whether our bit-by-bit strategy can satisfactorily account for all the workings of music or whether there are still important acoustical lessons in the way the pieces are put together. We took a major step in that direction in the last chapter by studying intervals, scales, and tuning. We now must look at how we build up the larger structures of music, finding again that there are new and important concepts in how things are combined that were not present in the elemental ingredients. Our perspective will be mainly that of the familiar European tradition; I must leave to someone more knowledgeable the application of similar thoughts to Asian and African music.

We shall ask similar questions first of melody, then of harmony: How do they achieve a sense of purpose, form, organization, or progress in music? This will require discussion of such concepts as tonality, chord functions in harmony, and consonance and dissonance. In the latter part of the chapter, we consider even larger structures in music—the overall form and organization of entire pieces and the evolution of musical styles over the centuries. We shall find light shed on all these questions by elementary concepts of information theory.

19.1 MELODIES AND MODES

Even without the complication of overt harmony, a good melody can do a great deal to establish musical form and organization. The melody has a rhythm and tempo that help set a mood, divide a piece into phrases and sections, and tell you what is beginning, middle, and end. Equally important are the choice of scale from which the notes of the melody are taken and the different roles assigned to different notes. These often result in implied harmonies even when none actually are sounding.

FIGURE 19.1 Use of the diatonic scale in several distinct modes, with tonic and dominant notes designated *T* and *D*. (a) Theme from second movement of Haydn's *Symphony no. 94* (''Surprise Symphony''), transposed to C major. (b) Subject of Bach's *Fugue in A Minor* for organ (S. 551). (c) Theme of a miniature for organ by Flor Peeters (Op. 55, no. 10) in the Dorian mode starting from D.

The persistence with which a melody returns to a particular note or avoids it, the length of time it is dwelt on, whether it is approached casually and stepwise or with a sudden leap—all serve to define a unique role for that note. More specifically, these influences usually result in singling out one note in the scale to serve as a home base, a center about which all the musical events revolve. When this happens, we call that tonal center a *tonic note.* We also say that this way of using the scale makes a *mode.*

We illustrate these terms in Figure 19.1, where we see that several different modes all can be made from the same diatonic scale. Out of seven possible choices of tonic note in the diatonic scale, two account for most of the familiar music of our culture. If the diatonic scale is represented by the white keys on a piano, having the tonic on C gives the *major mode*, and the *minor mode* has its tonic on A. Only the simplest melodies are limited strictly to the diatonic scale, however, and both modes are enriched by occasional use of chromatic notes (that is, extra or ''accidental'' sharps and flats).

---

*Technically, the natural minor mode defined above usually is altered to two slightly different forms for different purposes (harmonic minor and melodic minor, the latter having different patterns for ascending and descending). But we will not pursue that here.

Another important role is that of the *dominant note.* The dominant serves as a secondary center; the melody may linger around it for a while before returning to its primary center at the tonic. The interval between tonic and dominant was not the same in all the Medieval church modes. But in the familiar major and minor modes the dominant is a fifth above the tonic, G and E, respectively, in the examples above. For these modes the acoustical rela-

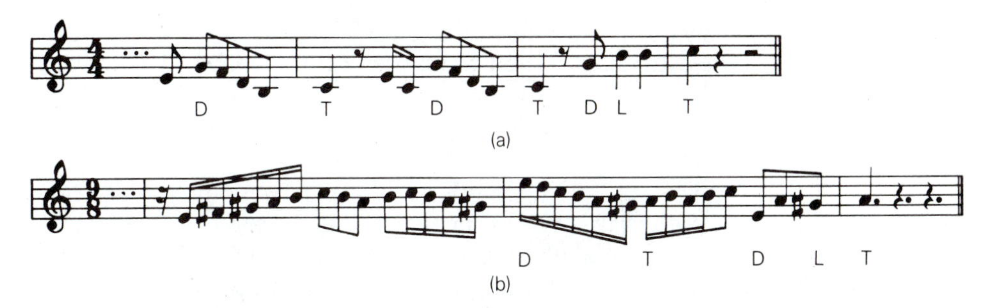

FIGURE 19.2 Use of the leading tone to approach the tonic, especially at the end of a phrase. (a) End of first movement of a Mozart piano sonato (K. 333), transposed to C major. Note how dominant *D* and leading tone *L* both help establish the final tonic *T* as a satisfying end. (b) End of the Prelude from Bach's *English Suite no. 6* for harpsichord transposed to A minor. Notice how the G is raised by a sharp sign so that it is only a semitone from the tonic and thus more effective as a leading tone.

tion between dominant and tonic is quite close—the most consonant interval other than an octave—so they sound as if they are giving complementary parts of the same message. They serve to strengthen each other in ways that become much more explicit with the introduction of harmony.

Another characteristic role is that of *leading tone*, a tone immediately adjacent to and leading directly into another more important tone. Implicit in this role is that the melody would sound disappointing and unfinished if it stopped on the leading tone. The note just below the tonic most often acts this way. In fact, this is such a useful way of reinforcing the tonic's own role that the subtonic note in the minor mode often is raised a half step so that it will lean more effectively toward the tonic (Figure 19.2). Furthermore, the leading tone often is played or sung somewhat sharp, making it less than 100¢ below the tonic.

These modes can be transposed to let any note on the keyboard serve as tonic; this may be necessary to make a melody lie within easy range for a particular voice or instrument. This is where the terminology of *keys* enters: A piece in the major mode is said to be in the key of C major if it uses C for tonic, but in E♭ major if E♭ serves as tonic and starting point for the *TTSTTTS* scale pattern. Similarly, the key of A minor means the minor mode with tonic A, and the key of F♯ minor has F♯ for tonic.

---

*We shall not require learning it here, but most musicians know an easy rule for relating the tonic of a mode to the number of sharps or flats that appear in its *key signature*: C major and A minor have none, and in all other cases counting around the circle of fifths to the transposed tonic tells how many sharps (in the C–G–D . . . direction) or flats (in the C–F–B♭ . . . direction).

Deliberate centering on a particular key, with strong differentiation of roles for different notes and a clearly identifiable tonic, is said to result in *tonality.* The addition of harmony may greatly strengthen a feeling of sustained tonality. When tonality is firmly established it is not merely a tonic note or tonic chord that serves as home base; it is a whole set of roles for several notes and chords that serves as an extended structural basis of the music.

The contrasting property of *atonality* requires a deliberate avoidance of any feeling that one particular note is the tonic, achieved by treating all notes in the scale alike. True atonality cannot be achieved with the seven notes of the diatonic scale; their uneven spacing makes it inescapable that some play different roles than others. But the 12 notes of the chromatic scale have no intrinsic structure, so they can be used to write atonal music. Arnold Schönberg (1874–1951) laid the foundations of atonal composition, and it has since been practiced by numerous composers of *serial music.* Various arbitrary rules take precedence over "pleasant" sounds in composing serial music; but that is only one particular way of achieving atonality. Schönberg himself made the distinction, and predicted that once atonality was well established, composers would get along fine without serialism.

---

*Knowing some readers will consider it reactionary, I offer the following personal opinion: I believe there are acoustical reasons why serial composition has remained an ivory-tower activity and never been accepted as mainstream music other than by a small avant-garde. I believe structure and contrast, recognizable symmetry and organization, are what cater to the abilities of ear and brain and make music esthetically pleasing. The deliberate avoidance of the most fundamental kind of structure in music (special acoustical relations of one note with another) seems bound to reduce it to uninteresting noise as far as most people are concerned. It is true that serial composers substitute other kinds of structure, but they are kinds directed toward the eye of someone studying the score, not the ear of a listener. Perhaps this should be judged not as music but as a new and distinct art form.

Chromaticism, the liberal use of the entire chromatic scale, need not, however, lead to atonality. It can instead add further interesting details to tonality and even strengthen it. This is the role played by chromatic tones in most of the classical music of the seventeenth through nineteenth centuries and in much popular music today.

19.2 CHORDS AND HARMONIC PROGRESSIONS

The elemental ingredients of harmony are chords, the most important of which involve three or four notes. But we should not think of a chord as just a collection of notes; it is more significant to regard a chord as made up of the intervals bounded by those notes. A *major triad,* for example, consists of a minor third immediately above a major third; if we reverse the thirds we get a *minor triad.* Just as the musical roles of simple two-note intervals depend on

FIGURE 19.3 Three contrasting chords. (a) A dissonant chord. (b) An F major triad, strongly consonant. (c) An F minor triad, somewhat less consonant.

their characteristic sound (as determined by frequency ratio), especially on their degrees of consonance or dissonance, so also for chords.

What makes some chords soothing, restful, or happy, and others sad, disturbing, or grating? Part of the answer, as with intervals, lies in how well the notes fit together to make a pattern that somehow appeals to us as recognizable, simple, or orderly. As a first crude approximation, we can say the nature of a chord is determined largely by the nature of its constituent intervals. A chord including a major second, as in Figure 19.3a, for instance, will generally sound rather dissonant because it contains that dissonant interval. Major and minor triads (Figure 19.3b) are considered consonant, because the intervals they contain (major third, minor third, and perfect fifth) are all consonant.

But this last example illustrates that listing the intervals in a chord is not sufficient—the major and minor triads contain exactly the same intervals yet sound very different. One is generally associated with joy and the other with sadness. So let us consider some other properties of a chord that may give clues to its musical character.

We said in the last chapter that intervals whose ratios occur in the harmonic series are especially important. They give your harmonic-pattern recognizer a chance to fit all the harmonics of both notes together into a unified pattern. Look back now at Figure 18.4a (page 405); we can further point out that the combined list of harmonics of both notes (200, 300, 400, 600, . . .) includes most of the same frequencies that make up the harmonic series based on 100 Hz. That is, there is an **implied fundamental** of 100 Hz, even though that frequency is not actually present. Likewise in Figure 18.4b, all the notes shown are harmonics of C_2. So C_2 is an implied fundamental, fitting like a glove with the actual sounding C_3 and G_3 and enhancing the feeling that they belong together.

Can we find implied fundamentals for chords with three or more notes? Let us try it first for a simple major triad. If justly tuned, the frequency ratios are 4:5:6. This alone might be enough for our brains to recognize a harmonic pattern, and it is all the easier if each tone has a complex waveform. Frequencies 400, 800, 1200, 1600, . . . from the first tone, 500, 1000, 1500, . . . from the second, and 600, 1200, 1800, . . . from the third, for instance, give a combined list of 400, 500, 600, 800, 1000, 1200, 1500, 1600, 1800, . . . that quite strongly suggests a missing fundamental at 100 Hz. We conclude that the three tones all cooperate to give an implied fundamental, called the *root* of the chord, which is critically important in classical harmonic analysis.

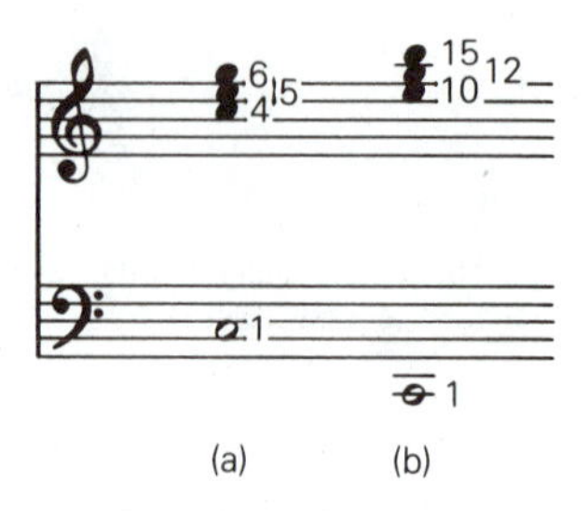

FIGURE 19.4 (a) The implied fundamental of a C_5 major triad is C_3. (b) The implied fundamental of an E_5 minor triad is C_2.

Now let us try the same thing with a minor triad in just tuning. If the lower pair are to have frequency ratio 5:6 and the upper pair 4:5, the smallest integers that will express this for the entire triad at once are 10:12:15. If, for convenience, we consider fundamentals 1000, 1200, and 1500 Hz, the combined list of harmonics is 1000, 1200, 1500, 2000, 2400, 3000, 3600, 4000, 4500, 4800, 5000, ... The largest number that includes all of these among its multiples is 100, so (at least as a mathematical exercise) we might again say there is an implied fundamental of 100 Hz.

But there are two important differences. First, a good representation of all the harmonics of 100 Hz is present in the major-triad example, whereas the selection is quite sparse for the minor triad. It is easy to believe that our pattern-recognition apparatus can pick up the 4:5:6 relation, aided by the presence of many other harmonics of the implied fundamental. But it would require much finer discrimination to pick out 10:12:15 (and be sure it was not, for instance, 9:11:14 instead), with only occasional additional representatives of the implied fundamental to help tie down the conclusion.

Second, consider in musical notation the identity of the implied fundamental (Figure 19.4). For the major triad it is just two octaves below one of the three original notes, and we have a strong root closely cooperating with the sounding notes. But for the minor triad a 1:10 ratio places the implied fundamental three octaves plus a major third ($\frac{1}{8} \times \frac{4}{5}$) below the lowest original note; it is not only farther away but also an entirely different note that clashes with the sounding notes! We must conclude that (1) there is probably much less tendency to perceive an identifiable root for a minor chord; (2) insofar as we admit the possibility of this weak implied fundamental, it only makes things more complicated instead of reinforcing a simple pattern; and so (3) the acoustical basis for consonance in minor chords is considerably weaker than in major chords.

The same conclusions still follow when we try alternative viewpoints. One is to backtrack to individual intervals and look for implied fundamentals there. For the major triad, both thirds imply the same fundamental discussed above, and the fifth implies another an octave higher; these all help give the same impression of a strong root (Figure 19.5a). But for the minor triad, the three intervals give three different implied fundamentals, one of which clashes with the original chord again (Figure 19.5b).

APPENDIX

A Written Music

There will be some users of this book who are not accustomed to reading music. They should not be discouraged, for it is not necessary to read music proficiently for this purpose. Something roughly equivalent to "Run, Dick, Run" will be enough music-reading skill to decipher most of the examples and exercises, except in Chapter 19.

This appendix provides a brief introduction to a few of the basic features of written music. Use it as necessary, especially along with Figure E. To give your ear some definite associations with what you see on the page, you should make a serious effort to get access to a piano, especially in connection with Chapter 7. Use Figure E to find the right keys and actually play some of the examples, even if only slowly and haltingly.

Written music allows the composer to give the performer detailed instructions about which sounds are to be produced and how they are to succeed one another with the passage of time. Some instructions apply generally to entire passages of music. These include tempo and meter (discussed in Section 7.1), as well as dynamic markings telling how loud or soft to play. The Italian words *forte* (loud) and *piano* (soft) are usually abbreviated with *f* and *p*. Greater degrees of loudness are indicated by *ff* (*fortissimo*) and *fff*, and greater softness by *pp* (*pianissimo*) and *ppp*. In between come *mf* (*mezzoforte*) and *mp* (*mezzopiano*), where *mezzo* means "moderately" (literally, "half" or "medium"). Changes in loudness can be gradual increases (*crescendo*, or <) or decreases (*decrescendo*, or >). Or they may be sudden as indicated by *subito piano* (suddenly soft, and continuing that way) or *sfz* (*sforzando*, a sudden strong accent followed by a return to the previous normal level).

Especially important to us now is the code that tells about each individual note. It is natural to place each symbol in succession across the page in the same direction as you read words (in our culture from left to right), to indicate their order in time. Our description of some notes as sounding high in pitch and others low suggests coding that information by placing symbols higher or lower on the page. (The word *note*, incidentally, is used both for the symbol and for the sound it represents.) But pitch height must be measured from some reference level, and this is conveniently represented by a line drawn across the page, as in Figure A.1.

You can see that it is rather inconvenient to estimate just how many steps above or below the line are intended. It is natural to add another reference line, then another and another. Although history was not actually so simple,

FIGURE A.1 A twelfth-century manuscript illustrating an early stage in the development of written music. (Courtesy of the British Library)

it is a useful fiction to contemplate a gradual accumulation of 11 lines (Figure A.2, left), standardized so that each line and each space between is home for one of the letter-named notes of the diatonic scale (white keys only on the piano). For our purposes it is purely a matter of historical accident which of these happens to be called A; but, of course, the reason the names begin again with A following G instead of H is that two notes an octave apart sound so much alike. This 11-line group has C_4 on the middle line; hence "middle C."

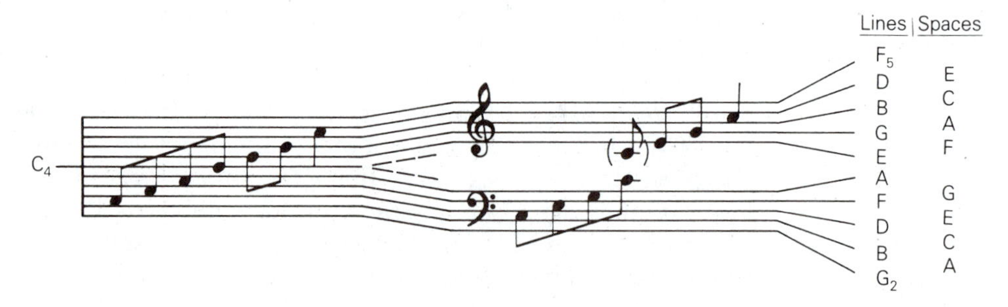

FIGURE A.2 A C major arpeggio (chord played one note after another), written first on an 11-line staff, and again on the modern double staff. Note the treble and bass clef signs to distinguish the two staves. The line for middle C has disappeared, and it is now written with a ledger line either above the bass clef or below the treble clef.

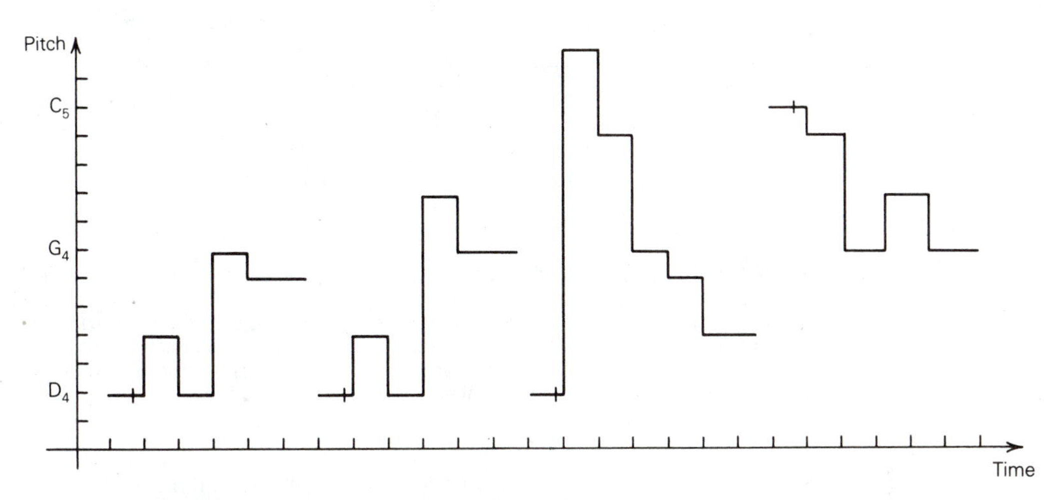

FIGURE A.3 The familiar tune "Happy Birthday," coded in standard music notation (above) and as a graph of pitch versus elapsed time (below). The fermata or "hold" (𝄐) and the ritard ("slow down") cause the pitch changes on the graph to fall later than the times marked on the axis when we would hear clicks from a metronome set to the original tempo. Notice the slight vertical distortion: The whole step D to E ("birth-day" in the first phrase) and the half step G to F♯ ("to you") are both represented on the musical staff by the same distance, yet they are not the same size interval.

But so many lines make it hard again to estimate just which pitch is intended—when playing a piece at a fast tempo, it is hard to make snap judgments whether a note symbol is on the fourth or the fifth line, for instance. So purely for ease of reading, we imagine the top five and bottom five lines separated (Figure A.2, center); each group is called a *staff*. The middle-C line disappears except when that note actually occurs; then it reappears briefly as a *ledger line*. (The same trick also is used to accommodate notes above or below the outermost lines; see Figure E.) By using the *bass clef* (𝄢) and *treble clef* (𝄞) signs, we can identify either group of five lines by itself. They correspond roughly to the ranges of low men's and average women's voices, respectively.

Consider the example of Figure A.3. The musical notes on the staff are reminiscent of the graph of a functional relationship. As a first approximation, you sweep your eye steadily from left to right and play high or low notes according to whether you encounter little marks higher or lower on the staff. But to insist that time elapse uniformly with distance toward the right would mean that rapid notes must be crowded together, while long-sustained notes simply waste a lot of space. Therefore we adopt code symbols to stand for longer and shorter notes (Figure A.4), and allow these symbols to be more or less evenly spaced as long as they follow one another in the correct order. We

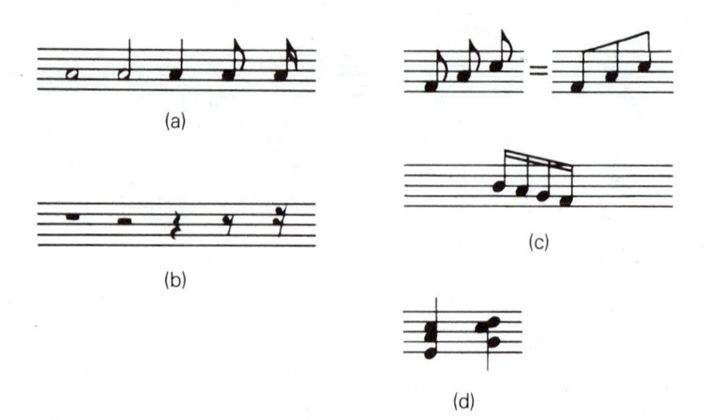

FIGURE A.4 From left to right in (a) are the symbols for the whole, half, quarter, eighth, and sixteenth note; each one is held only half as long as the one to its left. Underneath each in (b) is the symbol for a rest of the same duration. Any of these symbols with a dot immediately following (as occurs in Figures A.3 and A.6) is held half again as long as ordinarily. Groups of successive eighth or sixteenth notes may be joined together with bars (c) instead of each having individual flags on their stems. Notes to be sounded simultaneously as a chord are shown in (d).

also need *rest* symbols to indicate similar lengths of time during which no note is to be played.

We also allow some nonuniformity vertically. The real reasons for this distortion are quite subtle; suffice it to say that musicians have agreed to let equal vertical distances represent equal numbers of *white* keys on a piano keyboard, whereas it is only the white and black keys taken together that represent a succession of approximately equal steps in pitch (see Figure E). So each line and each space on the staff represent one of the white keys. The performer will not strike a black key unless he sees one of two special signs; the *sharp* sign (♯) immediately preceding a note says "instead of this white key, play the black key just to its right; that is, the one next higher in pitch," and the *flat* (♭) sign says "shift left to the next lower black-key pitch." If a piece requires certain black keys to be used routinely, it would be tiresome to repeat these signs many times. So they can be put in a *key signature* at the beginning of a score (like the F♯ in Figure A.3) to indicate that every occurrence of that note throughout the piece, including all its cousins of the same name in other octaves, will be played on the black key instead of the white. Sometimes in the middle of such a piece the instruction "use the white key after all" is needed, and this is given by the *natural* sign (♮).

FIGURE A.5 One octave of the chromatic scale. The pitch interval from each note to the one following has several names: semitone, half-step, and minor second are most common. These 12 intervals are all approximately the same size; possible inequalities are considered in Chapter 18 in our study of tuning theory.

FIGURE A.6 A familiar nursery tune to serve as review for any reader now learning to read music.

Figure A.5 shows how to code the instruction, "Start at middle C and play each higher note in order from there to C_5, including both white and black keys."

See if you can recognize the familiar nursery tune of Figure A.6. If you have trouble reading the musical notation in any of the examples in the book, be sure to ask for help.

APPENDIX

B The Metric System

The United States is the last major industrial nation that has yet to convert fully to the metric system of measurement. This delay has put us at a disadvantage in world trade and has cost us many billions of dollars. This edition is for the 1990s when we should already have been well along with metrication. I feel I would do my readers a great disservice by using any other than metric units throughout the book.

In that spirit, I offer English equivalents in the last column of Table B.1 only for completeness and for occasional comparison with past experience. There is never any need in this book to use these English equivalents in calculations. I urge you to simply work all problems in the metric system, rather than converting to English and back. There is no better way to become comfortable with the metric system than to simply use it.

TABLE B.1 Units for Motion, Force and Energy

Symbol	Concept	Defining Relation	Metric Units	English Equivalents
x	Length		1 m = **meter**	3.28 ft
S	Surface area	$S = x_1 x_2$	$1\ m^2$	
V	Volume	$V = x_1 x_2 x_3$	1 l = liter $= 10^{-3}\ m^3 = 10^3\ cm^3$	1.06 qt
m	Mass		1 kg = **kilogram**	0.068 slug
D	Density	$D = m/V$	$1\ kg/m^3$	$1.93 \times 10^{-3}\ sl/ft^3$
t	Time		1 s = **second**	
f	Frequency	$f = 1/P$	1 Hz = hertz = 1 cycle/s	
v	Velocity	$v = x/t$	1 m/s	2.24 mph
a	Acceleration		$1\ m/s^2$	
F	Force	$F = ma$	1 N = newton $= 1\ kg\text{-}m/s^2$	0.225 lb
p	Pressure	$p = F/S$	$1\ N/m^2$	$1.46 \times 10^{-4}\ lb/in^2$
			1 atmosphere* $= 1.013 \times 10^5\ N/m^2$	$14.7\ lb/in^2$
W E	Work or Energy	$W = Fx$	1 J = joule = 1 N-m	0.74 ft-lb
P	Power	$P = W/t$	1 W = watt = 1 J/s	1.35×10^{-3} hp
I	Intensity	$I = P/S$	$1\ W/m^2$	
T	Temperature difference		1 K = **kelvin**	1.8 F degrees

*The atmosphere is not an official basic metric unit; nevertheless it is convenient and useful to retain this as an alternative.

B.1 UNITS FOR PHYSICAL MEASUREMENTS

Each physical quantity must be measured in appropriate units. A summary of information about these units is presented for reference in Table B.1. Notice that length, mass, time, and temperature have unique status as primary quantities, which do not depend on prior definitions of any other concepts. Once the length of a certain metal bar kept in Paris is arbitrarily declared to be the standard meter, for instance, then all other lengths are measured by comparison to that standard.

All other units are secondary or derived; they are merely names for certain combinations of meters, kilograms, and seconds. The definition of each derived unit comes from a physical relationship, such as the hertz from $f = 1/P$ for frequency and period, or the joule from $W = Fx$ for the work done by a force F exerted through a distance x.

Some of these derived quantities involve areas or volumes. The proper way to read m^2 or m^3 is *square meters* or *cubic meters*, respectively. Several use the slash symbol, which is read as *per*: kg/m^3 means kilograms per cubic meter, for instance. You should think carefully about how *per* in turn means "for each." To say that the mass density of aluminum is 2700 kg/m^3, for instance, means that any block of aluminum, regardless of size, has 2700 kilograms of mass *for each* cubic meter of volume. A block 2 m wide, 5 m long, and 1 m thick contains 10 cubic meters, each with 2700 kg, for a total mass of 27,000 kg. The same concept applies just as well to small blocks. A cube of aluminum 0.1 m (10 cm) on a side has a total volume of 0.001 m^3 and a mass of 2.7 kg, and it, too, has density 2700 kg/m^3 in the sense that enough such blocks to make a cubic meter (in this case 1000) would have a total mass of 2700 kg.

The last example brings up another point often misunderstood by students. The volume of the 10-centimeter cube is $(0.1 \text{ m}) \times (0.1 \text{ m}) \times (0.1 \text{ m}) = (0.1 \text{ m})^3 = (0.1)^3 (\text{m})^3 = 0.001 \text{ m}^3$, and that is *not* the same as 0.1 m^3. Neither is it the same as $(0.001 \text{ m})^3$—that is the volume of a cube 1 mm on a side, which is only a billionth of a cubic meter. Another tempting mistake is to say

TABLE B.2 Standard Prefixes

Prefix	Equivalent Multiplier	Example
n = nano	10^{-9}	ns
μ = micro	10^{-6}	μfd
m = milli	$10^{-3} = 0.001$	ms, mm
c = centi	$10^{-2} = 0.01$	cm
d = deci	$10^{-1} = 0.1$	dB
K = kilo	$10^3 = 1000$	KW, KHz
M = mega	10^6	MHz
G = giga	10^9	GHz

FIGURE B.1 Copyright © 1972 United Feature Syndicate, Inc. Reprinted by permission.

$V = (10 \text{ cm})^3 = 1000$ cubic centimeters and then divide by 100 (the number of centimeters in a meter) to get the wrong answer $V = 10 \text{ m}^3$. Avoid this by picturing 1-centimeter cubes stacked 100 across, 100 long, and 100 high to fill a cubic meter. Then you realize that it takes a million cubic centimeters to make a cubic meter, and you get the correct answer that 1000 cm^3 divided by 1,000,000 cm^3/m^3 is 0.001 m^3.

Table B.2 shows standard prefixes that are used in describing fractions and multiples of any of the basic units. The first two and last entries in the table are not used in this book, but you might encounter them in your other reading (Figure B.1).

Table B.3 shows some additional units used in electricity and magnetism. They are given here only for completeness of reference and are not actually used elsewhere in the book. Note that the coulomb is another primary quantity (it is the amount of electric charge carried by 6.2×10^{18} electrons), and all the rest are derived units.

TABLE B.3 Electric and Magnetic Units

Symbol	Concept	Defining Relation	Metric Units	
Q	Charge		1 C = coulomb	
I	Current	$I = Q/t$	1 A = ampere	= 1 C/s
V	Voltage or potential difference	$V = W/Q$	1 V = volt	= 1 J/C
R	Resistance or impedance	$R = V/I$	1 Ω = ohm	= 1 V/A
E	Electric field	$F_e = QE$		1 V/m
B	Magnetic field	$F_m = QvB$	1 T = tesla	= 1 V-s/m^2 = 10^4 gauss

B.2 SCIENTIFIC NOTATION AND COMPUTATION

Scientists often must use immensely large numbers or exceedingly small fractions, which are inconvenient to write out. They have adopted a shorthand notation, based on the trivial ease of multiplying or dividing by 10. For instance, $10 \times 10 \times 10 = 10^3 = 1000$. In general, any number n of factors of 10 multiplied together gives a 1 followed by n zeros; but rather than write out all those zeros when n is large we just leave it in the form 10^n. Thus a thousand is 10^3, a million is 10^6, a billion is 10^9, and so on. The proper way to say 10^9 aloud is "ten to the ninth power," or just "ten to the ninth."

Other numbers can be put in this form as well; for instance, $23{,}500{,}000 = 2.35 \times 10{,}000{,}000 = 2.35 \times 10^7$. In general, anything times 10^n stands for that thing with the decimal point moved n places to the right. We extend this by supposing that if positive exponents mean moving the decimal to the right, then negative exponents must mean moving it to the left. For example, $10^{-5} = 0.00001$ and $3.4 \times 10^{-3} = .0034$. Be sure you do *not* write 7^4 in place of 7×10^4, for it is an entirely different number.

If you need to multiply or divide numbers in scientific notation, one way is to write them out in ordinary notation and then do the arithmetic in sixth-grade longhand style. For example, 3×10^5 divided by 2×10^{-4} becomes $.0002\overline{)300{,}000.0000} = 1{,}500{,}000{,}000$. But it is easier to know that you can handle all the powers of 10 just by adding or subtracting their exponents. Multiplication is taken care of by $(A \times 10^m) \times (B \times 10^n) = (A \times B) \times 10^{m+n}$ and division by $(A \times 10^m)/(B \times 10^n) = (A/B) \times 10^{m-n}$. These apply regardless of whether m and n are positive or negative. The example above becomes $3 \times 10^5/2 \times 10^{-4} = (3/2) \times 10^{5-(-4)} = 1.5 \times 10^9$. If you will practice this a few times, you will find that it is a very easy way to handle large and small numbers. Be sure to avoid this common error: $10^3 \times 10^4$ is 10^7, not 100^7.

One final piece of advice: Do not calculate blindly. All the exercises and examples in the main text (this appendix being an exception) involve quantities with *physical meaning*. The numbers or letters serve only as shorthand symbols for that concrete meaning. Do not be content with mere manipulation of the symbols, but make sure you always think about the underlying physical meaning.

In particular, you should form the habit of keeping the units together with the numbers for all physical quantities. For example, suppose you are to find the density D of a block of wood if it has mass $m = 1.5$ kg and volume $V = 0.003\ \text{m}^3$. The answer, given by $D = m/V$, is *not* merely $1.5/0.003 = 500$; it is $(1.5\ \text{kg})/(0.003\ \text{m}^3) = 500\ \text{kg/m}^3$. The number alone without the units does not have the same physical meaning; it simply is not a correct answer.

EXERCISES

1. 8 KHz means how many Hz?
   26 cm means how many m?
   3 MW means how many W?
   0.6 ms means how many seconds?

2. A rectangle 3 cm wide and 4 cm long has how much surface area in cm^2? Show that this is $1.2 \times 10^{-3}\ \text{m}^2$.

3. If a force of 12 N acts through a distance $x = 0.2$ m, how much work is done? (Use information about units in Table B.1 to express this as a number of joules.)

4. Use information about units in Table B.1 to show that (3 N) $\times$ (4 m/s) = 12 W.

5. If power $P = 10^{-3}$ W is spread over a surface area $S = 10^2\ \text{m}^2$, what is the corresponding intensity I?

6. Write each of these numbers in scientific notation: (a) 10,000,000,000,000, (b) 0.00001, (c) 3600, (d) 0.036.

7. Write each of these numbers in ordinary notation: (a) 10^{18}, (b) 10^{-12}, (c) 3×10^4, (d) 2.7×10^{-5}.

8. Do the following multiplication and division problems, leaving your answers in scientific notation:
   (a) $(2 \times 10^3) \times (3 \times 10^4)$
   (b) $(2 \times 10^{-3}) \times (3 \times 10^4)$
   (c) $(4 \times 10^{10})/(2 \times 10^8)$
   (d) $(4 \times 10^8)/(2 \times 10^{10})$
   (e) $(4 \times 10^{-3})/(2 \times 10^{-5})$
   (Hint: The last answer is $2 \times 10^{+2}$. If you feel at all unclear about which numbers are large or small or why the answers come out as they do, carry out the operations in ordinary notation also and verify that you get the same answers both ways.)

APPENDIX

G Glossary

Absorption Loss of energy from a sound wave upon reflection; especially its conversion into heat by friction or viscosity.

Acceleration Rate of change of velocity.

Amplitude (1) Maximum change in any variable (such as displacement or velocity) during each cycle of any vibratory disturbance. (2) For sound, this most often refers to the maximum *pressure* change caused by the wave.

Antinode A point in a standing-wave pattern where the oscillation has a greater amplitude than at any neighboring point. (See also *node.*)

Articulation Precise control of the joining or separation of successive notes by a performer.

Basilar membrane Inner-ear structure whose vibrations stimulate the hair cells to produce nerve impulses carrying information about sound to the brain. (See Chapters 6 and 17.)

Bass Sounds of low pitch, such as below G_3 (that is, frequencies less than approximately 200 Hz, and especially below 100 Hz).

Beats (1) Slow rise and fall of perceived loudness when signals with slightly different frequencies are combined. (2) A series of accented musical events, evenly separated in time.

Binaural "Two-eared"; perception based on different signals arriving at left and right ears.

Brass Wind instruments using the lips as a double reed. Often but not necessarily made of brass.

Bridge A support defining one end of the active length of a stretched string; especially such a support mounted on a soundboard.

Cadence Stylized musical formula for reaching a point of rest or repose.

Cent A unit of measurement for interval size, such that 1200¢ makes one octave.

Chromatic scale A scale with 12 notes per octave, evenly spaced in pitch or approximately so.

Consonance A pleasant or restful feeling produced by an interval or chord, which depends *both* on the intrinsic acoustical properties of the interval or chord *and* on the context in which it is heard.

Critical band Any range of frequencies within which any two signals strongly stimulate a common portion of the basilar membrane.

Current Rate of flow of electric charge around a circuit.

Damping The gradual decay of vibrations as their energy is lost to friction, viscosity, or radiation.

Decibels Units for sound level differences. One bel corresponds to an intensity ratio of 10, and $1 \text{ dB} = 0.1 \text{ B}$ to an intensity ratio $\sqrt[10]{10} = 1.26$.

Diatonic scale A scale with seven notes per octave, unevenly spaced in pitch, especially in the *TTSTTTS* pattern of ascending whole tone and semitone steps.

Diffraction Characteristic behavior of all types of waves, in which they tend to spread outward in all directions after passing through an opening, or to fill in the region behind an obstacle rather than leaving a shadow.

Displacement Distance away from a fixed reference point, such as an equilibrium position.

Dissipation Conversion of any other form of energy into heat, especially through friction.

Dissonance An unpleasant or agitated feeling produced by an interval or chord; a feeling of tension calling for movement to a *consonance.*

Edgetone Vibration produced by fluid-flow instability when a narrow stream of air is directed against a sharp edge.

Energy Capacity for doing work. Includes *kinetic energy* (due to motion) and *potential energy* (due to location, distortion, or chemical state), both of which are involved in sound waves.

Enharmonic Relation between two notes (such as G♮ and A♭) that are distinguished in written notation but *not* distinguished in performance (at least for some instruments, especially piano); distinct from *inharmonic.*

Equilibrium Any system configuration for which all forces are precisely balanced; when placed in this configuration, the system will remain at rest.

Exponential decay A type of *damping* for which passage of equal amounts of time always reduces the amplitude by the same fraction, regardless of the initial amplitude.

Feedback Any mechanism by which the present value of some physical variable can influence its future evolution. Negative feedback tends to return the system to equilibrium, while positive feedback destabilizes.

Force A push or pull of any kind, with total strength measured in newtons.

Formant A broad resonance (especially in the human vocal tract) that strengthens those Fourier components within a definite frequency range.

Fourier analysis and synthesis Mathematical representations of the equivalence between a complex waveform and a mixture of pure sine waves or *Fourier components.* (See Chapter 8.)

Frequency Repetition rate of a vibration, measured in Hz (cycles per second).

Fundamental The lowest frequency of a harmonic series, or the slowest natural mode of a vibrating system.

Harmonic distortion Introduction of additional harmonics into a signal by transducer nonlinearity.

Harmonic series A set of vibrations whose frequencies are all integral multiples of one fundamental frequency.

Helmholtz resonator An air reservoir with a relatively small opening, whose lowest resonant frequency is determined primarily by its total volume and neck dimensions but is practically independent of other details of its shape.

Impedance A measure of how much force must be applied to a wave-carrying medium to produce a given amount of motion. (See Box 10.3.)

Inertia Tendency of any moving object to continue onward with constant speed; a direct consequence of the object's mass.

Inharmonic Adjective referring to a set of natural modes and their frequencies, which *lack* the regular pattern of a harmonic series. Distinct from *enharmonic.*

Instability Any process that acts to drive a system farther and farther away from an equilibrium state.

Intensity Physical measure of sound strength as a rate of flow of energy per unit area: $I = E/St$.

Interference Combination of signals of the same frequency from two or more sources, which may cooperate at some points in space (*constructive* interference) but cancel at others (*destructive* interference).

Interval (1) Perceived separation between two pitches. (2) Musical element formed by a particular size of pitch separation, such as the perfect fifth.

Just noticeable difference (JND) Minimum change required in any stimulus (such as sound level or frequency) before the difference is reliably detected by a human listener.

Key (1) Device struck by a finger to control a hammer, pipe valve, or plectrum when playing a piano, organ, or harpsichord. (2) Bar struck by a mallet to produce sound on a xylophone, marimba, or vibraharp. (3) Finger-operated device to cover or uncover a woodwind tone hole.

(4) Short for *key signature*, or written instruction to the performer on how many sharps or flats to use. (5) Tonal center indicated by melodic and harmonic structure, as in "this passage modulates from the key of C major into E minor."

Linear versus nonlinear A relation between any two variables is linear if it has, but nonlinear if it lacks, the property that doubling one variable always precisely doubles the other variable. (See Sections 2.4 and 16.4.)

Loudness Psychological impression of sound strength; measured in sones.

Loudness level Limited information about the loudness of any sound, involving only comparative judgment but no magnitude estimation. The loudness level (in phons) is the same number as the sound level of a 1000-Hz sine wave whose loudness equals that of the sound in question.

Masking Ability of one sound to obscure the presence of another, especially one weaker and higher in frequency.

Mass Amount of material in an object, as measured by its inertia, or resistance to being accelerated.

Missing fundamental Tendency for pitch judgment to correspond to the fundamental of a harmonic series even if the fundamental is not physically present in the spectrum.

Mode (1) Manner of using a scale that assigns distinct musical roles (for example, tonic and dominant) to different notes. (2) See *natural mode*.

Mute A device for reducing or muffling the ordinary sound output of any instrument.

Natural mode Any pattern of motion of an extended object or system for which all parts move in simple harmonic motion of the same frequency. (See Chapter 9 for extended discussion.)

Node A point in a standing-wave pattern at which a physical variable remains constant in spite of its time variation at neighboring points. Pressure nodes must be distinguished from displacement nodes, as they occur at different places.

Note (1) A single musical sound of definite pitch. (2) A name (for example, C_4) identifying that pitch. (3) A symbol (for example, ♪) representing it in written music.

Octave (1) The particular musical *interval* characterized by the perception that one pitch seems just like the other even though distinctly higher. (2) The 2:1 ratio of frequencies that produces this sensation.

Ohm's law (1) In acoustics, the statement that human perception of a sound is determined entirely by the amplitudes of its Fourier components and not by their relative phases. True only within limited circumstances. (2) In electric circuits, a relation among current, voltage, and resistance.

Overblowing Using higher wind pressure (or vent holes) to cause a pipe or wind instrument to speak at higher pitch.

Overtone Any Fourier component of a sound except the one of lowest frequency. (Not used in this book, but common among musicians.)

Pattern recognition Ability and tendency to perceive complex stimuli as forming standard patterns; for sound, especially, the assignment of pitch by fitting an actual Fourier spectrum as closely as possible to a harmonic series template.

Perception Conscious awareness and judgment of a stimulus (such as sound), and various physiological and psychological components of this process.

Percussion Those instruments played by striking (or in a few cases rubbing, plucking, etc.) any object to produce transient sounds. In broad usage, sometimes taken to include the piano.

Period Length of time to complete one cycle of a vibration, or of any other repetitive phenomenon.

Periodic wave A vibration that repeats precisely the same pattern over and over again, thus producing a steady tone.

Periodicity theory Model for pitch perception dependent on ability of the brain (as opposed to ears) to measure periods of repetitive waveforms. (See Section 17.2.)

Phase Relation in time between two signals of

the same frequency. *In phase* means both peak together; (180°) *out of phase* means one lags a half-cycle behind the other. Other cases between these extremes can be described also; for example, "90° out of phase" for a quarter-cycle lag.

Phons Numerical labels to identify the *loudness level* of any sound; defined by the sound level in dB of a 1000-Hz sine wave of equal loudness.

Pitch (1) Psychological impression of the height of a sound (bass or treble). (2) A note name to identify this high or low pitch within some scale.

Pitch-class A set of notes all called by the same name, each being separated from all the others by one or more full octaves.

Place theory Model for pitch perception in which each different pitch results from stimulation of a distinct part of the basilar membrane. (See Chapters 6 and 17.)

Power Rate of transfer of energy; $P = E/t$.

Precedence effect Tendency of human judgment to ascribe the location of a sound to the direction from which it first arrives, even when early reinforcement from other directions is louder.

Pressure Force per unit area, $p = F/S$; a measure of the density or concentration of a force distributed over a surface.

Propagation Movement (as of a sound wave) from one place to another, and behavior (such as reflection, diffraction, and so forth) characteristic of this movement.

Reed (1) A thin strip of metal or wood whose vibrations serve to control air flow into a reed instrument. (2) In a broader sense, any other object serving the same purpose, such as vocal cords or brass players' lips.

Refraction Change of direction of wave travel due to difference in its speed from one place to another; to be distinguished from *diffraction*, which takes place even when the wave speed is the same everywhere.

Resonance Vibration of especially large amplitude occurring when three conditions are fulfilled: (1) A system with a natural vibration mode frequency f_n is acted upon by (2) an alternating driving force (possibly from a sound wave) with frequency f_0, and (3) f_0 is nearly the same as f_n.

Restoring force Any force whose action is always in such a direction as to return an object to an equilibrium position.

Reverberation Combined effect of multiple sound reflections within a room, giving a smooth and gradual decay of the perceived sound after a source stops.

Reverberation time Time for a decaying sound to drop 60 dB below its initial level. Meaning is ambiguous unless the decay continues always at the same rate.

Rhythm Patterns of strong and weak beats and subbeats, which may be stressed both by the composer's choice of notes and the performer's dynamics and articulation.

Scale A discrete set of pitches that provides all the notes to be used in a particular piece, or in a particular type of music. (See Chapter 18.)

Scaling The orderly progression in size and proportion among strings in a piano, pipes in an organ rank, or members of an instrument family. Proper scaling maintains a unified tone quality.

Simple harmonic motion (SHM) Sinusoidal motion, such as that of a mass bouncing on the end of an ideal linear spring, or that of a natural mode.

Sine (or sinusoidal) wave The simplest, smoothest waveform, one produced by simple harmonic motion. Its Fourier spectrum consists entirely of the fundamental, with no higher harmonics.

Sones Units for measurement of *loudness.*

Sound (1) A longitudinal-wave disturbance in any compressible substance, but especially in air. (2) The perceptual object "heard" in the brain when these waves stimulate human ears.

Sound level Coded information about the *intensity* of a sound, expressed in decibels. Specifically, the sound level corresponds (Table 5.1) to the ratio I/I_0, where $I_0 = 10^{-12}\ \mathrm{W/m^2}$.

Spectrum The recipe of strengths of the sinusoidal components of a complex wave. (See Chapter 8.)

Standing wave A natural mode of a continuous system (for example, a piano string or the air in an organ pipe), characterized by its frequency of vibration and by its pattern of stationary loops and nodes.

Syncopation A deliberate shift or disturbance of an established pattern of accented beats.

Temperament (1) Adjustment of the pitch of one note so that the interval it forms with another note will be out of tune by a small, deliberately controlled amount. (2) A complete scheme for adjusting the pitch of all notes in a scale by deliberately tempering one or more intervals.

Tempo "Pulse rate" of music; how often successive notes or rhythmic beats occur.

Timbre Tone color; the psychological impression of what characterizes a tone besides its pitch and loudness.

Tonality Deliberate organization of musical structure around a recognizable home key.

Tone (1) Any steady musical sound of definite pitch. (2) Used specifically by some authors to mean "pure tone" or "simple tone," that is, a tone with pure sinusoidal waveform and no higher harmonics. (3) An esthetic judgment of timbre in performance, as in "The soloist played with a robust and satisfying tone." (4) A particular size of musical *interval*, the whole tone or major second.

Transducer Any device that converts wave signals from one energy form to another, as, for example, a microphone converts sound waves to electrical signals.

Transient (1) A sound that is not sustained, but quickly dies away. (2) The rapidly changing initial attack or final decay portions of an otherwise steady sound.

Traveling wave A wave disturbance that carries energy from one region to another far away (as opposed to a standing wave, whose energy always remains in the same vicinity).

Treble Sounds of high pitch, such as above C_5 (that is, frequencies greater than approximately 500 Hz and especially above 1000 Hz).

Tremolo Amplitude modulation at subaudible frequencies, applied to any sound wave. (See Section 8.3.)

Tuning (1) Adjusting the pitch of one note or instrument into a precise intervallic relation with another. (2) A complete scheme for adjusting the pitch of all notes in a scale by tuning one interval at a time.

Velocity Rate of motion from one place to another.

Vibrato Frequency modulation at subaudible frequencies, applied to any sound wave. (See Section 8.3.)

Voltage Electrical potential difference; a measure of the net electrical push acting upon the electrons in a circuit.

Waveform The continuous sequence of displacements or pressure differences making up one complete cycle of a complex vibration or wave.

Wavelength Distance from one wave crest to the next along their direction of travel.

Weight The downward gravitational *force* upon any object, related to its mass by weight $= Mg$.

Woodwinds Wind instruments using reed (such as oboe and clarinet) or edgetone (such as flute) sound-generation mechanisms, but excluding lip reeds. From an acoustical viewpoint, this is a poorly defined category and would be better abandoned.

Work The transfer of energy to an object whenever a force moves it through a distance; $W = Fd$.

Xylophone (1) Any percussion instrument involving the striking of wooden objects. (2) That specific instrument with wooden (or plastic) bars, tuned to produce fundamental and third harmonics, arranged like a piano keyboard.

APPENDIX

H Hints and Answers to Selected Exercises

A reminder: The appearance of many answers here emphasizes that the final answers are always much less important than a clear presentation of the reasoning with which you obtain them. Make it a habit always to show *how* you get your answers.

CHAPTER 1

2. Roughly 69 cm.
3. 100 ms.
4. Roughly 3 C° either way.
5. If you have any trouble getting $p \cong 10^3$ atm, study Appendix B!
6. Dozens.
7. Maximum 1.000003 atm.
8. 4 atm; 20 cm^2 (both tires together).
9. 100 ms.
10. 200 ms.
11. 10^{-3} N.
12. 300 N/m^2.

CHAPTER 2

1. Almost 4 ms.
2. Remember the meaning of the prefix M!
5. I did not say 2.000 m!
6. 540 km.
8. 20 cm.
10. 700 Hz.
13. 30 cm.
15. 50 kg.
16. 1.6 Hz.
19. 2.5×10^5 bulbs.
20. 10 cm.

CHAPTER 3

2. How long a time for a wave round trip?
3. 1.4 m.
7. (c) 1600 Hz.
8. (c) 1.2 m.
9. 2% (remember Chapter 1).

CHAPTER 4

1. How does each λ compare to D?
2. Yes; largely. Explain.
3. D versus λ.
4. 1.7 KHz.
5. None.
6. Nearly 150 km/hr.
7. (b) 25 cm, 75 cm, . . . what next?
8. (a) 287 Hz, 573 Hz, . . .
   (b) 143 Hz, 430 Hz, . . .
9. One is constructive, the other is not.
11. One is 437 Hz.
12. Give three answers!

CHAPTER 5

1. 10 W/m^2.
2. 108 J; remember, 1 kg weighs 9.8 N.
3. (b) 7×10^{-10} J.
5. See Table 5.2.
6. (b) 3×10^{-4} W/m^2.
7. (b) 56 dB.
8. Last: 1000, not 30.
9. 70, 70, 73, . . .
10. 20.
11. 97 dB.
12. (a) 18 dB, (d) very little. Explain.
13. 76, 85, 83 dB.
14. (b) 100.
15. (b) 40–100.
16. 95%; half or less of original.
17. 50; 82 dB. Explain!
18. More than 300 m. Consider 6 dB at a time.

CHAPTER 6

4. Nearly 1 atm; 1 mm; 0.1 mm.
6. Use Figure 6.8 and divide the audible range into several parts. Estimates of approximately 2500 to 3000 are reasonable.
7. 60%.
8. (b) 2000 Hz.
9. Think in terms of octaves!
11. (c) Nearly 80 phons.

13. (b) Nearly 50 sones.
14. (c) 83 dB.
16. 12 dB, 27 dB, . . .
17. Choose frequencies such as 50–200 Hz to represent bass and 1–5 KHz for treble.
19. 12–15%.
20. (a) 58 dB.

CHAPTER 7

1. (b) 0.2 s.
3. 150.
5. 100 ms versus 125 ms.
10–18. See Figure E.
19. Examples: 16, 15, 14, respectively.
20. Approximately $\frac{1}{3}$ semitone.

CHAPTER 8

1. Same answer for both.
2. $P = 10$ ms.
3. No and yes. Explain.
4. 196 Hz and . . .
5. 250 Hz and . . .
8. Which sounds have periodic waveforms?

CHAPTER 9

1. In each case, only one changes. Why?
5. 6; two degenerate pairs.
6. How much are the springs stretched?
9. What would it mean for a mode frequency to approach zero?
10. $E^{\flat}_4$, D_5, and so on.
11. A_4.
13. The bends do not make much difference, so you can use your knowledge of the modes of a straight bar.
15. (a) None. Why?
17. Last: no. Why?
19. Downward or upward all together.
20. Examples: 3 strong, 5 absent.
22. 2 s.
23. Notice the antinode of mode 1 is a node for many others.
25. 1 KHz.

CHAPTER 10

1. One of them is 0.4 m.
2. 400 Hz.
3. 750 Hz.
5. Frequency ratio 1.414; half an octave.
6. 300 m/s.
7. 200 Hz.
8. 864 N.
15. Approximately one octave; $J_s \cong \frac{1}{2}J_b$; $J_s \cong 2J_b$.
16. Review the rules in Sections 9.6 and 9.7.
20. Which mode is practically missing from the recipe?
21. (a) Three possible places; one is the center. Where are the other two? (c) Octave plus fifth.
22. Remember the damping rule from Section 9.7; G_4.
23. Pluck and pickup *each* suppress certain harmonics.
25. Review Figure 10.7a.

CHAPTER 11

1. (d) Peaks shift left, then lower, then right and higher.
3. How strong a force does the bridge exert on the body?
5. (a) 4.8.
7. 0.63 gm/m.
8. (a) Any impedance change generates partial reflections. (c) No. (d) Yes, adversely. Explain.
10. Recall the rule of Section 9.7.
12. Remember $W = FD$ and Table 5.1.
15. How does damping affect resonance peaks?
16. No.
18. What happens to the air resonance?
20. P4, 2 cm, 6 cm.
21. 26 or 27 cm.
22. P4 up.
23. 12, 6, 10, 30.

CHAPTER 12

2. Look at parts (b) and (c).
3. 300 Hz.
4. 50 Hz.
5. 1 and 5 KHz.
6. 0.2 and 1.8 KHz.
7. (b) 10 cm.
8. 171 and 162 Hz.
9. 1600 Hz.
10. 2 cm.
12. (a) 4.
13. 1.7 KHz; 20 KHz.
14. (b) $5\frac{1}{3}$ is what fraction of 16?
15. G_4.
16. 4 cm up; 26 cm up.
17. Approximately 40 cm down.
19. Where is a node of the desired mode?

CHAPTER 13

3. Remember: You do not want the reed's own resonance peak to be too high and narrow.

4. A waveform including more sudden changes will involve stronger high harmonics.
5. (b) 5.1 m.
6. Less than 60 cm. Examine the bell.
7. More than 70 cm. Look at the reed.
8. (b) A_5.
9. Where are the nodes?
10. Roughly 40 and 15.
12. 2.22 m, 6.6 cm, 8.9 cm.
17. Rises approximately 0.3 semitone. Why?
18. There is no way it can work. Why?
19. 7.4 cm.

CHAPTER 14

1. Ask whether your calculated answer is realistic!
2. (a) *t-sh*.
4. Try it!
5. A continuous spectrum can be shaped by formants just as well as can a harmonic series.
6. What frequency ranges are involved?
10. Which harmonic is at 2.2 KHz?
11, 12, 13. Use Figure 14.12.
14. No and yes. Explain why.
16. Could argue for either *ah* or short *a*.
17. Approximately 3 cm. (Recall Figure 12.3.)
18. 0.015 atm.

CHAPTER 15

1. (c) gets the best diffusion in *all* directions.
3. Too long.
4. 10 Hz.
5. First find the percentage you are willing to let in, remembering Table 5.1.
6. 60, 6000; 2 or 3, 240.
8. (a) 1.6 s, meaning not less than 1.4 or more than 1.8.
10. (b) 0.84 s.
12. 0.4 s.
13. (a) 2.4 s. (b) 1.1 s. (c) 1 s and less. Even with a proper ceiling, 200 people will make the room dead.
14. (c) 1.1 s. (d) See Figure 15.8c.
15. 640 m^2.
16. 1.4 s.
17. 400 m^2.
18. (b) 14 m.
19. (a) 109 dB. (c) 103 dB.
20. 1200 m^3.
22. Tens of microseconds.
23. For what frequencies can the head make a shadow?
24. (b) 1.6 s.
25. 1000 m^2; intermediate case 1100 m^2 and 1.45 s.

CHAPTER 16

1. Less than 1 cm.
4. See Figure 16.9.
6. (b) 40 dB.
7. (a) 1.9 mm.
8. (a) 3.5 mm.
9. 0.17 semitone; 1.8 s.
11. (a) Roughly 1 KHz.
12. 500–1000 Hz; 2–3 m.
13. (b) Several meters.
14. Electrical power approximately 2 W and 6 W.
15. (b) 15 cm maximum.
    (c) 1 m minimum.
17. 6 ms.
18. 1.4 million; 5.6 billion; 1.4 square micrometers.

CHAPTER 17

3. See Figure 17.3.
4. (a) 200 Hz and 176 Hz; see Figure E for names.
6. C_4; to pick a definite frequency (within 2 or 3 Hz), insist that too-high and too-low votes balance out.
7. Two, approximately $2\frac{1}{2}$ semitones apart. What pitch names?
8. (a) 224 Hz; see Figure E.
9. See Figure 17.9.
10. (b) 2.5 KHz.
11. See Figure 17.10.
12. What *are* the harmonic frequencies? Then use Figure 17.10.
14. Roughly 100 Hz.
15. Several seconds.
18. (e) $C_6^\sharp$, C_2, $B_4^\flat$.
    (a)–(d) Some are in the cracks.
19. Yes, around E_3. What frequencies?
21. 50 sones. Remember Figures 6.12 and 6.13.
22. 5 sones.
23. See Figure 17.17.
24. Try to think of your own first. Then see Plomp, p. 105.

CHAPTER 18

1. (c) 8:3.
2. (a) Reduce the ratio to lowest terms. (b) Express it

in terms of other intervals already listed.

3. There are several reasons; see Exercise 12 in Chapter 17 for one.
4. 8:7.
5. (b) 5:2.
6. (a) 9:8 or 204¢.
7. (c) Recall Figures 12.2 and 12.3.
8. A_5, 10 Hz.
11. Slendro has clusters around 245¢, 490¢, three more; Pelog around 130¢, 415¢, several more.
12. 434¢, 608¢, ...
13. (e) All within 1.5¢.
16. Five distinct sizes.
18. (b) 7¢.
19. Show why the fifths are 695¢, then prepare a column as in Figure 18.8a. Compare the results with multiples of 63¢.
22. 12 Hz, E_6.
23. (a) 996¢, 16:9.
24. (a) 582.5¢.

CHAPTER 19

2. C_0; six different tones.
3. (a) C_3; C_3 twice and C_4.
4. (a) C_1. (b) $E_0^{\flat}$.

Index

THE CHROMATIC SCALE

Pictured in the middle are the 88 keys of a standard piano keyboard, with names of the white keys as recommended by the Acoustical Society of America and the U.S. Standards Association. Helmholtz's names still are used sometimes and are indicated for the C's only. Each black key can be called by two names, either ♯ of the white key to its left or ♭ of the one to its right. Even with each key and above it is the corresponding note as written on the staff.

Frequencies of these notes are given for the equal-tempered chromatic scale tied to A440. These are only an approximation to the frequencies used in most performance, even for the piano. Caution: See Chapter 18 for the importance of nonequal temperaments.

Lines below the piano keyboard indicate the usual playing ranges of several common orchestral instruments. For many, the highest playable notes depend strongly on the skill of the individual player. In all cases, higher harmonics also are present when notes with these fundamentals are played.

THE HARMONIC SERIES SLIDER

Photocopy and cut out the slider on the edge of this page. You can use it to locate quickly and easily all harmonic relations among notes in the chromatic scale. It should always be placed with its marked edge along the "H-S Slider" line on the piano keyboard, so that it is over the backs of the keys with black and white keys included equally. Here are directions for several of its applications. (Make allowances for the binding crease!)

1. To find the harmonics of any given fundamental, move the slider left or right until the Fundamental mark is centered in that key. Then next to each harmonic number you will find the note corresponding to that member of the harmonic series. Example: Place Fundamental on C_2, find 2 opposite C_3, 3 by G_3, 5 slightly off center from E_4, and so forth. Any harmonic marker that is not centered on a key (for example, 5 slightly off, 7 and 11 much worse) indicates that an equal-tempered instrument does not have a note accurately corresponding to that harmonic.

2. To find the fundamental corresponding to some given harmonic, move the slider to align the given harmonic number with its note, then look at the left end to see the fundamental to which it belongs. Example: If A_5 is sixth harmonic of some note, place 6 opposite A_5, and then observe that D_3 is its fundamental.

3. To find the frequency ratio corresponding to the justly tuned version of any musical interval (Chapter 18), locate the two notes on the keyboard that form the interval. Upon the lower of these notes place first the Fundamental mark, then the 2, the 3, and all others in turn. Each time, check the upper note. The first time you find a harmonic marker well-centered on that key, the two markers tell you a small-integer ratio of frequencies that may

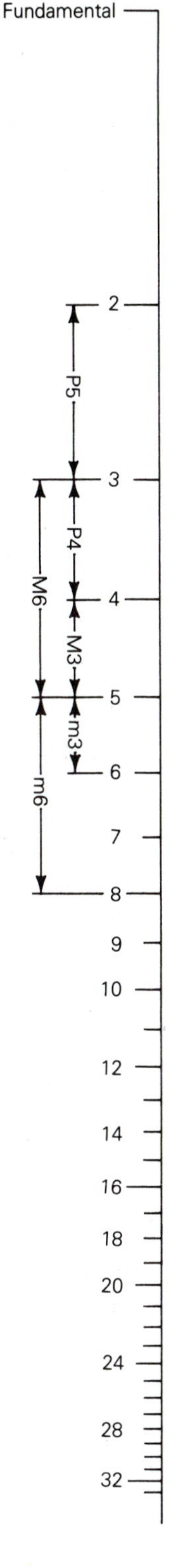

serve as a standard for that interval. For instance, for C_4 and F_4 you should find that this ratio is 3 to 4. But for F_4 and B_4, 5 to 7 is only a crude approximation (or rather, the equal-tempered tritone is only a crude approximation of 5 to 7), and only 19 to 27 comes very close. This procedure can, of course, also be reversed to go from ratio to notated interval.

4. To find combination tones (Chapter 17), locate the keys corresponding to the two notes played. As in (3), move the slider from right to left until the first time you find two marks centered on these notes. Subtract the corresponding harmonic numbers and find the number corresponding to the difference; opposite it will be the difference tone. Similarly, opposite the sum of the two numbers will be found the sum tone. For example, F_4 and D_5 fit the 3 to 5 ratio fairly well; the slider then indicates the difference tone $B^\flat_3$ opposite harmonic number 2 and the sum tone $B^\flat_5$ opposite 8. (This is strictly true only for just tuning; for equal temperament, the difference tone is in the crack between $B^\flat_3$ and B_3, as you can see either by subtracting the frequencies or by noting that 19 to 32 fits the diagram closer than 3 to 5.)

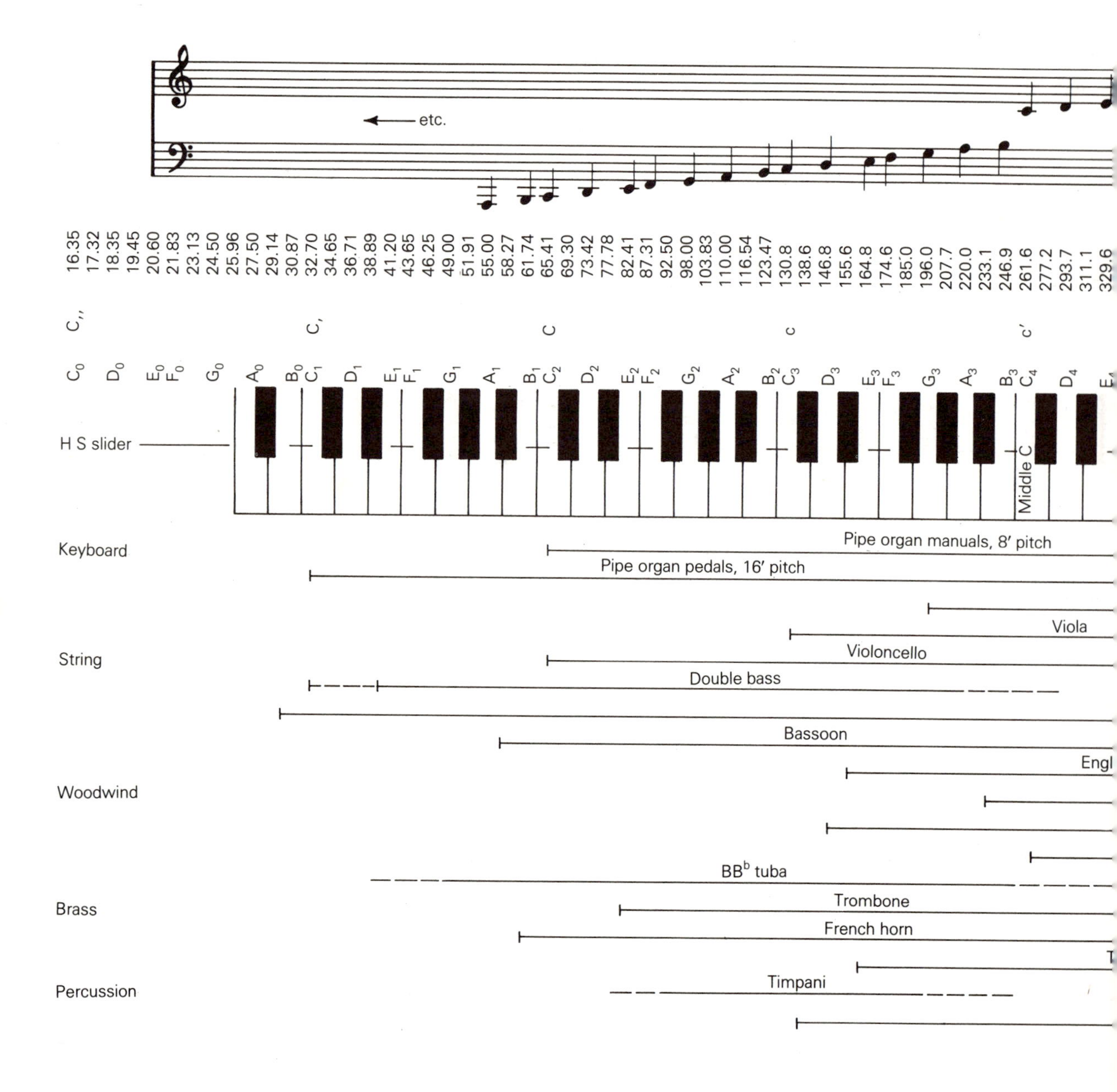

etc.
H S slider
Middle C
Keyboard
Pipe organ manuals, 8′ pitch
Pipe organ pedals, 16′ pitch
String
Viola
Violoncello
Double bass
Woodwind
Bassoon
Brass
BB♭ tuba
Trombone
French horn
Percussion
Timpani